Lecture Notes in Mathematics

Edited by A. Dold and B. Eckmann

928

Probability Measures on Groups

Proceedings of the Sixth Conference Held at Oberwolfach, Germany, June 28 – July 4, 1981

Edited by H. Heyer

Springer-Verlag
Berlin Heidelberg New York 1982

Editor
Herbert Heyer
Universität Tübingen, Mathematisches Institut
Auf der Morgenstelle 10, 7400 Tübingen, Fed. Rep. of Germany

AMS Subject Classifications (1980): 60B15, 60J15, 60J45, 60K05, 43A05, 60A10, 60B11, 60J65, 43A22, 43A25

ISBN 3-540-11501-3 Springer-Verlag Berlin Heidelberg New York
ISBN 0-387-11501-3 Springer-Verlag New York Heidelberg Berlin

Printing and binding: Beltz Offsetdruck, Hemsbach/Bergstr.
2141/3140-543210

INTRODUCTION

This Sixth Conference in the series "Probability Measures on Groups" held at the Mathematisches Forschungsinstitut Oberwolfach extends a well-established tradition: Participants of ten different countries contributed to an inspiring atmosphere and helped promoting the exchange of ideas on recent advances and new directions of research in the field. As usual the meeting received a favorable reception from various areas of stochastics and analysis covering probability theory and potential theory on algebraic-topological structures as well as their interrelations with the structure theory of locally compact groups, Banach spaces and Banach lattices.

The main subjects of discussion can be described as follows

(i) Probability measures on groups, semigroups and hypergroups.

(ii) Stochastic processes with values in groups and their potential theory.

(iii) Connections between probability theory on groups and abstract harmonic analysis.

(iv) Applications of probability theory on algebraic-topological structures to quantum physics.

To stress a few highlights of the recent progress in the theory we mention that profound work has been done during the last three years in the theory of probability on nilpotent Lie groups and on Banach spaces. The long standing problem of embedding infinitely divisible probability measures on a locally compact group received new impulses. Among the important and still challenging directions of research the problem of stability for probability measures on locally compact groups and locally convex vector spaces deserves special attention.

We give a short survey of the work published in this volume, organized by the underlying structure.

1. *Semigroups, hypergroups.*

Asymptotic equidistribution on a locally compact semigroup is the subject of F. K i n z l's contribution. A. M u k h e r j e a and his coauthor A. N a k a s s i s apply the theory of nonhomogeneous Markov chains with a countable state space to the convergence of convolution products of probability measures on countable semigroups. W. R. B l o o m studies the rational embeddability of infinitely divisible probability measures on a hypergroup and uses his results in dealing with the special hypergroup of double cosets.

2. *General locally compact groups.*

G. J. S z é k e l y introduces a new notion of prime probability on an Abelian locally compact group and characterizes all groups which admit such prime probability measures. I. Z. R u z s a extends the convergence principle to groups which are not necessarily second countable, at least in the Abelian case. In M. M c C r u d d e n's work a variant of the rational embedding of infinitely divisible probability measures on a locally compact group is considered which covers also the case of non strongly root compact groups.
W. H a z o d presents the actual state of affairs in the theory of stable probability measures on groups emphasizing stability with respect to a contracting group of automorphisms. In his detailed exposition C. J. S t o n e develops the potential theory of recurrent symmetric homogeneous processes with independent increments taking their values in an Abelian locally compact group. Main topics of the treatment are the balayage and equilibrium problems corresponding to an arbitrary relatively compact Borel set.

3. *Special locally compact groups.*

There is the paper by W. W o e s s on local limit theorems for random walks on certain countable groups, in particular on free groups. A number of contributors choose Lie groups as the domain of their probabilistic studies. E. S i e b e r t gives an extended analysis of convolution hemigroups of probability measures on a Lie group and of the corresponding problem of nonhomogeneous evolution equations. A dichotomy theorem for random walks on homogeneous spaces with an application to Harris recurrence in the case of the affine group is the topic taken up by L. Elie in her paper. E. Le Page studies sequences of independent, identically distributed random variables taking values in the special linear group and proves results of central limit type.

4. *Vector groups.*

Here we first mention the extensive paper of D. J. H. Garling on functional analytic and probabilistic methods in the theory of stable Banach spaces. Another topic of high actuality are the semistable measures. A. Janssen and A. Tortrat prove zero-one laws for such measures. Their papers are related to each other. The first author's general zero-one law applies to semistable and selfdecomposable measures on a locally convex vector space. Banach lattices of Poisson type are the underlying structure of E. Dettweiler's work on Poisson measures and on the central limit theorem for cone-valued random vectors.

5. *Interrelations and applications.*

Ph. Feinsilver investigates representations of the Heisenberg-Weyl algebra and deduces from them the orthogonal polynomials corresponding to the Bernoulli, Gamma, Poisson and Gauss generators. The aim of the paper by J. Erwen and B.-J. Falkowski is the analysis of the two-cocycles appearing in the Parthasarathy-Schmidt representation of infinitely divisible positive definite functions on the group SU(1,1). The main problem reduces to the computation of a certain first cohomology group and to the proof of the non-triviality of a certain two-cycle. J. A. Goldstein in his contribution shows how the well-known connection between the Poisson process and the abstract telegraph equation yields interesting results in the theory of equipartition of energy. Finally we mention the two papers by T. Drisch and by P. D. F. Ion and W. von Waldenfels on problems in quantum physics. T. Drisch considers quantum mechanical limit distributions, discusses within this framework the uniqueness and continuity theorems for the Fourier transform and a theorem of Bochner type, and applies these tools to central limit theory and stability. Basic to the analysis of Ion and von Waldenfels is the fact that the noise processes under consideration are considered as linear functionals on certain tensor algebras. In order to deepen the understanding of time ordered moments as linear functionals of an algebra, the authors propose an enlargement of the algebra with the aid of shuffle or Hurwitz products. The studies presented open the possibility of discussing the stochastic differential equation of the two-level atoms in a heat bath of photons.

We conclude this introduction with some comments on a unified setting of probability measures on homogeneous spaces. Studying probability measures on a locally compact group which are biinvariant with respect to a compact subgroup can be built up on various basic structures which are related to each other. The most general set up for investigations on the central limit theorem and on problems of divisibility for biinvariant probability measures on the group appears to be the notion of a hypergroup. Related structures are the delphic semigroups in the sense of D.G. Kendall and J.F.C. Kingman, and the generalized convolution algebras in the sense of K. Urbanik and J. Gilewski. Approaches of a more specific nature are related to generalized translation spaces and Gelfand pairs. Examples of Gelfand pairs are the Euclidean spaces viewed as homogeneous spaces defined by the motion group, hyperbolic spaces, spheres, cubes and homogeneous trees. Although there exists a rich harmonic analysis for general Gelfand pairs, very little genuine probability theory is available except in the above mentioned special cases.

Herbert Heyer

C O N T E N T S

The authors D. J. H. Garling, J. A. Goldstein, C. J. Stone and A. Tortrat kindly provided their papers, although they did not participate at the conference.

PAPERS GIVEN AT THE CONFERENCE
BUT NOT PUBLISHED IN THIS VOLUME

E. Dettweiler	Branching processes with a continuum of states.
G. Forst	Completely self-decomposable probabilities on $\mathbb{R}_+$ and $\mathbb{Z}_+$.
Y. Guivarc'h	Brownian motion on some coverings of compact manifolds.
F. Hirsch	Quotients of negative definite functions on $\mathbb{R}^n$.
G. Högnäs	Measures on the semigroup of singular matrices.
J. Kisyński	Exponential moments for convolution semigroups of probability measures on Lie groups.
R. G. Laha	Random fields over groups.
E. Lukacs	Recent advances in the theory of characteristic functions.
H. Rindler	Almost invariant sets and unique invariant means.
E. Siebert	Gaussian convolution semigroups on a Lie group.
D. W. Stroock	The Littlewood conjecture.

LIST OF PARTICIPANTS

W. R. Bloom	Perth, Western Australia
Y. Derriennic	Brest, France
E. Dettweiler	Tübingen, West Germany
T. Drisch	Dortmund, West Germany
L. Elie	Paris, France
B.-J. Falkowski	Neubiberg, West Germany
P. Feinsilver	Carbondale, USA
G. Forst	København, Denmark
L. Gallardo	Nancy, France
P. Gerl	Salzburg, Austria
Y. Guivarc'h	Rennes, France
W. Hazod	Dortmund, West Germany
H. Heyer	Tübingen, West Germany
F. Hirsch	Cachan, France
G. Högnäs	Abo, Finland
A. Janssen	Dortmund, West Germany
E. Kaniuth	Paderborn, West Germany
F. Kinzl	Salzburg, Austria
J. Kisyński	Warsaw, Poland
R. G. Laha	Bowling Green, USA
E. le Page	Rennes, France
E. Lukacs	Washington D. C., USA
M. McCrudden	Manchester, United Kingdom
A. Mukherjea	Tampa, USA
H. Rindler	Wien, Austria
I. Z. Ruzsa	Budapest, Hungary
G. Schlichting	München, West Germany
L. Schmetterer	Wien, Austria
E. Siebert	Tübingen, West Germany
D. W. Stroock	Boulder, USA
G. J. Székely	Budapest, Hungary
W. von Waldenfels	Heidelberg, West Germany
W. Woess	Salzburg, Austria

INFINITELY DIVISIBLE MEASURES ON HYPERGROUPS

Walter R. Bloom

1. Introduction

Let K be a hypergroup (convo) in the sense of Jewett [9] admitting a left Haar measure. Denote by $M(K)$ the algebra of bounded Radon measures on K, and by $M^1(K)$ the subsemigroup of probability measures. There has already been some development of probability theory on K, with a study of the Fourier mapping on $M^1(K)$ in [1], and an examination of the convergence behaviour of convolution products of probability measures leading to a result of Kawada-Ito type in [2]. In this paper we introduce the notion of root compactness for hypergroups and indicate its role in the study of divisibility properties of probability measures. The results obtained here are for the main part analogous to those for locally compact groups.

We begin in Section 2 by considering some properties of double coset hypergroups, and in Section 3 we introduce the notion of conjugacy class for hypergroups. The main sections of the paper are 4 and 5, which deal with root compactness and infinite divisibility. Firstly a relationship between root compactness and divisibility both in $M^1(K)$ and K is obtained. Then, using a version of the shift compactness theorem for hypergroups, the closure of the set of infinitely divisible measures in $M^1(K)$ is shown. It is also noted that the infinitely divisible measures on a root compact hypergroup are rationally embeddable. Finally, in Section 6, we consider the above ideas in the context of a specific example, the (hypergroup) dual of $SU(2)$.

For a general reference to and the background of the corresponding results for probability measures on locally compact groups the reader is referred to [8]. The notation and terms used in the present paper will be found in [1] and [2].

2. Double coset hypergroups

Double coset hypergroups were introduced by Jewett ([9], Section 14); we recall the definition. Let K be a hypergroup and let H be a compact subhypergroup. The double cosets of H are the sets $H * \{x\} * H$, where $x \in K$; we write HxH in place of $H * \{x\} * H$. The collection of double cosets of H will be denoted by $K /\!/ H$. It will be given the quotient topology with respect to the natural projection $\pi : K \to K /\!/ H$, defined by $\pi(x) = HxH$. For $\mu \in M(K)$ write the image of μ under π as $\pi_*(\mu)$, that is, $\pi_*(\mu)(g) = \mu(g \circ \pi)$ for all $g \in C_c(K /\!/ H)$. Alternatively one may write

$$\pi_*(\mu) = \int_K \varepsilon_{HxH} \, d\mu(x) .$$

It follows easily, using [9], Theorem 14.1A, that $\pi_*(\omega_H * \varepsilon_x * \omega_H) = \varepsilon_{HxH}$ for all $x \in K$, where ω_H denotes the normalised Haar measure of H. In fact π_* is an isomorphism from the Banach $\sim$-algebra

$$M_H(K) = \{\mu \in M(K) : \omega_H * \mu * \omega_H = \mu\}$$

onto $M(K /\!/ H)$. A convolution structure can be defined on $M(K /\!/ H)$ via

$$\mu * \nu = \pi_*\left(\pi_*^{-1}(\mu) * \pi_*^{-1}(\nu)\right)$$

under which $K /\!/ H$ is a hypergroup, with identity H and involution given by $(HxH)^- = Hx^-H$.

2.1 Lemma Let $f \in C_c(K /\!/ H)$. Then

(a) $\operatorname{supp}(f \circ \pi) = \pi^{-1}(\operatorname{supp}(f))$;

(b) $\pi_*(\mu)(A) = \mu(\pi^{-1}(A))$ for all σ-finite $\pi_*(\mu)$-measurable $A \subset K /\!/ H$;

(c) $\operatorname{supp} \pi_*(\mu) = \pi(\operatorname{supp}(\mu))$.

Proof The proof of (a) is easy, and that of (b) follows from [7], (12.46), using the properties that π is continuous, onto and satisfies $\pi^{-1}(C)$ is compact in X for every compact subset C of $K /\!/ H$ (for the latter assertion see [9], Theorem 14.1C, Lemma 13.1B and Lemma 13.2A). It remains to prove (c). This is a consequence of the implications : $\pi_*(\mu)(U) = 0$ if and only if $\mu(\pi^{-1}(U)) = 0$ if and only if $\operatorname{supp}(\mu) \subset \pi^{-1}(U)^c = \pi^{-1}(U^c)$ if and only if $\pi(\operatorname{supp}(\mu)) \subset U^c$, together with the fact that π is a closed map, where the statements above are understood to hold for all open subsets U of $K /\!/ H$. //

As a consequence of Lemma 2.1 we have, using the above notation:

2.2 Lemma Let $A, B \subset K$. Then

$$\pi(A) * \pi(B) = \pi(A * H * B) .$$

Proof For each $x, y \in K$ we have, using the definition of the convolution on $K /\!/ H$,

$$\varepsilon_{HxH} * \varepsilon_{HyH} = \pi_*\big((\omega_H * \varepsilon_x * \omega_H) * (\omega_H * \varepsilon_y * \omega_H)\big)$$

and, considering the supports of these measures,

$$\begin{aligned}\pi(x) * \pi(y) &= \{HxH\} * \{HyH\} \\ &= \operatorname{supp}\big(\pi_*(\omega_H * \varepsilon_x * \omega_H * \varepsilon_y * \omega_H)\big) \\ &= \pi(H * \{x\} * H * \{y\} * H) \\ &= \pi(\{x\} * H * \{y\}) .\end{aligned}$$

The result now follows. //

3. Conjugacy classes and invariance

For $x,y \in K$ write $x \sim y$ if there exist $z_1, z_2, \ldots, z_n \in K$ such that $y \in Z_n * \{x\} * Z_n^-$, where $Z_n = \{z_1\} * \{z_2\} * \ldots * \{z_n\}$.

3.1 Lemma $\sim$ is an equivalence relation on K.

The proof of Lemma 3.1 is clear. We write

$$C_x = \{y \in K : y \sim x\}$$

for the conjugacy class of K containing x. Alternatively,

$$C_x = \cup\{Z_n * \{x\} * Z_n^- : Z_n = \{z_1\} * \{z_2\} * \ldots * \{z_n\},$$
$$z_i \in K,\ i = 1,2,\ldots,n,\ n \in N\} .$$

Note that when K is a group then the definition reduces to the usual one.

Let $H \subset K$. A subset F of K is called H-invariant if $\{x\} * F * \{x^-\} \subset F$ for all $x \in H$. If $H^- = H$ then this is equivalent to demanding that $\{x\} * F * \{x^-\} = F$ for all $x \in H$; indeed if $x \in H$, in which case $x^- \in H$, we have

$$F \subset \{x\} * \{x^-\} * F * \{x\} * \{x^-\} \subset \{x\} * F * \{x^-\} \subset F .$$

It follows easily that for such H, F being H-invariant implies that $\{x\} * F = F * \{x\}$ for all $x \in H$.

3.2 Theorem Each conjugacy class C_y is K-invariant.

Proof Let $x \in K$ and consider $u \in \{x\} * C_y * \{x^-\}$. Then $u \in \{x\} * \{z\} * \{x^-\}$ for some $z \in C_y$. This implies that $z \sim u$ and, together with $z \sim y$, we have $u \in C_y$. Thus $\{x\} * C_y * \{x^-\} \subset C_y$ and, by the comments preceding the theorem, C_y is K-invariant. //

In contrast to the case for groups, a character of a commutative hypergroup need not be constant on the conjugacy classes. For example if $K = \{e,a\}$ with $\varepsilon_a * \varepsilon_a = \beta\varepsilon_e + (1-\beta)\varepsilon_a$, where $\beta \in (0,1)$ is given, then $C_e = K$, whereas the character χ given by $\chi(a) = -\beta$ is obviously not constant on K.

A hypergroup K is called class compact if all of its conjugacy classes are relatively compact. Note that even commutative hypergroups need not be class compact; an example will be given in Section 6.

3.3 Lemma Suppose that K is class compact and that it has a compact invariant neighbourhood V of e. Then every compact subset C of K is contained in a K-invariant compact set.

Proof First we have

$$C \subset V * C = \cup\{V * \{c\} : c \in C\}$$

and, since C is compact, there exist $c_1, c_2, \ldots, c_n \in C$ such that $C \subset \bigcup_{i=1}^{n} V * \{c_i\}$. Let C_i denote the conjugacy class containing c_i. Then $C \subset V * C'$, where $C' = \bigcup_{i=1}^{n} C_i$ is compact. We show that the compact set $V * C'$ is K-invariant. Indeed, for $x \in K$,

$$\{x\} * V * C' * \{x^-\} \subset \{x\} * V * \{x^-\} * \{x\} * C' * \{x^-\} \subset V * C',$$

using the invariance of V and C'. This completes the proof. //

4. Root compactness

4.1 Definition Let $n \in \mathbb{N}$. Call a hypergroup K n-root compact (written $K \in R_n$) if the following condition holds: For every compact $C \subset K$ there exists compact $C_n \subset K$ such that all finite sets $\{x_1, x_2, \ldots, x_n\}$ in K with $x_n = e$ satisfying

$$\{x_i\} * C * \{x_j\} * C \cap \{x_{i+j}\} * C \neq \phi$$

for $i + j \leq n$ are contained in C_n. Write $R = \bigcap_{n=1}^{\infty} R_n$ for the class of all root compact hypergroups.

The class of root compact hypergroups provides a useful framework within which to study divisibility properties of the semigroup of probability measures, which can then be related to divisibility in the underlying hypergroup; see Theorem 4.4 below. Also for a root compact hypergroup K we can show that the set of infinitely divisible measures is closed in $M^1(K)$, and that the infinitely divisible measures are rationally embeddable. As a first step we present two preliminary results for product sets.

4.2 Lemma Let $\mu, \nu \in M^1(K)$. Then $\mu * \nu(A * B) \geq \mu(A)\nu(B)$.

Proof We have, using [9], Lemma 3.1E,

$$\begin{aligned} \mu * \nu(A * B) &= \mu * \nu(\xi_{A*B}) \\ &= \int_K \int_K \xi_{A*B}(x * y)\, d\mu(x)\, d\nu(y) \\ &\geq \int_B \int_K \xi_{A*B}(x * y)\, d\mu(x)\, d\nu(y) . \end{aligned}$$

Since, by definition, $\operatorname{supp}(\varepsilon_x * \varepsilon_y) \subset A * B$ for each $x \in A$, $y \in B$, we have for such x,y,

$$\xi_{A*B}(x * y) = \int \xi_{A*B}\, d\, \varepsilon_x * \varepsilon_y = 1 .$$

It follows that $\xi_{A*B}(x * y) \geq \xi_A(x)$ for each $x \in K$, $y \in B$, and

$$\mu * \nu(A * B) \geq \int_B \int_K \xi_{A*B}(x * y)\, d\,\mu(x)\, d\,\nu(y) \geq \mu(A)\nu(B) . \quad //$$

4.3 Lemma Let $\mu \in M^1(K)$. Then for each $x \in K$ and compact $C \subset K$,

$$\varepsilon_x * \mu(C) \leq \mu(\{x^-\} * C) .$$

Proof Using [9], Lemma 4.2H,

$$\varepsilon_x * \mu(C) = \varepsilon_x * \mu(\xi_C) = \mu(\varepsilon_{x^-} * \xi_C)$$

and, since $\operatorname{supp}(\varepsilon_{x^-} * \xi_C) \subset \{x^-\} * C$ and $0 \le \varepsilon_{x^-} * \xi_C \le 1$, we have

$$\mu(\varepsilon_{x^-} * \xi_C) \le \mu(\{x^-\} * C) \ . \quad //$$

We cannot expect to obtain equality in Lemma 4.3, even in the case where K is compact and μ is its normalised Haar measure; see [9], Lemma 3.3C and the remarks preceding it.

Now for each $N \subset M^1(K)$, $C \subset K$, write

$$R(n,N) = \{\mu \in M^1(K) : \mu^n \in N\} ,$$

$$C^{1/n} = \{x \in K : \{x\}^n \subset C\} .$$

4.4 Theorem Let K be a hypergroup and consider the following conditions:

(i) $K \in \mathcal{R}_n$;

(ii) $R(n,N)$ is relatively compact for each relatively compact $N \subset M^1(K)$;

(iii) $C^{1/n}$ is compact for every compact $C \subset K$.

Then (i) => (ii) => (iii) .

If, in addition, K is class compact and has a compact invariant neighbourhood of e , then (iii) => (i) .

Proof (i) => (ii). Let $N \subset M^1(K)$ be relatively compact. By Prohorov's theorem there exists for $\varepsilon \in (0,1/3)$ compact $C \subset K$ such that $\mu(C) \ge 1 - \varepsilon/2$ for all $\mu \in N$. Let $\mu \in N$ and $\nu \in R(n,\{\mu\})$. Now

$$\nu^{n-i} * \nu^i(C) = \mu(C) \ge 1 - \varepsilon/2 \ , \ 1 \le i \le n \ . \tag{4.5}$$

If $\nu^i(\varepsilon_x * \xi_C) < 1 - \varepsilon$ for all $x \in K$ then, using [9], Lemma 4.2H,

$$\nu^{n-i} * \nu^i(C) = \int_K \varepsilon_x * \nu^i(\xi_C)\, d\nu^{n-i}(x)$$

$$= \int_K \nu^i(\varepsilon_{x^-} * \xi_C)\, d\nu^{n-i}(x)$$

$$\le 1 - \varepsilon ,$$

a contradiction of (4.5). Hence there exists x_i such that $\nu^i(\varepsilon_{x_i} * \xi_C) \ge 1 - \varepsilon$, and it follows from the proof of Lemma 4.3 that $\nu^i(\{x_i\} * C) \geq 1 - \varepsilon$.

Now consider $1 \leq j \leq n - i$. By Lemma 4.2,

$$\nu^{i+j}(\{x_i\} * C * \{x_j\} * C) \geq \nu^i(\{x_i\} * C)\nu^j(\{x_j\} * C) \geq (1-\varepsilon)^2 .$$

Put $x_n = e$. Then, for all $i + j \leq n$,

$$\begin{aligned}\nu^{i+j}(\{x_i\} * C * \{x_j\} * C \cap \{x_{i+j}\} * C) &\geq (1-\varepsilon)^2 - \nu^{i+j}([\{x_{i+j}\} * C]^c) \\ &\geq (1-\varepsilon)^2 - \varepsilon > 1 - 3\varepsilon > 0 .\end{aligned}$$

Thus $\{x_i\} * C * \{x_j\} * C \cap \{x_{i+j}\} * C \neq \phi$ for all $i + j \leq n$. Since, by assumption, $K \in \mathcal{R}_n$, there exists compact $C_n \subset K$ such that $\{x_1, x_2, \ldots, x_n\} \subset C_n$. We note that C_n is independent of ν ; for all $\nu \in R(n,N)$,

$$\nu(C_n * C) \geq \nu(\{x_1\} * C) \geq 1 - \varepsilon .$$

By Prohorov's theorem, $R(n,N)$ is relatively compact in $M^1(K)$.

(ii) => (iii). Let $C \subset K$ be compact. Then, by Prohorov's theorem,

$$N = \{\nu \in M^1(K) : \operatorname{supp}(\nu) \subset C\}$$

is relatively compact in $M^1(K)$. By assumption, $R(n,N)$ is relatively compact in $M^1(K)$, from which it follows that $\{\varepsilon_x : \{x\}^n \subset C\}$ is relatively compact in $M^1(K)$. Since $\varepsilon_x(C') \geq \frac{1}{2}$ if and only if $x \in C'$ we deduce, using Prohorov's theorem again, that $C^{1/n}$ is relatively compact. It is easy to see that $C^{1/n}$ must be closed, so that it is also compact.

(iii) => (i). Suppose K is class compact and has a compact invariant neighbourhood of e . By Lemma 3.3, every compact subset of K is contained in a K-invariant compact set. Let C be a compact subset of K , and let $\{x_1, x_2, \ldots, x_n\} \subset K$ with $x_n = e$ satisfy

$$\{x_i\} * C * \{x_j\} * C \cap \{x_{i+j}\} * C \neq \phi , \; i + j \leq n .$$

Since the same is true if C is replaced by any larger set, we may assume that $e \in C$, $C = C^-$ and C is K-invariant.

Firstly, $x_i \in \{x_1\}^i * C^{3(i-1)}$ for all $i = 1,2,\ldots,n$ (we define $C^0 = \{e\}$) ; indeed, if $x_k \in \{x_1\}^k * C^{3(k-1)}$ then, putting $i = k$, $j = 1$, we have

$$\{x_1\}^k * C^{3(k-1)} * C * \{x_1\} * C \cap \{x_{k+1}\} * C \neq \phi ,$$

from which it follows that

$$x_{k+1} \in \{x_1\}^k * C^{3k-2} * \{x_1\} * C * C^- = \{x_1\}^{k+1} * C^{3k} ,$$

the equality following using the invariance of C .

Now, taking $i = n$, we have

$$e = x_n \in \{x_1\}^n * C^{3(n-1)}$$

and hence

$$\{x_1^-\}^n \subset \{x_1\}^n * C^{3(n-1)} * \{x_1^-\}^n = C^{3(n-1)} .$$

Thus $x_1 \in \left(C^{3(n-1)}\right)^{1/n}$. Since $e \in C \cap \left(C^{3(n-1)}\right)^{1/n}$, we have

$$x_i \in \{x_1\}^i * C^{3(i-1)} \subset \left[\left(C^{3(n-1)}\right)^{1/n}\right]^n C^{3(n-1)} ;$$

this latter set is independent of the choice of $i \in \{1,2,\ldots,n\}$ and, by assumption, is compact. This completes the proof of this part, and hence of the theorem. //

Root compactness has several permanence properties. For locally compact groups these are detailed in [8], Section 3.1. It is clear that if $K \in R_n$ and H is a subhypergroup of K then $H \in R_n$. For the remainder of this section we shall consider how the membership of K in R_n is related to that of the double coset hypergroup $K // H$.

4.6 Theorem Let K, K' be hypergroups, and let $\pi : K \to K'$ be a continuous onto map such that $\pi^{-1}(D)$ is compact in K for every compact subset D of K' , and such that $\pi(B * C) \subset \pi(B) * \pi(C)$ for all compact sets $B, C \subset K$. Then $K' \in R_n$ implies that $K \in R_n$.

In the case H is a compact normal subhypergroup of K, the natural map $\pi : K \to K // H$ satisfies the stated conditions, and then the reverse implication holds as well.

Proof First note that by assumption on π, $\pi(a) \in \pi(a) * \pi(e)$ for all $a \in K$. Since π is onto we can choose $a \in K$ such that $\pi(a) = e$. It follows that $\pi(e) = e$. Now let $C \subset K$ be compact. Then so is $\pi(C)$ and, by assumption, there exists compact $D_n \subset K'$ containing all finite sets $\{y_1, y_2, \ldots, y_n\}$ in K' with $y_n = e$ that satisfy

$$\{y_i\} * \pi(C) * \{y_j\} * \pi(C) \cap \{y_{i+j}\} * \pi(C) \neq \phi , \quad i + j \le n .$$

Put $C_n = \pi^{-1}(D_n)$.

Suppose that $\{x_1, x_2, \ldots, x_n\} \subset K$ satisfies

$$\{x_i\} * C * \{x_j\} * C \cap \{x_{i+j}\} * C \neq \phi , \quad i + j \le n ,$$

where $x_n = e$. Then

$$\phi \neq \pi\left(\{x_i\} * C * \{x_j\} * C \cap \{x_{i+j}\} * C\right) \subset \{\pi(x_i)\} * \pi(C) * \{\pi(x_j)\} * \pi(C) \cap \{\pi(x_{i+j})\} * \pi(C)$$

and $\pi(x_n) = e$. By the choice of D_n, $\{\pi(x_1), \pi(x_2), \ldots, \pi(x_n)\} \subset D_n$. Thus $\{x_1, x_2, \ldots, x_n\} \subset C_n$, a compact set, so that $K \in \mathcal{R}_n$, and this proves the first part of the theorem.

Now consider the case $K' = K // H$, where H is a compact normal subhypergroup of K and π is the natural map. By [9], Theorem 14.1C and Lemma 13.1C, and Lemma 2.2, π satisfies the conditions stated in the theorem. Suppose that $K \in \mathcal{R}_n$ and choose compact $D \subset K // H$. Write $C = \pi^{-1}(D)$ and C_n for a compact set that contains every $\{x_1, x_2, \ldots, x_n\}$ satisfying $x_n = e$ and

$$\{x_i\} * H * C * H * \{x_j\} * H * C * H \cap \{x_{i+j}\} * H * C * H \neq \phi , \quad i + j \le n .$$

For $\{x_1, x_2, \ldots, x_n\}$ satisfying $x_n = e$ and

$$\{Hx_iH\} * D * \{Hx_jH\} * D \cap \{Hx_{i+j}H\} * D \neq \phi , \quad i + j \le n ,$$

we deduce from Lemma 2.2 that, for all $i + j \le n$,

$$\pi\left(\{x_i\} * H * C * H * \{x_j\} * H * C\right) \cap \pi\left(\{x_{i+j}\} * H * C\right) \neq \phi$$

and

$$H * \{x_i\} * H * C * H * \{x_j\} * H * C * H \cap H * \{x_{i+j}\} * H * C * H \neq \phi .$$

Thus, using the normality of H ,

$$\{x_i\} * (H * C * H) * \{x_j\} * (H * C * H) \cap \{x_{i+j}\} * (H * C * H) \neq \phi , \quad i + j \leq n ,$$

and, by the choice of C_n , $\{x_1, x_2, \ldots, x_n\} \subset C_n$. Hence $\{\pi(x_1), \pi(x_2), \ldots, \pi(x_n)\} \subset \pi(C_n)$, a compact set, and the theorem is proved. //

5. Infinitely divisible measures

5.1 Definition A measure $\mu \in M^1(K)$ is called infinitely divisible if for every $n \in \mathbb{N}$ there exists $\mu_n \in M^1(K)$ such that $\mu_n^n = \mu$. The set of infinitely divisible measures in $M^1(K)$ will be denoted by $I(K)$.

For root compact hypergroups, $I(K)$ is weakly closed in $M^1(K)$. To prove this we need a shift compactness theorem for hypergroups.

5.2 Theorem Let (μ_α), (ν_α) and (λ_α) be nets of measures in $M^1(K)$ with $\mu_\alpha = \nu_\alpha * \lambda_\alpha$ for all $\alpha \in \Lambda$.

(a) If (ν_α) and (λ_α) are tight then so is (μ_α) .

(b) If (μ_α) and (λ_α) are tight then so is (ν_α) .

(c) If (μ_α) is tight then there exists $(x_\alpha) \subset K$ such that $(\varepsilon_{x_\alpha} * \lambda_\alpha)$ is tight.

The assertions (a)-(c) continue to hold with "tight" replaced by "uniformly tight".

Proof (a) Let $\varepsilon > 0$ be given. There exists a compact set C_ε and $\alpha_0 \in \Lambda$ such that $\nu_\alpha(C_\varepsilon) \geq 1 - \varepsilon$ and $\lambda_\alpha(C_\varepsilon) \geq 1 - \varepsilon$ for all $\alpha \geq \alpha_0$. By Lemma 4.2 we have that

$$\mu_\alpha(C_\varepsilon * C_\varepsilon) \geq (1-\varepsilon)^2 \geq 1 - 2\varepsilon$$

for all $\alpha \geq \alpha_0$, which shows that (μ_α) is tight.

(b) We first show that for compact sets $B, C \subset K$,

$$\mu_\alpha(B) \le \nu_\alpha\left(B * C^-\right) + \lambda_\alpha\left(C^c\right) .$$

We have

$$(5.3) \quad \mu_\alpha(B) = \nu_\alpha * \lambda_\alpha(B) = \int_C \int_K \xi_B(x*y)\,d\nu_\alpha(x)\,d\lambda_\alpha(y) + \int_{C^c} \int_K \xi_B(x*y)\,d\nu_\alpha(x)\,d\lambda_\alpha(y) .$$

Now if $y \in C$ then $\{x\} * \{y\} \cap B \neq \phi$ implies that $x \in B * \{y^-\} \subset B * C^-$. Hence the first term on the righthand side of (5.3) becomes

$$\int_C \int_{B*C^-} \xi_B(x*y)\,d\nu_\alpha(x)\,d\lambda_\alpha(y) \le \int_C \nu_\alpha(B*C^-)\,d\lambda_\alpha(y)$$

$$\le \nu_\alpha(B*C^-) .$$

Also it is clear that

$$\int_{C^c} \int_K \xi_B(x*y)\,d\nu_\alpha(x)\,d\lambda_\alpha(y) \le \lambda_\alpha\left(C^c\right) ,$$

whence the desired inequality follows.

Now, given $\varepsilon > 0$, choose compact C_ε and $\alpha_0 \in \Lambda$ such that $\mu_\alpha\left(C_\varepsilon\right) \ge 1 - \varepsilon$ and $\lambda_\alpha\left(C_\varepsilon^c\right) \le \varepsilon$ for all $\alpha \ge \alpha_0$. Then, by the above working,

$$\nu_\alpha\left(C_\varepsilon * C_\varepsilon^-\right) \ge \mu_\alpha\left(C_\varepsilon\right) - \lambda_\alpha\left(C_\varepsilon^c\right) \ge 1 - 2\varepsilon$$

for all $\alpha \ge \alpha_0$, so that $\left(\nu_\alpha\right)$ is tight.

(c) Choose $\left(\varepsilon_n\right) \subset R^+$ with $\sum_{n=1}^{\infty} \varepsilon_n < \infty$. For each n there exists compact C_n and $\alpha(n) \in \Lambda$ such that $\mu_\alpha\left(C_n^c\right) < \varepsilon_n$ for all $\alpha \ge \alpha(n)$. Put $\Lambda_n = \{\alpha : \alpha \ge \alpha(n)\}$, $\Delta_n = \Lambda_n \setminus \Lambda_{n+1}$ for each $n \in \mathbf{N}$, and write $\Delta_\infty = \bigcap_{n=1}^{\infty} \Lambda_n$. We may assume that $\alpha(n) < \alpha(n+1)$ for each n, and also that $\Lambda_1 = \Lambda$. Then $\Lambda_{n+1} \subset \Lambda_n$ for all $n \in \mathbf{N}$ and $\bigcup_{n=1}^{\infty} \Delta_n = \Lambda \setminus \Delta_\infty$.

Choose $\left(\delta_n\right) \subset R^+$ satisfying $\lim_n \delta_n = 0$ and $\sum_{n=1}^{\infty} \varepsilon_n / \delta_n < 1/2$.

Write

$$B_{\alpha,n} = \{x \in K : \varepsilon_x * \lambda_\alpha(C_n) \geq 1 - \delta_n\}$$

and $F_\alpha = \bigcap_{n=1}^{m} B_{\alpha,n}$, $\alpha \in \Delta_m$, $m \in \mathbb{N} \cup \{\infty\}$. Now $F_\alpha \neq \phi$ for all $\alpha \in \Lambda$. Indeed, for $\alpha \in \Lambda_n$,

$$1 - \varepsilon_n \leq \mu_\alpha(C_n) = \int_{B_{\alpha,n}} \varepsilon_x * \lambda_\alpha(C_n)\,d\nu_\alpha(x) + \int_{B^c_{\alpha,n}} \varepsilon_x * \lambda_\alpha(C_n)\,d\nu_\alpha(x)$$

$$\leq \nu_\alpha(B_{\alpha,n}) + (1 - \delta_n)\nu_\alpha(B^c_{\alpha,n}) ,$$

so that $\nu_\alpha(B^c_{\alpha,n}) \leq \varepsilon_n/\delta_n$ for all $\alpha \in \Lambda_n$, $n \in \mathbb{N}$. It follows that

$$\nu_\alpha(F^c_\alpha) \leq \sum_{n=1}^{\infty} \varepsilon_n/\delta_n < 1/2$$

for all $\alpha \in \Lambda$. In particular $F_\alpha \neq \phi$, so that we can choose $x_\alpha \in F_\alpha$ for each $\alpha \in \Lambda$.

We now have that $(\varepsilon_{x_\alpha} * \lambda_\alpha)$ is tight. To show this, let $\varepsilon > 0$ and choose n such that $\delta_n \leq \varepsilon$. Then $\alpha \geq \alpha(n)$ implies that $\alpha \in \Delta_m$ for some $m \geq n$ or $\alpha \in \Delta_\infty$. In either case $x_\alpha \in B_{\alpha,n}$ and we have $\varepsilon_{x_\alpha} * \lambda_\alpha(C_n) \geq 1 - \delta_n \geq 1 - \varepsilon$, as required.

The proof of (a)-(c) with "tight" replaced by "uniformly tight" is similar. //

5.4 Theorem If $K \in \mathcal{R}$ then $I(K)$ is weakly closed in $M^1(K)$.

Proof Let $n \in \mathbb{N}$. We first prove that if (μ_α) is tight then (ν_α) satisfying $\nu^n_\alpha = \mu_\alpha$ for each $\alpha \in \Lambda$ is also tight. Let $\varepsilon \in (0, 1/3)$ be given . There exists compact $C \subset K$ and $\alpha_0(\varepsilon)$ such that $\mu_\alpha(C) \geq 1 - \varepsilon$ for all $\alpha \geq \alpha_0(\varepsilon)$. We have that $\nu^i_\alpha * \nu^{n-i}_\alpha = \mu_\alpha$ for $i = 1,2,\ldots,n$ and, by Theorem 5.2(c), there exist $x_{\alpha,i}$ such that $(\varepsilon_{x_{\alpha,i}} * \nu^i_\alpha)$ is tight for each $i = 1,2,\ldots,n$. Hence we can assume that C and $\alpha_0(\varepsilon)$ satisfy

(5.5) $$\varepsilon_{x_{\alpha,i}} * \nu^i_\alpha(C) \geq 1 - \varepsilon \quad \text{for all } \alpha \geq \alpha_0(\varepsilon) , \ i = 1,2,\ldots,n$$

(with $x_{\alpha,n} = e$) . Then, for each $i,j \in \{1,2,\ldots,n\}$, (5.5) gives

$$\nu_\alpha^{i+j}\left(\{x^-_{\alpha,i}\} * C * \{x^-_{\alpha,j}\} * C\right) \geq \nu_\alpha^{i}\left(\{x^-_{\alpha,i}\} * C\right)\nu_\alpha^{j}\left(\{x^-_{\alpha,j}\} * C\right)$$

$$\geq \varepsilon_{x_{\alpha,i}} * \nu_\alpha^i(C)\, \varepsilon_{x_{\alpha,j}} * \nu_\alpha^j(C)$$

$$\geq (1-\varepsilon)^2 > 1/3 \quad ,$$

where we have made use of Lemma 4.3, and similarly

$$\nu_\alpha^{i+j}\left(\{x^-_{\alpha,i+j}\} * C\right) \geq 1 - \varepsilon > 2/3 \, .$$

Thus, for all $\alpha \geq \alpha_0(\varepsilon)$ and $i + j \leq n$, $i,j \in \{1,2,\ldots,n\}$,

$$\{x^-_{\alpha,i}\} * C * \{x^-_{\alpha,j}\} * C \cap \{x^-_{\alpha,i+j}\} * C \neq \phi \, .$$

Since $K \in \mathcal{R} \subset \mathcal{R}_n$ we have the existence of compact $C_n \subset K$ such that $x^-_{\alpha,i} \in C_n$ for all $\alpha \geq \alpha_0(\varepsilon)$, $i \in \{1,2,\ldots,n\}$. Then, for $\alpha \geq \alpha_0(\varepsilon)$,

$$\nu_\alpha\left(C_n * C\right) \geq \nu_\alpha\left(\{x^-_{\alpha,1}\} * C\right) \geq \varepsilon_{x_{\alpha,1}} * \nu_\alpha(C) \geq 1 - \varepsilon \, ,$$

which says that (ν_α) is tight.

Now let $(\mu_\alpha) \subset I(K)$ with $\mathcal{T}_w$-lim $\mu_\alpha = \mu$. Then, by [8], 1.2.20 (2), (μ_α) is tight and, by the above working, so is any (ν_α) satisfying $\nu_\alpha^n = \mu_\alpha$ for all α . By [8], 1.2.20 (1), (ν_α) admits a subnet $(\nu_{\alpha(\beta)})$ converging to ν say. From this it follows that $\nu^n = \mu$ and, since this holds for any $n \in \mathbb{N}$, it must be the case that $\mu \in I(K)$. //

Finally, with no change to the proof of [8], Theorem 3.1.32, we can write down immediately:

5.6 Theorem If $K \in \mathcal{R}$ then every $\mu \in I(K)$ is rationally embeddable.

6. Examples

A list of locally compact groups satisfying a root compactness property is given in [8], Section 3.1. In view of Theorem 4.6 we have that $\mathbb{R}^+$ with the hypergroup structure given by Naimark (see [9], Example 9.5) is not in $\mathcal{R}_n$ for any $n \geq 2$. Indeed, if it were for some n then, since $SL(2,\mathbb{C}) /\!/ SU(2) \cong \mathbb{R}^+$ (see [9], Section 15.5), we could deduce that $SL(2,\mathbb{C}) \in \mathcal{R}_n$, which is not the case. We also note that if a hypergroup has "too many" self adjoint elements ($x \in K$ is called self adjoint if $x^{-} = x$) then it cannot belong to $\mathcal{R}_2$.

6.1 **Theorem** Let K be a hypergroup such that its set H of self adjoint elements is not relatively compact. Then $K \notin \mathcal{R}_2$

Proof We see that, for each $x \in H$,

$$\{x\} * \{e\} * \{x\} * \{e\} \cap \{e\} * \{e\}$$

$$= \{x\} * \{x\} \cap \{e\} \neq \phi \ .$$

If $K \in \mathcal{R}_2$ we would deduce that $H \subset C_2$ for some compact set C_2 , contradicting the assumption on H . //

An interesting example of a hypergroup satisfying the assumption of the theorem is the dual of $SU(2)$. The set of continuous unitary irreducible representations of $SU(2)$ is given by $\{T^{(0)}, T^{(1)}, T^{(2)}, \ldots\}$, where $T^{(n)}$ has dimension $n + 1$; see [6], (29.13) . We identify this set with $\mathbb{N}' = \mathbb{N} \cup \{0\}$. For any $m,n \in \mathbb{N}'$ the tensor product $T^{(m)} \otimes T^{(n)}$ is unitarily equivalent to

$$T^{(|m-n|)} \oplus T^{(|m-n|+2)} \oplus \ldots \oplus T^{(m+n)}$$

(see [6], (29.26) for details). Following [10], Section 5, we define a convolution structure on $M(\mathbb{N}')$ via

$$\varepsilon_m * \varepsilon_n = \sum_{i=0}^{m \wedge m'} \frac{|m - m'| + 2i + 1}{(m+1)(n+1)} \varepsilon_{|m-m'|+2i} \ ,$$

where $m \wedge m' = \frac{1}{2}(m + m' - |m - m'|)$. With this convolution $\mathbb{N}'$ becomes a commutative hypergroup (with the discrete topology) and, since all the $T^{(n)}$ are self conjugate, the elements of $\mathbb{N}'$ are self adjoint. It follows from Theorem 6.1 that $\mathbb{N}' \notin \mathcal{R}_2$. Note also that $\mathbb{N}'$ is not class compact $(C_0 = 2\mathbb{N}')$.

Although a complete description of $I(\mathbb{N}')$ is not known it is the case that if $\mu \in I(\mathbb{N}')$ and $\mu \neq \varepsilon_0$ then $\mathrm{supp}(\mu) = \mathbb{N}'$ or $\mathrm{supp}(\mu) = 2\mathbb{N}'$. To see this, first note that for such μ , $\mathrm{supp}(\mu)$ must be infinite. Indeed for each $n \in \mathbb{N}$, $\mathrm{supp}(\mu_n) \neq \{0\}$ (where $\mu_n^n = \mu$) , and $\mathrm{supp}(\mu) = \bigcup_{n=1}^{\infty} \mathrm{supp}(\mu_n^n)$ shows that $\mathrm{supp}(\mu)$ cannot be finite. It follows that $\mathrm{supp}(\mu_n)$ is infinite for each n .

Now if $m \in \mathrm{supp}(\mu_2)$ then, by the definition of the convolution, $\{0,2,\ldots,2m\} \subset \mathrm{supp}(\mu)$, so that $2\mathbb{N}' \subset \mathrm{supp}(\mu)$. If there exists n for which

$2n+1 \in \mathrm{supp}(\mu)$ then $2m+1 \in \mathrm{supp}(\mu_4)$ for some m, so that $2N' \cup \{2m+1\} \subset \mathrm{supp}(\mu_4^2)$, from which it follows that $\mathrm{supp}(\mu) = N'$.

Although we have not gone into much detail here on N' as the dual of $SU(2)$, much work has in fact been done on its space of probability measures; see [3], [4] and [5].

REFERENCES

[1] Walter R. Bloom and Herbert Heyer, The Fourier transform for probability measures on hypergroups, Rend. Mat. (to appear).

[2] Walter R. Bloom and Herbert Heyer, Convergence of convolution products of probability measures on hypergroups, Rend. Mat. (to appear).

[3] Pierre Eymard et Bernard Roynette, Marches aléatoires sur le dual de SU(2). Analyse harmonique sur les groupes de Lie (Sém. Nancy-Strasbourg, 1973-1975), pp.108-152. Lecture Notes in Math. Vol.497, Springer, Berlin, Heidelberg, New York, 1975.

[4] L. Gallardo et V. Ries, La loi des grands nombres pour les marches aléatoires sur le dual de SU(2), Studia Math. 66 (1979), 93-105.

[5] Léonard Gallardo, Théorèmes limites pour les marches aléatoires sur le dual de SU(2), Ann. Sci. Univ. Clermont Math. 67 (17), (1979), 63-68.

[6] Edwin Hewitt and Kenneth A. Ross, Abstract harmonic analysis, Vol.II, Springer, Berlin, Heidelberg, New York, 1970.

[7] Edwin Hewitt and Karl Stromberg, Real and abstract analysis, Springer, Berlin, Heidelberg, New York, 1965.

[8] Herbert Heyer, Probability measures on locally compact groups, Springer, Berlin, Heidelberg, New York, 1977.

[9] Robert I. Jewett, Spaces with an abstract convolution of measures, Adv. in Math. 18 (1975), 1-101.

[10] Kenneth A. Ross, Hypergroups and centers of measure algebras, Symposia Mathematica 22 (1977), 189-203.

School of Mathematical and Physical Sciences,
Murdoch University,
Perth,
Western Australia 6150,
AUSTRALIA.

POISSON MEASURES ON BANACH LATTICES

Egbert Dettweiler

Probability theory on Banach spaces has shown an intimate relationship between the validity of probabilistic theorems in Banach spaces and the underlying geometry of the spaces (cf. [4] and [19] as general references on this subject). Therefore it seemed to be a natural conjecture that the study of random variables taking their values in the cone of a Banach lattice would lead to similar connections to the lattice structure.

Our results on this subject are divided into two parts. The first part is of preparatory character and illustrates the different situation which arises if one considers only random vectors with values in the cone E_+ of a Banach lattice E. So one gets very weak conditions (without any independence assumption) that an increasing sequence of random vectors converges a.s. For Banach lattices with order continuous norm the infinitely divisible probability measures on E_+ are of the form $\mu = \varepsilon_a * e_o(F)$ where $a \in E_+$ and $e_o(F)$ is the limit of a sequence of Poisson measures $e(G_n)$ on E_+ such that the exponents G_n increase to the so-called Lévy measure F of μ.

The problem of characterizing the infinitely divisible measures by their Lévy measures is closely related to the geometric structure of the Banach lattice. A weak solution of this problem is already given at the end of chapter I. If $T:E \longrightarrow G$ is a cone p-absolutely summing positive operator into a second Banach lattice G then any measure F on E_+ with $\sup_{\|x'\|\leq 1} \int \langle x, |x'|\rangle^p F(dx) < \infty$ and $\sup_{\|x'\|\leq 1} \int \langle x, |x'|\rangle F(dx) < \infty$ is transformed into a Lévy measure T(F). The proof is based on a lemma of Rosenthal ([14]) which also plays a key rôle for the second chapter.

In the second chapter we introduce the notions of a Banach lattice of Poisson type and of Poisson cotype p resp. ($1 \leq p < \infty$). A Banach lattice E is said to be of Poisson type p if there exists a constant $C > o$ such that for all finite sequences (x_i) in E_+ and all finite sequences (ξ_i) of independent Poissonian random variables the following inequality holds:

$$\mathbb{E}\|\sum \xi_i x_i\|^p \leq C\ (\sum \mathbb{E}\xi_i^p \|x_i\|^p + \|\sum \mathbb{E}\xi_i x_i\|^p)$$

Similarly, E is said to be of Poisson cotype p if the converse inequality holds. The lemma of Rosenthal implies that every abstract L^p space is simultaneously of Poisson type and cotype p. It is shown that the Banach lattices of Rademacher type p ($1\leq p\leq 2$) (cf. [4]) are of Poisson type p but not conversely, since every abstract L^1 space is of Poisson type p (and Poisson cotype p) for all $p\geq 1$.

The Banach lattices of Poisson cotype p are exactly those Banach lattices having an equivalent p-superadditive lattice norm. For Banach lattices of Poisson type p it seems to be more difficult to get a complete geometric characterization by elementary properties of the norm. We can only prove that a Banach lattice of Poisson type p, with the additional property that c_o and ℓ^1 are not finitely representable (in the lattice sense, cf. [7]), always has an equivalent q-subadditive lattice norm for all $q<p$.

The Banach lattices of Poisson type turn out to be the natural domain for the solution of the central limit problem for cone valued random vectors since on these Banach lattices the infinitely divisible measures can be completely characterized by simple integrability properties of their Lévy measures. To any Lévy measure F one can associate a Poisson process $\xi = (\xi_S)$ indexed by the Borel subsets of E_+ with finite F-measure. This Poisson process defines a sort of stochastic integral and it is shown that the space of all E-valued functions f such that f is ξ-integrable is a Banach lattice. With the aid of this stochastic integral it is possible to prove a general central limit theorem in giving conditions for the convolution products of a uniformly infinitesimal triangular sequence (μ_{nj}) of probability measures on E_+ to converge towards a given infinitely divisible measure on E_+ (cf. [19]).

This paper contains results of the Habilitationsschrift of the author (Poisson Measures on Banach Lattices. Habilitationsschrift, Tübingen 1981, unpublished).

I. General convergence theorems for sequences of Poisson measures

§o - Preliminaries

The basic notations and definitions of the theory of topological vector spaces and Banach lattices are taken from [10], [15] and [16]. For the subject of probability theory on Banach spaces the reader is refered to [4] and [19].

A cylindrical measure on a locally convex vector space E is by definition a projective family of measures on the finite dimensional quotient spaces of E. If E is an ordered locally convex space then a cylindrical measure μ is said to be a cylindrical measure on the cone E_+ of E (cf. [3]) if for all positive linear operators v on E with finite rank the measure $v(\mu)$ is concentrated on $v(E)_+$. $\check{\mathbb{P}}(E_+)$ denotes the set of all cylindrical measures on E_+. For any $\mu \in \check{\mathbb{P}}(E_+)$ we can define the Laplace transform $L\mu : E'_+ \to \mathbb{R}_+$ of μ, given by

$$L\mu(x') := \int e^{-\langle x,x'\rangle}\mu(dx) \qquad \text{(cf. [2])}.$$

A Radon measure on a Hausdorff space S is a locally bounded regular Borel measure (cf. [18] as a reference for the general theory of Radon measures on topological spaces). $M^b_+(S)$ and $\mathbb{P}(S)$ denote the spaces of all positive bounded Radon measures and all probability (Radon) measures on S resp. If $G \in M^b_+(E)$ where E is a Banach space then the Poisson measure $e(G)$ is defined by $e(G) := e^{-\|G\|}(\sum_{k\geq 0} G^k/k!)$. G^k denotes the kth power of G with respect to the convolution product and G^0 is defined as the Dirac measure ε_0.

If $(\Omega, \mathfrak{A}, P)$ is a probability space then an S-valued random variable X always means a measurable map from Ω into S such that the distribution μ of X is a Radon measure on S. If S = E is a Banach space this is equivalent to the requirement that X is Bochner measurable, i.e. that X is the a.s. limit of a sequence of step functions. In this case we also say that X is an E-valued random vector. The space of (equivalence classes of) E-valued random vectors with the topology of convergence in probability is denoted by $L^0(\Omega, E)$ and $L^p(\Omega, E)$ (or simply L^p_E) denotes the subspace of all $X \in L^0(\Omega, E)$ with $\|X\|_p := (\int \|X\|^p dP)^{1/p} < \infty$.

If there is no danger of confusion we will in general not explicitely mention the underlying probability space if we consider E-valued random vectors. The symbol $\mathbb{E}$ always denotes the mathematical expectation with respect to an (often unspecified) probability space.

§1 - Some remarks on the convergence of positive random vectors

Let (X_k) be an independent sequence of random vectors in a Banach space E and put $Y_n := \sum_{k=1}^{n} X_k$. Then the theorem of Ito and Nisio (cf. [4] for a general version) asserts that for the sequence (Y_n) the notions of a.s. convergence, convergence in probability, and weak convergence are all equivalent. If E is a Banach lattice and (X_k) is a sequence of positive random vectors, then it is not hard to prove this equivalence principle without the assumption of independence (cf. [5]).

1.1 Proposition. Let (Y_n) be an (a.s.) increasing sequence of random vectors in the cone E_+ of a Banach lattice E and let (μ_n) denote the corresponding sequence of distributions. Then the following assertions are equivalent:

(a) (Y_n) converges a.s.
(b) (Y_n) converges in probability.
(c) (μ_n) converges weakly.
(d) (μ_n) is uniformly tight relative to the norm topology.
(e) (μ_n) is uniformly tight relative to the weak topology.

Proof. We only have to show that (e) implies (a). The assumption (e) means that there exists an increasing sequence (K_j) of weakly compact subsets of E_+ such that $P[Y_n \in K_j] \geq 1-1/j$ for all $n, j \in \mathbb{N}$. We put

$$A_j := \bigcap_n \bigcup_{i \geq n} [Y_i \in K_j] \quad \text{and} \quad A := \bigcup_j A_j .$$

Then $P(A)=1$. Let $\omega \in A$ be a fixed element and assume $\omega \in A_j$. This implies that a certain subsequence of $(Y_n(\omega))$ converges weakly in K_j. Since $(Y_n(\omega))$ is increasing, the whole sequence $(Y_n(\omega))$ converges strongly, i.e. (Y_n) converges a.s. ⌋

If E is a Banach lattice with order continuous norm one can add a further and extremely weak condition to the equivalent conditions of prop.1.1.

1.2 Proposition. Let (Y_n) be an increasing sequence of positive random vectors in a Banach lattice E with order continuous norm. Then the following holds.

(a) If there exists an E_+-valued random vector Y such that for all $x' \in E'_+$ $P[\langle Y_n, x'\rangle \leq \langle Y, x'\rangle] = 1$ then (Y_n) converges a.s.
(b) If the associated sequence $(L\mu_n)$ of Laplace transforms converges pointwise to the Laplace transform of a Radon measure μ on E_+ then (Y_n) converges a.s.

Proof. In a Banach lattice with order continuous norm the solid hull of a compact subset of E_+ is weakly compact. Since Y is an E_+-valued random vector, this shows that for all $\varepsilon > o$ there exists a weakly compact, convex and solid set K with $P[Y \in K] \geq 1-\varepsilon$. For all $x_1', \dots, x_n' \in E_+'$ let $\pi_{x_1', \dots, x_n'}$ denote the projection defined by

$\pi_{x_1', \dots, x_n'}(x) := (\langle x, x_1' \rangle, \dots, \langle x, x_n' \rangle)$ for all $x \in E$, and for any subset C of a Banach lattice we denote by s(C) the solid hull of C. Then the assumption in (a) implies

$$\mu_n(\pi^{-1}_{x_1', \dots, x_m'}(s(\pi_{x_1', \dots, x_m'}(K))) \geq 1 - \varepsilon$$

for all $n \in \mathbb{N}$ and every finite sequence $(x_1', \dots, x_m')$ in E_+'. The family

$$(E_+ \cap \pi^{-1}_{x_1', \dots, x_m'}(s(\pi_{x_1', \dots, x_m'}(K))))_{(x_1', \dots, x_m') \in E_+'^{\mathbb{N}}}$$

forms a decreasing family of closed sets in E_+ with intersection $K \cap E_+$, where the last assertion follows from the separation theorem. Therefore we have proved $\mu_n(K) \geq 1-\varepsilon$ for for all $n \in \mathbb{N}$. Since $\varepsilon > o$ was arbitrary the sequence (μ_n) is uniformly tight for the weak topology and the assertion of (a) follows from proposition 1.1. The assertion (b) can be proved in a similar way. One only has to observe that the convergence of the Laplace transforms is equivalent to the weak convergence of the marginal distributions $(\pi_{x_1', \dots, x_m'}(\mu_n))$ for all $x_1', \dots, x_m' \in E_+'$. $\rfloor$

§2 - Poisson measures and Lévy measures

In the following let E always denote a Banach lattice with order continuous norm. For any $\mu \in \mathbb{P}(E_+)$ the set $S := \{x \in E_+ : \mu(E_+ + x) = 1\}$ is an increasing and directed set and one has $\mu(\bigcap_{x \in S}(E_+ + x)) = 1$. Hence S is order bounded . Therefore a:=sup S exists in E_+ and one has $\mu(E_+ + a) = 1$ and $\mu(E_+ + x) < 1$ for all $x \notin [o, a]$. Let us call a the support point of μ. Now let μ be an infinitely divisible measure on E_+ with Lévy measure F. If a is the support point of μ, then $\mu * \varepsilon_{-a}$ is an infinitely divisible measure on E_+ with support point o and with the the same Lévy measure as μ. Let us denote this unique infinitely divisible measure with Lévy measure F and support point o by $e_o(F)$. Hence any infinitely divisible measure λ on E_+ with Lévy measure F has the unique decomposition $\lambda = \varepsilon_b * e_o(F)$ with $b \in E_+$.

The measure $e_o(F)$ can be obtained also as a limit of a sequence $(e(G_n))$ of Poisson measures such that (G_n) increases to F. For this we need the following lemma.

2.1 Lemma. Let $(\mu_i)_{i\in I}$ be a uniformly tight family of probability measures on E_+ such that for all $i\in I$ the measure μ_i has the decomposition $\mu_i = \lambda_i * \nu_i$ with $\lambda_i, \nu_i \in \mathbb{P}(E_+)$. Then the families $(\lambda_i)_{i\in I}$ and $(\nu_i)_{i\in I}$ are uniformly tight relative to the system of all weakly compact solid subsets of E_+.

Proof. An inspection of the proof of a theorem of Parthasarathy, Ranga Rao and Varadhan (cf. [13], ch.III, theorem 2.2) shows the existence of families (a_i) and (b_i) in E_+ such that the families $(\lambda_i * \varepsilon_{a_i})$, $(\nu_i * \varepsilon_{b_i})$ and $(\varepsilon_{a_i+b_i})$ are uniformly tight. Therefore we have $a_i + b_i \in K_1$ for a certain compact set $K_1 \subset E_+$ for all $i\in I$. Consequently, (a_i) and (b_i) are families in the weakly compact set $s(K_1)$. Now we choose for an arbitrary $\varepsilon > 0$ a compact $K_2 \subset E_+$ such that $\lambda_i(K_2 - a_i) \geq 1-\varepsilon$. Then $\lambda_i(s(K_1+K_2)) \geq 1-\varepsilon$ since $E_+ \cap (K_2 - a_i) \subset s(K_1+K_2)$. This proves the uniformly tightness of (λ_i). ⌋

2.2 Proposition. Suppose that we are given a uniformly tight sequence (μ_n) in $\mathbb{P}(E_+)$ with $\mu_n = e(G_n) * \nu_n$ for all $n\in\mathbb{N}$, where (ν_n) is a sequence in $\mathbb{P}(E_+)$ and (G_n) is an increasing sequence in $M_+^b(E_+)$. Then the sequence of Poisson measures $(e(G_n))$ converges and $\lim e(G_n) = e_o(F)$ with $F = \sup G_n$.

Proof. Since (G_n) is increasing there exists an increasing sequence (Y_n) of random vectors such that $e(G_n)$ is the distribution of Y_n. Now the above lemma and proposition 1.1 show that $(e(G_n))$ converges. If one computes the Laplace transform of $\lim e(G_n)$ then it is not hard to see that $\lim e(G_n) = e_o(F)$. ⌋

For sequences of Poisson measures on a Banach space it is usually necessary to shift the measures in order to get convergence. One can show with a more detailed analysis of the Laplace transforms that in the lattice case this shifting is normally unnecessary even if the sequence (G_n) is not increasing.

2.3 Proposition. Let $(e(G_n))$ be a sequence of Poisson measures on the cone E_+ such that $(e(G_n) * \varepsilon_{a_n})$ converges for a certain sequence (a_n) in E_+. Then $(e(G_n))$ converges if and only if

$$\text{(u)} \quad \lim_{s\to 0} \sup_{n\in\mathbb{N}} \sup_{x'\geq 0, \|x'\|\leq 1} \int_{[\|x\|\leq s]} \langle x, x'\rangle G_n(dx) = 0 .$$

If $\lim e(G_n) * \varepsilon_{a_n} = e_o(F)$ then condition (u) is automatically satisfied. If E is weakly sequentially complete then necessarily $(e(G_n))$ converges to $e_o(F)$ relative to the weak topology on E.

Proof. The convergence of $(e(G_n)*\varepsilon_{a_n})$ implies

$$\lim_{n,m\to\infty} \sup_{\substack{x'\geq 0\\ \|x'\|\leq 1}} | L(e(G_n)*\varepsilon_{a_n})(x') - L(e(G_m)*\varepsilon_{a_m})(x')| = 0.$$

Therefore (a_n) is a Cauchy sequence if and only if $\lim_{n,m\to\infty} A_{n,m} = 0$ where $A_{n,m} := \sup_{\substack{x'\geq 0\\ \|x'\|\leq 1}} \left|\int (1-e^{-\langle x,x'\rangle})G_n(dx) - \int (1-e^{-\langle x,x'\rangle})G_m(dx)\right|$.

Now the following inequality holds:

$$A_{n,m} \leq \sup_{\substack{x'\geq 0\\ \|x'\|\leq 1}} \left(\int_{[\|x\|\leq s]} (1-e^{-\langle x,x'\rangle})G_n(dx) + \int_{[\|x\|\leq s]} (1-e^{-\langle x,x'\rangle})G_m(dx)\right)$$

$$+ \sup_{\substack{x'\geq 0\\ \|x'\|\leq 1}} \left|\int_{[\|x\|>s]} (1-e^{-\langle x,x'\rangle})G_n(dx) - \int_{[\|x\|>s]} (1-e^{-\langle x,x'\rangle})G_m(dx)\right| .$$

If condition (u) holds then $A_{n,m}$ becomes small for n,m large enough since the convergence of $(e(G_n)*\varepsilon_{a_n})$ implies that for all $s>0$ the sequence $(\mathrm{Res}_{[\|x\|>s]} G_n)$ converges (since $(\mathrm{Res}_{[\|x\|>s]} G_n)$ is uniformly tight). Similarly, one can show that the convergence of (a_n) implies (u). Suppose now that $\lim e(G_n)*\varepsilon_{a_n} = e_0(F)$. Then for $s<1/2$ and $\|x'\|\leq 1$, $x'\in E'_+$ we have the inequality

$$\frac{1}{2}\int_{[\|x\|\leq s]} \langle x,x'\rangle G_n(dx) \leq \int_{[\|x\|\leq s]} (1-e^{-\langle x,x'\rangle})G_n(dx) + \langle a_n,x'\rangle$$

$$\leq \left|\int (1-e^{-\langle x,x'\rangle})G_n(dx) + \langle a_n,x'\rangle - \int (1-e^{-\langle x,x'\rangle})F(dx)\right|$$

$$+ \left|\int_{[\|x\|>s]} (1-e^{-\langle x,x'\rangle})G_n(dx) - \int_{[\|x\|>s]} (1-e^{-\langle x,x'\rangle})F(dx)\right| ,$$

and the same argument as in the first part of the proof shows the validity of (u). Now suppose that E is weakly sequentially compact, and take a fixed $x'\in E'_+$. Lemma 2.1 implies that the sequence $(e(\pi_{x'}(G_n)))$ is relative compact in $\mathbb{P}(\mathbb{R}_+)$. Let $(e(\pi_{x'}(G_{n_k})))$ be a convergent subsequence. By the first part of the proposition for this subsequence condition (u) holds (in the Banach lattice $\mathbb{R}$). Now it is not difficult to prove with the aid of condition (u) that necessarily $\lim e(\pi_{x'}(G_{n_k})) = e_0(\pi_{x'}(F))$. This shows that the whole sequence $(e(\pi_{x'}(G_n)))$ converges. Hence $(\langle a_n,x'\rangle)$ is a Cauchy sequence for all $x'\in E'_+$ and by assumption on E the sequence (a_n) converges weakly. This together with the relative compactness of $(e(G_n))$ with respect to the weak topology on E yields the asserted convergence of $(e(G_n))$ since we already know that the sequence of Laplace transforms converges pointwise to the Laplace transform of $e_0(F)$. $\lrcorner$

For infinitely divisible measures on E_+ the support can easily be obtained from the support of the Lévy measure. For $A \subset E_+$ let us denote by $SG^-(A)$ the closed subsemigroup of E_+ generated by A.

2.4 Proposition. $\operatorname{supp}(e_o(F)) = SG^-(\operatorname{supp}(F) \cup \{o\})$.

Proof. Let (G_n) be an increasing sequence in $M_+^b(E_+)$ with $\sup G_n = F$. Then there exists a sequence (ν_n) in $\mathbb{P}(E_+)$ with limit ε_o such that $e_o(F) = e(G_n) * \nu_n$ for all $n \in \mathbb{N}$.

1. Let $a \in \operatorname{supp}(e_o(F))$, $a \neq o$. For every fixed o-neighborhood U we choose a (convex, circled) o-neighborhood V with $4V \subset U$. By assumption we have for a certain $\delta > o$ $e_o(F)(V+a) \geq \delta$ and also $\nu_n(V^c) \leq \delta/2$ for n large enough. Hence we get

$$e_o(F)(V+a) = \int e(G_n)(V+a-x)\ \nu_n(dx)$$

$$\leq \int_V e(G_n)(V+a-x)\ \nu_n(dx) + \delta/2 \leq e(G_n)(2V+a) + \delta/2,$$

i.e. $e(G_n)(2V+a) > o$. This shows that there is a $b \in 2V+a$ with $b \in \operatorname{supp}(e(G_n))$. Consequently, $G_n^k(V+b) > o$ for a suitable $k \in \mathbb{N}$ and there is a $c \in V+b$ with

$$c \in \operatorname{supp}(G_n^k) = (\underbrace{\operatorname{supp} G_n + \ldots + \operatorname{supp} G_n}_{k \text{ times}})^{-} .$$

So we get k elements $x_1,\ldots,x_k \in \operatorname{supp} G_n \subset \operatorname{supp} F$ such that

$$\sum_{j=1}^k x_j \in V + c \subset 2V + b \subset 4V + a \subset U + a .$$

Since $\sum_{j=1}^k x_j \in SG(\operatorname{supp} F)$ the arbitrariness of U shows $a \in SG^-(\operatorname{supp} F \cup \{o\})$.

2. If $o \notin \operatorname{supp} F$ then $F(V)=o$ for a suitable o-neighborhood V and F is a bounded measure. But then $o \in \operatorname{supp}(e_o(F))$ since $e_o(F)(V) \geq e^{-\|F\|} > o$. Now we suppose $o \in \operatorname{supp} F$ and take $a \in SG^-(\operatorname{supp} F)$. Let U be an arbitrary o-neighborhood and V a convex circled o-neighborhood with $2V \subset U$. Then we have for all $n \in \mathbb{N}$

$$e_o(F)(U+a) \geq \int_V e(G_n)(U+a-x)\nu_n(dx) \geq e(G_n)(V+a)\nu_n(V) ,$$

and we have only to prove that $e(G_n)(V+a) > o$ for all $n \in \mathbb{N}$ large enough. If $a=o$ then surely $G_n(V) > o$ and therefore $e(G_n)(V) > o$ for all n large enough. If $a \neq o$ we may suppose $o \notin V+a$. By assumption on a there exist k non-zero elements $x_1,\ldots,x_k \in \operatorname{supp} F$ such that

$$a - \sum_{j=1}^k x_j \in W := 1/2\ V .$$

If n is large enough then surely $x_j \in \operatorname{supp} G_n$ for all j. Therefore $\sum_{j=1}^k x_j \in \operatorname{supp}(G_n^k)$ and hence $\sum_{j=1}^k x_j \in \operatorname{supp}(e(G_n))$.

Since $W + \sum_{j=1}^{k} x_j \subset V+a$ we get $e(G_n)(V+a) > o$. ┘

<u>Remark</u>. The above proof shows that $\mathrm{supp}(e_o(F)) = SG^-(\mathrm{supp}\, F)$ for an unbounded Lévy measure F. The bounded Lévy measures are exactly those measures with $e_o(F)(\{o\}) > o$.

Let $(e(G_n))$ be a sequence of Poisson measures on E_+ such that (G_n) increases. If one does not know in advance that one of the equivalent conditions of prop.1.1 holds, then one can look for conditions on the measure $F = \sup G_n$ which ensure the convergence of $(e(G_n))$. This is equivalent to the problem to determine those measures F on E_+ which are actually Lévy measures. A solution of this problem depends on the geometric structure of the Banach lattice as will be shown in the next chapter. If there is no restriction on the given Banach lattice on can get only results for special measures. This will be illustrated in the following proposition.

2.5 <u>Proposition</u>. Let E be an arbitrary Banach lattice. If (x_k) is a sequence in $E_+ \setminus \{o\}$ such that for a sequence (a_k) of positive numbers with the property $\sum([a_k]!)^{-1} < \infty$ the series $\sum a_k x_k$ converges, then $F = \sum \varepsilon_{x_k}$ is a Lévy measure.

<u>Proof</u>. Let (ξ_k) be a sequence of independent Poisson distributed random variables with parameter 1 defined on a probability space $(\Omega, \mathfrak{A}, P)$. Prop. 1.1 shows that F is a Lévy measure if and only if $\sum \xi_k x_k$ converges in probability. We show that for all $\varepsilon > o$ there exists an $A \in \mathfrak{A}$ such that $P(A) \geq 1-\varepsilon$ and

$$\lim_{\substack{n,m\to\infty \\ n<m}} \int_A \left\| \sum_{k=n}^{m} \xi_k x_k \right\| dP = o .$$

Clearly, this implies the convergence in probability. Let A be an event of the form $A := [\xi_k \leq n_k \text{ for all } k \in \mathbb{N}]$ where (n_k) is a sequence in $\mathbb{N}$. One obtains

$$P(A) = \prod_{k\geq 1} e^{-1}\Big(\sum_{j=0}^{n_k} \frac{1}{j!}\Big) \geq \prod_{k\geq 1}\Big(1 - \frac{1}{(n_k+1)!}\Big).$$

This implies that the probability of A is near to 1 if $\sum_{k\geq 1}(n_k+1)!^{-1}$ is small. For all $k,j \in \mathbb{N}$ we now define

$$n_k(j) := j + [a_k] \quad \text{and} \quad A_j := [\xi_k \leq n_k(j) \text{ for all } k] .$$

Since $\sum_{k\geq 1} n_k(j)!^{-1} \leq j^{-1} \sum_{k\geq 1} [a_k]!^{-1}$, we get for $\varepsilon > o$ and j large enough $P(A_j) \geq 1-\varepsilon$. Furthermore we have

$$\int_{A_j} \left\| \sum_{k=n}^{m} \xi_k x_k \right\| dP \leq \left\| \sum_{k=n}^{m} n_k(j) x_k \right\| \leq j \left\| \sum_{k=n}^{m} x_k \right\| + \left\| \sum_{k=n}^{m} a_k x_k \right\| .$$

The convergence of $\sum a_k x_k$ now implies $\lim\limits_{n,m\to\infty} \int_{A_j} \|\sum_{k=n}^{m} \xi_k x_k\| dP = 0.$ ∎

Remark. It will follow from the results of the next chapter that for many Banach lattices the convergence of $\sum x_k$ is already sufficient for F to be a Lévy measure.

2.6 Corollary. Let E be an abstract L^p space of infinite dimension ($1<p<\infty$). For all q with $1\leq q<p$ there exists a Lévy measure F on $E_+\setminus\{0\}$ concentrated on the unit ball B such that $\int \|x\|^p F(dx)<\infty$ and $\int \|x\|^q F(dx) = \infty$. If E is a Banach lattice containing a Banach sublattice lattice isomorphic to c_o then there exists a Lévy measure F concentrated on $B\subset E_+$ with $\int \|x\|^p F(dx) = \infty$ for all $1\leq p<\infty$.

Proof. In the first case we choose a sequence (y_k) in E_+ of disjoint elements of norm 1 and put $x_k := k^{-1/q} y_k$ and $F:=\sum \varepsilon_{x_k}$. Then

$$\int \|x\|^p F(dx) = \sum k^{-p/q} < \infty \ , \quad \int \|x\|^q \ F(dx) = \sum k^{-1} = \infty \ ,$$

and F is a Lévy measure by prop.2.5 if one defines $a_k := k^{(p-q)/(p+q)q}$ for all $k\in\mathbb{N}$. For the second assertion we assume w.l.o.g. $E = c_o$. If (e_k) denotes the unit vector basis we put

$$x_k := (\log k)^{-2}\ e_k \quad \text{and} \quad a_k := \log k\ .$$

Then $\sum a_k x_k \in c_o$ and $F=\sum \varepsilon_{x_k}$ is a Lévy measure by prop.2.5, but for all $1\leq p<\infty$ we have $\int \|x\|^p \ F(dx) = \sum (\log k)^{-2p} = \infty.$ ∎

§3 - A characterization of cone p-absolutely summing operators by Lévy measures

A σ-finite measure F on E_+ is called a weak Lévy measure if there is an infinitely divisible cylindrical measure μ on E_+ such that for all $x_1',\dots,x_n' \in E_+'$ the measure $\pi_{x_1',\dots,x_n'}(F)$ restricted to $\mathbb{R}_+^n\setminus\{0\}$ is the Lévy measure of $\pi_{x_1',\dots,x_n'}(\mu)$. A weak Lévy measure is said to be of type p ($1\leq p<\infty$) if for all $x'\in E_+'$ the condition $\int \langle x,x'\rangle^p F(dx)<\infty$ holds. Clearly, in general a weak Lévy measure is not a Lévy measure, but one can ask wether certain operators into a second Banach lattice transform weak Lévy measures into Lévy measures.

An operator T from E into a Banach space F is called cone p-absolutely summing if T transforms every weakly p-summable sequence in E_+ into an absolutely p-summable sequence in F. One knows (cf. [2], [17]) that T is cone p-absolutely summable if and only if T transforms every cylindrical measure on E_+ of type p into a Radon measure on $(F'', \sigma(F'',F'))$ of order p, and we will show in the following that T is already cone p-absolutely summing if (and only if) T transforms every weak Lévy measure being simultaneously of type p and type 1 into a Lévy measure. A key to this result is the following lemma of H.P. Rosenthal ([14], lemma 1):

3.1 Lemma. Let $1 \leq p < \infty$ and let $\xi_1, \ldots, \xi_n$ be independent positive (real) random variables belonging to L^p. Then there exists a constant C only depending on p such that

$$\int \left(\sum_{i=1}^{n} \xi_i\right)^p dP \leq C\left(\sum_{i=1}^{n} \int \xi_i^{\,p}\, dP + \left(\sum_{i=1}^{n} \int \xi_i\, dP\right)^p\right).$$

3.2 Theorem. Let $T: E \to G$ be a weakly continuous positive linear operator from E into a Banach lattice G with order continuous norm. Let $1 \leq p < \infty$ be given. Then the following assertions are equivalent:

(a) T is cone p-absolutely summing.

(b) For any sequence (X_k) of independent E_+-valued random vectors such that

$$\sum \int \langle X_k, x'\rangle^p\, dP < \infty \quad \text{and} \quad \int \langle \sum X_k, x'\rangle dP < \infty \quad \text{for all } x' \in E'_+$$

the series $\sum T \circ X_k$ converges in probability.

(c) T transforms any weak Lévy measure which is both of type p and type 1 into a Lévy measure.

(d) For all sequences (x_k) in E_+ and all sequences (ξ_k) of independent Poisson distributed (real) random variables with parameter $\lambda_k \leq 1$ for all $k \in \mathbb{N}$ the series $\sum \xi_k T(x_k)$ converges in probability if

$$\sum \lambda_k \langle x_k, x'\rangle^p < \infty \quad \text{and} \quad \sum \lambda_k \langle x_k, x'\rangle < \infty \text{ for all } x' \in E'_+ .$$

(e) There is a constant $C > 0$ (only depending on p) such that

$$\int \Big\| \sum_{k=1}^{n} \xi_k T(x_k) \Big\|^p dP \leq C\Big(\sup_{\substack{x' \in E'_+ \\ \|x'\| \leq 1}} \sum_{k=1}^{n} \lambda_k \langle x_k, x'\rangle^p + \Big\|\sum_{k=1}^{n} \lambda_k x_k\Big\|^p\Big)$$

for every finite sequence $(x_k)_{1\leq k\leq n}$ in E_+ and every finite sequence $(\xi_k)_{1\leq k\leq n}$ of independent Poisson distributed random variables with with parameter $\lambda_k \leq 1$.

Proof. We prove the implications (a)$\Rightarrow$(b), (b)$\Rightarrow$(c), (c)$\Rightarrow$(d), (d)$\Rightarrow$(a), (a)$\Rightarrow$(e), (e)$\Rightarrow$(d).

(a)$\Rightarrow$(b): If T is cone p-absolutely summing then there exists a constant $D > 0$ such that

$$\int \Big\| \sum T \circ Y_k \Big\|^p dP \leq D \sup_{x' \in E'_+,\, \|x'\| \leq 1} \int \left(\sum \langle Y_k, x'\rangle\right)^p dP$$

(cf. [17], p.197) for every finite sequence (Y_k) of E_+-valued random vectors. Therefore the assertion follows from lemma 3.1 together with prop. 1.1.

(b)$\Longrightarrow$(c): If F is a weak Lévy measure then $F = \sum G_k$ where (G_k) is a sequence in $M_+^b(E_+)$. Let (X_k) be an independent sequence of E_+-valued random vectors X_k with distribution $e(G_k)$. For all $k \in \mathbb{N}$ one has

$$\int \langle X_k, x' \rangle \, dP = \int \langle x, x' \rangle \, G_k(dx) \text{ and}$$

$$\int \langle X_k, x' \rangle^p \, dP \leq C \left(\int \langle x, x' \rangle^p \, G_k(dx) + \left(\int \langle x, x' \rangle G_k(dx) \right)^p \right)$$

where the latter inequality follows from lemma 3.1 by approximating every G_k by measures of the form $\sum_{j=1}^{n} \lambda_j \varepsilon_{x_j}$ with $\lambda_j \leq 1$ and $x_j \in E_+$ for $1 \leq j \leq n$ (see ch.II for a general proof of this approximation procedure). Now (b) implies that $\sum T \circ X_k$ converges in probability and the corresponding limit distribution clearly is infinitely divisible with T(F) as Lévy measure.

(c)$\Longrightarrow$(d): This follows immediately if one takes $F := \sum \lambda_k \varepsilon_{x_k}$ as weak Lévy measure.

(d)$\Longrightarrow$(a): First we show that for all $\varepsilon > o$ there exists a $\delta > o$ such that for all finite sequences (λ_k) in $[o,1]$ and (x_k) in E_+ the condition

$$\sup_{x' \in E'_+, \|x'\| \leq 1} \sum \lambda_k \langle x_k, x' \rangle^p + \left\| \sum \lambda_k x_k \right\|^p \leq \delta$$

implies

$$\int \inf(1, \left\| \sum \xi_k T(x_k) \right\|^p) dP \leq \varepsilon ,$$

where (ξ_k) is an independent sequence of Poisson distributed random variables with parameter λ_k.

Suppose that this assertion does not hold. Then there is an $\varepsilon > o$ and for all $n \in \mathbb{N}$ there are finite sequences $(\lambda_{k,n})$ in $[o,1]$ and $(x_{k,n})$ in E_+ such that

$$\sup_{x' \in E'_+, \|x'\| \leq 1} \sum_k \lambda_{k,n} \langle x_{k,n}, x' \rangle^p + \left\| \sum_k \lambda_{k,n} x_{k,n} \right\|^p \leq 2^{-n}$$

and $\qquad \int \inf(1, \left\| \sum_k \xi_{k,n} T(x_{k,n}) \right\|^p) \, dP \geq \varepsilon$ for all $n \in \mathbb{N}$,

and we may and do suppose that the whole double sequence $(\xi_{k,n})$ of Poisson distributed random variables with parameters $\lambda_{k,n}$ is independent. Now from assumption (d) it follows that the series $\sum_{k,n} \lambda_{k,n} T(x_{k,n})$ converges in probability and this is a contradiction to $\int \inf(1, \left\| \sum_k \xi_{k,n} T(x_{k,n}) \right\|^p) dP \geq \varepsilon$ for all $n \in \mathbb{N}$.

For $\varepsilon := 1/(2e)$ now we choose a $\delta > o$ such that the above condition holds. Let (x_k) be an arbitrary finite sequence in E_+ such that $\sup_{x'\geq 0, \|x'\|\leq 1} \sum_k \langle x_k, x'\rangle^p \leq \delta/2$. For any sequence (λ_k) of positive numbers with $\sum \lambda_k = 1$ and all $x' \in E'_+$ with $\|x'\| \leq 1$ we have

$$(\sum_k \lambda_k \langle \lambda_k^{-1/p} x_k, x'\rangle)^p \leq \sum_k \lambda_k \langle \lambda_k^{-1/p} x_k, x'\rangle^p \leq \delta/2 \text{ , and hence}$$

$$e^{-1} \sum_k \lambda_k \inf(1, \lambda_k^{-1} \|T(x_k)\|^p) \leq \int \inf(1, \|\sum_k \xi_k T(\lambda_k^{-1/p} x_k)\|^p) dP \leq 1/(2e).$$

For the special choice $\lambda_k := \|T(x_k)\|^p / \sum_i \|T(x_i)\|^p$ we get $\sum_k \|T(x_k)\|^p \leq 1/2$. Therefore T is cone p-absolutely summing.

(a)$\Longrightarrow$(e) follows from lemma 3.1 together with the inequality in the proof of (a)$\Longrightarrow$(b), and the last implication (e)$\Longrightarrow$(d) is clear.|

3.3 <u>Corollary</u>. A Banach lattice E with order continuous norm is isomorphic to an abstract L^1 space if and only if every weak Lévy measure of type 1 is already a Lévy measure.

<u>Proof</u>. The theorem implies that the identity on E is cone absolutely summing. Hence E is isomorphic to an abstract L^1 space ([16], ch.IV, theorem 2.7). The converse assertion is clear.|

<u>Remarks</u>. 1. If the Banach lattice G in the theorem has not necessary order continuous norm, then the theorem remains valid if one replaces the condition (c) by the stronger condition that any weak Lévy measure of type p and type 1 will be transformed by T into a Lévy measure of an infinitely divisible measure of order p.

2. One can get similar characterizations of the p-absolutely summing operators on Banach spaces. An operator $T:E \longrightarrow G$ from a Banach space E into a Banach space G is p-absolutely summing for $1 \leq p < \infty$ if and only if every weak Lévy measure of type p (and in addition of type 2 if $p > 2$) is transformed into a Lévy measure. Here [14], theorem 3 has to be used instead of lemma 3.1. There is no restriction on G necessary since in the Banach space case that theorem can be proved in using only symmetric random vectors and symmetric infinitely divisible measures.

II. Banach lattices of Poisson type and the central limit problem

§1 - Banach lattices of Poisson type and cotype

The lemma of Rosenthal (I.3.1) implies that for a sequence (ξ_k) of independent positive random variables the series $\sum \xi_k$ converges in L^p if (and only if) the two series $\sum \int \xi_k^p dP$ and $\sum \int \xi_k dP$ converge. This result carries over to the case that the random variables take their values in the cone of an L^p space.

1.1 Proposition. Let $E = L^p(X,\Sigma,\mu)$ where (X,Σ,μ) is a σ-finite measure space and $1 \leq p < \infty$. Then there exists a constant $C > 0$ such that for every finite sequence $(X_k)_{1\leq k\leq n}$ of independent E_+-valued and p-integrable random vectors on some probability space $(\Omega,\mathcal{A},P)$ the following inequality holds:

$$\int \| \sum_{k=1}^n X_k \|^p \, dP \leq C \left(\sum_{k=1}^n \int \| X_k \|^p \, dP + \| \sum_{k=1}^n \int X_k \, dP \|^p \right) .$$

Proof. Since the inequality only depends on the distributions μ_k of the X_k, we may suppose that the random vectors X_k are the projections on the probability space $(E^n, \mu_1 \otimes \ldots \otimes \mu_n)$. Let $Z_k : (E,\mu_k) \to E$ denote for $1 \leq k \leq n$ the identity on E. By assumption Z_k is p-integrable and hence can be approximated in L^p_E by a sequence $(Z_{k,l})$ of simple random vectors (of the form $Z = \sum_{j=1}^m 1_{A_j} z_j$ with $A_j \in \mathcal{B}(E)$ and $z_j \in E_+$). If we define $X_{k,l} := Z_{k,l} \circ X_k$ for $1 \leq k \leq n$ and all $l \in \mathbb{N}$ then for every fixed $l \in \mathbb{N}$ the sequence $(X_{k,l})$ is independent and $\sum_{k=1}^n X_k$ is the limit of $(\sum_{k=1}^n X_{k,l})$ in L^p_E. Hence it is sufficient to prove the asserted inequality for finite sequences $(X_k)_{1\leq k\leq n}$ of independent simple random vectors in E_+.

Let us suppose $X_k = \sum_{i\in I_k} 1_{A_{k,i}} x_{k,i}$ for $k=1,\ldots,n$ where $A_{k,i} \in \mathcal{A}$ and $x_{k,i} \in E_+ = L^p(X,\Sigma,\mu)_+$ for all i in a finite index set I_k. With the aid of lemma I.3.1 and Fubinis theorem we obtain

$$\int \| \sum_{k=1}^n X_k \|^p dP = \int_\Omega \int_X \Big(\sum_{k=1}^n \sum_{i\in I_k} 1_{A_{k,i}}(\omega) x_{k,i}(t) \Big)^p \mu(dt) \, P(d\omega)$$

$$= \int_X \int_\Omega \Big(\sum_{k=1}^n \sum_{i\in I_k} 1_{A_{k,i}}(\omega) x_{k,i}(t) \Big)^p P(d\omega) \, \mu(dt)$$

$$\leq C \Big[\int_X \Big(\sum_{k=1}^n \int_\Omega \Big(\sum_{i\in I_k} 1_{A_{k,i}}(\omega) x_{k,i}(t) \Big)^p P(d\omega) \Big) \mu(dt) + \int_X \Big(\int_\Omega \sum_{k=1}^n \sum_{i\in I_k} 1_{A_{k,i}}(\omega) x_{k,i}(t) P(d\omega) \Big)^p \mu(dt) \Big]$$

$$= C \Big[\sum_{k=1}^n \Big(\sum_{i\in I_k} P(A_{k,i}) \| x_{k,i} \|^p \Big) + \| \sum_{k=1}^n \sum_{i\in I_k} P(A_{k,i}) x_{k,i} \|^p \Big]$$

$$= C \Big[\sum_{k=1}^n \int \| X_k \|^p \, dP + \| \sum_{k=1}^n \int X_k \, dP \|^p \Big] .$$

Remark. The above theorem can be proved directly for an arbitrary abstract L^p space without using any concrete representation of these spaces. One only has to utilize the fact that a Banach lattice is isomorphic to an abstract L^p space if and only if it is p-concave and p-convex (see [10], p.59).

1.2 Proposition. Let E be an abstract L^1 space. Then for all $1 \leq p < \infty$ there exists a constant $C > o$ only depending on p such that the inequality of proposition 1.1 holds for every finite sequence (X_k) of independent E_+-valued and p-integrable random vectors.

Proof. This follows easily from lemma I.3.1 since for an abstract L^1 space the norm is additive on the positive cone.∎

The above two propositions can be viewed as first motivation for the following definition.

1.3 Definition. A Banach lattice E is called a Banach lattice of Poisson type p ($1 \leq p < \infty$) if there exists a constant $C > o$ only depending on p such that for all finite sequences (ξ_k) of independent Poisson distributed real random variables with parameters $\lambda_k \leq 1$ and for all finite sequences (x_k) in E_+ the following inequality holds:

$$\int \| \sum \xi_k x_k \|^p \, dP \leq C \left(\sum \lambda_k \|x_k\|^p + \| \sum \lambda_k x_k \|^p \right).$$

E is called a Banach lattice of Poisson cotype p ($1 \leq p < \infty$) if for a certain constant $c > o$ the converse inequality holds, i.e.

$$c\left(\sum \lambda_k \|x_k\|^p + \| \sum \lambda_k x_k \|^p \right) \leq \int \| \sum \xi_k x_k \|^p \, dP .$$

Remarks. 1. For the Poisson distribution π_λ with $\lambda \leq 1$ one has $\int t^p \pi_\lambda(dt) \leq k_p \lambda$ for a certain constant $k_p > o$. Therefore the defining inequality in the definition of a Banach lattice of Poisson type p is equivalent to the inequality of proposition 1.1 for the special sequences $(\xi_k x_k)$ of random vectors.

2. The definition of Poisson cotype is somewhat redundant, since the convexity of $\|\cdot\|^p$ implies that $\| \sum \lambda_k x_k \|^p \leq \int \| \sum \xi_k x_k \|^p \, dP$ always holds. Hence E is of Poisson cotype p if and only if for a certain constant $c > o$ one always has

$$c\left(\sum \lambda_k \|x_k\|^p \right) \leq \int \| \sum \xi_k x_k \|^p \, dP .$$

This inequality is surely satisfied if the norm satisfies a lower p-estimate (see [10], p.82-84), i.e. if there is a constant $k > o$ such that for every positive finite sequence (x_k) one has $\sum \|x_k\|^p \leq k \| \sum x_k \|^p$.

3. It is not difficult to prove that a Banach lattice E is of Poisson

type p ($1 \leq p < \infty$) if and only if for all sequences (ξ_k) of independent Poisson variables with parameters $\lambda_k \leq 1$ and all sequences (x_k) in E_+ the conditions

$$\|\sum \lambda_k x_k\| < \infty \quad \text{and} \quad \sum \lambda_k \|x_k\|^p < \infty$$

imply

$$\int \|\sum \xi_k x_k\|^p \, dP < \infty ,$$

where we use the notation $\|\sum a_k\| := \sup_n \|\sum_{k=1}^n a_k\|$ for any series $\sum a_k$ in E_+ even if $\sum a_k$ is not convergent. Similar, E is of Poisson cotype p if and only if

$$\int \|\sum \xi_k x_k\|^p \, dP < \infty \quad \text{always implies} \quad \sum \lambda_k \|x_k\|^p < \infty .$$

Let E be an abstract L^p space of infinite dimension with $p > 1$. If $1 \leq q < p$ then by corollary I.2.6 there exists a sequence (x_k) in E_+ of pairwise orthogonal elements such that $\sum \|x_k\|^p = \|\sum x_k\|^p < \infty$ but $\sum \|x_k\|^q = \infty$. This shows that E is not of Poisson cotype q since

$$\left(\int \|\sum \xi_k x_k\|^q \, dP\right)^{1/q} \leq \left(\int \|\sum \xi_k x_k\|^p \, dP\right)^{1/p} < \infty .$$

Conversely, if $q > p$ then E cannot be of Poisson type q. To show this let (y_k) be a disjoint sequence in E_+ with $\|y_k\| = 1$ for all $k \in \mathbb{N}$. Since $p(p-1)^{-1}(q-1) > p + (p-1)^{-1}(q-p)$ there exists a sequence (c_k) in $]0,1[$ such that

$$\sum c_k{}^{p(p-1)^{-1}(q-1)} < \infty \quad \text{but} \quad \sum c_k{}^p c_k{}^{(p-1)^{-1}(q-p)} = \infty .$$

Now we define $\lambda_k := c_k{}^{(p-1)^{-1}(q-p)}$ and $x_k := c_k y_k$. Then we have

$$\sum \lambda_k \|x_k\|^q = \sum c_k{}^{p(p-1)^{-1}(q-1)} < \infty \text{ , and similar}$$

$$\|\sum \lambda_k x_k\|^p < \infty \text{ , but } \sum \lambda_k \|x_k\|^p = \sum c_k{}^{(p-1)^{-1}(q-p)} c_k{}^p = \infty .$$

Since E is of Poisson cotype p we get

$$\int \|\sum \xi_k x_k\|^p \, dP = \infty \quad \text{and hence also} \quad \int \|\sum \xi_k x_k\|^q \, dP = \infty ,$$

i.e. E is not of Poisson type q.

The question whether a Banach lattice of Poisson type p ($p > 1$) or of Poisson cotype p resp. is also of Poisson type q for $q < p$ or of Poisson cotype q for $q > p$ resp. seems to be more difficult. A partial answer is contained in the following result.

1.4 <u>Theorem</u>. Let E be a Banach lattice of Rademacher type p ($1 \leq p \leq 2$). Then E is also of Poisson type p. If E is a Banach lattice of Rademacher cotype p ($2 \leq p < \infty$) then E is also of Poisson cotype p.

Proof. Let (x_k) be a finite sequence in E_+ and let (ξ_k) be a finite sequence of independent Poisson variables on some probability space with $\mathbb{E}\xi_k = \lambda_k \leq 1$ for all k. If E is of Rademacher type p then there exists a constant $C>0$ such that

$$\int \| \sum (\xi_k - \lambda_k) x_k \|^p \, dP \leq C^p (\sum \int |\xi_k - \lambda_k|^p \, dP \|x_k\|^p .$$

Furthermore, $\int |\xi_k - \lambda_k|^p \, dP \leq \int (\xi_k + \lambda_k)^p \, dP \leq D^p \lambda_k$

for a certain constant $D>0$ since $\lambda_k \leq 1$ for all k. Now the following estimate shows that E is of Poisson type p:

$$\begin{aligned}(\int \| \sum \xi_k x_k \|^p dP)^{1/p} &\leq (\int \| \sum (\xi_k - \lambda_k) x_k \|^p dP)^{1/p} + \| \sum \lambda_k x_k \| \\ &\leq C (\sum \int |\xi_k - \lambda_k|^p dP \, \|x_k\|^p)^{1/p} + \| \sum \lambda_k x_k \| \\ &\leq CD (\sum \lambda_k \|x_k\|^p)^{1/p} + \| \sum \lambda_k x_k \| .\end{aligned}$$

If E is of Rademacher cotype p ($p \geq 2$) we denote by (ε_k) a finite Bernoulli sequence independent of the sequence (ξ_k). Then the following estimate holds:

$$\begin{aligned}\sum \lambda_k \|x_k\|^p &\leq \sum \int \xi_k{}^p dP \, \|x_k\|^p = \sum \int |\varepsilon_k \xi_k|^p dP \, \|x_k\|^p \\ &\leq K \int \| \sum \varepsilon_k \xi_k x_k \|^p dP \leq K \int \| \sum \xi_k x_k \|^p \, dP ,\end{aligned}$$

where the constant $K>0$ exists by the assumption that E is of Rademacher cotype p. $\rfloor$

Since an L^1 space of infinite dimension is not of Rademacher type $p>1$ the above proposition shows that for Banach lattices the notion of Poisson type is more general than the notion of Rademacher type.

The Banach lattices of Poisson cotype p can be completely characterized. They are exactly those Banach lattices possessing a lower p-estimate (see [10], p.82). For the proof we need the following lemma.

1.5 Lemma. Let E be a p-concave Banach lattice (see [10], p.56) for some $p \geq 1$. Then there exists a constant $K>0$ such that for every finite sequence (x_k) in E_+ and every finite sequence (ζ_k) of positive random variables with $\mathbb{E}\,\zeta_k{}^p \leq C$ for all k the following inequality holds:

$$\int \| \sum \zeta_k x_k \|^p \, dP \leq KC \| \sum x_k \|^p .$$

Proof. A Banach lattice E is p-concave if and only if every positive linear operator $T:c_o \to E$ is p-absolutely summing. For any p-absolutely summing operator T from a Banach space F into E let $\pi_p(T)$ as usual denote the smallest constant $k \geq o$ such that

$$(\sum \|Tx_k\|^p)^{1/p} \leq k \sup_{x' \in E', \|x'\| \leq 1} (\sum |\langle x_k, x' \rangle|^p)^{1/p}$$

holds for every finite sequence (x_k) in F. For every $d \in \mathbb{N}$ let further ℓ_d^∞ denote the space $\mathbb{R}^d$ with the ℓ^∞ norm. We will first prove that every positive linear operator $T:c_o \to E$ is p-absolutely summing if and only if there exists a constant $K>o$ such that for all $d \in \mathbb{N}$ and every positive operator $T:\ell_d^\infty \to E$ one has $\pi_p(T) \leq K\|T\|$.

Let us suppose that there is a sequence (d_n) in $\mathbb{N}$ and a sequence (T_n) of positive operators $T_n:\ell_{d_n}^\infty \to E$ with $\|T_n\| = 1$ and $\pi_p(T_n) \geq n2^n$. In virtue of the natural imbeddings $\ell_{d_n}^\infty \to \bigoplus_{k \in \mathbb{N}} \ell_{d_k}^\infty \to c_o$ every operator T_n can be viewed as an operator defined on c_o. If we define $T := \sum 2^{-n} T_n$ we get $\|T\| \leq 1$ and $\pi_p(T) \geq 2^{-n}\pi_p(T_n) \geq n$ for all n. But this contradicts the assumption that T as a positive operator on c_o is p-absolutely summing.

Now let $T:\ell_d^\infty \to E$ be a positive operator such that $T((c_i)_{1 \leq i \leq d}) = \sum_{i=1}^d c_i x_i$ for all $(c_i)_{1 \leq i \leq d} \in \ell_d^\infty$ $(x_i \in E_+)$. Then $\|T\| = \|\sum_{i=1}^d x_i\|$. The factorization theorem of Pietsch (cf. [9], p.64) yields the existence of a sequence $(a_i)_{1 \leq i \leq d}$ in $\mathbb{R}_+$ with $\sum_{i=1}^d a_i = 1$ such that

$$\|\sum_{i=1}^d c_i x_i\| \leq \pi_p(T)(\sum_{i=1}^d |c_i|^p a_i)^{1/p} \leq K\|\sum_{i=1}^d x_i\|(\sum_{i=1}^d |c_i|^p a_i)^{1/p}$$

for all $(c_i)_{1 \leq i \leq d}$, and the assertion of the lemma follows by integration.∎

1.6 <u>Theorem</u>. The following assertions are equivalent:

(a) E is a Banach lattice of Poisson cotype p.

(b) There is a constant $C > o$ such that for all finite disjoint sequences (x_k) in E_+ one has the inequality

$$(*) \qquad \sum \|x_k\|^p \leq C \int \|\sum \xi_k^1 x_k\|^p \, dP \quad ,$$

where (ξ_k^1) denotes an independent sequence with $\xi_k^1(P) = \pi_1$ for all k.

(c) E satisfies a lower p-estimate.

<u>Proof</u>. We have only to prove the non-trivial implication (b)⟹(c). First we show that for all $p \geq 1$ the inequality (*) does not hold for the Banach lattice c_o. If (*) is valid then also

$$(\sum \|x_k\|^p)^{1/p} \leq D \left(\int \|\sum \xi_k^1 x_k\|^q\right)^{1/q} \quad \text{for all } q > p.$$

Let (e_i) denote the canonical basis in c_o and let (a_i) be a sequence in $\mathbb{R}_+$ such that

$$\sum a_i^p = \infty \text{ and } \sum a_i^q < \infty \ . \text{ Define } \ b_i := a_i^q (\sum a_j^q)^{-1}.$$

Then we have for every finite sequence (t_i) in $\mathbb{R}_+$

$$\| \sum_i t_i a_i e_i \|_{c_o}^q = \sup_i (t_i a_i)^q \leq \sum_i t_i^q a_i^q = (\sum_j a_j^q) \sum_i t_i^q b_i .$$

If we integrate this inequality we get

$$\int \| \sum \xi_i^1 (a_i e_i) \|_{c_o}^q \ dP \leq K(\sum a_i^q)$$

for a certain constant $K>o$ and all n. This shows that in c_o the inequality (✱) cannot hold since $\sum \|a_i e_i\|^p = \sum a_i^p = \infty$.

Since (✱) does not hold in c_o it is not hard to show that c_o is not finitely representable in E in the lattice sense. For if c_o is finitely representable in E then the first part of the proof shows that there is a sequence (K_n) in $\mathbb{R}_+$ with $\lim K_n = \infty$ and for all $n \in \mathbb{N}$ a finite sequence $(c_{i,n})$ in c_o with $c_{i,n} = \sum_j a_{i,j,n} e_j$ (where $a_{i,j,n}$=o only for a finite number of indices) such that

$$\sum_i \|c_{i,n}\|_{c_o}^p \geq K_n \int \| \sum_i \xi_i^1 c_{i,n} \|_{c_o}^p \ dP \quad \text{for all } n \in \mathbb{N}.$$

Since by assumption c_o is finitely representable in E there exists for all $n \in \mathbb{N}$ a disjoint sequence $(x_{j,n})$ in E_+ such that for all sequences (t_i) in $\mathbb{R}_+$

$$\| \sum_i t_i \sum_j a_{i,j,n} e_j \|_{c_o} \leq \| \sum_i t_i \sum_j a_{i,j,n} x_{j,n} \| \leq \lambda \| \sum_i t_i \sum_j a_{i,j,n} e_j \|_{c_o}$$

for an arbitrary given constant $\lambda > 1$. With $y_{i,n} := \sum_j a_{i,j,n} x_{j,n}$ we get for all $n \in \mathbb{N}$

$$\sum_i \|y_{i,n}\|^p \geq K_n \lambda^{-p} \int \| \sum_i \xi_i^1 y_{i,n} \|^p dP$$

and this is a contradiction to assumption (b).

Since c_o is not finitely representable in E there exists (s. [10], theorem 1.f.12) a $q \geq 1$ such that E satisfies a lower q-estimate. We put

$$q_* := \inf \{ q \geq 1 \ : \ \text{E satisfies a lower q-estimate} \}$$

and

$$p_* := \inf \{ p \geq 1 \ : \ (✱) \text{ holds in E} \} ,$$

and we will show $q_* = p_*$. If E satisfies a lower q-estimate then surely (✱) holds for p=q and hence $q_* \geq p_*$. Now suppose $q_* > p_*$. Then there exist p,q with $q \geq q_* > p \geq p_*$ such that (✱) holds for p whereas E satisfies no lower p-estimate. Since E has a lower q-estimate we may suppose that every positive operator $T:c_o \to E$ is q-absolutely summing (cf. [10], theorem 1.f.7) and we can apply lemma 1.5. We get

$$(\sum_i \|x_i\|^q)^{1/q} \leq (\sum_i \|x_i\|^p)^{1/p} \leq c_p(\int \|\sum_i \xi_i^1 x_i\|^p \, dP)^{1/p}$$
$$\leq c_p(\int \|\sum_i \xi_i^1 x_i\|^q dP)^{1/q} \leq D\|\sum_i x_i\|$$

for every finite sequence (x_i) in E_+. Therefore E has a lower p-estimate and hence $q_* > p_*$ is not possible.

Similar one can show that (*) is valid for p_* if and only if E satisfies a lower q_*-estimate. Therefore the implication (b)$\Longrightarrow$(c) always holds.

1.7 Corollary. A Banach lattice E is isomorphic to an abstract L^p space if (and only if) there are constants $C,D > o$ such that for all finite disjoint sequences (x_i) in E_+ the following inequality holds:

$$C(\sum_i \|x_i\|^p) \leq \int \|\sum_i \xi_i^1 x_i\|^p \, dP \leq D(\sum_i \|x_i\|^p) .$$

Since any Banach lattice of Rademacher type p $(1 \leq p \leq 2)$ satisfies an upper p-estimate, we also get as a corollary the following characterization of abstract L^p spaces for $1 \leq p \leq 2$ including for the Banach lattice case the theorem of Kwapien which asserts that a Banach space of Rademacher type and cotype 2 is necessarily isomorphic to a Hilbert space.

1.8 Corollary. If E is a Banach lattice of Rademacher type p and of Poisson cotype p $(1 \leq p \leq 2)$ then E is isomorphic to an L^p space.

The last corollary poses the problem whether a Banach lattice of Poisson type p and Poisson cotype p must be isomorphic to an abstract L^p space. Clearly, by proposition 1.2 one has to exclude the case that E contains a copy of ℓ^1. The solution of this problem is open, but we will give some results going in this direction.

1.9 Proposition. Let E be a Banach lattice of Poisson type p $(1 < p < \infty)$ such that c_o and ℓ^1 are not finitely representable in E in the lattice sense. Then E satisfies an upper q-estimate for all $q < p$.

Proof. If $q < p$ then ℓ^q is not of Poisson type p for $q > 1$ and hence ℓ^q can not be finitely representable in E. Since ℓ^1 is not finitely representable there exists a $p_1 > 1$ such that E satisfies an upper p_1-estimate. Define $p_E := \sup\{r > 1 : \text{E satisfies an upper r-estimate}\}$. $p_E = \infty$ would imply that
$q_{E'} := \inf\{s > 1 : \text{E' satisfies a lower s-estimate}\} = 1$, and hence ℓ^1 would be finitely representable in E' by a theorem of Krivine ([7]), and this would contradict the assumption that c_o is not finitely

representable in E. Therefore we have $1 < p_E < \infty$ and again Krivine's theorem implies that ℓ^{p_E} is finitely representable in E. ∎

The next proposition shows that the Banach lattices , which can be simultaneously of Poisson type and cotype p for more than one number p, are in a certain sense close to the L^1 spaces. Define

$$p^* := \sup\{p \geq 1 : \text{E is of Poisson type p}\} \quad \text{and}$$
$$p_* := \inf\{p \geq 1 : \text{E is of Poisson cotype p}\}.$$

If E is of Poisson type p and of Poisson cotype p ($p > 1$) then clearly $p_* \leq p^*$. Suppose $p_* < p^*$. By Krivine's theorem ℓ^{p_*} is finitely representable in E. Since E is of Poisson type $p>1$, this shows that either $p_* = 1$ of $p_* = p^*$. Thus we have proved:

1.1o <u>Proposition</u>. Let E be a Banach lattice of Poisson type p and cotype p for some $p > 1$. Then $p_* = p^*$ or ℓ^1 is finitely representable in E in the lattice sense.

§2 - <u>The central limit problem</u>

Let E be a Banach lattice as before and let F be a measure of the form $F = \sum_{i=1}^{n} \lambda_i \varepsilon_{x_i}$ ($\lambda_i \in \mathbb{R}_+$, $x_i \in E_+ \setminus \{o\}$). If (ξ_i) is an independent sequence of Poisson variables with $\mathbb{E}\xi_i = \lambda_i$ then to any function $f: E_+ \longrightarrow G$, where G denotes a second Banach lattice, we can associate the stochastic integral $\int f d\xi := \sum_{i=1}^{n} \xi_i f(x_i)$. $\int f d\xi$ is a G-valued random vector whose distribution is the Poisson measure $e(\sum_{i=1}^{n} \lambda_i \varepsilon_{f(x_i)})$. If G is a Banach lattice of Poisson type p then we get

$$\left\| \int |f| d\xi \right\|_{L^p_E} \leq C\left(\left(\int \| |f(x)| \|^p F(dx)\right)^{1/p} + \left\| \int |f(x)| F(dx) \right\|\right),$$

where the constant C only depends on p. We now want to define such a stochastic integral for more general measures F. For this we first construct to a given measure F a certain Poisson process (ξ_A) indexed by the sets of finite F-measure. The existence of these Poisson processes will be a consequence of the following general result.

2.1 <u>Proposition</u>. Let E be a locally convex vector space, T an abelian semigroup and $(\mu_t)_{t \in T}$ a semigroup in $\mathbb{P}(E)$. Let further $\mathbb{H}$ be a lattice semiring on the positive cone of some vector lattice G (cf. [3]). We put $\mathbb{H}^{(n)} := \{\vec{y} \in \mathbb{H} : \vec{y} = (y_1, \ldots, y_n),\ y_i \in G_+\}$ for all n and identify $\mathbb{H}^{(1)}$ with the subset $S := \{y \in G_+ : (y) \in \mathbb{H}\}$ of G_+. Suppose further that we are given a map $\lambda: S \longrightarrow T$ which is additive in the following sense: for all $n \in \mathbb{N}$ and for all $\vec{y} = (y_1, \ldots, y_n) \in \mathbb{H}^{(n)}$ one has $\lambda(\sum_{i=1}^{n} y_i) = \sum_{i=1}^{n} \lambda(y_i)$. Then there is a probability space $(\Omega, \mathfrak{A}, P)$

and a stochastic process $(\xi_y)_{y\in S}$ on Ω with values in E such that

(i) $\xi_y(P) = \mu_{\lambda(y)}$ for all $y\in S$, and

(ii) for all $\bar{y} = (y_1,\dots,y_n)\in \mathbb{H}$ the sequence (ξ_{y_i}) is independent and satisfies $\sum_{i=1}^{n} \xi_{y_i} = \xi_{\sum_{i=1}^{n} y_i}$.

Now suppose that the following additional conditions hold: E is a Banach space, T is a topological semigroup with zero, $\lim t_n = o$ (where (t_n) is a sequence in T) always implies $\lim \mu_{t_n} = \varepsilon_o$, and $(y_n)\downarrow o$ implies $\lim \lambda(y_n) = o$ for every sequence (y_n) in S. Then the process $(\xi_y)_{y\in S}$ is σ-additive: For all sequences (y_k) in S such that $(y_1,\dots,y_n)\in \mathbb{H}$ for all n and $y := \sum_k y_k \in S$ the series $\sum_k \xi_{y_k}$ converges almost surely. If E is a Banach lattice, $T = \mathbb{R}_+$, (μ_t) a semigroup in $\mathbb{P}(E_+)$, and if $\int \|x\|^p \mu_r(dx) < \infty$ for a certain $r\in T$, then in the case $y = \sum_k y_k \in S$ the series $\sum_k \xi_{y_k}$ even converges in L^p.

<u>Proof</u>. We put $F:=(E',\sigma(E',E'^*))$, $F^*:=(E'^*,\sigma(E'^*,E'))$, $F_y:=F$ for all $y\in S$, and $F_{\bar{y}}:=F_{y_1}\times\dots\times F_{y_n}$ for all $\bar{y}=(y_1,\dots,y_n)\in \mathbb{H}$. On $F^*_{\bar{y}}$ we define the probability measure $\mu_{\bar{y}} := \mu_{\lambda(y_1)}\otimes\dots\otimes\mu_{\lambda(y_n)}$. For $\bar{y},\bar{z}\in\mathbb{H}$ with $\bar{y} = (y_1,\dots,y_n) < \bar{z} = (z_1,\dots,z_m)$ (cf. [3]) there exists by definition of a lattice semiring a partition $(I_i)_{i=o,1,\dots,n}$ of $\{1,\dots,m\}$ such that $y_i = \sum_{j\in I_i} z_j$ for $1\le i\le n$. We define the canonical imbedding $\chi_{\bar{y},\bar{z}}: F_{\bar{y}}\to F_{\bar{z}}$ by $\chi_{\bar{y},\bar{z}}(u_1,\dots,u_n) = (v_1,\dots,v_m)$ with $v_j=u_i$ for $j\in I_i$ $(1\le i\le n)$ and $v_j=o$ for $j\in I_o$. Then the transposed map $\pi_{\bar{y},\bar{z}}: F^*_{\bar{z}}\to F^*_{\bar{y}}$ of $\chi_{\bar{y},\bar{z}}$ is given by $\pi_{\bar{y},\bar{z}}(t_1,\dots,t_m) = (\sum_{j\in I_1} t_j,\dots,\sum_{j\in I_n} t_j)$ for all $(t_1,\dots,t_m)\in F^*_{\bar{z}}$. Now we define

$$F_{\mathbb{H}} := \varinjlim_{\bar{y}\in\mathbb{H}} F_{\bar{y}} \quad\text{and}\quad F^*_{\mathbb{H}} := \varprojlim_{\bar{y}\in\mathbb{H}} F^*_{\bar{y}} .$$

Then $F^*_{\mathbb{H}}$ is the topological dual of $F_{\mathbb{H}}$ (cf. [15] , ch.IV, theorem4.5). Finally we denote by $\chi_{\bar{y}}$ and $\pi_{\bar{y}}$ resp. the canonical imbedding of $F_{\bar{y}}$ into $F_{\mathbb{H}}$ and the canonical projection of $F^*_{\mathbb{H}}$ onto $F^*_{\bar{y}}$ resp.

On the set algebra

$$\mathfrak{Z}(F^*_{\mathbb{H}}) := \bigcup_{\bar{y}\in\mathbb{H}} \pi_{\bar{y}}^{-1}(\mathfrak{B}(F^*_{\bar{y}}))$$

we can define the set function $\tilde{P}$ in setting $\tilde{P}(A) := \mu_{\bar{y}}(B)$ if $B\in\mathfrak{B}(F^*_{\bar{y}})$ and $A=\pi_{\bar{y}}^{-1}(B)$. Then it is not hard to show (cf. [3]) that $\tilde{P}$ is continuous at $\emptyset$ and hence can be extended to a probability measure P on the σ-algebra $\mathfrak{A}$ generated by $\mathfrak{Z}(F^*_{\mathbb{H}})$, i.e. we have defined our probability space $(\Omega,\mathfrak{A},P)$.

For all $y \in S$ we now put $\xi_y := \pi_{(y)}$. Then we have surely $\xi_y(P) = \pi_{(y)}(P) = \mu_{(y)} = \mu_{\lambda(y)}$, and for $\bar{y} = (y_1,\dots,y_n) \in \mathbb{H}$ the sequence $(\xi_{y_i})_{1 \le i \le n}$ is independent. Furthermore we have

$$\xi_{\sum_{i=1}^n y_i} = \pi_{(\sum_{i=1}^n y_i)} = \pi_{(\sum_{i=1}^n y_i),(y_1,\dots,y_n)} \circ \pi_{(y_1,\dots,y_n)}$$
$$= \pi_{(\sum_{i=1}^n y_i),(y_1,\dots,y_n)}((\xi_{y_1},\dots,\xi_{y_n})) = \sum_{i=1}^n \xi_{y_i} ,$$

i.e. (i) and (ii) are satisfied.

Now suppose that the additional assumptions of the theorem hold. By the construction of the process $\xi = (\xi_y)_{y \in S}$ we have

$$\mu_{\lambda(y)} = \mu_{\lambda(\sum_{k=1}^n y_k)} * \mu_{\lambda(y - \sum_{k=1}^n y_k)} \quad \text{and} \quad \xi_y = \sum_{k=1}^n \xi_{y_k} + \xi_{y - \sum_{k=1}^n y_k}$$

for any sequence (y_n) in S such that $(y_1,\dots,y_n) \in \mathbb{H}$ for all n and $y = \sup_n \sum_{k=1}^n y_k \in S$. Since by assumption $\lim_n \mu_{\lambda(y - \sum_{k=1}^n y_k)} = \varepsilon_o$ the theorem of Ito and Nisio implies the a.s. convergence of $\sum \xi_{y_k}$.

If $T = \mathbb{R}_+$ and if all measures μ_t are concentrated on E_+ then we have for every constant $c > o$ and $t > s > o$

$$\int_{[\|x\|>c]} \|x\|^p \mu_t(dx) = \int_{[\|x+y\|>c]} \|x+y\|^p \mu_s(dx)\mu_{t-s}(dy) \ge \int_{[\|x\|>c]} \|x\|^p \mu_s(dx).$$

Hence we get

$$\overline{\lim_{n\to\infty}} \sup_{s \le r} \int_{[\|x\|>n]} \|x\|^p \mu_s(dx) \le \lim_{n\to\infty} \int_{[\|x\|>n]} \|x\|^p \mu_r(dx) = o$$

if $\int \|x\|^p \mu_r(dx) < \infty$. Since

$$\int \|\sum_{k=1}^n \xi_{y_k}\|^p \, dP = \int \|x\|^p \mu_{\lambda(\sum_{k=1}^n y_k)}(dx)$$

the L^p-convergence of the series $\sum \xi_{y_k}$ now follows. ┘

Remarks. 1. Let (X,Σ,μ) be a measure space, and let G be the Banach lattice of all bounded Σ-measurable functions. Then Σ defines in a natural way a lattice semiring $\mathbb{H}$ such that S is the set of characteristic functions of sets in Σ. The proposition will be applied to this case with $E = \mathbb{R}_+$ and $T = \mathbb{R}_+$. The mapping $\lambda : S \to \mathbb{R}_+$ is given by $\lambda(1_A) := \mu(A)$ for $A \in \Sigma$.

2. The following example may indicate that it can be useful to have the above existence theorem in its stated general form. Let $T := M_+(X, \mathbb{Z}_+)$ be the semigroup of all integer-valued positive measures on some locally compact second countable state space X, let $E = G = M(X)$ be the space of all Radon measures on X, and let $\mathbb{H}$ be the lattice semiring in $M_+(X,\mathbb{Z}_+)$ containing the n-tuples $(\nu_1,\dots,\nu_n)$

of simple measures with disjoint supports. Then the proposition shows that to any semigroup $(\mu_t)_{t\in T}$ of probability measures on T there exists a process $\xi = (\xi_s)_{s\in S}$ with the properties stated in the proposition. Here S is the subset of $M_+(X,\Sigma_+)$ consisting of all simple measures.

2.2 <u>Corollary</u>. Let (X,Σ,μ) be a measure space and put $\Sigma_o := \{S\in\Sigma : \mu(S)<\infty\}$. Then there is a process $\xi = (\xi_S)_{S\in\Sigma_o}$ on some probability space $(\Omega,\mathfrak{A},P)$ with the properties

(a) for all $S\in\Sigma_o$ the random variable ξ_S is Poisson distributed with $\mathbb{E}\,\xi_S = \mu(S)$,

(b) for every disjoint sequence (S_k) in Σ_o with $S = \bigcup S_k \in \Sigma_o$ one has $\sum \xi_{S_k} = \xi_S$ a.s. and (ξ_{S_k}) is an independent sequence.

2.3 <u>Definition</u>. Any process $\xi = (\xi_S)_{S\in\Sigma_o}$ with the two properties of the above corollary will be called a <u>Poisson process with respect to μ</u>.

Let $\mathcal{E}(X,E)$ denote the set of all simple functions from (X,Σ,μ) into a Banach lattice E which are μ-integrable, and let ξ be a fixed Poisson process with respect to μ defined on some probability space $(\Omega,\mathfrak{A},P)$. On $\mathcal{E}(X,E)$ we define a positive linear operator T with values in $L^o(\Omega,E)$ in the following way. For $f = \sum_{i=1}^{n} 1_{S_i} x_i \in \mathcal{E}(X,E)$, where $(S_i)_{1\le i\le n}$ is a disjoint sequence in Σ_o and $(x_i)_{1\le i\le n}$ is a sequence in E, we put $Tf := \sum_{i=1}^{n} \xi_{S_i} x_i$. On $T(\mathcal{E}(X,E))$ we define the following norm $\|\cdot\|_{p,\xi}$:

$$\|Tf\|_{p,\xi} := \left(\int \Big\| \sum_{i=1}^{n} \xi_{S_i} |x_i| \Big\|^p dP\right)^{1/p} ,$$

and denote by $L^{p,\xi}(\Omega,E)$ the closure of $T(\mathcal{E}(X,E))$ with respect to $\|\cdot\|_{p,\xi}$. Because of

$$\int \Big\|\sum_{i=1}^{n} \xi_{S_i} x_i\Big\|^p dP \le \int \Big\|\sum_{i=1}^{n} \xi_{S_i} |x_i|\Big\|^p dP$$

$L^{p,\xi}(\Omega,E)$ can be viewed as a subspace of $L^p(\Omega,E)$.

Similarly, we define on $\mathcal{E}(X,E)$ a norm $\|\cdot\|_{p,1}$ by

$$\|f\|_{p,1} := \left(\int \Big\|\sum_{i=1}^{n} 1_{S_i} |x_i|\Big\|^p d\mu\right)^{1/p} + \Big\|\int \Big(\sum_{i=1}^{n} 1_{S_i} |x_i|\Big) d\mu\Big\|$$

$$= \left(\int \|f\|^p d\mu\right)^{1/p} + \Big\|\int |f|\, d\mu\Big\| ,$$

and we denote by $L^{p,1}(X,E)$ the closure of $\mathcal{E}(X,E)$ with respect to the norm $\|\cdot\|_{p,1}$. If μ is a bounded measure then clearly $L^{p,1}(X,E)$ is isomorphic to $L^p(X,E)$. Directly from definition 1.3 we now get:

2.4 Theorem. If E is a Banach lattice of Poisson type p ($1 \leq p < \infty$) then the operator $T: \mathcal{E}(X,E) \longrightarrow L^{p,\xi}(\Omega,E)$ can be extended to a continuous operator on $L^{p,1}(X,E)$ (which again will be denoted by T). If in addition E is also of Poisson cotype p then T is an isomorphism.

If E is of Poisson type p we use the notation $\int f d\xi := Tf$ for all $f \in L^{p,1}(X,E)$, and we will call f ξ-integrable and $\int f d\xi$ the ξ-integral of f.

2.5 Proposition. If (X,Σ,μ) is a σ-finite measure space then every μ-measurable function $f: X \longrightarrow E$ with $\int \|f\|^p d\mu < \infty$ and $\int |f| \, d\mu \in E_+$ belongs to $L^{p,1}(X,E)$. Hence these functions are ξ-integrable if E is of Poisson type p.

Proof. Since μ is σ-finite there is a disjoint sequence (B_n) in Σ_o with $\mu((\bigcup B_n)^c) = o$. Let $f: X \longrightarrow E$ be a μ-measurable function with $\int \|f\|^p d\mu < \infty$ and $\int |f| \, d\mu \in E_+$ and let (δ_k) be a fixed null sequence of positive numbers. For every $k \in \mathbb{N}$ there exists an $n_k \in \mathbb{N}$ such that

$$\int_{\bigcup_{n>n_k} B_n} \|f\|^p \, d\mu < \delta_k^{\,p} \quad \text{and} \quad \Big\| \int_{\bigcup_{n>n_k} B_n} |f| \, d\mu \Big\| < \delta_k \;,$$

where the latter estimate holds because of $\int |f| \, d\mu \in E_+$. Since f is μ-measurable and since the restriction of μ to every B_n is a bounded measure there exists for all $n,k \in \mathbb{N}$ a function $g_{n,k} \in \mathcal{E}(B_n,E)$ such that

$$\Big(\int_{B_n} \|f-g_{n,k}\|^p d\mu\Big)^{1/p} < 2^{-n}\delta_k \quad \text{and} \quad \Big\| \int_{B_n} |f-g_{n,k}| \, d\mu \Big\| < 2^{-n}\delta_k .$$

Let us define $f_k := \sum_{n=1}^{n_k} g_{n,k}$. Then we get

$$\Big(\int \|f-f_k\|^p d\mu\Big)^{1/p} \leq \sum_{n=1}^{n_k} \Big(\int_{B_n} \|f-g_{n,k}\|^p d\mu\Big)^{1/p} + \delta_k < 2\delta_k \quad \text{and}$$

similar $\| \int |f-f_k| \, d\mu \| < 2\delta_k$. Hence $\|f-f_k\|_{p,1} < 4\delta_k$ holds for all $k \in \mathbb{N}$, which shows $f \in L^{p,1}(X,E)$. ┘

With the aid of proposition 2.5 it is not difficult to obtain the following characterization of the Banach spaces of Poisson type and cotype.

2.6 Theorem. (a) If (and only if) E is of Poisson type p ($1 \leq p < \infty$) then every Radon measure F on $E_+ \setminus \{o\}$ with $\int \|x\|^p F(dx) < \infty$ and $\int x F(dx) \in E_+$ is a Lévy measure of an infinitely divisible measure of order p.
(b) If (and only if) E is of Poisson cotype p ($1 \leq p < \infty$) then the Lévy measure F of every infinitely divisible measure of order p fulfills the conditions $\int \|x\|^p F(dx) < \infty$ and $\int x F(dx) \in E_+$.

With the aid of the stochastic integral with respect to a

Poisson process we can now prove the following limit theorem for sequences of infinitely divisible measures which is a first step to the solution of the central limit problem (cf. [19]).

2.7 Proposition. Let E be a Banach lattice of Poisson type p ($1 \leq p < \infty$) and suppose that F_n ($n \in \mathbb{N}$) and F are (not necessarily bounded) Radon measures on E_+ such that

$$\int \inf(1,\|x\|^p)F_n(dx) < \infty \quad \text{and} \quad \int_{B_+} x\, F_n(dx) \in E_+ \quad \text{for all } n \in \mathbb{N} \text{ and}$$

$$\int \inf(1,\|x\|^p)F(dx) < \infty \quad \text{and} \quad \int_{B_+} x\, F(dx) \in E_+ \ ,$$

where B_+ denotes the positive part of the unit ball of E. Then the sequence $(e_o(F_n))$ converges weakly towards $e_o(F)$ if the following three conditions hold:

(L1) $\lim_{n\to\infty} \operatorname{Res}_{rB_+^c} F_n = \operatorname{Res}_{rB_+^c} F$ for all $r > o$,

(L2) $\lim_{r\to o} \overline{\lim_{n\to\infty}} \int_{rB_+} \|x\|^p F_n(dx) = o$,

(L3) $\lim_{r\to o} \overline{\lim_{n\to\infty}} \|\int_{rB_+} x\, F_n(dx)\| = o$.

Proof. We will show that every subsequence of $(e_o(F_n))$ possesses a subsequence converging to $e_o(F)$.

If ξ is a Poisson process with respect to (X,Σ,μ) then for any ξ-integrable function f and all $A \in \Sigma$ we will denote by $\int_A f d\xi$ the ξ-integral of the function $f \cdot 1_A$. Now let ξ_n ($n \in \mathbb{N}$) and ξ be Poisson processes belonging to F_n and F resp. For $o<r<1$ we choose a compact subset K of $rB_+^c \cap B_+$. Then $\int x\, d\xi_n$ has the following independent decomposition:

$$\int x d\xi_n = \int_{rB_+} x d\xi_n + \int_{(rB_+^c \cap B_+)\setminus K} x\, d\xi_n + \int_K (x-f(x))d\xi_n + \int_{B_+^c} x\, d\xi_n + \int_K f(x)\, d\xi_n \ ,$$

where $f: E_+ \to E_+$ denotes a simple function of the form $f = \sum_{i=1}^m 1_{B_i} x_i$ such that $x_i \in B_i$ ($1 \leq i \leq m$) and $(B_i)_{1 \leq i \leq m}$ is a disjoint sequence of F-continuity sets.

We will now estimate the single terms of the above decomposition of $\int x\, d\xi_n$.

1. By the assumptions (L2) and (L3) there exists for all $\varepsilon > o$ an $r > o$ and an $n_o := n_o(\varepsilon, r)$ such that for all $n \geq n_o$ the following inequality holds since E is of Poisson type p:

$$\int \| \int_{rB_+} x \, d\xi_n \|^p \, dP \leq C(\int_{rB_+} \|x\|^p F_n(dx) + \| \int_{rB_+} x \, F_n(dx) \|^p) < \varepsilon .$$

2. Because of (L1) we can find for every $\varepsilon > o$ a compact set K such that for all $n \in \mathbb{N}$ $F_n((rB_+^c \cap B_+) \setminus K) < \varepsilon$. Since E is of Poisson type p this gives us

$$\int \| \int_{(rB_+^c \cap B_+) \setminus K} x \, d\xi_n \|^p \, dP < 2C\varepsilon .$$

3. To every compact K we can choose a simple function f such that $\|f(x) - x\| < \varepsilon$ for all $x \in K$, and we get

$$\int \| \int_K (f(x)-x) d\xi_n \|^p \, dP \leq C \; (\varepsilon^p F_n(K) + \varepsilon^p F_n(K)^p).$$

4. Because of (L1) the sequence $(e(\mathrm{Res}_{B^c} F_n))$ converges to $e(\mathrm{Res}_{B^c} F)$ and one only has to observe that $e_o(\mathrm{Res}_A^+ F)$ is the distribution of $\int_A x d\xi$ for all $A \in \mathfrak{B}(E_+)$ if ξ is a Poisson process with respect to a Lévy measure F.

5. For all $n \in \mathbb{N}$ the random integral $\int_K f d\xi_n$ has the distribution $e(\sum_{i=1}^m F_n(B_i)\varepsilon_{x_i})$. Since the sets B_i $(i=1,\dots,m)$ are all F-continuity sets, the sequence $(e(\sum_{i=1}^m F_n(B_i)\varepsilon_{x_i})$ converges to $(e(\sum_{i=1}^m F(B_i)\varepsilon_{x_i})$.

Now let (δ_k) denote a fixed null sequence of positive numbers. Since we are only considering Radon measures the probability measures $e_o(F_n)$ $(n \in \mathbb{N})$ and $e_o(F)$ are concentrated on a closed separable sublattice of E. Hence we can assume w.l.o.g. that E itself is separable and hence that the topology of weak convergence can be described by the Prohoroff metric. Then the above estimates show that for all $k \in \mathbb{N}$ we can find an $n_k \in \mathbb{N}$ such that the distance of $e_o(F_{n_k})$ and $e_o(F)$ with respect to the Prohoroff metric is smaller than δ_k. This proves the convergence of $(e_o(F_n))$ to $e_o(F)$. $\rfloor$

2.8 Lemma. Let $(\mu_{nj})_{n\geq 1; j=1,\dots,k_n}$ be a uniformly infinitesimal triangular array of probability measures on the cone E_+ of a Banach lattice E such that the following two conditions hold:

(i) For all $r > o$ the sequence $(\mathrm{Res}_{rB_+^c} \sum_{j=1}^{k_n} \mu_{nj})$ converges weakly to $\mathrm{Res}_{rB_+^c} F$ where F is a Lévy measure on E_+.

(ii) $\lim_{r\to o} \overline{\lim_{n\to\infty}} \| \int_{rB_+} x \, (\sum_{j=1}^{k_n} \mu_{nj})(dx) \| = o$.

Then $\lim_{n\to\infty} e(\sum_{j=1}^{k_n} \mu_{nj}) = e_o(F)$ implies $\lim_{n\to\infty} \mathop{*}_{j=1}^{k_n} \mu_{nj} = e_o(F)$.

Proof. Suppose $\lim_{n\to\infty} e(\sum_{j=1}^{k_n} \mu_{nj}) = e_o(F)$. By Le Cam's theorem ([8], th.3) there is a sequence (x_n) in E such that $(\mathop{*}_{j=1}^{k_n} \mu_{nj} * \varepsilon_{x_n})$ is uniformly tight. For the asserted convergence of the sequence $(\mathop{*}_{j=1}^{k_n} \mu_{nj})$ to $e_o(F)$ it therefore suffices to show that the sequence $(\prod_{j=1}^{k_n} \hat{\mu}_{nj})$ of Fourier transforms converges for all $k \in \mathbb{N}$ uniformly on kB^o to the Fourier transform of $e_o(F)$. We will prove

$$\lim_{n\to\infty} \sup_{x' \in kB^o} \left| \log \prod_{j=1}^{k_n} \hat{\mu}_{nj}(x') - \log e(\widehat{\sum_{j=1}^{k_n} \mu_{nj}})(x') \right| = o.$$

This limit relation holds because of the following estimates:

$$\left|\log(\prod_{j=1}^{k_n} \hat{\mu}_{nj})(x') - \log e(\widehat{\sum_{j=1}^{k_n} \mu_{nj}})(x')\right|$$
$$\leq K \sup_{1\leq j\leq k_n} |1 - \hat{\mu}_{nj}(x')| \sum_{j=1}^{k_n} |1 - \hat{\mu}_{nj}(x')| \quad \text{and}$$

$$\sum_{j=1}^{k_n} |1 - \hat{\mu}_{nj}(x')| \leq \sum_{j=1}^{k_n} \int_{B_+} |\langle x, x'\rangle| \mu_{nj}(dx) + 2(\sum_{j=1}^{k_n} \mu_{nj})(B_+^c)$$
$$\leq \langle \int_{B_+} x \, (\sum_{j=1}^{k_n} \mu_{nj})(dx), |x'| \rangle + 2(\sum_{j=1}^{k_n} \mu_{nj})(B_+^c)$$
$$\leq \left\| \int_{B_+} x \, (\sum_{j=1}^{k_n} \mu_{nj})(dx) \right\| \|x'\| + 2(\sum_{j=1}^{k_n} \mu_{nj})(B_+^c) < \infty .$$

2.9 <u>Theorem</u>. Let E be a Banach lattice of Poisson type p $(1\leq p<\infty)$ and let $(\mu_{nj})_{n\geq 1; j=1,\ldots,k_n}$ be a uniformly infinitesimal triangular array of probability measures on E_+. Let further F be a Radon measure on E_+ with $\int \inf(1, \|x\|^p) F(dx) < \infty$ and $\int_{B_+} x \, F(dx) \in E_+$ and suppose that the following three conditions hold:

(CL1) For all $r>o$ the sequence $(\mathrm{Res}_{rB_+^c} (\sum_{j=1}^{k_n} \mu_{nj}))$ converges weakly to $\mathrm{Res}_{rB_+^c} F$.

(CL2) $\lim_{r\to o} \overline{\lim_{n\to\infty}} \sum_{j=1}^{k_n} \int_{rB_+} \|x\|^p \mu_{nj}(dx) = o$.

(CL3) $\lim_{r\to o} \overline{\lim_{n\to\infty}} \left\| \sum_{j=1}^{k_n} \int_{rB_+} x \, \mu_{nj}(dx) \right\| = o$.

Then $(\mathop{*}_{j=1}^{k_n} \mu_{nj})$ converges weakly to $e_o(F)$.

<u>Proof</u>. By proposition 2.7 the sequence $(e(\sum_{j=1}^{k_n} \mu_{nj}))$ converges to $e_o(F)$. There-fore the sequence $(\mathop{*}_{j=1}^{k_n} \mu_{nj})$ also converges to $e_o(F)$ by Lemma 2.8.

2.1o <u>Proposition</u>. Let E be a Banach lattice of Poisson type p. Then for any sequence (X_k) of independent E_+-valued random vectors with the properties $\sum_k \int \|X_k\|^p dP < \infty$ and $\sum_k \int X_k \, dP \in E_+$ the series $\sum_k X_k$

converges a.s.

Proof. We only have to show that $\sum X_k$ converges weakly. We put $\mu_k := X_k(P)$, $F_n := \sum_{k=1}^{n} \mu_k$, $F := \sum_{k=1}^{\infty} \mu_k$, and $\lambda_n := \ast_{k=1}^{n} \mu_k$. Then clearly the conditions (L2) and (L3) of Proposition 2.7 are satisfied, and (L1) holds since

$$\sum_k P[\|X_k\| \geq \varepsilon] \leq \varepsilon^{-p} \sum_k \int \|X_k\|^p dP < \infty .$$

Hence $(e(F_n))$ converges to $e_o(F)$. By Le Cam's theorem it remains to show that the sequence $(\hat{\lambda}_n)$ of Fourier transforms converges uniformly on the sets kB^o for all $k \in \mathbb{N}$. But this can be proved as in the proof of theorem 2.9. In this special situation we do not need the assumption of uniform infinitesimality. ⌋

Problems. 1. Do there exist Banach lattices not being of Poisson type p such that the central limit theorem 2.9 holds? 2. Do there exist Banach lattices not being of Poisson type p such that proposition 2.1o holds? Both problems depend on the solution of the following problem: Suppose that E is a Banach lattice such that every Radon measure F on $E_+ \setminus \{o\}$ with the integrability properties $\int \|x\|^p F(dx) < \infty$ and $\int x F(dx) \in E_+$ is the Lévy measure of an infinitely divisible measure μ, where we do not demand that μ is of order p. Does this already imply that E is of Poisson type p?

References

[1] E. DETTWEILER: Infinitely divisible measures on the cone of an ordered locally convex vector space. Annales scientifiques de l'Université de Clermont, N° 61, 14e fascicule, 11-17 (1976).

[2] E. DETTWEILER: The Laplace transform of measures on the cone of a vector lattice. Math. Scand. 45, 311-333 (1979).

[3] E. DETTWEILER: Cylindrical measures on vector lattices and random measures. Mh. Math. 89, 185-2o4 (198o).

[4] J. HOFFMANN-JØRGENSEN: Probability in Banach spaces. In: Ecole d'été de probabilités de Saint-Flour VI-1976. 1-186. Lecture Notes in Mathematics 598. Berlin-Heidelberg-New York: Springer 1977.

[5] T. KAMAE, U. KRENGEL, G.L. O'BRIEN: Stochastic inequalities on partially ordered spaces. The Annals of Probability 5, 899-912 (1977).

[6] J.L. KRIVINE: Théorèmes de factorisation dans les espaces réticulés. Séminaire Maurey-Schwartz, 1973-1974, Exposes 22-23 Ecole Polytechnique, Paris.

[7] J.L. KRIVINE: Sous espaces de dimension finite des espaces de Banach réticulés. Ann. of Math. 1o4, 1-29 (1976).

[8] L. LE CAM: Remarques sur le théorème limite central dans les espaces localement convexes. Les Probabilités sur les structures algebriques, Colloq. CNRS Paris 197o, 233-249.

[9] J. LINDENSTRAUSS, L. TZAFRIRI: Classical Banach Spaces I, Sequence Spaces. Berlin-Heidelberg-New York: Springer 1977.

[1o] J. LINDENSTRAUSS, L.TZAFRIRI: Classical Banach Spaces II, Function Spaces. Berlin-Heidelberg-New York: Springer 1979.

[11] B. MAUREY: Type et cotype dans les espaces munis de structures locales inconditionelles. Séminaire Maurey-Schwartz 1973-74, Exposes 24-25, Ecole Polytechnique, Paris.

[12] B. MAUREY, G. PISIER: Séries de variables aléatoires vectorielles indépendentes et propriétés géométriques des espaces de Banach. Studia Math. 58, 45-9o (1976).

[13] K.R. PARTHASARATHY: Probability measures on metric spaces. New York-London: Academic Press 1967.

[14] H.P. ROSENTHAL: On the subspaces of L^p ($p>2$) spanned by sequences of independent random variables. Israel J.Math.8,273-3o3 (197o).

[15] H.H. SCHAEFER: Topological vector spaces. 3rd printing, Berlin-Heidelberg-New York: Springer 1971.

[16] H.H. SCHAEFER: Banach lattices and positive operators. Berlin-Heidelberg-New York: Springer 1974.

[17] L. SCHWARTZ: Probabilités cylindriques et applications radonifiantes. J. Fac. Sci. Univ. Tokyo, Sect. I A Math., Vol. 18 no 2, 139-286 (197o).

[18] L. SCHWARTZ: Radon measures on arbitrary topological spaces and cylindrical measures. Oxford: Oxford Univ. Press 1973.

[19] W.A. WOYCZYNSKI: Geometry and martingales in Banach spaces II. Adv. in Probability, vol 4, 267-517. New York: Decker 1978.

E. Dettweiler, Mathematisches Institut der Universität Tübingen, Auf der Morgenstelle 1o, D-74oo Tübingen, BRD

First Elements of a Theory of Quantum Mechanical Limit Distributions

T. Drisch

The aim of this little report is to inform about certain basic concepts, methods and results of a "noncommutative" analogy to the theory of limit distributions as a part of a quantum mechanical probability theory, introduced by C. Cushen/R. Hudson 1971 [2] and based on the "classical" formulation of quantum mechanics characterised by the names of E. Schrödinger, W. Heisenberg, J. v. Neumann. Nevertheless, no quantum mechanical prerequisites are necessary, because an intuitive approach to the theory is also possible by emphasising the strict formal similarity to measure theoretic probability theory. On the other hand, the methods emerge mainly from the theory of group representations. Especially, the theorems on limits of (the products of the rows of) triangular arrays of normalised positive definite kernels on locally compact groups (as in [13]) regains an interpretation as limit theorems for sums of independent quantum mechanical random variables, a programmatic claim, for which only a few hints can be given. Therefore, consider this report just as an invitation to look at the first elements of a theory of quantum mechanical limit theorems (which, by the way, has a more regular behaviour as the classical counterpart, reflecting, for instance, the fact that the algebra of bounded operators has in many respects better manners compared with an algebra of continuous functions).

§ 1 Quantum Mechanical Probability Spaces and Random Variables

The most elementary classical description of a quantum mechanical system is given by an infinite dimensional complex separable Hilbert space H whose rays $\{\alpha\varphi : \alpha \in \mathbb{C}, |\alpha| = 1\}$ $(\varphi \in H, |\varphi| = 1)$ are the pure states of the system (which we will identify with the orthogonal projections $Q_{\mathbb{C}\varphi}$), whereas the composed states are convex combinations $\sum \alpha_n Q_{\mathbb{C}\varphi_n}$ of the pure states giving the set of positive nuclear operators with trace norm $\|\rho\|_1 := \text{trace}\,|\rho| = 1$ (the density matrices), and inducing a nonnegative normed functional P_ρ : $M \longmapsto \text{trace}\, Q_M \rho$ on the orthocomplemented complete lattice $\mathcal{B}_H$ of all closed subspaces of H which is σ- additive on sequences of mutually orthogonal elements of $\mathcal{B}_H$. σ- additive bounded real functionals on $\mathcal{B}_H$ are called Gleason-measures, referring to the Theorem of Gleason ([6], [8]):

> Let H be a (not necessary separable) Hilbert space over $\mathbb{R}$ or $\mathbb{C}$ or the quaternions, whose dimension is a not real measurable cardinal number $\neq 2$. Then each Gleason measure P on $\mathcal{B}_H$ is induced by a selfadjoint nuclear operator ρ_P on H, which is uniquely determined by P, and the map $P \mapsto \rho_P$ is an isomorphic isometry from the Banach space $M^b(H)$ of Gleason measures on $\mathcal{B}_H$ (with respect to the norm of total variation) to the Banach space $T_r(H)$ of selfadjoint nuclear operators on H (with the trace norm) which preserves positivity (i.e. maps the positive cone of $M^b(H)$ on the positive cone of $T_r(H)$).

A cardinal I is said to be real measurable if I is uncountable and if there exists a positive measure $\mu \neq 0$ on the power set of I with $\|\mu\|<\infty$, $\mu(\{\alpha\}) = 0$ for each $\alpha \in I$. It is known that, on the basis of ZF (axioms of Zermelo-Fraenkel) and AC (axiom of choice) and GCH (generalised continuum hypothesis), the existence of real measurable cardinals cannot be proved. The norm of total variation is given by $\|P\| := \sup\{\sum|P(L_j)| \;:\; L_j$ mutually orthogonal closed subspaces of H$\}$.

The theorem gives the legitimation for considering the tripel $(H, \mathcal{B}_H, P)$ - P a positive normed Gleason measure on $\mathcal{B}_H$ - as quantum mechanical probability space. $\mathcal{B}_H$ is not distributive reflecting the peculiarity of incompatibility of quantum mechanical measurements (expressed by Heisenbergs uncertainty principle). The commutative analogy is a discrete probability space $(\Omega, \mathcal{P}(\Omega), \sum\alpha_n \varepsilon_{\omega_n})$ (where $\mathcal{P}(\Omega)$ is the power set of Ω, ε_ω the point measure in $\omega \in \Omega$); we gain the general (nondiscrete) concept by replacing $\mathcal{B}_H$ by the projectors of a W*- algebra of operators on H (which are the experimentally verifiable statements on the given object, called the logic of the object).

In the commutative case the experiments are described by real random variables X on the measurable space $(\Omega, \mathcal{A})$, the phase space, which are - according to a theorem of Varadarajan [16] - the same as the σ- homomorphisms from the Boolean σ- lattice $\mathcal{B}_{\mathbb{R}}$ of real Borel sets into A, identifying X with $X^{-1}: \mathcal{B}_{\mathbb{R}} \longrightarrow \mathcal{A}$. In strict formal analogy we call σ- homomorphisms $E: \mathcal{B}_{\mathbb{R}} \longrightarrow \mathcal{B}_H$ (whose definition includes that $B \cap B' = \emptyset$ implies $E(B) \perp E(B')$) quantum mechanical random variables and identify them via the spectral theorem with the selfadjoint operators on H, the observables (a σ- homomorphism $\mathcal{B}_{\mathbb{R}} \longrightarrow \mathcal{B}_H$ is obviously the same as a (projector-valued) spectral measure on $\mathcal{B}_{\mathbb{R}}$). In analogy to the distribution $P_X: B \longmapsto P(X^{-1}(B))$ (B real Borel set) we define, given an observable A on H with spectral measure E_A and a positive normed Gleason measure P on $\mathcal{B}_H$, a probability measure

$$P_A : B \longmapsto P(E_A(B)(H)) = \text{trace } E_A(B)\rho_P$$

on $\mathcal{B}_{\mathbb{R}}$ which describes the distribution of the experimental observations of A when the quantum mechanical system is in the state P. P_A allows the definition of the usual probabilistic notions; for example, the expectation of A with respect to P is given by

$$E_P(A) := \int t \, dP_A(t)$$

(if the identity is P_A - quasi integrable).

§ 2 Quantum Mechanical Distributions and Quantum Mechanical Independence

The logic of the system contains no specific information characterising the system - in our most elementary case we have only one logic, namely $\mathcal{B}_H$ (H complex separable, $\dim H = \infty$). The relevant information gives a locally compact group G, describing the symmetries of the system and inducing a group of automorphisms of the logic. (For example, considering the motion of a free particle, G is the group of Euklidean motions of $\mathbb{R}^3$ in the nonrelativistic and the inhomogeneous Lorentz group in the relativistic case). The automorphisms of the special logic $\mathcal{B}_H$ can be described in a bijective manner modulo phase factors (i.g. complex numbers of modulus 1) by unitary or antiunitary operators T on H via $L \longmapsto T(L)$ for $L \in \mathcal{B}_H$ ([17], th. 7.27); therefore, in the case that all T are unitary (for example, if G is divisible or a connected Lie group), the automorphisms group of the logic $\mathcal{B}_H$ is isomorphic to the projective group $\mathcal{U}(H)/\mathbb{T} \cdot id_H$ of H, where $\mathcal{U}(H)$ denotes the group of unitary operators on H and $\mathbb{T}$ the torus $\{ z \in \mathbb{C} : |z| = 1\}$. Thus, using a section

$\mathfrak{U}(H)/_{\mathbb{T}\cdot id_H} \longrightarrow \mathfrak{U}(H)$ ([17], th. 1o.4, th. 1o.5), we arrive at <u>projective representations</u> of G in $\mathfrak{U}(H)$, systematically studied by V. Bargmann [1] and G. W. Mackey [1o], the phase factors being expressed by <u>multipliers</u> $\omega : G \times G \longrightarrow \mathbb{C}$ (i.e. functions with

$$\omega(g, e) = \omega(e, g) = 1, \ \omega(g, g^{-1}) = 1,$$
$$\omega(g, g')\ \omega(gg', g'') = \omega(g', g'')\ \omega(g', g'')$$

for all $g, g', g'' \in G$, satisfying

(*) $T_g\, T_{g'} = \omega(g, g')\, T_{g g'}$).

We will assume that G is second countable, that ω is continuous and that $g \longmapsto T_g$ is strongly continuous (see [12] for an exposition of the multiplier problem), and we will concentrate us on the following <u>canonical situation</u> for the pair (G, ω) :

> There exists (up to unitary equivalence) <u>only one</u> irreducible (strongly continuous) projective ω- representation of G in $\mathfrak{U}(H)$.

(The existence of irreducible ω- representations is always guaranteed ([17], 1o.17)). The first example of this situation was given by J. v. Neumann, considering the <u>kinematic</u> of a quantum mechanical particle not underlying electromagnetic forces, which can be described by a <u>canonical pair</u> (p, q) on H (i.e. two unbounded operators on a common dense domain of definition D with $p(D) \subset D$, $q(D) \subset D$, such that p, q and $p^2 + q^2$ are essentially selfadjoint and that the <u>canonical commutation relations</u> $i\,(pq - qp)\big|_D = id_D$ are fulfilled). Assuming the particle as one-dimensional, the symmetry group is $\mathbb{R}^2$ (homogenity of time and space), the projective representation is defined by

$$S_{x,y} \quad := \quad \exp\,(i(xp + yq)) \qquad\qquad x,y \in \mathbb{R},$$

having as multiplier

$$m((x,y),(x',y')) \quad := \quad \exp\Big(\tfrac{i}{2}\ \det\begin{pmatrix} x & x' \\ y & y' \end{pmatrix}\Big) \ ;$$

the equations (*) are called, in this special case, the <u>Weyl-Relations</u>. Converaely, the infinitesimal generators of (a projective representation of $\mathbb{R}^2$, satisfying) the Weyl-Relations are - according to a theorem of Dixmier [4] - a canonical pair. Now, von Neumann's Uniqueness Theorem asserts that the pair ($\mathbb{R}^2$, m) satisfies the canonical situation [11].

The characterising property of the multiplier m implying the uniqueness of the m - representation of $\mathbb{R}^2$ can be formulated for multipliers ω of abelian groups G as follows: Fixing one argument we obtain a character $\chi_g : g' \longmapsto \omega(g, g')\,\overline{\omega(g', g)}$ of G, and $g \longmapsto \chi_g$ is a continuous homomorphism from G into the dual $\hat{G}$. As a consequence of Mackey's Imprimitivity Theorem we have the Theorem of Fell ([14], th. 7.1/th. 7.4):

> If $g \longmapsto \omega_g$ is an algebraic-topological isomorphism from G onto $\hat{G}$, than (G, ω) fulfils the canonical situation.

The typical (abelian) example for the validity of this theorem is $G = K \oplus \hat{K}$ - K abelian locally compact second countable group - and $\omega^K((k, \hat{k}), (k', \hat{k}')) := \overline{\langle k', \hat{k}\rangle}$. We know as a consequence of the Imprimitivity Theorem or the Naimark-Ambrose-Godement Theorem that the irreducible ω^K - representation is of the form $S_{(k, \hat{k})} = V_k W_{\hat{k}}$ with irreducible representations V of K and W of $\hat{K}$ obeying $V_k W_{\hat{k}} = \langle k, \hat{k}\rangle W_{\hat{k}} V_k$ (taking, for example, V as left regular representation and W as multiplication by the corresponding character); see [10], § 11. In v. Neumann's special case the last description corresponds to the canonical pair $(p_o, q_o) := (-i\frac{d}{dx}, M_y)$ on $D := \mathcal{S}(\mathbb{R})$, M_y being multiplication by y, $\mathcal{S}(\mathbb{R})$ being the space of rapidly decreasing $\mathcal{C}^\infty$- functions (Schrödinger representation of canonical pairs), and we have

$$S_{(x,y)} = \exp\left(-\frac{i}{2}xy\right) e^{ixp_o} e^{iyq_o} = \exp\left(\frac{i}{2}xy\right) e^{iyq_o} e^{ixp_o}.$$

Given a pair (G, ω) in the canonical situation, fix an irreducible projective ω - representation $S : G \longrightarrow \mathcal{U}(H)$ (which plays the role of coordinisation of the sample space H). Then, for each ω - representation $T : G \longrightarrow \mathcal{U}(\tilde{H})$ there exists a unitary operator U_T from $\tilde{H}$ to $H \otimes H_T$ (H_T complex separable Hilbert space describing the multiplicity of T) with

$$T_g = U_T^{-1}\,(S_g \otimes id_{H_T})\,U_T \qquad g \in G.$$

Given a Gleason measure P on $\mathcal{B}_H$ with defining nuclear operator ρ_P, there exists a unique nuclear operator ρ_T on H with

$$\text{trace } A\rho_T = \text{trace } (U_T^{-1}(A \otimes id_{H_T})U_T)\rho_P$$

for each bounded operator A on H; the Gleason measure P_T on $\mathcal{B}_H$ associated with ρ_T is called the <u>quantum mechanical distribution</u> of T. Obviously, we have $E_P(T_g) = E_{P_T}(S_g)$ in analogy to the <u>transformation theorem</u> of the Kolmogoroff probability theory (S_g replacing the characters of $\mathbb{R}$) or - more general:

$$E_P\,(f(T_g)) = E_{P_T}\,(f(S_g))$$

for each bounded Borel measurable function f: $\mathbb{R} \longrightarrow \mathbb{R}$.

Let T, T' be two commuting ω-representations of G (on the same Hilbert space $\tilde{H}$); then $(G \oplus G, \omega \oplus \omega)$ being in the canonical situation, we obtain a unitary operator $U_{T,T'} : \tilde{H} \longrightarrow H \otimes H \otimes H_{T,T'}$ with

$$T_g = U_{T,T'}^{-1} \quad (S_g \otimes id_H \otimes id_{H_{T,T'}})\; U_{T,T'}\,,$$

$$T_g' = U_{T,T'}^{-1} \quad (id_H \otimes S_g \otimes id_{H_{T,T'}})\; U_{T,T'}\,,$$

defining the <u>common distribution</u> $P_{T,T'}$ of T and T' in the same way as the distribution P_T of T is defined. We call T and T' <u>independent</u>, if the associated nuclear operators fulfil $\rho_{T,T'} = \rho_T \otimes \rho_{T'}$. The integration of this concept in a more general version of independence as well as its heuristic meaning is given in [7].

Obviously, $\rho_T \otimes \rho_{T'}$ induces the <u>Gleason product measure</u> on H $\otimes$ H.

Now, the concept of <u>a family of independent ω-representations</u> is at hand; for the realisation of infinite families of independent observables (involving infinite tensor products) see also [7] (the analogy to the Kolmogoroff construction for realizing families of independent random variables).

Finally, let us say a family of canonical observables $((p_\alpha, q_\alpha))_{\alpha \in I}$ to be <u>independent</u> if the corresponding family of m-representations is independent.

§ 3 Quantum Mechanical Fourier Transformation

Let us consider a pair (G, ω) in the canonical situation, and a fixed irreducible projective ω- representation S of G on the separable complex Hilbert space H. Given a Gleason measure P on $\mathcal{B}_H$, the Fourier transform of P is defined by

$$\hat{P}(g) := E_P(S_g) \qquad g \in G$$

where the right side of this equation is given by splitting S_g in the hermitian and the antihermitian part. In the special case of the kinematic of a one-dimensional particle the defining equation reads as

$$\hat{P}(x, y) = E_P(e^{i(xp + yq)}) \qquad x, y \in \mathbb{R},$$

enlightening the strict formal analogy to the usual ("commutative") case.

It is not difficult to see, that $\hat{P}$ is continuous, hermitian and ω- positive definite in the case of positivity of P, i.e. for all sequences $g_1, \ldots, g_n \in G$, $z_1, \ldots, z_n \in \mathbb{C}$ it yields:

$$\sum_{j,k} \omega(g_j, g_k^{-1}) \hat{P}(g_j g_k^{-1}) z_j \bar{z}_k \geq 0.$$

Conversely we have the quantum mechanical theorem of Bochner (for a proof of the theorems of this paragraph, see [5]):

> $F : G \longrightarrow \mathbb{C}$ is the Fourier transform of a positive Gleason measure P, if F is continuous in e, hermitian and ω- positive definite. In this case, $F(e) = P(H) = \| P \|$.

Towards the uniqueness theorem, the main step is the following fundamental lemma for the canonical situation (in the following we assume G being unimodular):

> A continuous ω - positive definite function is square integrable (with respect to Haar measure on G).

As a corollary, we gain the quantum mechanical formula of Plancherel:

There exists $d > o$ such that for all positive Gleason measures P, P' on $\mathcal{B}_H$:

$$\int \hat{P}(g)\, \overline{\hat{P}'(g)}\, dg = d \text{ trace } \rho_P\, \rho_{P'}.$$

(dg a fixed Haar measure on G). Given a finite-dimensional subspace K of H, and denoting the orthogonal projection on K by Q_K, the last corollary reads with $\rho_P = Q_K$ as the quantum mechanical Fourier Inversion Theorem:

$$P(K) = \frac{1}{d} \int \hat{P}(g)\, \overline{\text{trace } S_g Q_K}\, dg.$$

Especially we arrive at the Uniqueness Theorem of Fourier Transformation, regarding, that P is determined by the one-dimensional subspaces, H being separable and P being σ- additive.

For a formulation of the most important analytic theorem of Fourier transformation (from the view point of the theory of limit distributions), the continuity theorem, we remember that the functional analytic dual of the Banach space T(H) of nuclear operators on H in the trace norm is the Banach space L(H) of bounded operators on H in the operator norm using the duality $(\rho, A) \longmapsto \text{trace } A\rho$ for $\rho \in T(H)$, $A \in L(H)$. Regarding the formulation of Gleason's theorem, given in § 1, L(H) is the dual of the Banach space $M^b_{\mathbb{C}}(H)$ of complex (bounded) Gleason measures on $\mathcal{B}_H$ in the norm of total variation, by the duality

$$(P, A) \longmapsto E_P(A) \qquad P \in M^b_{\mathbb{C}}(H),\ A \in L(H).$$

Therefore we said a sequence $P_1, P_2, \ldots$ of Gleason measures on $\mathcal{B}_H$ to be weakly convergent with limit P, if

$$\lim E_{P_n}(A) = E_P(A) \qquad \text{for all } A \in L(H).$$

Then, the fundamental lemma allows the proof of the quantum mechanical continuity theorem:

Let $P_1, P_2, \ldots$ be a sequence of positive normed Gleason measures on $\mathcal{B}_H$.

(A) If $\hat{P}_1, \hat{P}_2, \ldots$ converges pointwise to a function $F : G \longrightarrow \mathbb{C}$, continuous in e, then F is the Fourier transform of a positive normed Gleason measure P, and the measures P_n are weakly convergent to P.

(B) If $P_1, P_2, \ldots$ converges weakly to P, then $\hat{P}_1, \hat{P}_2, \ldots$ converges to $\hat{P}$ in the topology of compact convergence.

Therefore, a sequence of ω - representations $T_1, T_2, \ldots$ converge in distribution to P, if and only if the characteristic functions $\hat{P}_{T_1}, \hat{P}_{T_2}, \ldots$ converges pointwise to $\hat{P}$, obtaining the following proof strategy:

Let G_ω be the Heisenberg group of G, i.e. the set $G \times \mathbb{T}$ with the product topology and the product

$$(g,z)(g',z') := (gg', \omega(g,g')zz').$$

G_ω is a unimodular (G supposed to be unimodular) locally compact second countable group and a central extension of $\mathbb{T}$, and the Haar measure on G_ω is the product of the Haar measures of G and $\mathbb{T}$ ([17], th.10.8). The ω - positive definite functions F on G are in a bijective correspondence to the positive definite functions F_ω on G_ω with the property $F_\omega(g,z) = zF_\omega(g,1)$ (given by $F_\omega(g,z) := zF(g)$). Now, taking a limit theorem for sequences of positive definite functions on G_ω, we obtain - via continuity theorem - a limit theorem for quantum mechanical distributions (remembering the theorem of Bochner).

We see that even the abelian case involves a certain amount of noncommutative theory (the Heisenberg group), whereas, on the other side, the prohabilistic("measure theoretic") interpretation of the limit theorems of positive definite functions don't require duality theory (in the case that G is abelian, the canonical situation implies naturally the selfduality of G).

According to the aim of the little report I will not enumerate the resulting theorems, but give certain hints on some at first glance perhaps a little surprising results compared with the ordinary theory, restricting myself for convenience on the motivating (most elementary) example of the canonical pairs.

§ 4 Some aspects of quantum mechanical limit distributions

Let (p,q),(p',q') be two commuting pairs; then, by Stone's Theorem, we obtain for $\alpha,\alpha' \in \mathbb{R}$ a pair of new observables

$$\bar{p} := \alpha p + \alpha' p', \qquad \bar{q} := \alpha q + \alpha' q'$$

as the infinitesimal generators of

$$e^{i\alpha p} \cdot e^{i\alpha' p'} \quad \text{resp.} \quad e^{i\alpha q} \cdot e^{i\alpha' q'} .$$

Let P be a positive normed Gleason measure on $\mathcal{V}_H$. Let

$$f_{p,q}(x,y) := E_P\,(e^{i(xp+yq)}) \qquad x,y \in \mathbb{R}$$

be the characteristic function of (p,q), which can be interpreted, by the transformation theorem, as the Fourier transform of the quantum mechanical distribution with respect to a fixed canonical pair. This interpretation proves the quantum mechanical convolution theorem:

If (p,q),(p',q') are independent (in the sense of § 2), then:

$$f_{\bar{p},\bar{q}}(x,y) = f_{p,q}(\alpha x,\alpha y)\, f_{p',q'}(\alpha' x,\alpha' y)$$

(obviously valid also for other independent projective representations not having an interpretation as sums of the associated generators).

As the last ingredient for a theory of limit distributions we have the notion of the quantum mechanical normal distribution as the Gleason measure P_Γ corresponding by Bochner's Theorem to the hermitian function $n_\Gamma(x,y) := \exp(-\frac{1}{2}(x,y)\Gamma\binom{x}{y})$ (Γ being a positive symmetric real 2 x 2 - matrix), fixing a canonical pair (p,q), for convenience the Schrödinger pair (p_0,q_0). Now, the necessary and sufficient condition of m-positive definiteness of n_Γ is equivalent to $\det\Gamma \geq \frac{1}{4}$. A short calculation shows, that $\Gamma = (\Gamma_{jk})$ is the covariance matrix of the pair (p_0,q_0) in the state P_Γ:

$$\Gamma_{11} = E_{P_\Gamma}(p_0^{\,2}), \quad \Gamma_{22} = E_{P_\Gamma}(q_0^{\,2}),$$

$$\Gamma_{12} = \Gamma_{21} = E_{P_\Gamma}(\tfrac{1}{2}(p_0 q_0 + q_0 p_0)) .$$

Thus, $\det\Gamma \geq \frac{1}{4}$ implies $E_{P_\Gamma}(p_0^{\,2})\, E_{P_\Gamma}(q_0^{\,2}) \geq \frac{1}{4}$, a reformulation of the uncertainty principle.

$\det\Gamma = \frac{1}{4}$ is equivalent to the purity of P_Γ [15], being in this case the ground state of the Hamiltonian H_Γ of the harmonic oszillator, i.e. the projector on $\mathbb{C}\varphi_0$, φ_0 being the first eigenvalue of H_Γ. If

$\det \Gamma > \frac{1}{4}$, the density matrix associated to P_Γ is the equilibrium state of H_Γ [2].

Perhaps it is noteworthy that, because there exists no sharp distribution (with $E_p(p_o^2) = E_p(q_o^2) = 0$), we have no possibility to formulate a law of large numbers.

Concerning the central limit theorem, we observe first that if a sequence of covariance matrices $\Gamma^1, \Gamma^2, \ldots$, such that $\frac{1}{n} \sum_{k=1,\ldots,n} \Gamma^k$

is convergent, defines as limit also a covariance matrix Γ (i.e. $\det \Gamma \geq \frac{1}{4}$), by using the following lemma:

> Let $A_1, \ldots, A_n$ be positive definite symetric real $n \times n$ - matrices, let $\alpha_1, \ldots, \alpha_n$ be positive real numbers with $\sum \alpha_k = 1$. Then:
>
> $$\det \sum_{k=1}^{n} \alpha_k A_k \geq \prod_{k=1}^{n} (\det A_k)^{\alpha_k} .$$

Second, we observe by using the convolution theorem that

$f_{\frac{1}{\sqrt{2}}(p+q),\ \frac{1}{\sqrt{2}}(p'+q')}$ is also m-positive definite such that the proof strategy of the end of § 3 can be applied. We arrive at the quantum mechanical limit theorem, whose formulation for identically distributed canonical pairs is given in [2] :

> Let (p_1,q_1) $(p_2,q_2),\ldots$ be inde-pendent canonical pairs with expectation 0 and finite variances (i.e. $E_p(p_n^2) < \infty$, $E_p(q_n^2) < \infty$), such that the arithmetical means of the covariance matrices Γ^k are convergent to Γ. Write:
>
> $$\sigma_k^2(x,y) \; := \; E_p(xp+yq)^2 \quad (= \; (x,y)\,\Gamma^k \binom{x}{y} \;) ,$$
>
> $$s_n^2 \; := \; \sum_{k=1}^{n} \sigma_k^2 \qquad (x,y \in \mathbb{R},\ k,n \in \mathbb{N}) .$$
>
> Let P_k be the distribution of (p_k,q_k) in the sense of § 2, using the associated m-representations and the fixed pair (p_o,q_o); write:
>
> $$L_\varepsilon^n(x,y) \; := \; \frac{1}{s_n^2(x,y)} \sum_{k=1}^{n} \int_{|t| \geq \varepsilon s_n(x,y)} t^2 \, d(P_k)_{xp_o+yq_o}(t)$$
>
> $$\varepsilon > 0, \quad (x,y) \neq (0,0) .$$

Then we obtain the following equivalences:

(i) $\lim_n L_\varepsilon^n(x,y) = 0$ for all $\varepsilon > 0$, $(x,y) \neq 0$

(Lindeberg-Feller Condition).

(ii) The sequence $(p_1,q_1),(p_2,q_2),\ldots$ converges in distribution to P_Γ and $S_n(x,y) \to \infty$, $\sigma_n'(x,y)\, S_n^{-1}(x,y) \to 0$ for all $(x,y) \neq (0,0)$.

(iii) $\lim_n \sum_{k=1} (f_{p_k,q_k}(\frac{x}{\sqrt{n}},\frac{y}{\sqrt{n}}) - 1) = -\frac{1}{2}(x,y)\Gamma\binom{x}{y}$

(accompanying law).

I will notice, by the way, that we can establish a theorem on accompanying laws, the concept of uniform infinitesimality being obvious.

Finally, I will mention the result of Urbanik [15] on stable quantum mechanical distributions (which is, as well as the problem of infinite divisibility, beyond the scope of the proof strategy, the sums having other multipliers). A Gleason probability P is said to be stable if (in analogy to the "commutative" definition) there exists a sequence of independent identical distributed canonical pairs (p_n,q_n) and a norming sequence of strict positive numbers α_n such that the sums $(\overline{p}_n,\overline{q}_n) := (\alpha_n(p_1+\ldots+p_n), \alpha_n(q_1+\ldots+q_n))$ are convergent in distribution to P. The convolution theorem implies, assuming $\alpha_1 = 1$, and using the "trivial" part (B) of the continuity theorem (valid obviously also in this a little more general situation):

$$\hat{P}(x,y) = \lim f(\alpha_n x, \alpha_n y)^n \qquad x,y \in \mathbb{R},$$

f being the characteristic function of (p_1,q_1). Now, by the Lemma of Schur, the product of ω-positive definite functions is positive definite [2], showing that $\hat{P}^2$ is the characteristic function of a classical stable distribution μ on $\mathbb{R}^2$.

Analysing $\hat{P}^2$ in the case of symmetry (i.e. the distributions of all (p_n,q_n) are invariant under the inversion $(J(\varphi)(t) := \varphi(-t)$ on $L^2(\mathbb{R})$ (this means: J commutes with the distributions)) implying

μ as symmetric, Urbanik obtains the following result, based on the fundamental lemma on square - integrability:

A symmetric quantum mechanical distribution is stable if and only if it is a ground state.

Recently, Czkwianianc generalised this theorem on symmetric semistable distributions [3], based on a representation of the characteristic functions of the classical analogies on $\mathbb{R}^n$. (He used the following definition of semistability:

The convergence of the (p_n, q_n) is only considered on a subsequence $(\bar{p}_{n_j}, \bar{q}_{n_j})$ satisfying

$\lim_j \frac{n_{j+1}}{n_j} = r$ for a real number r with $1 \leq r < \infty$.)

References:

1 V. BARGMANN, On Unitary Ray Representations of Continuous Groups. Annals Math. 59, 1-46 (1954).

2 C.D. CUSHEN, R.L. HUDSON, A Quantum Mechanical Central Limit Theorem. J. Appl. Prob. 8, 454-469 (1971).

3 E. CZKWIANIANC, Symmetric Semistable Measures on $\mathbb{R}^n$ and Symmetric Semistable Distribution Operators in Quantum Mechanics. Reports Math. Physics 17, 89-99 (198o).

4 J. DIXMIER, Sur la relation i (PQ - QP) = 1. Compositio Math. 13, 263-27o (1958).

5 T. DRISCH, A Generalisation of Gleason's Theorem. Int. J. Theor. Physics 18, 239-243 (1979).

6 T. DRISCH, Die Sätze von Bochner und Lévy für Gleason-Maße. Arch. Math. 34, 6o-68 (198o).

7 T. DRISCH, Zur Realisierung unabhängiger kanonischer Paare. Arch. Math. 34, 357-37o (198o).

8 A. GLEASON, Measures on the Closed Subspaces of a Hilbert Space. J. Math. Mech. 6, 885-893 (1957).

9 G.W. MACKEY, Unitary Representations of Group Extensions I. Acta Math. 99, 265-311 (1958).

1o G.W. MACKEY, Unitary Group Representations in Physics, Probability and Number Theory. Benjamin, Reading (Mass.)1978.

11 J. v. NEUMANN, Die Eindeutigkeit der Schrödingerschen Operatoren. Math. Ann. 1o4, 57o-578 (1931).

12 K.R. PARTHASARATHY, Multipliers on Locally Compact Groups. Springer Lecture Notes Math. 93, Heidelberg 1969.

13 K.R. PARTHASARATHY, K. SCHMIDT, Positive Definite Kernels, Continuous Tensor Products and Central Limit Theorems. Springer Lecture Notes Math. 272, Heidelberg 1975.

14 B. SIMON, Topics in Functional Analysis. In: R. STREATER, Mathematics of Contemporary Physics, London - New York 1972.

15 K. URBANIK, Stable Symmetric Probability Laws in Quantum Mechanics. In: Springer Lecture Notes 472, Heidelberg 1975.

16 V.S. VARADARAJAN, Probability in Physics and a Theorem on Simultaneous Observability. Comm. Pure App. Math. 15, 189-217(1962).

17 V.S. VARADARAJAN, Geometry of Quantum Theory I, II. Van Nostrand, Princeton 1968.

Thomas Drisch

Abt. Mathematik

der Universität

Postfach 5oo 5oo

46 Dortmund (BRD)

SUR LE THEOREME DE DICHOTOMIE

POUR LES MARCHES ALEATOIRES SUR LES ESPACES HOMOGENES

Laure ELIE

INTRODUCTION

Soit G un groupe localement compact à base dénombrable et soit μ une mesure de probabilité adaptée et adaptée (cf. 1-2) sur G. Si nous considérons la marche aléatoire gauche de loi μ sur G, nous savons que soit tout état de G est transient, soit tout état de G est récurrent et dans ce dernier cas la marche sur G est récurrente Harris de mesure invariante la mesure de Haar à gauche m_G sur G. Dans [4], B. Hennion et H. Roynette ont étudié cette dichotomie pour les marches aléatoires induites sur les espaces homogènes, mais cherchant à obtenir un résultat analogue à celui existant pour les marches sur les groupes, ils ont été amenés à faire des hypothèses,

- soit sur la loi μ, du type le semi-groupe fermé T_μ engendré par le support de μ est G entier.
- soit sur l'espace homogène : plus précisément si G opère à gauche sur l'espace homogène M, ils réclamaient l'existence d'une mesure ρ_M sur M
  . relativement invariante sous l'action de G (i.e. $g\rho_M$ est proportionnelle à ρ_M pour tout g de G)
  . et excessive pour la marche induite de loi μ (i.e. $\mu * \rho_M \leq \rho_M$).

Ils obtenaient alors la dichotomie suivante : soit tout état est transient, soit tout état est récurrent et dans ce cas la marche induite est récurrente Harris de mesure invariante m. Dans le premier cas ($T_\mu = G$), m est une mesure quasi-invariante sous l'action de G (i.e. gm est équivalente à m pour tout g de G) et dans le deuxième cas $m = \rho_M$.

Ils prouvaient en particulier qu'une telle dichotomie existait sur les espaces homogènes M = G/H pour lesquels G et H étaient des groupes unimodulaires. Par contre ils mettaient en évidence que sur l'espace homogène $\mathbb{R}$ du groupe affine $\mathbb{R}^{+*} \times \mathbb{R}$, un tel résultat n'était pas possible pour certaines mesures .

Dans [5] D. Revuz, à l'aide de résultats de théorie ergodique, a simplifié la preuve du théorème de dichotomie de H. Hennion et B. Roynette. Ce sont ses arguments qui ont inspiré notre démarche dans ce papier.

Ici nous étudions pour une marche induite sur l'espace homogène M la décomposition de M en classes ergodiques et ensembles transients. Nous en déduisons en particulier que si $T_\mu^{-1}T_\mu$ (ou $T_\mu T_\mu^{-1}$) opère transitivement sur M,

• soit tout état de M est transient,

• soit il existe un ensemble F absorbant tel que la marche induite soit récurrente Harris sur M de mesure invariante m et tel que si $\overline{F}$ désigne la fermeture de F, tout état de $\overline{F}^c$ est transient. De plus, la mesure invariante m est la restriction à F d'une mesure quasi-invariante sous l'action de G et est à une constante multiplicative près l'unique mesure sur M vérifiant $\mu * m \leq m$.

Comme F n'est pas toujours égal à M, certains états sont transients, d'autres récurrents.

L'hypothèse ci-dessus sur T_μ est vérifiée de façon générale dès que les fonctions μ (ou $\hat{\mu}$) harmoniques bornées sur G sont constantes ($\hat{\mu}$ désignant l'image de μ par $g \to g^{-1}$) car alors (cf. prop. IV-3 de [1]) $T_\mu^{-1} T_\mu$ (ou $T_\mu T_\mu^{-1}$) est G entier. Elle est en particulier toujours vérifiée sur l'espace homogène $\mathbb{R}$ du groupe affine, exemple que nous traitons au paragraphe 6.

D'autre part, nous verrons que le théorème de dichotomie de [4] peut s'étendre au cas où il existe sur l'espace homogène M une mesure m

• quasi-invariante sous l'action de G

• et excessive pour la marche induite de loi μ.

1.- HYPOTHESES ET NOTATIONS

1.1.- Soit G un groupe localement compact à base dénombrable et H un sous-groupe fermé de G. Nous appelons M l'espace homogène G/H sur lequel G opère à gauche et π l'application canonique continue et ouverte sur M. Par définition, pour $(g,x) \in G \times M$, $g \cdot x = \pi[g\pi^{-1}(x)]$ et si ν est une mesure sur M, $g\nu$ est la mesure définie pour A borélien de M par

$$g\nu(A) = \varepsilon_g * \nu(A) = \nu(g^{-1}A)$$

Nous savons [2] qu'il existe sur M une mesure <u>quasi-invariante</u> λ (c'est-à-dire telle que pour tout g de G, $g\lambda$ soit équivalente à λ), que deux mesures quasi-invariantes quelconques sont équivalentes et qu'une partie borélienne A de M est de λ-mesure nulle si et seulement si $\pi^{-1}(A)$ est négligeable dans G pour une mesure de Haar sur G. Nous désignerons par m_G une mesure de Haar à gauche sur G.

1.2.- Soit μ une mesure de probabilité sur G. Nous ferons sur μ les <u>hypothèses</u> suivantes :

- μ est <u>adaptée</u>, i.e. le sous-groupe fermé engendré par le support de μ est G entier.

- μ est <u>étalée</u>, i.e. il existe un entier n tel que μ^n ne soit pas étrangère à m_G.

1.3. Définition. Si μ est une mesure de probabilité sur G, T_μ désigne le semi-groupe fermé engendré par le support de μ et S_μ l'ensemble des points g de G pour lesquels il existe une mesure de Haar m sur G et un entier p tels que la restriction de m à un voisinage de g soit majorée par μ^p.

1.4.- La marche aléatoire gauche de loi μ sur G est la chaîne de Markov X d'espace d'états G et de probabilité de transition $Q(g,.) = \mu * \varepsilon_g$ $(g \in G)$. Comme G opère à gauche sur M, cette marche induit sur M une chaîne de Markov Y de probabilité de transition $P(x,.) = \mu * \varepsilon_x$ $(x \in M)$, que l'on appelle marche aléatoire sur M (cf.[6]). On note $U(x,.) = \sum_{n\geq 0} P^n(x,.) = \sum_{n\geq 0} \mu^n * \varepsilon_x$ $(x \in M)$ le noyau potentiel associé.

Si A est un borélien de M, nous notons

$$h_A(x) = P_x\,[\overline{\lim}\ \{Y_n \in A\}]$$

Cette fonction est harmonique bornée et est donc continue puisque μ est supposée étalée.

Un état x de M est dit transient si il existe un voisinage V de x tel que $h_V(x) = 0$ et il est dit récurrent si pour tout voisinage V de x, $h_V(x) = 1$. Il est prouvé dans [4] que tout élément de M est soit transient, soit récurrent et que l'ensemble R des états récurrents est un fermé absorbant (i.e. $P(x,R) = 1$ pour tout x de R). De plus si x est un état transient, il existe un voisinage V de x tel que la fonction $U\,1_V$ soit bornée.

1.5.- Contraction associée à P : Comme λ est une mesure quasi-invariante sur M, la mesure λP définie si A est un borélien de M par $\lambda P(A) = \int_M \lambda(dx)\, P(x,A)$ est absolument continue par rapport à λ et nous noterons par T la contraction positive induite par P sur $L^1(M,\lambda)$: pour tout $f \in L^1(M,\lambda)$, $Tf = \frac{d((f\lambda)P)}{d\lambda}$ est la dérivée de Radon-Nykodim de $f\lambda P$ par rapport à λ.

Nous désignerons par C la partie conservative (définie à une classe d'équivalence près) de T et par $D = C^c$ la partie dissipative. Si f est un élément positif de $L^1(M,\lambda)$, nous savons que,

$$\sum_0^\infty T_n f = 0 \quad \text{ou} \quad +\infty \text{ sur } C$$

$$\sum_0^\infty T_n f < \infty \quad \text{sur } D.$$

L'opérateur adjoint T^* de T est la contraction positive sur $L^\infty(M,\lambda)$ définie si $f \in L^\infty(M,\lambda)$ par la classe d'équivalence dans $L^\infty(M,\lambda)$ de Pf. Nous savons (prop. 2.8 de [6]) que $T^* 1_D = 0$ λ-p.s. sur C et donc que $P\,1_C = 1$ λ-p.s. sur C.

On peut en fait construire un ensemble $C_1 = C$ λ-p.s. tel que $P1_{C_1} = 1$ partout sur C_1, en d'autres termes tel que C_1 soit <u>absorbant</u>. Il suffit (cf. prop.2 de [3]) de considérer la suite $h_o = 1_C$, $h_{n+1} = \inf(h_n, Ph_n)$; cette suite décroissante bornée converge vers h telle que $Ph \geq h$. L'ensemble $C_1 = \{h = 1\}$ est absorbant et comme $h_n = 1_C$ λ-p.s. pour tout n, $C_1 = C$ λ-p.s. Dans toute la suite nous choisirons comme représentant de la partie conservative cet ensemble C_1 que nous noterons C.

2.-<u>ETUDE DE LA PARTIE DISSIPATIVE</u>

La partie dissipative D de T est composée en grande partie d'états transients, comme l'exprime la proposition suivante:

2-1.- <u>Proposition</u> : Soit $D_1 = \{x \in M, \exists V$ voisinage de $x, \lambda(V \cap D^c) = 0\}$. L'ensemble D_1 est un ouvert vérifiant $\mathring{D} \subset D_1 \subset \mathring{\bar{D}}$ (si $\mathring{D}$ désigne l'intérieur de D et $\bar{D}$ sa fermeture) et tous les états de D_1 sont transients. De plus si K est un compact de M inclus dans D_1, le potentiel $U1_K$ est borné.

La preuve de cette proposition reposera sur plusieurs lemmes simples.

2-2.- <u>Lemme</u> : L'ensemble D s'écrit $\bigcup_{n=o}^{\infty} D_n$ avec $h_{D_n} = 0$.

<u>Preuve</u> : Soit f un élément strictement positif de $L^1(M,\lambda)$. Il existe une suite de boréliens D_n de M croissant vers D tels que $\lambda(D_n)$ soit fini et tels que $\sum_o^{\infty} T^p f$ soit inférieur à n sur D_n. Alors par dualité

$$< \sum_o^{\infty} T^p f \, , \, 1_{D_n} >_\lambda = < f \, , \sum_o^{\infty} T^{*p} 1_{D_n} >_\lambda = < f \, , \, U1_{D_n} >_\lambda \leq n.\lambda(D_n).$$

D'où $U1_{D_n} < \infty$ λ-p.s. Par suite $h_{D_n} = 0$ λ-p.s. et (cf. 1-4) comme h_{D_n} est continue, $h_{D_n} = 0$ partout.

2-3.- <u>Lemme</u> : Si E est un borélien de M s'écrivant $\bigcup_{n=o} E_n$ avec $h_{E_n} = 0$, alors les états de l'ouvert $S_\mu^{-1}E$ sont transients.

<u>Preuve</u> : On construit comme dans le lemme 2 de [4] une fonction borélienne f sur M telle que $f > 0$ sur E et telle que $Uf \leq 1$ partout. Si μ_a^n désigne la partie absolument continue par rapport à m_G de μ^n et si $P_a^n f(x) = \mu_a^n * \varepsilon_x(f)$, la fonction $h = \sum_{n \geq 1} \frac{1}{2^n} P_a^n f$ est continue, strictement positive sur l'ouvert $S_\mu^{-1}E$ et vérifie $Uh \leq Uf \leq 1$. Considérons un point x de l'ouvert $S_\mu^{-1}E$ et V un voisinage compact de x inclus dans $S_\mu^{-1}E$, la fonction h est minorée par une constante γ strictement positive sur V et par suite $U1_V \leq 1/\gamma$. Le point x est donc transient.

2-4.- Lemme : Si E est un borélien de M, la fonction $U1_E$ est bornée sur l'ensemble $(S_\mu^{-1}E)^c$ et par suite la fonction $U1_{(E\cap(S_\mu^{-1}E)^c)}$ est bornée partout.

Preuve : Soient pour tout entier n, μ_a^n et μ_s^n les parties absolument continue et singulière relativement à m_G de μ^n. Comme μ est étalée il existe une constante k et un entier p tels que pour tout $n \geqslant p$, $\|\mu_s^n\| \leqslant k^n$. Soit y un élément de $(S_\mu^{-1}E)^c$. Alors $S_\mu y \cap E = 0$. Par suite $U_a 1_E(y) = \sum_n \mu_a^n * \varepsilon_y(E) = 0$ et

$U1_E(y) = U_s 1_E(y) = \sum_n \mu_s^n * \varepsilon_y(E)$ est majoré par une constante k' ne dépendant que de k et de p. Il en résulte que $U1_{(E\cap(S_\mu^{-1}E)^c)}$ est bornée par k' sur $(S_\mu^{-1}E)^c$ et donc partout d'après le principe du maximum.

2-5.- Preuve de la proposition 2-1 : Comme C est absorbant, $\lambda(S_\mu C \cap D) = 0$. Par suite $\lambda(S_\mu^{-1}D \cap C) = 0$ et l'ouvert $S_\mu^{-1}D$ est inclus dans l'ouvert D_1. D'après les lemmes 2-2 et 2-3, tous les états de $S_\mu^{-1}D$ sont transients. Considérons donc un élément x de $D_1 \cap (S_\mu^{-1}D)^c$. Alors

$$U1_{D_1}(x) \leqslant U1_D(x) + U1_{D_1\cap C}(x).$$

D'après le lemme 2-4, $U1_D(x)$ est fini. Comme λ est quasi-invariante et comme $\lambda(D_1 \cap C) = 0$, la fonction $U_a 1_{D_1\cap C}$ est nulle et le potentiel $U1_{D_1\cap C} = U_s 1_{D_1\cap C}$ est une fonction bornée. Il en résulte que $U1_{D_1}(x)$ est fini. Par suite $h_{D_1}(x) = 0$ et l'état x est transient.

Tout point de D_1 est donc transient et admet (cf. prop.2 de [4]) un voisinage V tel que $U1_V$ soit borné. Il est alors aisé d'en déduire que si K est un compact de M inclus dans D_1 , $U1_K$ est borné.

2-6.- Remarque : La proposition 2-1 repose seulement sur le fait que l'ensemble D s'écrit $\bigcup_{n=0}^{\infty} D_n$ avec $h_{D_n} = 0$.

3.- CLASSES ERGODIQUES

3-1.- Définition : Soit (E,ν) un espace mesuré, T une contraction de $L^1(E,\nu)$, C sa partie conservative. La contraction T sera dite ergodique si pour tout f de $L_+^1(E,\nu)$ l'ensemble $C_f = \left\{ \sum_{n=0}^{\infty} T^n f = \infty \right\}$ est ν-p.s. égal à l'ensemble vide ou à C.

On vérifie alors (cf. [6]) que si T est ergodique et si f est un élément de $L^\infty(E,\nu)$ nul sur C^c vérifiant $T^* f = f$ ν-p.s. sur C, alors f est ν-p.s. constante sur C.

3-2.- Définition : Soient μ une mesure de probabilité sur G, M un espace homogène sur lequel G opère à gauche et P la probabilité de transition sur M définie par $P(x,.) = \mu * \varepsilon_x$; soit F un ensemble absorbant pour P. Nous dirons que la restriction de P à F est <u>ergodique</u> si la restriction à F de la contraction associée à P définie en 1-5 est ergodique.

Nous allons expliciter des ensembles absorbants de M sur lesquels P est ergodique. Remarquons que des ensembles absorbants naturels à étudier sont les ensembles $T_\mu x$ $(x \in M)$; et précisément nous avons :

3-3.- <u>Proposition</u> : Soient μ une mesure de probabilité adaptée et étalée sur G, M un espace homogène sur lequel G opère à gauche. Si x est un élément de M, la restriction à l'ensemble absorbant $T_\mu x$ de P est ergodique.

<u>Preuve</u> : Remarquons tout d'abord que $T_\mu T_\mu^{-1} = S_\mu S_\mu^{-1}$ (cf. déf. 1-4). En effet de la relation $T_\mu S_\mu \subset S_\mu \subset T_\mu$, il résulte que $T_\mu S_\mu S_\mu^{-1} T_\mu^{-1} \subset S_\mu S_\mu^{-1} \subset T_\mu T_\mu^{-1}$, et donc puisque $S_\mu S_\mu^{-1}$ contient l'élément neutre, $S_\mu S_\mu^{-1} = T_\mu T_\mu^{-1}$. De même $T_\mu^{-1} T_\mu = S_\mu^{-1} S_\mu$.

Notons $F = T_\mu x$, x étant fixé dans M. L'ensemble F est absorbant, donc si T est la contraction définie en 1-5, $T^* 1_{F^c} = P 1_{F^c} = 0$ sur F. Donc si f est un élément de $L_+^1(M,\lambda)$ nul sur F^c, il en est de même de Tf et nous pouvons considérer la restriction de T à F. Nous noterons C_x sa partie conservative. Il s'agit de montrer que pour $f \in L_+^1(M,\lambda)$ nul sur F^c, l'ensemble $C_f = \left\{ \sum_{n=0}^{\infty} T^n f = \infty \right\}$ est λ-p.s. égal à l'ensemble vide ou à C_x. Comme C_f satisfait $P 1_{C_f} = 1_{C_f}$ λ-p.s. sur C_x (cf. [6] p. 112), il suffit de prouver que si A est un borélien de M inclus dans l'ensemble absorbant C_x tel que $P 1_A = 1_A$ λ-p.s. sur C_x, alors $A = \emptyset$ ou C_x λ-p.s. Nous allons utiliser le lemme suivant.

3-4.- <u>Lemme</u> : Si A est un borélien de M inclus dans un ensemble absorbant E et tel que

$$(*) \qquad P 1_A = 1_A \quad \lambda\text{-p.s. sur } E,$$

et si A et E-A sont de λ-mesure positive, alors il existe $y \in A$ et $z \in E-A$ tels que

$$\lambda(S_\mu S_\mu^{-1} y \cap S_\mu S_\mu^{-1} z \cap E) = 0.$$

<u>Preuve du lemme</u> : Prouvons pour commencer que si A vérifie (*), nous avons

$$\forall n \in \mathbb{N} \quad \forall k \in \mathbb{N} \qquad P^n \hat{P}^k 1_A \leq 1_A \quad \lambda\text{-p.s. sur } E$$

$$P^n \hat{P}^k 1_{(E-A)} \leq 1_{(E-A)} \quad \lambda\text{-p.s. sur } E$$

où $\hat{P}$ est la probabilité de transition sur M associée à $\hat{\mu}$ (image de μ par $g \to g^{-1}$).

En effet comme $P^k 1_A = 1_A$ λ-p.s. sur E, nous obtenons en remontant à G (cf. 1-4 pour les notations), $Q^k 1_{\pi^{-1}(A)} = 1_{\pi^{-1}(A)}$ m_G-p.s. sur $\pi^{-1}(E)$.

Par dualité sur G, nous en déduisons, si $\hat{Q}$ désigne la probabilité de transition sur G associée à $\hat{\mu}$,

$$\langle Q^k 1_{\pi^{-1}(A)}, 1_{\pi^{-1}(E-A)} \rangle_{m_G} = \langle 1_{\pi^{-1}(A)}, \hat{Q}^k 1_{\pi^{-1}(E-A)} \rangle_{m_G}$$

et donc que

$$\hat{Q}^k 1_{\pi^{-1}(E-A)} = 0 \quad m_G\text{-p.s.} \quad \text{sur } \pi^{-1}(A).$$

D'où

$$\hat{P}^k 1_{(E-A)} \leqslant 1_{(E-A)} \qquad \lambda\text{-p.s.} \quad \text{sur } E.$$

Comme E est absorbant, nous en concluons que sur E, nous avons si $n \in \mathbb{N}$,

$$P^n \hat{P}^k 1_{(E-A)} = P^n(1_E \hat{P}^k 1_{(E-A)}) \leqslant P^n 1_{E-A} = 1_{E-A} \qquad \lambda\text{-p.s.}$$

D'où la deuxième inégalité. Si A vérifie (*), il en est de même de E-A et l'autre inégalité s'obtient de manière identique en considérant cet ensemble E-A .

Supposons donc que A et E-A soient de λ-mesure positive, alors il existe $y \in A$ et $z \in E-A$ tels que

$$\forall n \in \mathbb{N}, \forall k \in \mathbb{N}, \quad P^n \hat{P}^k 1_A(z) = P^n \hat{P}^k 1_{(E-A)}(y) = 0 .$$

Alors $\qquad \lambda(A \cap S_\mu S_\mu^{-1} z) = \lambda((E-A) \cap S_\mu S_\mu^{-1} y) = 0 .$

D'où $\quad \lambda(S_\mu S_\mu^{-1} y \cap S_\mu S_\mu^{-1} z \cap E) \leqslant \lambda(A \cap S_\mu S_\mu^{-1} z) + \lambda((E-A) \cap S_\mu S_\mu^{-1} y) = 0 .$

3-5.- <u>Fin de la preuve de 3-3</u> : Il résulte du lemme 3-4 que si A et C_x-A sont de λ-mesure positive, alors il existe y et z appartenant à F tels que

$$\lambda(S_\mu S_\mu^{-1} y \cap S_\mu S_\mu^{-1} z \cap C_x) = 0 .$$

Comme $T_\mu T_\mu^{-1} = S_\mu S_\mu^{-1}$ et comme y et $z \in T_\mu x$, on en déduit que $\lambda(T_\mu x \cap C_x) = 0$, et donc que $\lambda(C_x) = 0$, ce qui contredit le fait que A et E-A soient de λ-mesure positive. La restriction de P à $T_\mu x$ est donc ergodique.

La proposition suivante donne des conditions pour lesquelles P est ergodique sur M entier.

3-6. <u>Proposition</u> : Si $T_\mu^{-1} T_\mu$ (ou $T_\mu T_\mu^{-1}$) opère transitivement sur M, P est ergodique.

<u>Preuve</u> : Soit C la partie conservative de la contraction T associée à P. Si P n'est pas ergodique, il existe d'après la preuve de la proposition 3-3, deux éléments y et z de C tels que $\lambda(S_\mu S_\mu^{-1} y \cap S_\mu S_\mu^{-1} z \cap C) = 0$. Alors si $T_\mu T_\mu^{-1}$ opère transitivement sur M, $S_\mu S_\mu^{-1} y = S_\mu S_\mu^{-1} z = M$ et $\lambda(C) = 0$, ce qui contre dit la non-ergodicité de P. D'où la proposition dans ce cas. Si $T_\mu^{-1} T_\mu$ opère transitivement sur M, l'ensemble ouvert $S_\mu y \cap S_\mu z$ est non vide et donc de λ-mesure positive. De plus comme C est

absorbant, les ensembles $S_\mu y$ et $S_\mu z$ sont λ-p.s. inclus dans C. Par suite

$$0 < \lambda(S_\mu y \cap S_\mu z \cap C) \leqslant \lambda(S_\mu S_\mu^{-1} y \cap S_\mu S_\mu^{-1} z \cap C) ,$$

ce qui est contradictoire.

4.- RECURRENCE HARRIS

4-1.- Définition : La marche induite de loi μ sur l'espace homogène M est dite récurrente Harris sur un ensemble absorbant F si il existe une mesure invariante m_F σ-finie portée par F telle que $m_F(A) > 0$ entraine $h_A = 1$ sur F.

Elle sera dite transiente si pour tout compact de M, le potentiel $U1_K$ est borné.

Nous allons étudier la marche induite sur M en nous plaçant sur des ensembles absorbants où P est ergodique.

4-2.- Théorème : Soient μ une mesure de probabilité adaptée et étalée sur G et M un espace homogène sur lequel G opère à gauche. Soient F un ensemble absorbant de M sur lequel la restriction de P à F est ergodique et C_F la partie conservative associée.

Si $\lambda(C_F) = 0$, F est réunion dénombrable d'ensembles transients F_n (i.e. $h_{F_n} = 0$).

Si $\lambda(C_F) > 0$, la marche induite sur M de loi μ est récurrente Harris sur C_F de mesure invariante m_F. La mesure m_F est la restriction à C_F d'une mesure sur M quasi-invariante sous l'action de G. De plus $F-C_F$ est réunion dénombrable d'ensembles transients.

Ce théorème admet plusieurs corollaires que nous donnons tout de suite. Le premier permet d'obtenir la décomposition de M en ensembles absorbants sur lesquels la marche induite est récurrente Harris et en ensembles transients et on retrouve le résultat de [9].

4-3.- Corollaire : Soient μ une mesure de probabilité adaptée et étalée sur G et M un espace homogène sur lequel G opère à gauche. Alors M est réunion de deux ensembles disjoints C' et D' vérifiant

- $C' = \bigcup_{p \in \mathbb{N}} C_p$; pour tout p de $\mathbb{N}$, l'ensemble C_p est absorbant , la marche de loi μ est récurrente Harris sur C_p de mesure invariante m_p et m_p est la restriction à C_p d'une mesure quasi-invariante sur M.

- $D' = \bigcup_{p \in \mathbb{N}} E_p$; pour tout p de $\mathbb{N}$, E_p est transient (i.e. $h_{E_p} = 0$).

De plus $D'_1 = \{ x \in M, \exists V$ voisinage de $x, \lambda(V \cap C') = 0 \}$ est exactement l'ensemble des points transients. Il satisfait $\mathring{D}' \subset D'_1 \subset \mathring{\overline{D'}}$ et $D'_1 \cap D' = D'$ λ-p.s.

4-4.- Remarque : Même sur un espace homogène compact , il peut exister effectivement une infinité dénombrable d'ensembles absorbants sur lesquels la marche induite est récurrente Harris. J. Rosenberg en donne un exemple dans [8] en considérant un groupe G semi-simple de centre non fini , KAN sa décomposition d' Iwasawa, et l'espace homogène compact $K \simeq G/AN$.

Preuve du corollaire 4-3 : Puisque, pour tout x de M, $T_\mu x$ n'est pas d'intérieur vide, il existe une suite $(x_n)_{n\in N}$ telle que $M = \bigcup_n T_\mu x_n$. La restriction de P à l'ensemble $T_\mu x_n$ est ergodique (prop. 3-3) et donc d'après le théorème 4-2, si C_n est un ensemble absorbant représentant la partie conservative de la restriction de P à $T_\mu x_n$, la mar- induite est récurrente Harris sur C_n de mesure invariante m_n et $T_\mu x_n - C_n$ est réunion dénombrable d'ensembles transients. Remarquons de plus que si i et j sont deux entiers distincts, les ensembles C_i et C_j sont égaux ou disjoints λ-p.s., puisque les mesures m_i et m_j sont les restrictions à C_i et C_j de mesure quasi-invariante sur M. Par suite si $(n_p)_{p\in N}$ est une sous-suite telle que les ensembles C_{n_p} soient deux à deux disjoints λ-p.s. et telle que $\bigcup_p C_{n_p} = \bigcup_n C_n$ λ-p.s., il nous suffit de poser $C' = \bigcup_p C_{n_p}$ et $D' = M - C'$ pour obtenir la décomposition réclamée.

D'après la proposition 2-1 et la remarque 2-6 les points de D'_1 sont transients. Soit alors x n'appartenant pas à D'_1. Si V est un voisinage de x, $\lambda(V\cap C')>0$. Soit $I = \{p\in N \text{ tel que } \lambda(V\cap C_p) > 0\}$. Puisque la marche induite estrécurrente Harris sur C_p la fonction $h_{V\cap C_p}$ est égale à 1 sur C_p. Par suite $h_{V\cap C'} = 1$ sur $\bigcup_{p\in I} C_p$. De plus comme $\overset{\circ}{D'} \subset D'_1$, x est adhérent à C'. Montrons qu'en fait x est adhérent à $\bigcup_{p\in I} C_p$. Sinon il existerait $V_1 \subset V$ tel que $V_1 \cap (\bigcup_{p\in I} C_p) = \emptyset$ et $\lambda(V_1\cap C')$ serait nul, ce qui est impossible. Comme la fonction $h_{V\cap C'}$ est continue, elle est égale à 1 sur $\overline{\bigcup_{p\in I} C_p}$ et $h_V(x)$ est égal à 1 ; le point x est donc récurrent.

Lorsque P est ergodique, nous obtenons immédiatement le corollaire suivant:

4-5.- Corollaire : Soient μ une mesure de probabilité adaptée et étalée sur G et M un espace homogène sur lequel G opère à gauche. Supposons P ergodique et soit C sa partie conservative. Alors

- si $\lambda(C) = 0$, la marche induite de loi μ sur M est transiente.
- si $\lambda(C) > 0$, la marche est récurrente Harris sur C de mesure invariante m, et m est la restriction à C d'une mesure quasi-invariante sur M. L'ensemble des points récurrents est exactement le support de m. Le potentiel de tout compact inclus dans le complémentaire du support de m est borné.

4-6.- Remarque : Si $T_\mu^{-1}T_\mu$ (ou $T_\mu T_\mu^{-1}$) opère transitivement sur M, P est ergodique et les conclusions du corollaire 4-5 sont vérifiées.

La preuve du théorème 4-2 va reposer sur le lemme suivant :

4-7.- Lemme : Soient F un ensemble absorbant de M sur lequel la restriction de P à F est ergodique et C_F la partie conservative associée. Supposons $\lambda(C_F) > 0$. Alors la marche induite sur M restreinte à F est $\lambda_{|C_F}$-irréductible, c'est à dire que pour tout borélien A de M inclus dans F et vérifiant $\lambda(A \cap C_F) > 0$, le potentiel $U1_A$ est strictement positif sur F.

Preuve : Commençons par montrer que la marche induite restreinte à F est $\lambda_{|C_F}$-essentiellement irréductible, c'est à dire que pour tout borélien A de M inclus dans F et vérifiant $\lambda(A \cap C_F) > 0$, $U1_A > 0$ λ_{C_F}-p.s.

Soit $B = \{ x \in C_F, \ U1_A(x) = 0 \}$. Alors par dualité,

$$\left\langle \sum_{n \geq 0} P^n 1_A , 1_B \right\rangle_\lambda = \left\langle 1_A , \sum_{n \geq 0} T_n 1_B \right\rangle_\lambda = 0 ;$$

et $\sum_{n \geq 0} T_n 1_B = 0$ λ-p.s. sur $A \cap C_F$. Comme $A \cap C_F$ est de λ-mesure positive et comme P est ergodique, on en déduit que $\sum_{n \geq 0} T_n 1_B = 0$ λ-p.s. sur C_F et donc que $1_B = 0$ λ-p.s. . D'où l'assertion.

Pour montrer que la marche sur F est $\lambda_{|C_F}$ - irréductible, il suffit maintenant (cf. [6] p.78) de vérifier que pour x appartenant à un borélien inclus dans C_F de λ-mesure positive, les mesures $\lambda_{|C_F}$ et $\sum_{n \geq 0} 2^{-n} P^n(x,.)$ ne sont pas étrangères. Or $\sum_{n \geq 0} 2^{-n} P^n(x,.) \geq \phi m_G * \varepsilon_x$ où ϕ est une fonction borélienne sur G strictement positive sur S_μ et la mesure $\phi\, m_G * \varepsilon_x$ est la restriction à $S_\mu x$ d'une mesure quasi-invariante sur M. L'ensemble C_F est absorbant (cf. 1-5) et donc $\lambda(S_\mu C_F \cap C_F^c)$ est nul. Par suite pour tout x de C_F les mesures $\lambda_{|C_F}$ et $\sum_{n \geq 0} 2^{-n} P^n(x,.)$ ne sont pas étrangères. D'où le lemme.

4-8.- Preuve du théorème 4-2 : Le cas où $\lambda(C_F) = 0$ découle du lemme 2-2 . Supposons donc $\lambda(C_F) > 0$. Comme la marche induite sur F est $\lambda_{|C_F}$-irréductible, soit le noyau U restreint à F est propre, soit la marche est récurrente Harris sur un ensemble absorbant E de $\lambda_{|C_F}$- mesure pleine. Si U restreint à F était propre, il existerait une suite $F_p \nearrow F$ telle que $U1_{F_p}$ soit borné. Alors pour $f \in L^1(M,\lambda)$ nulle sur F^c, $\sum_{n \geq 0} T_n f$ serait λ-p.s. fini sur F_p pour tout p et donc sur F, ce qui entrainerait

$\lambda(C_F) = 0$. Par suite la marche est récurrente Harris sur E absorbant et admet une mesure invariante m_F portée par E telle que $\lambda_{|C_F} \ll m_F$ (cf.[5] p.74). De plus si A inclus dans C_F vérifie $\lambda(A) = 0$, alors $\sum_{n\geq 0} P^n 1_A = 0$ λ-p.s. et donc $m_F(A) = 0$. Par conséquent la restriction à C_F de m_F est équivalente à $\lambda_{|C_F}$. Montrons qu' en fait m_F est portée par C_F. Supposons donc qu'il existe un ensemble B inclus dans $F-C_F$ tel que $m_F(B)$ soit > 0. D' après le lemme 2-2, l'ensemble $F-C_F$ est réunion dénombrable d'ensembles transients D_n et il existe un entier n tel que $m_F(B\cap D_n) > 0$. Mais alors $h_{B\cap D_n} = 1$ sur E ce qui contredit le fait que D_n soit transient. En conclusion, la mesure m_F est équivalente à $\lambda_{|C_F}$ et la marche est récurrente Harris sur C_F. D'où le théorème.

Etudions maintenant plus précisement le cas où $T_\mu^{-1}T_\mu$ opère transitivement sur M.

4-9.- Proposition : Si $T_\mu^{-1}T_\mu$ opère transitivement sur M, alors

- soit la marche induite de loi μ sur M est transiente,
- soit il existe un ensemble absorbant C sur lequel la marche est récurrente Harris et partant d'un élément x de C^c la marche atteint C avec une probabilité strictement positive.

Preuve : Il s'agit de montrer (d'après 4-6) la dernière assertion, c'est à dire que h_C est strictement positive sur M. Les ensembles $A = \{ h_C = 0 \}$ et $B = \{ h_C = 1 \}$ sont absorbants et l'ensemble B contient C. Supposons que A soit non vide et considérons un élément x de A et un élément y de C. Comme A et B sont absorbants, $\lambda(S_\mu x\cap A^c) = \lambda(S_\mu y\cap B^c) = 0$. Par suite $\lambda(S_\mu x\cap S_\mu y) = 0$, et l'ensemble ouvert $S_\mu x\cap S_\mu y$ est vide. L' ensemble $T_\mu^{-1}T_\mu = S_\mu^{-1}S_\mu$ n'opère donc pas transitivement sur M.

Et réciproquement nous avons

4-10.- Proposition : Si il existe un ensemble absorbant C sur lequel la marche induite de loi μ soit récurrente Harris et tel que h_C soit > 0, alors $T_\mu^{-1}T_\mu$ opère transitivement sur M.

Preuve : Soient x et y deux éléments de M. Comme $h_C(x)$ est > 0, $\lambda(T_\mu x\cap C) > 0$; il en résulte, puisque la marche est récurrente Harris sur C et puisque $h_C(y)$ est strictement positif que $h_{(T_\mu x\cap C)}(y)$ est strictement positif. On en déduit que $\lambda(T_\mu x\cap T_\mu y\cap C)$ est > 0 et donc que $T_\mu x\cap T_\mu y$ est non vide. D'où la proposition.

Le corollaire suivant va nous prouver que la condition $T_\mu^{-1}T_\mu$ opère transitivement sur M est une bonne condition d'ergodicité pour P.

4-11.- <u>Corollaire</u> : Si l' espace homogène M est compact, P est ergodique sur M si et seulement si $T_\mu^{-1} T_\mu$ opère transitivement sur M.

<u>Preuve</u> : Si P est ergodique et M compact, on est dans la deuxième alternative du corollaire 4-5;montrons que la fonction h_C est strictement positive. Comme h_C est continue, l'ensemble $B = \{ h_C = 0 \}$ est un compact de M inclus dans $\overline{C}^c$ et d'après la proposition 2-1, le potentiel $U1_B$ est borné. L'ensemble B étant absorbant, on en conclut que B est vide. Il résulte de la proposition 4-10 que $T_\mu^{-1} T_\mu$ opère transitivement sur M, ce qui avec 3-6 prouve le corollaire.

En fait lorsque M est compact,P est quasi-compact [7] et vérifie donc la condition de Doeblin relativement à la mesure invariante m définie en 4-5 (cf.[6] p.178). Il en résulte immédiatement que puisque $m(C^c) = 0$, le potentiel $U1_{C^c}$ est borné et donc dans ce cas $h_C = 1$ partout.

5.- MESURES INVARIANTES ET EXCESSIVES

5-1.- Soit ν une mesure excessive pour P sur M . Nous pouvons considérer la contraction T_ν de $L^1(M,\nu)$ définie pour $f \in L^1(M,\nu)$ par $T_\nu f = \frac{d(f\nu P)}{d\nu}$. Si C_ν et D_ν désignent les parties conservative et dissipative de T_ν , alors $T_\nu D_\nu = P1_{D_\nu} = 0$ ν-p.s. sur C_ν. Mais comme l'a fait remarquer D.Revuz dans [5], ν étant excessive, T_ν à ν-p.s. même partie dissipative et conservative que la contraction de $L^1(M,\nu)$ qui à f associe Pf. Or si f est un élément strictement positif de $L^1(M,\nu)$, l'ensemble $\{ \sum_{n\geq 0} P^n f = \infty \}$ est un ensemble absorbant ; il en résulte que $P1_{C_\nu} = 0$ ν-p.s. sur D_ν. Par conséquent $P1_{C_\nu} = 1_{C_\nu}$ ν-p.s.

La contraction T définie en 1-5 est la contraction T_λ où λ est une mesure quasi-invariante. Remarquons que si C est la partie conservative de $T = T_\lambda$, alors si ν est absolument continue par rapport à λ , $C = C_\nu$ ν-p.s.

5-2.- <u>Proposition</u> : Soient μ une mesure de probabilité adaptée et étalée sur G et M un espace homogène sur lequel G opère à gauche. Supposons que $T_\mu^{-1} T_\mu$ (ou $T_\mu T_\mu^{-1}$) opère transitivement sur M. Si la marche induite de loi μ sur M n'est pas transiente, il existe un ensemble absorbant C sur lequel la marche est récurrente Harris de mesure invariante m. Alors m est à une constante multiplicative près l'unique mesure sur M excessive pour P.

<u>Preuve</u> : D' après 4-6, il suffit de vérifier que m est l'unique mesure excessive pour P. Soit donc m' une mesure excessive. Comme C est absorbant, la restriction m'_C de m' à C est excessive; mais puisque la marche de loi μ est récurrente Harris

sur C, m'_C est invariante et proportionnelle à m. Par suite la restriction m'_D de m' à $D = C^c$ est excessive. Supposons que m'_D est non nulle. On peut vérifier en adaptant les lemmes II-2 et II-3 de [1] que m'_D majore une mesure quasi-invariante sur un ouvert absorbant Q. La restriction m'_Q de m'_D à Q est encore excessive et notons m'^a_Q la partie absolument continue de m'_Q par rapport à la mesure quasi-invariante λ. Comme la mesure $\mu * m'^a_Q$ est encore absolument continue par rapport à λ, nous en déduisons que la mesure m'^a_Q est excessive et est la restriction à l'ouvert Q d'une mesure quasi-invariante. Posons $\nu = m + m'^a_Q$. Alors si C_ν est la partie conservative de T_ν, nous savons (5-1) que $P1_{C_\nu} = 1_{C_\nu}$ ν-p.s. Par suite $P1_{C_\nu} = 1_{C_\nu}$ λ-p.s. sur l'ensemble absorbant $C \cup Q$. Il résulte alors du lemme 3-4 et de la preuve de 3-6 que C_ν est ν-p.s. égal à l'ensemble vide ou à $C \cup Q$. Mais ceci contredit le fait que $C_\nu = C$ ν-p.s. (cf. 5-1) et donc la mesure m'_D est nécessairement nulle.

5-3.- Remarque : On peut démontrer de manière analogue que sur chaque ensemble $T_\mu x$ tel que $\lambda(C_x) > 0$, il existe une unique mesure excessive qui est invariante et portée par C_x.

La proposition suivante généralise le théorème de dichotomie de [4].

5-4.- Proposition : Soient une mesure de probabilité adaptée et étalée μ sur G et M un espace homogène sur lequel G opère à gauche. Si il existe sur M une mesure λ quasi-invariante sous l'action de G et excessive pour P, alors

- soit la marche induite de loi μ sur M est transiente et tous les états sont transients;
- soit la marche est récurrente Harris sur M entier de mesure invariante λ et tous les états sont récurrents.

Preuve : Elle repose sur les idées de [5]. Comme λ est excessive, si C est la partie conservative de $T = T_\lambda$, $P1_C = 1_C$ λ-p.s. (cf. 5-1). En remontant au groupe G on en déduit que $Q1_{\pi^{-1}}(C) = 1_{\pi^{-1}}(C)$ m_G-p.s. (cf. paragraphe 1 pour les notations). En raisonnant alors comme dans le lemme 4 de [4] puisque sur G, Q et $\hat{Q}$(associé à $\hat{\mu}$) sont en dualité, on en conclut que $\pi^{-1}(C) = \emptyset$ ou G m_G-p.s. et donc que C est l'ensemble vide ou M λ-p.s. On montre de la même manière que P est ergodique.
Par conséquent il résulte du corollaire 4-5 que si $C = \emptyset$ λ-p.s., la marche est transiente et que si $C = M$ λ-p.s., la marche est récurrente Harris de mesure invariante λ. L'ensemble des points récurrents est le support de λ, c'est donc M. De plus pour tout borélien A de M tel que $\lambda(A) > 0$, la fonction h'_A vaut 1 sur C et donc partout, puisqu'elle est continue. La marche est récurrente Harris sur M entier.

6.- EXEMPLE DU GROUPE AFFINE

6-1. Nous représentons le groupe affine G_1 de la droite réelle par le produit semi-direct $\mathbb{R}^{+*} \times \mathbb{R}$ muni du produit $(a,b)(a',b') = (aa', b+ab')$. Nous désignerons par a et b les projections respectives de G sur $\mathbb{R}^{+*}$ et $\mathbb{R}$ et tout élément g de G_1 s' écrira $(a(g),b(g))$. Le groupe G opère à gauche sur l'espace homogène $\mathbb{R}$ par l'application qui à (g,x) associe $g.x = a(g)x + b(g)$. Si μ est une mesure de probabilité sur G_1, nous allons étudier la marche induite sur $\mathbb{R}$ par la marche aléatoire gauche de loi μ sur G_1.

Une mesure de probabilité μ sur G_1 sera dite avoir un moment d'ordre β $(\beta \in \mathbb{R}^+)$ si la fonction d^β est μ-intégrable où pour $g \in G_1$,

$$d(g) = \operatorname{Log}(a(g)) + \operatorname{Log}^+(b(g)) .$$

Si μ est adaptée, étalée et admet un moment d'ordre 1, alors soit les fonctions μ-harmoniques bornées sont constantes , soit les fonctions $\hat{\mu}$-harmoniques bornées sont constantes (cf. 11). Par suite pour une telle mesure μ , $T_\mu^{-1}T_\mu$ (ou $T_\mu T_\mu^{-1}$) est égal à G (prop IV.3 de [1]). Cette dernière propriété est encore vraie si μ n'admet pas de moment d'ordre 1, puisque dans la condition $T_\mu^{-1}T_\mu$ (ou $T_\mu T_\mu^{-1}) = G$, n'intervient que le support de μ . En conséquence, pour toute probabilité μ étalée et adaptée sur G_1 ,la probabilité de transition P sur l'espace homogène $\mathbb{R}$ définie par $P(x, \ ) = \mu * \varepsilon_x$ est <u>ergodique</u> (cf.3-6).

6-2.- <u>Théorème</u> : Soit μ une mesure de probabilité sur G_1 adaptée, étalée et telle que $\operatorname{Log}(a(g))$ soit μ-intégrable. Posons

$$\alpha = \int_{G_1} \operatorname{Log}(a(g)) \, d\mu(g) .$$

Alors la marche induite sur $\mathbb{R}$ de loi μ est

- récurrente Harris sur $\mathbb{R}$ entier de mesure invariante m de masse finie si μ admet un moment d'ordre 1 et si $\alpha < 0$,
- récurrente Harris sur $\mathbb{R}$ entier de mesure invariante m de masse infinie si μ admet un moment d'ordre $2+\varepsilon$ $(\varepsilon \in \mathbb{R}^+)$ et si $\alpha = 0$,
- transiente si $\alpha > 0$.

De plus lorsque la marche est récurrente Harris, elle est apériodique.

<u>Preuve</u> :

-<u>Cas 1</u>: $\alpha < 0$.

Si $\alpha < 0$ et si μ admet un moment d'ordre 1 on sait [1] qu'il existe sur $\mathbb{R}$ une mesure μ-invariante m de masse finie et m est absolument continue par rapport à la mesure de Lebesgue λ sur $\mathbb{R}$. Remarquons que λ est quasi-invariante et même relativement invariante sous l'action de G_1.

Alors si on considère la contraction T_m, sa partie conservative C_m est égale m-p.s. à $\mathbb{R}$ puisque $1 \in L^1_+(\mathbb{R},m)$. Or si C est la partie conservative de $T = T_\lambda$, $C = C_m$ m-p.s. et par conséquent $\lambda(C) > 0$. On est donc dans la deuxième alternative du corollaire 4-5 : la marche induite est récurrente Harris sur C de mesure invariante m et la mesure m est la restriction à C d'une mesure quasi-invariante sur $\mathbb{R}$. De plus on montre (corollaire 5.35 de [10]) que pour tout x de $\mathbb{R}$, la marche partant de atteint C et donc que $h_C = 1$ partout. Par suite la marche induite est récurrente Harris sur $\mathbb{R}$ entier.

- <u>Cas 2</u> : $\alpha = 0$.

Si X est la marche aléatoire gauche de loi μ sur G_1, $b(X)$ est la marche sur $\mathbb{R}$. Considérons la suite des temps d'arrêt $(\tau_k)_{k\in\mathbb{N}}$ suivante :

$$\tau_0 = 0, \ldots\ldots, \quad \tau_k = \inf \{ n > \tau_{k-1}, a(X_n) < a(X_{\tau_{k-1}}) \}.$$

Ces temps d'arrêt sont p.s. finis et $(X_{\tau_k})_{k\in\mathbb{N}}$ est une marche aléatoire sur de loi ρ définie pour tout borélien A de G_1 par $\rho(A) = P_e(X_{\tau_1} \in A)$, e désignant l'élément neutre de G_1.

Alors si μ admet un moment d'ordre $2+\varepsilon$, ρ admet un moment d'ordre 1 (5.46 de [10]) et bien sûr $\int \operatorname{Log}(a(g))\, d\rho(g) < 0$. De plus ρ est adaptée et étalée (2.26 de [10]) et (cas 1 ci-dessus) la marche $(b(X_{\tau_k}))_{k\in\mathbb{N}}$ est récurrente Harris sur $\mathbb{R}$ entier. Il en résulte que la marche $(b(X_n))_{n\in\mathbb{N}}$ est récurrente Harris sur $\mathbb{R}$ entier de mesure invariante m_1. Cette mesure m_1 est de masse infinie; en effet si ce n'était pas le cas, pour tout $f \in L^\infty(\mathbb{R},m_1)$, la fonction $\phi(g) = \varepsilon_g * m_1(f)$ serait μ-harmonique bornée sur G_1. Or si $\alpha=0$, ces fonctions sont constantes (cf. [11]) et la mesure m_1 satisferait $\varepsilon_g * m_1 = m_1$ pour tout g de G_1, ce qui contredirait le fait que m_1 soit de masse finie.

De plus dans les deux cas ci-dessus la marche induite de loi μ est récurrente Harris apériodique sur $\mathbb{R}$, c'est à dire n'admet pas plusieurs classes cycliques. En effet si ce n'était pas le cas, il existerait un entier p tel que la marche de loi μ^p sur $\mathbb{R}$ admette plusieurs classes ergodiques, ce qui contredirait l'ergodicité de la probabilité de transition associée à μ^p.

BIBLIOGRAPHIE

[1] R. AZENCOTT : Espaces de Poisson des groupes localement compacts, Lecture Notes N° 148, Springer Verlag, 1970.

[2] N. BOURBAKI : Livre VI, Intégration chapitre VII, Hermann.

[3] Y. DERIENNIC : Lois "zéro ou deux" pour les processus de Markov. Application aux marches aléatoires. Ann. Institut Henri Poincaré 12 (2), (1976), pp. 111-129;

[4] H. HENNION et B. ROYNETTE : Un théorème de dichotomie pour une marche aléatoire sur un espace homogène. Astérisque N°74 (1980), pp. 99-122.

[5] D. REVUZ : Sur le théorème de dichotomie de Hennion-Roynette. Préprint (à paraître dans un ecture otes consacré aux journées sur les marches à Nancy, mai 1981).

[6] D. REVUZ : Markov chains. North Holland publishing company, 1975.

[7] A. BRUNEL et D. REVUZ : Quelques applications probabilistes de la quasi-compacité, A.I.H.P. 10 (1974) p.301-337.

[8] J.ROSENBERG:Bull. Soc. Math. France, Mémoire 54 (1977).

[9] W. WINKLER : Doeblin's and Harris'theory of Markov Processes. Z. Wahrschein. 31 (1975) p.79-88.

[10] L. ELIE : Comportement asymptotique du noyau potentiel sur les groupes; A paraître aux Ann. Sci. Ecole Norm. Sup.

[11] A. RAUGI : Fonctions harmoniques et théorèmes limites pour les marches aléatoires sur les groupes . Bull. Soc. Math.France, Mémoire 54 (1977).

[12] F. SPITZER : Principle of Random Walks, Van Nostrand; New-York (1972).

Laure ELIE

Université Paris VII

U.E.R. de Mathématiques

2, Place Jussieu

75251 PARIS Cedex 05

FRANCE

CONTINUOUS COHOMOLOGY, INFINITELY DIVISIBLE POSITIVE DEFINITE FUNCTIONS AND CONTINUOUS TENSOR PRODUCTS FOR SU(1,1)

Joachim Erven

und

Bernd-Jürgen Falkowski

§ 0 Introduction

We investigate the infinitely divisible positive definite (I.D.P.) functions on SU(1,1). Applying the theory of Parthasarathy and Schmidt (cf [5]) with some modifications the problem can be reduced to the computation of a certain first cohomology group and the proof of the nontriviality of a certain 2-cocycle. This may be of interest in itself since it provides an explicit generator for $H^2(SU(1,1);R)$.

Finally we describe the construction of a continuous tensor product according to [8] and realize this construction by exhibiting a family of I.D.P. functions on SU(1,1).

§ 1 The Structure of I.D.P. Functions on SU(1,1)

First of all we need to recall some definitions:

(1.1) Definition:

Let G be a topological group and $g \to U_g$ be a unitary representation of G in a Hilbert space H. Then a continuous map δ: G $\to$ H is a 1 - cocycle if it satisfies

$$U_{g_1}\ \delta(g_2) = \delta(g_1 g_2) - \delta(g_1) \quad \forall g_1, g_2 \in G.$$

If in addition δ is analytic it is called an analytic 1-cocycle.

Remarks:

(i) If we set $\delta(g) := U_g v - v$ for some fixed $v \in H$ then we obtain a trivial 1-cocycle (coboundary).

(ii) It follows from the cocycle equation that the range of an analytic cocycle is contained in the space of analytic vectors of the representation.

(iii) Pinczon and Simon have shown (cf. [6]) that analytic cocycles differ from continuous ones only by coboundaries.

(1.2) Definition:

A continuous map $S:G\times G \to \mathbb{R}$ is an <u>additive 2-cocycle</u> if it satisfies

a) $S(g_1,g_2) + S(g_1g_2,g_3) = S(g_1,g_2g_3) + S(g_2,g_3) \quad \forall g_1,g_2,g_3 \in G$

b) A continuous map $\sigma : G\times G \to S^1$ is a <u>multiplicative 2-cocycle</u> if it satisfies

$$\sigma(g_1,g_2)\ \sigma(g_1g_2,g_3) = \sigma(g_1,g_2g_3)\ \ \sigma(g_2,g_3) \quad \forall g_1,g_2,g_3 \in G$$

Remarks:

(i) If we set $S(g_1,g_2) := b(g_1) + b(g_2) - b(g_1g_2)$ for some continuous map $b: G \to \mathbb{R}$ we obtain a trivial additive 2-cocycle (coboundary).

(ii) In analogy to (i) we obtain a trivial multiplicative 2-cocycle by setting
$\sigma(g_1,g_2) := \beta(g_1)\ \beta(g_2)\ \overline{\beta(g_1g_2)}$
for some continuous map $\beta : G \to S^1$.

We shall only be interested in 2-cocycles which satisfy the normalization conditions

a) $S(g,g^{-1}) \equiv S(g,e) \equiv S(e,g) \equiv 0 \quad \forall g \in G$

b) $\sigma(g,g^{-1}) \equiv \sigma(g,e) \equiv \sigma(e,g) \equiv 1 \quad \forall g \in G$

We shall also need two further definitions from [3]:

(1.3) Definition:

A continuous function $f:G \to \mathbb{C}$ is called <u>σ-positive</u> if

(i) $$\sum_{i=1}^{n} \sum_{j=1}^{n} \alpha_i \bar{\alpha}_j \ \ \sigma(g_j^{-1},g_i) \ \ f(g_j^{-1}g_i) \geq 0$$

$\forall (\alpha_1,\ldots,\alpha_n) \in \mathbb{C}^n, \ \ \forall (g_1,\ldots,g_n) \in G^n,$

(ii) $f(e) = 1$ (i.e. f is normalized at e),

where σ is a multiplicative 2-cocycle.

Remark: If $\sigma \equiv 1$ then f is called <u>positive definite.</u>

(1.4) Definition:

A pair (f,σ) with f σ-positive, is called <u>infinitely divisible</u> if $\forall n \in N \; \exists \; (f_n,\sigma_n)$

with

(i) f_n is σ_n-positive,

(ii) $f_n^n \equiv f$,

(iii) $\sigma_n^n \equiv \sigma$.

Remark:

If $\sigma \equiv 1$, then we obtain the definition of an I.D.P. function.

As a technical device to deal with I.D.P. functions we are also going to need the following theorem stated in [3]. Because of its importance and since the relevant facts are rather widely spread in the literature we shall give a proof here.

(1.5) Theorem:

Let G be a connected, locally connected, locally compact, second countable group. Let (f,σ) be infinitely divisible. Then there exist a 1-cocycle δ such that

(i) $f(g) = \exp[-\frac{1}{2} < \delta(g) \; , \; \delta(g) > + \; i \; b(g)]$

(ii) $\sigma(g_1,g_2) = \exp i \; S(g_1,g_2)$

(iii) $S(g_1,g_2) = \text{Im} < \delta(g_2), \; \delta(g_1^{-1}) > + \; b(g_1) + b(g_2) - b(g_1 g_2)$

where $b : G \to R$ is continuous satisfying $b(g^{-1}) = -b(g) \quad \forall g \in G$

(N.B. $<.,.>$ denotes the inner product in H and Im the imaginary part of a complex number.)

Proof:

Utilize theorem (3.6) and (3.7) in [2] together with corollary (12.8) in [5] to ascertain the existence of the relevant logarithms. Then determine the precise nature of these logarithms using theorem (3.1) in [2]. q.e.d.

We now state a lemma which is analogous to lemma (4.1) in [3].

(1.6) Lemma:

Let b: SU(1,1) → R be continuous with $b(g^{-1}) = -b(g)$. Then there is a bijection η between pairs (f,b) with f I.D.P. and certain σ-positive (ψ,σ) given by η : (f,b) → (ψ,σ) where

(i) $\psi(g) = \exp[i\, b(g)]\ f(g)$

(ii) $\sigma(g_1,g_2) = \exp i[b(g_1) + b(g_2) - b(g_1 g_2)]$.

Proof:

The proof proceeds along the same lines as the proof of lemma (4.1) in [3] where again semi-simplicity of SU(1,1) is used.

q.e.d.

Combining theorem (1.5) and lemma (1.6) we are now in a position to state the main result of this section:

(1.7) Theorem:

Every I.D.P. function f on

$$SU(1,1):=\{\begin{bmatrix}\alpha & \beta\\ \bar{\beta} & \bar{\alpha}\end{bmatrix}:\ \alpha,\beta\in\mathbb{C};\ |\alpha|^2-|\beta|^2=1\}$$

is of the form

$$f(g) = \exp(-[\tfrac{1}{2} < \delta(g),\ \delta(g) > + i\ b(g)])$$

where δ is a 1-cocycle and the additive 2-cocycle S given by

$$S(g_1,g_2) := \mathrm{Im} < \delta(g_2),\ \delta(g_1^{-1}) > \text{ satisfies}$$

$$S(g_1,g_2) = b(g_1) + b(g_2) - b(g_1 g_2)$$

where b: G → R is continuous.

As a 2-cocycle associated with a 1-coboundary is easily seen to be trivial we thus have reduced the classification of I.D.P. functions on SU(1,1) to computing the nontrivial 1-cocycles of SU(1,1) and then investigating the nontriviality of the 2-cocycle S described above.

§ 2 The Non-trivial Cocycles for SU(1,1)

In this section we are going to describe all non-trivial cocycles of SU(1,1). To this end we need a modification of a theorem due to Parthasarathy and Schmidt. This reduces the cocycle problem for representations induced (in the sense of Mackey) from unitary representations of subgroups to the cocycle problem for the subgroups themselves. This theorem may also be used to give all cocycles with values in a pre-Hilbert space H obtained by inducing from non-unitary representations. Although, in general, one doesn't obtain the complete solution of the cocycle problem in this fashion, for SU(1,1) we were successful: In this case it turned out that the pre-Hilbert space H mentioned above is the space of continuous, complex-valued functions on S^1 and this contains the analytic vectors for all unitary representations of SU(1,1) (cf. [10] p. 303). Thus we obtain the complete solution of the analytic cocycle problem, which already is the general solution (see remark (1.1)).

In order to find a suitable subgroup for the inducing construction we give an Iwasawa décomposition of SU(1,1):

(2.1) Lemma:

Let $K := \{k(x) := \begin{bmatrix} e^{ix} & o \\ o & e^{-ix} \end{bmatrix} : x \in R\}$

$N := \{n(s) := \begin{bmatrix} 1+is & -is \\ is & 1-is \end{bmatrix} : s \in R\}$

$A := \{a(t) := \begin{bmatrix} \cosh t & \sinh t \\ \sinh t & \cosh t \end{bmatrix} : t \in R\}$

Then K is a compact, N a nilpotent, and A an abelian subgroup of SU(1,1). An Iwasawa decomposition is given by:

$g = \begin{bmatrix} \alpha & \beta \\ \bar{\beta} & \bar{\alpha} \end{bmatrix} \in SU(1,1)$ can be written as

$$g = k(x)\ n(s)\ a(t) \quad \text{where}$$

$t = \log|\alpha+\beta|, \quad x = \arg(\alpha+\beta), \quad s = \operatorname{Im} \alpha\bar{\beta}.$

Proof: Computation.

We now set $D:=ZNA$ where $Z=\{\pm I\}$ is the centre of $SU(1,1)$. Then it is shown in [1] that all unitary irreducible representations of $SU(1,1)$ may be induced from characters (not necessarily unitary) of D. Since D is a semi-direct product, a straight-forward computation shows that the only character of D having a non-trivial cocycle is given by

$$\chi(z\ n(s)\ a(t)):=e^{t}$$

where $z \in Z$ and $n(s)$ and $a(t)$ are as in (2.1). In order to perform the inducing construction we now need some notation and a lemma: Let $C_o(S^1)$ denote all continuous complex-valued functions on S^1 and consider

$$C_+^1(S^1):=\{f\in C_o(S^1): \int_0^{2\Pi} e^{-inx} \quad f(e^{ix})dx=0 \qquad \forall n < 1\}$$

$$C_-^1(S^1):=\{f\in C_o(S^1): \int_0^{2\Pi} e^{-inx} \quad f(e^{ix})dx=0 \qquad \forall n > -1\}$$

(2.2) Lemma:

Let $<.,.>$ be defined on $V:=C_+^1(S^1)\cup C_-^1(S^1)$ by:

$$<f_1,f_2>:= -\int_0^{2\Pi}\int_0^{2\Pi} \log[\frac{|e^{ix} - e^{ix'}|^2}{2}]\ f_1(e^{ix})\overline{f_2}(e^{ix'})dxdx'$$

Then $<.,.>$ is an inner product on V.

Proof(Sketch):

The difficult part is to show that $<.,.>$ is positive definite. We first note that $|e^{ix} - e^{ix'}| = 2(1 - \cos(x-x'))$.

Then for $f\in C_+^1(S^1)$ we use the Fourier expansion $\sum_{n=1}^{\infty} a_n e^{inx}$.

to evaluate the integral. For this one also needs to show that

$$\int_0^{2\Pi} \log(1-\cos x)\ \cos nx = \frac{-2\Pi}{n} \text{ (Induction!)}$$

The rest is computation. An analogous argument goes through for $f \in C_-^1(S^1)$. q.e.d.

The inducing construction now yields

(2.3) Theorem:

Let $V = C_+^1(S^1) \cup C_-^1(S^1)$, let $<.,.>$ be as in (2.2).
Then the unitary representation U of SU(1,1) induced from $\chi(d) = e^t$ is given in $(V,<.,.>)$ by

$$(U_g f)(z) := \frac{1}{|\alpha - \bar{\beta} z|^2} f(g^{-1} z)$$

where $g := \begin{bmatrix} \alpha & \beta \\ \bar{\beta} & \bar{\alpha} \end{bmatrix} \in SU(1,1)$, $z \in S^1$, and

$$g \cdot z := \frac{\alpha z + \beta}{\bar{\alpha} + \bar{\beta} z}.$$

U splits into the irreducible parts $U|C_+^1(S^1)$ and $U|C_-^1(S^1)$ (denoted by U^+ and U^-). Both irreducibles lie in the discrete series.

Proof: Computation; for details see [1].
The generalization of the theorem of Parthasarathy and Schmidt now gives

(2.4) Theorem:

The only non-trivial cocycles associated with irreducible unitary representations of SU(1,1) are given by

$$\delta_1(g)(z) := \frac{\bar{\beta} z}{\alpha - \bar{\beta} z} \qquad \text{(up to constant multiples!)}$$

$$\delta_2(g)(z) := \frac{\overline{\beta z}}{\bar{\alpha} - \overline{\beta z}}$$

They are associated with U^+ and U^- respectively.

Remark:

We note for reference purposes that

$$\delta(g)(z) := \frac{1}{|\alpha - \overline{\beta} z|^2} - 1$$

defines a non-trivial cocycle which is real and the direct sum of δ_1 and δ_2 described above. This cocycle is already mentioned in [9].

§ 3 The Non-triviality of Im $< \delta_1(g_2), \quad \delta_1(g_1^{-1}) >$

In this section we show that neither δ_1 nor δ_2 will give rise to I.D.P. functions via (1.7). For this it is clearly sufficient (since $\delta_1 = \overline{\delta_2}$) to show that there is no continuous function b: SU(1,1) → R satisfying

$$S(g_1,g_2) := \mathrm{Im} < \delta_1(g_2), \delta_1(g_1^{-1}) > = b(g_1) + b(g_2) - b(g_1 g_2) \quad (1)$$

Using the Fourier expansion for $\delta_1(g)$ we obtain:

$$S(g_1,g_2) = -8\pi^2 \ \mathrm{Im} \log\left[\frac{\alpha_1\alpha_2 + \beta_1\overline{\beta}_2}{\alpha_1\alpha_2}\right] \quad (2)$$

where log is defined on $\mathbb{C}\setminus(-\infty,0]$ with log 1 = 0.

We assume the existence of a function b satisfying (1) and derive a contradiction. If such a b exists it must be necessarily unique, since because of (1) two such functions b and b' can differ only by a continuous homomorphism which must be trivial since SU(1,1) is semi-simple.
Using the Iwasawa decomposition described in (2.1) it is clear from (2) that the following equation holds:

$$S(g,k) \equiv S(k,g) \equiv 0 \ \forall (g,k) \in SU(1,1) \times K \quad (3)$$

(3) and (1) together imply that b|K is a continuous homomorphism from K to R. K being compact this implies

$$b(k) \equiv 0 \quad \forall k \in K \quad (4)$$

Now (3) and (4) yield:

$$b(k_1 g k_2) = b(g) \qquad \forall g \in SU(1,1), \quad \forall k_1, k_2 \in K \tag{5}$$

In particular we get:

$$b(nak) = b(na) \qquad \forall (k,n,a) \in K\times N\times A \tag{6}$$

We investigate b further on the semi-direct product NA. According to (2.1)

$$d := \begin{bmatrix} \alpha & \beta \\ \overline{\beta} & \overline{\alpha} \end{bmatrix} \in NA \qquad \text{is given by}$$

$$\alpha = \cosh t - ise^{-t}, \quad \beta = \sinh t - ise^{-t}$$

where $s, t \in R$.

Using elementary properties of the complex logarithm and the fact that because of the special form of the elements of NA

$$\log\left[\frac{\alpha_1\alpha_2 + \beta_1\overline{\beta}_2}{\alpha_1\alpha_2}\right] - \log[\alpha_1\alpha_2 + \beta_1\overline{\beta}_2] + \log\alpha_1 + \log\alpha_2$$

is a well defined continous function on R^4 we obtain from the connectedness of R^4 that

$$\log\left[\frac{\alpha_1\alpha_2 + \beta_1\overline{\beta}_2}{\alpha_1\alpha_2}\right] = \log[\alpha_1\alpha_2 + \beta_1\overline{\beta}_2] - \log\alpha_1 - \log\alpha_2 \tag{7}$$

holds an NA.

Hence if $\alpha : NA \to \mathbb{C}$ denotes the continuous map assigning to any matrix in NA its (1,1)-element then it follows from (7) and (2) that

$$S(g_1,g_2) = \tilde{b}(g_1) + \tilde{b}(g_2) - \tilde{b}(g_1 g_2) \quad \forall g_1, g_2 \in NA$$

where $\tilde{b}(g) := 8\pi^2 \operatorname{Im} \log \alpha(g)$

Since NA is not semi-simple $\tilde{b}$ may differ from b on NA by a continuous homomorphism:

$$b(n(s)\ a(t)) = \tilde{b}(n(s)\ a(t)) \mp \gamma t \text{ where } \gamma \in R.$$

Now let $k = \begin{bmatrix} i & 0 \\ 0 & -i \end{bmatrix}$, $a = \begin{bmatrix} 1 & 0 \\ 0 & 1 \end{bmatrix}$.

The Iwasawa decomposition of $n(s)\ ak = \begin{bmatrix} i-s & -i \\ -s & -i-s \end{bmatrix}$

is given by $\tilde{k}\ \tilde{n}\ \tilde{a}$ with

$$\tilde{n} = \begin{bmatrix} 1-is & is \\ -is & 1+is \end{bmatrix} \text{ and } \tilde{a} = \begin{bmatrix} \cosh \tilde{t} & \sinh \tilde{t} \\ \sinh \tilde{t} & \cosh \tilde{t} \end{bmatrix}$$

where $\tilde{t} = \log\sqrt{4s^2 + 1}$.

An application of (5) and (6) to our special choice of k and a implies that

$b(n) = b(\tilde{n}\ \tilde{a})$ must hold for every $n \in N$.
Using (8) we easily derive a contradiction. Hence
$\text{Im} < \delta_1(g_2),\ \delta_1(g_1^{-1}) >$ is non-trivial and (1.7) can't be applied. We note the interesting fact that $S(g_1,g_2)$, as described above, provides a complete description of $H^2(SU(1,1);\ R)$, since $H^2(SU(1,1);\ R)$ is known to be one-dimensional, cf. [7]. Since $H^2(SL(2;\mathbb{C});\ R)$ is well-known to be trivial (see [4]) this also shows that the fact that $SL(2;\mathbb{C})$ is simply connnected, whilst $SU(1,1) \cong SL(2;R)$ is not, is of crucial importance in the proof that $H^2(SL(2;\mathbb{C});\ R)$ is trivial.

Remark:

It is clear that for every $\gamma \in R$, $\gamma \neq 0$, γs also is a continuous non-trivial 2-cocycle. Consequently $\exp i\gamma s$ is a multiplicative 2-cocycle for SU(1,1). A similar analysis of $\exp i\gamma s$ leads to the following result:

$\exp i\gamma s$ is a coboundary iff $8\pi^2\gamma \in Z$.

§ 4 Continuous Tensor Products (CTP's) and I.D.P. functions

In this section we describe the construction of continuous tensor products of group representations following [8] and realize the construction for SU(1,1).
In order to motivate the construction of a CTP we shall start with a description of v. Neumann's product of a countable number of Hilbert spaces.

(4.1) v. Neumann's Product and its Generalization

Let $\{H_i\}_{i\in N}$ be a sequence of Hilbert spaces and $\Omega := \{\Omega_i\}_{i\in N}$ be a sequence of unit vectors with $\Omega_i \in H_i$ $\forall i \in N$. In fact we shall only be interested in the situation where $H_i = H$ (some fixed Hilbert space) for every $i \in N$. Then Ω may be viewed as a section of the trivial Hilbert bundle with total space NxH. Let $\psi := \{\psi_i\}_{i\in N}$ be a sequence which differs from Ω in only finitely many places and suppose further that D is the set of finite formal linear combinations of such ψ's. We then equip D with a sesquilinear form as follows:

$$\text{Set } \langle \psi^{(1)}, \psi^{(2)} \rangle := \prod_{i=1}^{\infty} \langle \psi_i^{(1)}, \psi_i^{(2)} \rangle_{H_i}$$

and extend by linearity/antilinearity. Separation and completion then gives the required Hilbert space which is, of course, dependent on the "reference section" Ω. Thus it will be denoted by

$$\overset{\Omega}{\bigotimes_{i\in N}} H_i$$

If we wish to generalize this construction to a continuous product we need the analogue of the inner product

$$\prod_{i=1}^{\infty} \langle \psi_i^{(1)}, \psi_i^{(2)} \rangle_{H_i} = \exp\Big[\sum_{i=1}^{\infty} \log \langle \psi_i^{(1)}, \psi_i^{(2)} \rangle\Big] \text{ formally}$$

where the sum should be replaced by an integral somehow and where obviously the appearance of the logarithm is going to cause problems.

However, let's not worry about problems of existence at the moment and proceed with the formal construction of the continuous analogue. Naturally we consider $\{H_x\}_{x\in R}$ with $H_x = H \ \ \forall x\in R$ and a continuous map Ω, $x \to \Omega_x$, from R to the unit vectors in H(i.e. a section of the trivial Hilbert bundle with total space RxH).

We now consider continuous sections ψ which differ from Ω only on a compact set. Again we let D be the set of all finite formal linear combinations of such ψ's and try to equip this set with an inner product by setting

$$\langle\psi_1, \psi_2\rangle := \exp \int_R \log\langle\psi_1(x), \psi_2(x)\rangle dx$$

and extending by linearity/antilinearity.

It is clear that neither existence nor definiteness are by any means guaranteed.
However, let G be a (suitably well-behaved) topological group and suppose that $g \to U_g$ is a unitary representation of G in a Hilbert space H with cyclic vector Ω. (This will give our "reference section" by setting $\Omega_x := \Omega \ \forall x\in R$),
We then consider the group of all continuous functions from R to G with compact support("multiplication" is pointwise) and denote it by $C_e(R,G)$, where this is furnished with the UCC-topology.
Any γ in $C_e(R,G)$ then defines a section ψ by setting

$$\psi(x) := U_{\gamma(x)}\Omega.$$

In this situation it is possible to give a necessary and sufficient condition for the existence of the CTP: It is simply that the positive definite function

$$f(g) := \langle U_g\Omega, \Omega\rangle$$

should be infinitely divisible. We refer to [8] for the details of the proof. So we are led to consider

(4.2) A Family of I.D.P. functions on SU(1,1)

We now utilize the real non-trivial cocycle $\delta(g)(z) :=$

$\frac{1}{|\alpha-\bar{\beta}z|^2} - 1$ mentioned after (2.4) to obtain via (1.7):

Theorem:

For every $c>0$ the function f_c given by $f_c(g) :=$

$$\exp[-\frac{c}{8\pi^2} < \delta(g), \delta(g)>] = [\frac{4}{\text{trace } gg^* + 2}]^{2c}$$

is infintely divisible positive definite.

Now let Exp V denote the symmetric Fock space over V (as in (2.2)) and consider the span of all vectors of the form

$$\text{Exp } \delta(g) := 1 \oplus \delta(g) \oplus \frac{1}{\sqrt{2!}}(\delta(g) \otimes \delta(g)) \oplus \frac{1}{\sqrt{3!}}(\delta(g) \otimes \delta(g) \otimes \delta(g))\ldots$$

We set $\Omega:=\text{Exp } \delta(e)$ and define a cyclic representation by setting

$$U_g \text{Exp } \delta(g') := \frac{f_c(gg')}{f_c(g')} \text{ Exp } \delta(gg')$$

Then it is easily seen that this representation has f_c as "expectation value", i.e.

$$f_c(g) = <U_g \Omega, \Omega>$$

and thus we have constructed the CTP.

Remark:

The family of I.D.P. functions in the theorem above was described in [9] although the definiteness proof was given there in a laborious ad hoc fashion whilst here it is immediate from the theory.

[1] Erven, J.: Über Kozyklen erster Ordnung von SL(2;R). Dissertation, TU München, (1979)

[2] Falkowski, B.-J.: Factorizable and infinitely divisible PUA-representations of locally compact groups. J. of Math. Phys. (1974)

[3] Falkowski, B.-J.: Infinitely divisible positive definite functions on SO(3) Ⓢ R^3 p. 111-115 in: Probability Measures on groups. Lecture Notes in Mathematics, Vol. 706, Springer Verlag, Berlin(1979)

[4] Parthasarathy, K.R.: Multipliers on Locally Compact Groups. Lecture Notes in Mathematics, Vol. 93, Springer Verlag, (1969)

[5] Parthasarathy, K.R.: Schmidt, K.: Positive Definite Kernels, Continuous Tensor Products, and Central Limit Theorems of Probability Theory. Lecture Notes in Mathematics, Vol, 272, Springer Verlag, (1972)

[6] Pinzon, G.; Simon,J.: On the 1-Cohomology of Lie Groups. Letters in Math. Phys. 1 (1975)

[7] Stasheff, J.D.: Continuous cohomology of groups and classifying spaces. Bull. of the AMS, Vol. 84, No. 4, (1978)

[8] Streater, R.F.: Current commutation relations, continuous tensor products and infinitely divisible group representations. Rend. Sci. Int. Fisica E. Fermi XI, (1969)

[9] Vershik, A.M.; Gelfand, I.M.; Graev, M.I.: Representations of the group SL(2;R), where R is a ring of functions. Russ. Math. Surv., 28, (1973)

[10] Warner, G.: Harmonic Analysis on Semi-simple Lie Groups I, Springer Verlag (1972)

Fachbereich Informatik
Hochschule der Bundeswehr München
Werner-Heisenberg-Weg 39
8014 Neubiberg

Canonical Representation of the Bernoulli Process

P. Feinsilver
Department of Mathematics
Southern Illinois University
at Carbondale
Carbondale, Illinois 62901
USA

Moment Systems

Let p(dx) denote a measure on $\mathbb{R}$ such that $\int_{\mathbb{R}} e^{zx}p(dx) = e^{L(z)}$ where L(z) is analytic in a neighborhood of the origin in $\mathbb{C}$. Then we have $\int_{\mathbb{R}} f(x+y)p(dy) = e^{L(D)}f(x)$, D denoting d/dx. If we think of L as the generator of a translation-invariant stochastic process w(t), then we can also write $\int_{\mathbb{R}} f(x+y)p(dy) = \langle f(w(1))\rangle_x$. Our point of view is to consider the measure p(dx) as a measure on the Heisenberg (or Weyl) algebra H generated by D and x. We will see how this leads to a canonical transformation or representation of the usual Bernoulli measures on $\mathbb{R}$. Presently, we continue the discussion for the standard representation D, x and for general L.

Let L generate a "flow" on H via $x(t) = e^{tL}xe^{-tL}$, $D(t) = e^{tL}De^{-tL} = D$. Then the <u>moment polynomials</u> are given by $x(t)^n 1$, $n \geq 0$. Setting $C = x(t)$, $h_n(x,t) = C^n 1$ we have:

(1) $Ch_n = h_{n+1}$ and $Dh_n = nh_{n-1}$

(2) $CDh_n = nh_n$

(3) $\dfrac{\partial h_n}{\partial t} = L(D)h_n$

Using the fundamental relation $[D,x] = 1$ we can compute $C = x(t) = x + tL'(D)$, $L' = \partial L/\partial D$. This shows that for functions f analytic around the origin, we can define $\langle f(w(t))\rangle_x$ by $f(x + tL'(D))1$, i.e., the t is linearized. It is not, then, required that the measures or distributions p_t satisfying $e^{tL(z)} = \int_{\mathbb{R}} e^{zx} p_t(dx)$, i.e., the convolution family containing $p = p_1$, actually be probability measures. That is, L need not be the generator of an infinitely divisible process in order for us to define a continuously parametrized family of "expectations." However, the family of functions f is restricted. In any case, all that is required to determine a sequence $h_n(x,t)$ of "evolved powers" is a function L(z) analytic in a neighborhood of the origin with $L(0) = 0$; $L'(0) = \mu$, $L''(0) = \sigma^2$ are the "mean" and "variance" of the corresponding distribution (which need not be a positive measure).

Orthogonal Systems

For the notable case $L = z^2/2$ it is well-known that $h_n(x,-t)$ are the Hermite polynomials. This case is interesting since the Hermite polynomials, besides being "evolved powers," are the orthogonal polynomials for the corresponding Gaussian measure. It is natural to ask if there is a simple transformation that, in general, relates the moment sequence $h_n(x,t)$ and the orthogonal polynomials, denoted in general by $J_n(x,t)$.

This problem had been considered by Meixner [3] from the viewpoint of generating functions and is discussed thoroughly from the Heisenberg algebra viewpoint in [2]. Briefly, the result is that a "simple" transformation, specifically, a smooth change of variable $z \to V(z)$, yields the appropriate orthogonal family only for Bernoulli generators or limiting cases: gamma, Poisson, or Gaussian distributions. The "moment systems" approach allows these to be defined in general by complex parameters so that only for certain real values of the parameters do the generators L actually correspond to families of probability measures.

The Bernoulli Case

Consider the standard random walk where a particle moves on the line with position after t steps $S_t = \sum_{j=1}^{t} X_j$,

$$X_j = \begin{cases} +1 \text{ with probability } p, \\ -1 \text{ with probability } \bar{p}. \end{cases}$$

The characteristic function, for integer $t > 0$, $\langle e^{zS_t}\rangle = (pe^z + \bar{p}e^{-z})^t$ yields the generator $L(z) = \log(pe^z + \bar{p}e^{-z})$, analytic in a neighborhood of 0. The canonical representation is given by a pair V, ξ satisfying $[V,\xi] = 1$ so that for the orthogonal polynomials $J_n(x,t)$:

(1) $J_{n+1}(x,t) = \xi(-t)J_n(x,t)$ where $\xi(-t) = e^{-tL}\xi e^{tL}$,
$J_0(x,t) = 1.$

(2) $V(D)J_n(x,t) = nJ_{n-1}(x,t)$

(3) $\xi(-t)V(D)J_n = nJ_n$

(4) $\dfrac{\partial J_n}{\partial t} + L(D)J_n = 0$

In fact $V(z) = L'(z) - L'(0)$ and $\xi = \xi(0) = x(V')^{-1}$ give the desired transformation $z \to V$, $x \to \xi$.

Let us consider the case $p = \bar{p} = 1/2$, the symmetric random walk, explicitly. In this case $L(z) = \log\cosh z$. The canonical representation is determined by the transformation $z \to V(z) = \tanh z$, $x \to \xi = x\cosh^2 z$. In terms of V, the canonical generator is $M(v) = L(V^{-1}(v)) = -\frac{1}{2}\log(1 - v^2)$. This is the generator of the canonical symmetric Bernoulli process on the Heisenberg algebra generated by V and ξ.

So we think of a random (operator) variable $\xi = x\cosh^2 z$. The generator $M(v)$ yields the moments $e^{tM(v)} = (1 - v^2)^{-t/2} = \sum_{k=0}^{\infty} v^{2k}\mu_{2k}(t)$, where $\mu_{2k}(t) = (2k)!(t)(t+2)\ldots(t+2(k-1))/2^k k!$. Taking the Fourier transform we find that in this representation the random walk measures correspond to the functions $p_t(\xi) = \int_{\mathbb{R}} e^{-i\xi v}(1+v^2)^{-t/2}dv/2\pi$ $= \pi^{-1/2}\Gamma(t/2)^{-1}(\xi/2)^{\nu}K_\nu(\xi)$, where $\nu = (t-1)/2$ and K_ν is MacDonald's (modified Bessel) function (note that p_t is actually a function of $|\xi|$, i.e., $p_t(\xi) = p_t(-\xi)$). For t an even integer, it is well-known that these functions can be expressed in elementary terms. Applying $p_t(\xi)$ to 1 yields an ordinary function. We substitute $v = \tan\theta$ and use the relation $\exp(a\xi)1 = \exp(xV^{-1}(a))$ to get $p_t(\xi)1 =$ $\int_{-\pi/2}^{\pi/2} e^{ix\theta}\cos^{t-2}\theta\, d\theta/2\pi$.

In the V,ξ representation the "moment polynomials" $x(t)^n 1$ become $\xi(t)^n 1$ and $\xi(-t)^n 1$ are the orthogonal polynomials for the original Bernoulli measures (for integer $t > 0$) -- namely they are the Krawtchouk polynomials. In terms of z and x, $\xi(-t) = x\cosh^2 z - t\sinh z\cosh z$. $\xi(-t)V(z)$ is a symmetric operator with respect to the (generalized) Bernoulli measures and can be written simply as $x\partial_{-2} + (x-t)\partial_{+2}\partial_{-2}$ where the difference operator $\partial_a = (e^{aD}-1)/a$. This is a discrete version of the confluent hypergeometric operator.

We have the theorem:

Representation Theorem for the Symmetric Bernoulli Process:

Corresponding to the generator $L(z) = \log\cosh z$,

1) The canonical variables are $V(z) = \tanh z$, $\xi = x\cosh^2 z$, and $\xi(t) = x\cosh^2 z + t\sinh z\cosh z$.
2) The canonical generator $M(v) = -\frac{1}{2}\log(1-v^2)$.
3) The number operator $\xi(-t)V(z) = x\partial_{-2} + (x-t)\partial_2\partial_{-2}$, $\partial_a = (e^{az}-1)/a$.
4) In the canonical variables the corresponding moment polynomials are the orthogonal family, the Krawtchouk polynomials.

Proof:

From Theorem 2, p. 23 of [2], $V(z) = L'(z) = \tanh z$. Then $W = 1/V' = \cosh^2 z$. Thus $\xi = xW = x \cosh^2 z$. $\xi(t) = e^{tL} x e^{-tL} \cosh^2 z = (x + tV)\cosh^2 z = x \cosh^2 z + t \sinh z \cosh z$.

The relation $V(z) = \tanh z = v$ may be inverted to yield z as a function of v. Substitution into L(z) gives M(v) as above (no. 2).

For 3), calculate $\xi(-t)V(z) = x \cosh z \sinh z - t \sinh^2 z$ using 1).

From $\partial_2 = (e^{2z} - 1)/2$, $\partial_{-2} = (1 - e^{-2z})/2$ check that $\partial_2 \partial_{-2} = \sinh^2 z$, $(\partial_2 + \partial_{-2})/2 = \sinh z \cosh z$ and that $(\partial_2 - \partial_{-2})/2 = \partial_2 \partial_{-2}$. Then

$$\begin{aligned}\xi(-t)V(z) &= x(\partial_2 + \partial_{-2})/2 - t\partial_2\partial_{-2} \\ &= x\partial_{-2} + x(\partial_2 - \partial_{-2})/2 - t\partial_2\partial_{-2} \\ &= x\partial_{-2} + (x - t)\partial_2\partial_{-2}.\end{aligned}$$

Concluding Remarks

It is an interesting question whether the approach illustrated above extends to other groups. Essentially, we are looking for canonical representations for (certain) processes invariant under a given group. The appropriate extension for a Lie group would determine the general form of the "Bernoulli" generators explicitly in terms of the Lie algebra. Finally, such an extension would yield a "probabilistic" approach to the special function theory related to the invariance group. A discussion of the theory for $\mathbb{R}^N$ may be found in [1].

References

1. Feinsilver, P. "Moment Systems and Orthogonal Polynomials in Several Variables," to appear in J. Math. Analysis and Appl.

2. Feinsilver, P. "Special Functions, Probability Semigroups, and Hamiltonian Flows," Springer-Verlag, Lecture Notes in Math 696, 1978.

3. Meixner, J. "Orthogonale Polynomsysteme mit einer besonderen Gestalt der erzeugenden Funktion," J. London Math. Soc. 9, 1934, p. 6-13.

CAPACITES, MOUVEMENT BROWNIEN ET PROBLEME DE L'EPINE DE LEBESGUE SUR LES GROUPES DE LIE NILPOTENTS

Léonard Gallardo
U. E. R. Sciences Mathématiques
Université de Nancy I (*)
C. O. 140, 54037 Nancy Cedex, France

In this paper, we show that one can develop a potential theory for brownian motion on every graded nilpotent simply connected Lie group and most of the classical results can be generalized in a natural way. In particular, we study the problem of Lebesgue thorn and the problem of recurrent and transient sets. We deduce also some asymptotic results concerning the area sweeped out by 2-dimensional brownian motion from the previous facts by considering the Heisenberg group.

Introduction

Le but de cet article est de montrer que la théorie classique du potentiel a un prolongement naturel à une large classe de groupes non abéliens : les groupes de Lie nilpotents simplement connexes gradués. Sur ces groupes, la diffusion dont le générateur infinitésimal est le laplacien de Kohn, a des propriétés remarquables, tout à fait analogues à celles du mouvement brownien de l'espace euclidien. Ceci est dû essentiellement à l'existence d'automorphismes qui jouent le rôle des dilatations de $\mathbb{R}^k$, par rapport auxquels le laplacien de Kohn est homogène et de degré 2 .

Dans les paragraphes 1 et 2, nous rappelons un certain nombre de résultats connus sur les groupes de Lie nilpotents gradués et sur le mouvement brownien $X(t)$ associé au laplacien de Kohn. Nous établissons aussi une loi du zéro-un (proposition (2.6)) qui se révélera utile dans la suite. Le paragraphe 3 est crucial ; dans le cas où le groupe a une dimension homogène $k \geqslant 3$ ($X(t)$ est alors transient), nous montrons (théorème (3.1)) que le noyau potentiel de $X(t)$ a une expression analogue au noyau newtonien classique (la norme euclidienne étant remplacée par une norme homogène) et il transforme les fonctions mesurables bornées à support compact en fonctions continues tendant

(*) et E.R.A. n° 839 du C.N.R.S.

vers zéro à l'infini. On est alors dans le cadre de la théorie générale de Hunt. Ceci nous permet de définir une notion de capacité dont nous étudions les propriétés au paragraphe 4. Nous donnons également une caractérisation géométrique de la capacité (théorème (4.9)) généralisant celle de Spitzer (cf. [24]). Dans le paragraphe 5, nous établissons un critère de régularité (test de Wiener, théorème (5.1)), d'où l'on peut déduire un critère du cône (critère de Poincaré). Enfin, dans le paragraphe 6, nous appliquons les résultats précédents au problème de l'épine de Lebesgue (théorème (6.2)) et dans le paragraphe 7, nous étudions de manière analogue le problème des ensembles récurrents. On obtient comme application un résultat sur le comportement quand $t \downarrow 0$ et quand $t \uparrow +\infty$ de l'intégrale stochastique $\int_0^t \beta_1(s)\, d\beta_2(s) - \beta_2(s)\, d\beta_1(s)$ (voir (6.8) et (7.5)).

1. Préliminaires

(1.1) Soit $(N,[.,.])$ une algèbre de Lie nilpotente réelle. Si on définit le produit $x.y$ de deux éléments de N par la formule de Campbell-Hausdorff :

$$x.y = x + y + \frac{1}{2}[x,y] + \dots ,$$

$(N,.)$ est alors un groupe de Lie nilpotent s. c. (s. c. : simplement connexe) et, à isomorphisme près, c'est le seul groupe nilpotent s. c. d'algèbre de Lie $(N,[.,.])$ (cf. [15]). Ainsi, à côté de la structure de groupe $(N,.)$, nous utiliserons constamment sa structure d'algèbre de Lie (et donc d'espace vectoriel) sous-jacente. On voit alors, par exemple, que $x^{-1} = -x$ $(x \in N)$, que le vecteur 0 est l'élément neutre de $(N,.)$, que la mesure de Haar de $(N,.)$ n'est autre que la mesure de Lebesgue de l'espace sous-jacent et que les automorphismes de $(N,.)$ coïncident avec les automorphismes de l'algèbre de Lie $(N,[.,.])$.

(1.2) Soit $N = N^1 \supset N^2 = [N,N] \supset \dots \supset N^\ell = [N,N^{\ell-1}] \supset N^{\ell+1} = \{0\}$ la série centrale descendante de N (l'entier ℓ est la classe de nilpotence de N). On dit que le groupe $(N,.)$ (resp. l'algèbre $(N,[.,.])$) est gradué (resp. graduée) s'il existe une décomposition $N = m_1 \oplus \dots \oplus m_\ell$, où m_i est un sous-espace vectoriel supplémentaire de N^{i+1} dans N^i $(i=1,\dots,\ell)$ et vérifiant les conditions :

$$[m_i,m_j] \subset m_{i+j} \text{ (resp. } = \{0\}) \text{ si } i + j \leq \ell \text{ (resp. si } i + j > \ell) .$$

Cette définition est celle de Goodman (cf. [10], p. 13). Dans [5] ou dans [22], les auteurs utilisent la dénomination d'algèbre de Lie stratifiée. Par exemple, les groupes abéliens $(\mathbb{R}^k,+)$ sont trivialement gradués.

(1.3) <u>Exemples de groupes nilpotents s. c. gradués</u> : a) Le groupe d'Heisenberg $\mathbb{H}_d$ est un groupe de Lie nilpotent s. c. de classe $\ell = 2$ avec $m_1 = \mathbb{R}^{2d}$, $m_2 = \mathbb{R}$ qu'on représente traditionnellement par $\mathbb{C}^d \oplus \mathbb{R}$ avec les coordonnées $(z,u) = (z_1,\ldots,z_d,u)$ où $z_j = x_j + iy_j$ et avec le produit :

$$(z,u)\cdot(z',u') = (z+z',u+u'+2\ \mathrm{Im} \sum_{j=1}^{d} z_j\, \bar{z}'_j)\ .$$

b) Le groupe nilpotent libre à n générateurs et de classe ℓ (cf. [22], p. 256).

c) N = la partie nilpotente dans la décomposition d'Iwasawa d'une algèbre de Lie semi-simple.

d) N = l'algèbre de Lie des matrices réelles $m \times m$ (a_{ij}) telles que $a_{ij} = 0$ si $i \geqslant j$ (avec le crochet $[M,N] = MN - NM$) .

(1.4) Supposons N gradué. Alors la dilatation δ_r de rapport $r > 0$, définie par :

$$\delta_r(x) = \sum_{i=1}^{\ell} r^i\, x^{(i)} \quad (x^{(i)} = \text{composante de } x \in N \text{ sur } m_i)\ ,$$

est un automorphisme de $(N,.)$ et l'on a : $\delta_r \circ \delta_{r'} = \delta_{rr'}$ $(r,r'>0)$, $\delta_{1/r} = \delta_r^{-1}$ et $\delta_1 = \mathrm{Id}_N$. De plus, le jacobien du changement de variables $y = \delta_r(x)$ vaut $\frac{dy}{dx} = r^k$ où l'entier $k = \sum_{i=1}^{\ell} i \dim(m_i)$ est <u>la dimension homogène de N</u> . Ce nombre k est aussi le degré de croissance polynomiale de N (cf. [11], p. 342).

(1.5) On appelle <u>norme homogène</u> sur le groupe gradué $(N,.)$, toute application continue $x \longmapsto |x|$ de N dans $\mathbb{R}_+$ vérifiant les conditions :

i) $|x| = 0$ si et seulement si $x = 0$

ii) $|x| = |-x|$

iii) $|\delta_r(x)| = r\,|x|$ pour tout $r > 0$.

Les normes homogènes vérifient une inégalité triangulaire :

$$|xy| \leqslant C_0(|x|+|y|) \quad (x,y \in N)$$

où C_0 est une constante $\geqslant 1$. Un exemple important de norme homogène est donné par :

$$|x|_\infty = \sup_{1 \leq i \leq \ell} \|x^{(i)}\|^{1/i}$$

où $\|x^{(i)}\|$ est une norme (quelconque) d'espace vectoriel sur m_i.

Les normes homogènes sont toutes équivalentes : étant données les normes $|.|_1$ et $|.|_2$, il existe une constante $C \geq 1$ telle que $C^{-1}|x|_1 \leq |x|_2 \leq C|x|_1$ pour tout $x \in N$.

Enfin, étant donnée une norme homogène $|.|$, <u>les boules de Koranyi</u> $B(0,r) = \{x \in N ; |x| < r\}$ $(r \in \mathbb{R}_+)$ forment un système fondamental de voisinages de 0 pour la topologie de $(N,.)$ qui n'est autre que la topologie naturelle d'espace vectoriel sous-jacent.

2. Mouvement brownien sur N

Dans toute la suite de l'article, $(N,.)$ est un groupe de Lie nilpotent s. c. <u>gradué</u> et nous utiliserons les notations du paragraphe 1 sans les rappeler.

(2.1) Considérons les champs de vecteurs invariants à gauche $X_1, \ldots, X_{p_1}$ sur $(N,.)$ correspondant à une base $e_1, \ldots, e_{p_1}$ de m_1 (i. e. : $X_i(f)(x) = \lim_{t \to 0} \frac{1}{t}(f(x.te_i) - f(x))$, $x \in N$ et f = fonction dérivable définie sur N). Alors l'opérateur du second ordre :

$$L = \frac{1}{2}(X_1^2 + \ldots + X_{p_1}^2)$$

est appelé <u>laplacien de Kohn</u> (ou sous-laplacien).

(2.2) D'après un théorème de Hunt (cf. [16]), il existe un semi-groupe fortement continu $(P^t)_{t>0}$ de générateur infinitésimal L et pour tout $t > 0$, une probabilité μ_t sur N telle que :

$$P^t f(x) = \int_N f(xy)\, \mu_t(dy) \quad (f \in C_K^\infty(N)) .$$

Si maintenant on définit une distribution p sur $N \times (0,+\infty)$ par (cf. [5], p. 178 ou [14], p. 454) :

$$\langle p, u \otimes v\rangle = \int_0^\infty \int_N u(x)\, v(t)\, \mu_t(dx)\, dt \quad (u \in C_K^\infty(N),\ v \in C_K^\infty(0,\infty)) ,$$

on voit facilement que p est solution de l'équation

$$\left(\frac{\partial}{\partial t} - L\right) p = 0 .$$

Mais l'opérateur $\frac{\partial}{\partial t} - L$ est hypoelliptique, car m_1 engendre l'algèbre de Lie N . Donc $p \in C^{\infty}(N \times (0,+\infty))$ et on a ainsi $\mu_t(dx) = p(t,x)\, dx = p_t(x)\, dx$; d'où $p_t(x) \geq 0$ $(\forall x \in N \text{ , } \forall t > 0)$ et de plus $p_t(x^{-1}) = p_t(x)$ puisque L est symétrique.

Si $N = m_1 = \mathbb{R}^k$, $(P^t)_{t>0}$ est évidemment le semi-groupe du mouvement brownien et $p_t(x) = (2\pi t)^{-k/2} \exp(-\|x\|^2/2t)$, mais en général l'expression explicite de $p_t(x)$ n'est pas connue (cf. [8] pour le cas du groupe d'Heisenberg et le cas du groupe libre de classe 2).

(2.3) L'opérateur L est homogène et de degré 2 par rapport aux dilatations (i. e. : $L(f \circ \delta_r) \circ \delta_{1/r} = r^2 Lf)$, donc le semi-groupe $(P_{r^2t})_{t>0}$ engendré par r^2L vérifie la relation $P_{r^2t}(f) = P_t(f \circ \delta_r) \circ \delta_{1/r}$. On en déduit que les densités $p_t(x)$ vérifient la relation d'homogénéité :

$$p_{r^2t}(\delta_r(x)) = r^{-k}\, p_t(x) \quad (\forall x \in N \text{ et } \forall t \text{ et } r > 0) ,$$

(où k est la dimension homogène de N) et la relation utile suivante :

$$(2.3.1) \qquad p_t(x) = t^{-k/2}\, p_1(\delta_{t^{-1/2}}(x)) .$$

On voit de plus qu'à x fixé, $x \neq 0$, on a $\lim_{t\to 0} p_t(x) = 0$ et qu'à $t > 0$ fixé, $x \longmapsto p_t(x)$ est à décroissance rapide quand $x \longmapsto \infty$.

(2.4) Des considérations précédentes, on voit que le semi-groupe $(P^t)_{t>0}$ définit sur N un processus de diffusion $X(t,\omega)$ (qu'on notera simplement $X(t)$ ou X_t) à accroissements indépendants pour la loi de groupe, si on définit la probabilité partant de x par :

$$P_x(X(t) \in dy) = p_t(x^{-1}y)\, dy .$$

Par analogie avec le cas abélien, la diffusion $X(t)$ est appelée <u>mouvement brownien (gauche) sur N</u> .

Dans [23], B. Roynette a donné une construction explicite de $X(t)$ à l'aide de p_1 mouvements browniens indépendants sur $\mathbb{R}$: $\beta_1(t)$, ... , $\beta_{p_1}(t)$. Les composantes de $X(t)$ sur une base (e_i) sont alors des sommes d'intégrales stochastiques relatives aux $\beta_i(t)$. On n'aura pas besoin ici de cette représentation, mais on notera cependant que la projection de $X(t)$ sur m_1 est un mouvement brownien usuel.

(2.5) La translation (shift) sur l'espace Ω des trajectoires de la diffusion est l'application $\theta_s : \Omega \longrightarrow \Omega$ telle que

$$X(t,\omega) = X(t-s,\theta_s(\omega)) \quad (s \in \mathbb{R}_+) .$$

Soit $\mathcal{J} = \{A \; ; \; \theta_s(A) = A \; , \; \forall s \in \mathbb{R}_+\}$ la tribu des événements invariants du mouvement brownien $X(t)$. Rappelons qu'on dit qu'une tribu est triviale relativement à une mesure μ si elle ne contient que deux éléments (i. e. : Ω et $\emptyset$) μ presque sûrement. On a alors la loi du zéro-un suivante :

(2.6) PROPOSITION. <u>La tribu</u> $\mathcal{J}$ <u>est triviale relativement à</u> P_x <u>pour tout</u> $x \in N$.

DEMONSTRATION. Soit $\Gamma \in \mathcal{J}$ et posons $Y = 1_\Gamma$ et $h(x) = E_x(Y)$. On a :

$$h(X_t) = E_{X_t}(Y) = E_x(Y \circ \theta_t / \mathcal{F}_t) \; P_x \; p.s \quad (\forall x \in N) ,$$

d'après la propriété de Markov ($\mathcal{F}_t$ est la filtration de $X(t)$) . On a donc

$$h(X_t) = E_x(Y/\mathcal{F}_t) \; P_x \; p.s \quad (\text{pour tout } x \in N)$$

puisque Y est invariante. On en déduit d'une part que $E_x(h(X_t)) = E_x(Y) = h(x)$ i. e. : $P^t h(x) = h(x)$. Donc h est harmonique pour la marche aléatoire de loi $\mu_t(dx) = p_t(x)\,dx$; il s'ensuit par un résultat de R. Azencott (cf. [1]) que $h \equiv c^{te}$. D'autre part, $h(X_t)$ est une martingale bornée par rapport à la filtration croissante de tribus $\mathcal{F}_t$. D'où $\lim\limits_{t\to+\infty} h(X_t) = Y$ P_x p.s pour tout $x \in N$. On a donc

$$Y = 1_\Gamma = c^{te} \quad P_x \; p.s \quad (\forall x \in N)$$

i. e. : $P_x(\Gamma) \equiv 0$ ou 1 ∎

3. Etude du noyau potentiel de X(t)

<u>Nous supposerons dans toute la suite de cet article que la dimension homogène</u> k <u>du groupe gradué</u> $(N,.)$ <u>est supérieure ou égale à</u> 3 (on voit de suite que $N = \mathbb{R}$ ou $\mathbb{R}^2$ si $k \leq 2$; ceci correspond au cas où le mouvement brownien est récurrent).

Soit $Vf(x) = \int_0^{+\infty} P^t f(x)\,dt$ le noyau potentiel de $X(t)$ et soit $\mathcal{B}_K(N)$ l'ensemble des fonctions mesurables bornées à support compact dans N . On a alors le théorème :

(3.1) THEOREME. 1) <u>La fonction</u> $x \longmapsto Vf(x)$ <u>est continue, bornée sur</u> N <u>et tend vers zéro quand</u> $x \longrightarrow \infty$ <u>pour toute</u> $f \in \mathcal{B}_K(N)$.
2) <u>Il existe une norme homogène</u> $x \longmapsto |x|$ <u>de classe</u> C^∞ <u>en dehors de zéro, telle que pour tout</u> $f \in \mathcal{B}_K(N)$, <u>on ait</u> :

$$Vf(x) = \int_N \frac{1}{|x^{-1}.y|^{k-2}} f(y)\, dy .$$

DEMONSTRATION.

1) $Vf(x) = \int_0^1 \int_N p_t(x^{-1}.y)\, f(y)\, dy\, dt + \int_1^{+\infty} \int_N p_t(x^{-1}.y)\, f(y)\, dy\, dt$.

La première intégrale est clairement continue et bornée ; la deuxième l'est aussi, car $p_t(x) \leqslant c^{te}\, t^{-k/2}$ $(x \in N)$ d'après (2.3.1). Le fait que $\lim_{x\to\infty} Vf(x) = 0$ résulte alors du théorème de convergence dominée de Lebesgue puisque $\lim_{x\to\infty} p_t(x) = 0$.

2) $Vf(x) = \int_N g(x^{-1}.y)\, f(y)\, dy$ où $x \longmapsto g(x) = \int_0^\infty p_t(x)\, dt$ est une fonction de classe C^∞ sur $N - \{0\}$. En effet, d'après (2.3) on a $p_t(x) = O(t^{-k/2})$ si $t \longrightarrow +\infty$ pour tout x fixé et $p_t(x) = O(t^N)$ pour tout entier N quand $t \longrightarrow 0$ (à $x \neq 0$ fixé) . De même si D est un opérateur différentiel homogène et de degré d sur N , on a : $Dp_t(x) = O(t^{-(k+d)/2})$ quand $t \longrightarrow +\infty$ (x fixé) et $Dp_t(x) = O(t^N)$ si $t \longrightarrow 0$. Ainsi pour $x \neq 0$, l'intégrale $\int_0^{+\infty} Dp_t(x)\, dt$ converge localement uniformément en x et on peut donc différentier g sous le signe intégrale.

De plus, on a $g(x) > 0$ $(\forall x \in N - \{0\})$ puisque pour tout $t > 0$ on a :

$$p_t(0) = \int_N p_{t/2}(x)\, p_{t/2}(x^{-1})\, dx = \int_N [p_{t/2}(x)]^2\, dx > 0 ,$$

donc pour x fixé, grâce à (2.3.1) et à la continuité de p_1 , on voit que $p_t(x) > 0$ pour t assez grand (on peut aussi remarquer que $Lg(x) = 0$ $(x \neq 0)$ donc, d'après un principe du maximum dû à Bony (cf. [3], p. 286), g ne peut s'annuler en un point sans être nul partout).

Enfin, on notera que $g(\delta_r(x)) = r^{2-k} g(x)$ $(\forall r > 0)$ grâce à (2.3), donc $g(x) \longrightarrow +\infty$ quand $x \longrightarrow 0$ (puisque $2 - k < 0$) . Posons alors :

$$|x| = [g(x)]^{-\frac{1}{k-2}} \quad \text{si } x \neq 0 \text{ et } |x| = 0 \text{ si } x = 0 ,$$

$x \longmapsto |x|$ possède alors toutes les propriétés voulues. ∎

(3.2) REMARQUES ET EXEMPLES. a) Dans le cas où $N = \mathbb{R}^k$ $(k \geqslant 3)$, on a $g(x) = |x|^{2-k}$ avec $|x| = C_k^{1/2-k} \|x\|$ où $\|x\|$ désigne la norme euclidienne et $C_k = 2\pi^{-k/2}\,\Gamma(k/_2 - 1)$.

b) Si $N = \mathbb{H}_d$ (cf. (1.4) a)), on a $g(z,u) = |(z,u)|^{-2d}$ (ici $k = 2d+2$) avec $|(z,u)| = C_d(|z|^4+u^2)^{1/4}$ ($C_d = C^{te} > 0$) et $|z| = \left(\sum_{j=1}^{d} z_j\,\bar{z}_j\right)^{1/2}$ (cf. [6]).

c) Dans le cas $N = \mathbb{R}^k$ ou $\mathbb{H}_d$, la norme homogène $|.|$ du théorème vérifie une vraie inégalité triangulaire : $|xy| \leqslant |x| + |y|$ (on le vérifie très aisément par un calcul élémentaire pour $\mathbb{H}_d$ (cf. [4], p. 371). Il serait intéressant de savoir si c'est toujours le cas.

Du théorème (3.1), on peut déjà déduire la transience de $X(t)$:

(3.3) PROPOSITION. <u>Soit</u> A <u>un ensemble borné</u> (i. e. : <u>relativement compact) dans</u> N. <u>Alors, quel que soit</u> $x \in N$, <u>on a</u> :

$$P_x(X(t) \in A \text{ infiniment souvent quand } t \uparrow +\infty) = 0 .$$

Pour la démonstration de (3.3), on aura besoin d'un lemme :

(3.3.1) LEMME. <u>Soient</u> K <u>un compact de</u> N <u>et un nombre</u> $a > 0$. <u>Il existe alors un compact</u> $\tilde{K}$ <u>contenant</u> K <u>tel que</u> :

$$\inf_{0 \leqslant s \leqslant a}\ \inf_{z \in K} P_z(X(s) \in \tilde{K}) = \alpha > 0 .$$

DEMONSTRATION. Soit $B = B(0,r)$ une boule de Koranyi, pour une norme homogène quelconque sur N, telle que :

$$\int_B p_1(y)\,dy = \alpha > 0$$

et soit $\tilde{K} = K \,.\, \delta_{a^{1/2}}(B)$. On a alors

$$P_z(X(s) \in \tilde{K}) = \int_{\tilde{K}} p_s(z^{-1}.x)\,dx =$$

$$= \int_{\tilde{K}} s^{-k/2}\, p_1(\delta_{s^{-1/2}}(z^{-1}.x))\,dx = \int_{\delta_{s^{-1/2}}(z^{-1}.\tilde{K})} p_1(y)\,dy \geqslant \alpha$$

car $z^{-1}.\tilde{K} \supset \delta_{a^{1/2}}(B)$, donc $\delta_{s^{-1/2}}(z^{-1}.\tilde{K}) \supset \delta_{(\frac{a}{s})^{1/2}}(B) \supset B$, puisque $\frac{a}{s} \geqslant 1$. D'où le lemme. ∎

DEMONSTRATION DE LA PROPOSITION (3.3). On va suivre, en l'adaptant, une idée de Port et Stone (cf. [20], p. 162).

Soit $T_{\bar{A}} = \inf\{t > 0 \; ; \; X(t) \in \bar{A}\}$ le temps d'entrée dans l'adhérence de A. On va appliquer le lemme avec $\bar{A} = K$ et $a = 1$:

$$\int_0^{t+1} P_x(X(s) \in \widetilde{K}) \, ds \geqslant \int_0^{t+1} P_x(X(s) \in \widetilde{K} \, , \, T_{\bar{A}} \leqslant s) \, ds =$$

$$= \int_0^{t+1} \int_0^{s} \int_{\bar{A}} P_x(T_{\bar{A}} \in dr \, , \, X(T_{\bar{A}}) \in dz) \, P_z(X(s-r) \in \widetilde{K}) \, ds$$

$$= \int_0^{t+1} \left(\int_r^{t+1} P_z(X(s-r) \in \widetilde{K}) \, ds \int_{\bar{A}} P_x(T_{\bar{A}} \in dr \, , \, X(T_{\bar{A}}) \in dz) \right)$$

$$\geqslant \int_0^{t} \left(\int_0^{1} P_z(X(u) \in \widetilde{K}) \, du \int_{\bar{A}} P_x(T_{\bar{A}} \in dr \, , \, X(T_{\bar{A}}) \in dz) \right)$$

$$\geqslant \alpha \, P_x(T_{\bar{A}} \leqslant t) \; .$$

En faisant $t \longrightarrow +\infty$, on en déduit :

$$\int_{\widetilde{K}} g(x^{-1}.y) \, dy \geqslant \alpha \, P_x(T_{\bar{A}} < +\infty) \; .$$

On a alors :

$$P_x(X(s) \in \bar{A} \text{ pour un } s > t) = \int_N p_t(x^{-1}.z) \, P_z(T_{\bar{A}} < \infty) \, dz$$

$$\leqslant \frac{1}{\alpha} \int_N p_t(x^{-1}.z) \int_{\widetilde{K}} g(z^{-1}.y) \, dy \, dz = \frac{1}{\alpha} \int_t^{+\infty} P^s \, 1_{\widetilde{K}}(z) \, ds \downarrow 0 \text{ si } t \uparrow +\infty \; ■$$

(3.4) REMARQUES. a) On déduit immédiatement de la proposition que pour toute norme homogène $|.|$, on a :

$$P_x(\lim_{t \to +\infty} |X(t)| = +\infty) \equiv 1 \; .$$

b) Le résultat de la proposition (3.3) est bien connu dans le cas des marches aléatoires (cf. [12], p. 118). D'ailleurs le mouvement brownien regardé aux instants entiers fournit une marche aléatoire $X(n)$ de loi $\mu = p_1(x)\,dx$; on a donc des inclusions du type :

$$(X(n) \in A \text{ i. s. quand } n \longrightarrow \infty) \subset (X(t) \in A \text{ i. s. quand } t \uparrow +\infty) \; .$$

On peut donc parfois déduire des résultats sur $X(t)$ lorsqu'ils sont connus pour les marches aléatoires.

c) Soit B un borélien quelconque. D'après la loi du zéro-un (voir 2.6), on a :

$P_x(X(t) \in B$ infiniment souvent quand $t \uparrow +\infty) = 0$ ou 1.

Nous verrons au paragraphe 7 un critère permettant de savoir si la probabilité précédente est 0 ou 1 .

4. Capacités - Propriétés et Applications

(4.1) D'après l'assertion 1 du théorème (3.1) et comme $X(t)$ est un processus autodual (car $p_t(x) = p_t(x^{-1})$), on est dans le cadre de la théorie générale de Hunt (cf. [2], p. 284). En particulier, étant donné un borélien borné $A \subset N$, il existe une mesure π_A de masse finie, portée par $A \cup A^r$ (où $A^r = \{x \in \partial A \,;\, P_x(T_A = 0) = 1\}$ est le bord régulier de A et $T_A = \inf\{t > 0 \,,\, X(t) \in A\}$) telle que :

$$P_x(T_A < +\infty) = V\pi_A(x) = \int_N \left(\frac{1}{|x^{-1}.y|}\right)^{k-2} d\pi_A(y)$$

et la masse $\pi_A(N) = C(A)$ est <u>la capacité (newtonienne) de A</u> qui vérifie les propriétés :

(4.2.) $\qquad C(A \cup B) + C(A \cap B) \leq C(A) + C(B)$,

(4.3) $\qquad C(A) \leq C(B)$ si $A \subset B$

et la propriété extrémale (cf. [2], p. 286) :

(4.4) $C(A) = \sup\{\mu(N) \,;\, \mu$ mesure finie à support dans A et $V\mu \leq 1\}$.

(4.5) On notera que si K est compact, on a $P_x(T_K < +\infty) = P_x(T_{\partial K} < +\infty)$, car les trajectoires de $X(t)$ sont continues, donc $V\pi_K(x) = V\pi_{\partial K}(x)$, d'où $\pi_K = \pi_{\partial K}$ et en particulier, $C(K) = C(\partial K)$.

(4.6) REMARQUE. On peut évidemment aussi définir des λ-capacités ($\lambda > 0$) . Soient $g^\lambda(x) = \int_0^{+\infty} e^{-\lambda t} p_t(x)\, dt$ et $\phi^\lambda(x) = E_x(e^{-\lambda T_A})$. Si A est un borélien borné, il existe une mesure finie π_A^λ portée par $A \cup A^r$ telle que :

$$\phi^\lambda(x) = \int_N g^\lambda(x^{-1}.y)\, d\pi_A^\lambda(y)$$

et $\pi_A^\lambda(N) = C^\lambda(A)$ est la <u>λ-capacité de A</u> qui vérifie aussi les propriétés (4.2) et (4.3) et qui est telle que :

$$\lim_{\lambda \to 0} C^\lambda(A) = C(A) \qquad \text{(cf. [17], p. 191).}$$

Le résultat qui suit a un grand intérêt technique que nous verrons par la suite :

(4.7) PROPOSITION. Soit A un borélien borné de N. On a

1) $C(x.A) = C(A) \quad (\forall x \in N)$

2) $C(\delta_r(A)) = r^{k-2}\, C(A) \quad (\forall r > 0)$

3) $C(A) = \lim_{x\to\infty} |x|^{k-2}\, P_x(T_A < +\infty)$.

DEMONSTRATION. Il suffit de démontrer l'assertion 3, les autres s'en déduisent grâce à l'invariance par translation à gauche et l'homogénéité du processus $X(t)$ par rapport aux dilatations. De (4.1), on déduit aussitôt l'encadrement suivant :

$$(4.7.0) \quad \inf_{y\in A} |x^{-1}.y|^{2-k}\, C(A) \leqslant P_x(T_A \leqslant +\infty) \leqslant \sup_{y\in A} |x^{-1}.y|^{2-k}\, C(A).$$

Il suffit donc de voir que $\frac{|x^{-1}.y|}{|x|} \longrightarrow 1$ quand $|x| \longrightarrow +\infty$, uniformément pour $y \in A$. Or on a :

$$\left|\frac{|x^{-1}.y|}{|x|} - 1\right| = \left|\left|\delta_{\frac{1}{|x|}}(x^{-1})\ \delta_{\frac{1}{|x|}}(y)\right| - \left|\delta_{\frac{1}{|x|}}(x^{-1})\right|\right|,$$

le résultat découle alors de l'inégalité de la moyenne suivante :

(4.7.1) LEMME. Soient $x \longmapsto |x|$ une norme homogène de classe $\mathcal{C}^2$ en dehors de $x = 0$, $x \longmapsto |x|_\infty$ la norme homogène définie en (1.6) et C une constante telle que $|x|_\infty \leqslant C\,|x|$ pour tout $x \in N$. Alors il existe une constante $\alpha > 0$ telle que $\big||xy| - |x|\big| \leqslant \alpha\,|y|$ pour tout $|y| \leqslant \frac{1}{2C}|x|$.

DEMONSTRATION. Si on remplace x et y respectivement par $\delta_r(x)$ et $\delta_r(y)$, les deux inégalités sont multipliées par r. On peut donc supposer $|x| = 1$ et $|y| \leqslant \frac{1}{2C}$; la fonction $y \longmapsto |xy|$ est alors $\mathcal{C}^2$ et il existe une constante $C' > 0$, telle que :

$$\big||xy| - 1\big| \leqslant C'\, \|y\|_\infty ,$$

où (avec les notations de (1.6) $\|y\|_\infty = \sup_{1\leqslant i\leqslant \ell} \|y^{(i)}\|$. On a donc :

$$\big||xy| - 1\big| \leqslant C'\, |y|_\infty \leqslant C'C\, |y|$$

car $\|y\|_\infty \leqslant |y|_\infty \leqslant \frac{1}{2}$. D'où le résultat en posant $\alpha = C'C$ ∎

(4.8) REMARQUE. Comme on vient de le faire pour le noyau newtonien, on peut également définir sur N un noyau et une capacité de Riesz d'indice $\alpha \in]0,2[$. Les processus qu'on obtient ainsi sont une généralisation des processus symétriques stables de $\mathbb{R}^k$ (cf. [7]). Voyons maintenant une caractérisation géométrique de la capacité généralisant celle de Spitzer dans le cas du mouvement brownien classique (cf. [24]).

(4.9) THEOREME. Soient A un borélien borné de N , μ la mesure de Lebesgue sur N et

$$V_t = \mu\left(\bigcup_{o \leq s \leq t} A.X(s)^{-1}\right)$$

le volume balayé par A entre les instants 0 et t dans son déplacement attaché à la trajectoire brownienne. On a alors :

$$\lim_{t\to+\infty} \frac{V_t}{t} = C(A) \quad P_o \text{ presque sûrement.}$$

DEMONSTRATION. Posons $V_{r,t} = \mu\left(\bigcup_{r\leq s\leq t} A.X(s)^{-1}\right)$. Le processus $V_{r,t}$ est sous-additif ($V_{r,u} \leq V_{r,t} + V_{t,u}$ pour tous $r < t < u$) . Donc, à condition que $E(V_{o,t}) = E(V_t)$ existe, la limite $\lim_{t\to+\infty} \frac{V_t}{t}$ existe presque sûrement (cf. [24], p. 905) et c'est une constante d'après la loi du zéro-un (2.6).

Il restera à montrer que $\lim_{t\to+\infty} \frac{E(V_t)}{t} = C(A)$.

(4.9.1) LEMME. Pour tout $t > 0$, on a

$$E(V_t) = \int_N P_x(T_A \leq t)\, dx < +\infty.$$

DEMONSTRATION DU LEMME. Pour montrer la convergence de l'intégrale, choisissons un compact K contenant $\bar{A}$ tel que l'on ait (cf. lemme (3.3.1)) :

$$\inf_{s\in[o,t]} \inf_{z\in\bar{A}} P_z(X(s) \in K) = \alpha > 0 .$$

On a alors

$$P_x(X(t) \in K) \geq P_x(X(t) \in K\ ,\ T_{\bar{A}} \leq t) =$$

$$= \int_0^t \int_{\bar{A}} P_z(X(t-u) \in K)\ P_x(X(T_{\bar{A}}) \in dz\ ,\ T_{\bar{A}} \in du) \geq \alpha\ P_x(T_{\bar{A}} \leq t)\ .$$

Ainsi on a

$$\int_N P_x(T_{\bar{A}} \leq t)\ dx \leq \frac{1}{\alpha} \int_N \int_K p_t(x^{-1}.\ y)\ dy\ dx = \frac{1}{\alpha}\ \mu(K) < +\infty\ .$$

Enfin l'inégalité entre $E(V_t)$ et l'intégrale résultent du fait que pour une trajectoire ω partant de x, on a $T_A(\omega) \leq t$ si et seulement si $x \in \bigcup_{0\leq s\leq t} A.X_s(\omega)^{-1}$ ■

Pour terminer, on utilise une méthode due à Getoor (cf. [9]) :

Soient $e_A(t) = E(V_t)$ et $e_A^*(\lambda) = \int_0^{+\infty} e^{-\lambda t}\ de_A(t)$ $(\lambda > 0)$ sa transformée de Laplace. Par un calcul élémentaire, on voit que

$$\lambda\ e_A^*(\lambda) = \int_N \phi^\lambda(x)\ dx = C^\lambda(A) \quad (\text{voir } (4.6)),$$

on a donc $\lambda\ e_A^*(\lambda) \longrightarrow C(A)$ quand $\lambda \longrightarrow 0$, d'où $e_A(t)/_t \longrightarrow C(A)$ quand $t \longrightarrow +\infty$ d'après le théorème taubérien. ■

5. Un critère de régularité

On généralise dans ce paragraphe un critère de régularité connu sous le nom de test de Wiener (cf. [18], p. 255 ou [13], p. 220).

(5.1) THEOREME. Soient $|x|$ la norme homogène du théorème (3.1) et $C_o \geq 1$ la constante de l'inégalité triangulaire (1.6) pour $|x|$. Soient A un borélien de N, α une constante $> C_o$, x un point de N et $A_n = \{y \in A\ ;\ \alpha^{-n-1} < |x^{-1}.y| \leq \alpha^{-n}\}$. Alors :

$P_x(T_A = 0) = 0$ ou 1 suivant que la série $\sum_n \alpha^{n(k-2)}\ C(A_n)$ converge ou diverge.

DEMONSTRATION. Par invariance par translation, on peut supposer $x = 0$. Grâce à (4.7.0), on a :

$$\alpha^{n(k-2)}\ C(A_n) \leq P_o(T_{A_n} < +\infty) \leq \alpha^{(n+1)(k-2)}\ C(A_n)\ .$$

Ainsi la convergence de la série $\sum_n \alpha^{n(k-2)}\ C(A_n)$ implique immédiatement, d'après Borel-Cantelli, que presque sûrement au plus un nombre fini d'événements $[T_{A_n} < +\infty]$ sont réalisés, donc $P_o(T_A = 0) = 0$.

Supposons maintenant que la série converge ; alors il existe une suite d'entiers $n_i \uparrow +\infty$ tels que $n_{i+1} - n_i \geqslant 2$ et vérifiant $\sum_i \alpha^{n_i(k-2)} C(A_{n_i}) = +\infty$. Désignons alors par θ (comme en 2.5) le "shift" sur les trajectoires. On a :

$$P_o(T_{A_{n_i}} < +\infty , T_{A_{n_j}} < +\infty) \leqslant P_o(T_{A_{n_i}} < +\infty , T_{A_{n_j}} \circ \theta_{T_{A_{n_i}}} < +\infty) +$$

$$+ P_o(T_{A_{n_j}} < +\infty , T_{A_{n_i}} \circ \theta_{T_{A_{n_j}}} < +\infty)$$

($T_{A_{n_i}} \circ \theta_{T_{A_{n_j}}}$ désigne le premier temps d'entrée dans A_{n_i} après être entré dans A_{n_j}).

Supposons $j > i$ pour fixer les idées. Si $x \in \bar{A}_{n_i}$ et $y \in \bar{A}_{n_j}$ on a $|x^{-1}.y| = |y^{-1}.x| \geqslant \frac{|x|}{C_o} - |y| \geqslant \frac{\alpha^{-(n_i+1)}}{C_o} - \alpha^{-n_j}$

i. e. : $|x^{-1}.y| \geqslant \alpha^{-n_j}\left(\frac{\alpha}{C_o} - 1\right)$ car $n_j - (n_i + 1) \geqslant 1$. Ainsi

$$P_o(T_{A_{n_i}} < +\infty , T_{A_{n_j}} \circ \theta_{T_{A_{n_i}}} < +\infty) =$$

$$= \int_{A_{n_i}} P_x(T_{A_{n_j}} < +\infty).P_o(X(T_{A_{n_i}}) \in dx , T_{A_{n_i}} < +\infty)$$

$$\leqslant \sup_{x \in A_{n_i}} P_x(T_{A_{n_j}} < +\infty).P_o(T_{A_{n_i}} < +\infty)$$

$$\leqslant \sup_{x \in A_{n_i}} \int_{\bar{A}_{n_j}} \left(\frac{1}{|x^{-1}.y|}\right)^{k-2} \pi_{A_{n_j}}(dy).P_o(T_{A_{n_i}} < +\infty)$$

$$\leqslant \left(\frac{1}{\frac{\alpha}{C_o} - 1}\right)^{k-2} P_o(T_{A_{n_i}} < +\infty).\alpha^{n_j(k-2)} C(A_{n_j})$$

$$\leqslant C^{te} P_o(T_{A_{n_i}} < +\infty).P_o(T_{A_{n_j}} < +\infty) ,$$

où $C^{te} = \left(\frac{1}{\frac{\alpha}{C_o} - 1}\right)^{k-2} > 0$. On a une majoration du même type pour

$P_o(T_{A_{n_j}} < +\infty , T_{A_{n_i}} \circ \theta_{T_{A_{n_j}}} < +\infty)$. On a ainsi montré :

$$P_o(T_{A_{n_i}} < +\infty \, , \, T_{A_{n_j}} < +\infty) \leq C^{te} \, P_o(T_{A_{n_i}} < +\infty) \, P_o(T_{A_{n_j}} < +\infty) \, .$$

D'après une réciproque du lemme de Borel-Cantelli due à J. Lamperti (cf. [19], p. 59), une infinité d'événements $(T_{A_{n_i}} < +\infty)$ est réalisée avec une probabilité > 0, donc avec probabilité un d'après la loi du zéro-un de Blumenthal. Par la continuité des trajectoires et la transience de $X(t)$, on en déduit que $P_o(T_A = 0) = 1$, d'où le résultat. ■

On peut maintenant généraliser le critère du cône de Poincaré :

(5.2) DEFINITION. Un cône nilpotent $\mathfrak{C}$ de sommet 0 est un borélien de N d'intérieur non vide, stable par les dilatations et tel que $0 \in \partial \mathfrak{C}$. Plus généralement, un cône $\mathfrak{C}$ de sommet $a \in N$ est un ensemble tel que $a^{-1}.\mathfrak{C}$ soit un cône de sommet 0.

(5.3) EXEMPLES DE CÔNES. a) Soit B un borélien borné d'intérieur non vide et ne contenant pas 0, alors l'ensemble $\mathfrak{C} = \bigcup_{r \geq 0} \delta_r(B)$ est un cône de sommet 0 engendré par B. Réciproquement, il n'est pas difficile de voir que tout cône de sommet 0 est de ce type.

b) Soient $|.|$ une norme homogène, $\alpha > 0$ une constante et P et D deux sous-espaces vectoriels supplémentaires dans N, alors $\mathfrak{C} = \{x = z + v \, ; \, z \in P \, , \, v \in D \text{ et } |z| \leq \alpha \, |v|\}$ est un cône de sommet 0. Si D est une droite, nous dirons que $\mathfrak{C}$ est d'axe D. On voit alors que D est stable par dilatations si et seulement si elle est incluse dans un sous-espace m_i. Dans ce cas, posons $v = ue$ avec $e \in D$ et $u \in \mathbb{R}$. L'ensemble :

$$\mathfrak{C}' = \{x = z + ue \, ; \, u > 0 \text{ et } |z| \leq \alpha \, u^{1/i}\}$$

est alors un cône (à une nappe) d'axe D et de sommet 0.

(5.4) COROLLAIRE (Critère de Poincaré). Soit A un borélien de N et $a \in N$. S'il existe un cône $\mathfrak{C}$ de sommet a et un voisinage V de a tel que $V \cap \mathfrak{C} \subset A$, alors $a \in A^r$ (i. e. : $P_a(T_A = 0) = 1$).

DEMONSTRATION. On applique le test de Wiener à $\mathfrak{C}$ et on constate que $C(A_n)$ est proportionnel à $\alpha^{-n(k-2)}$ car tous les A_n se déduisent de l'un d'entre eux par les dilatations et on peut alors appliquer l'assertion 2) de la proposition (4.7). La série du test de Wiener est donc divergente, d'où la conclusion. ■

6. L'épine de Lebesgue nilpotente

Dans ce paragraphe, nous étudions dans quelle mesure le mouvement brownien peut entrer instantanément dans un ensemble plus effilé qu'un cône de sommet O .

(6.1) NOTATIONS. On rappelle que N est gradué, de dimension homogène $k \geq 3$ et de classe de nilpotence ℓ . Soient $|.|_1$ une norme homogène (quelconque), D une droite de N , P un supplémentaire de D dans N et h une fonction positive définie sur $\mathbb{R}_+$ avec $h(0) = 0$. L'ensemble

$$A_h = \{x = z + ue \ ; \ z \in P \ , \ e \in D \ , \ u > 0 \ \text{ et } \ |z|_1 \leq h(u)\}$$

est appelé <u>épine</u> d'axe D , de "rayon" h et de sommet O . On a alors le théorème suivant :

(6.2) THEOREME. <u>On suppose que</u> $D \subset m_\ell$ <u>et que</u> $\frac{h(u)}{u^{1/\ell}} \downarrow 0$ <u>quand</u> $u \downarrow 0$. <u>Alors</u> :

1) <u>si</u> $k - 2 - \ell > 0$, $P_0(T_{A_h} = 0) = 0$ <u>ou</u> 1 <u>suivant que l'intégrale</u> $\int_{0^+}\left(\frac{h(u)}{u^{1/\ell}}\right)^{k-2-\ell} \frac{du}{u}$ <u>converge ou diverge</u>.

2) <u>si</u> $k - 2 - \ell = 0$ (i. e. : $N = \mathbb{R}^3$ <u>si</u> $\ell = 1$ <u>et</u> $N = \mathbb{H}_1$ <u>si</u> $\ell = 2$) , $P_0(T_{A_h} = 0) = 0$ <u>ou</u> 1 <u>suivant que l'intégrale</u> $\int_{0^+}\left|\log \frac{h(u)}{u^{1/\ell}}\right|^{-1} \frac{du}{u}$ <u>converge ou diverge</u>.

(6.3) REMARQUES. a) Ce résultat contient le cas classique de $\mathbb{R}^k$ ($\ell = 1$) (cf. [18] ou [21], p. 68).

b) Deux groupes se singularisent par le fait que le critère intégral de régularité est différent du cas général ; ce sont $\mathbb{R}^3$ et le premier groupe d'Heisenberg $\mathbb{H}_1$ (voir lemme (6.4)).

c) Dans le cas où $h(u) = C^{te} u^{\frac{1}{\ell}}$, A_h est un cône (voir (5.3) b)). Si l'on veut un ensemble plus effilé qu'un cône, il est raisonnable de mettre l'hypothèse $\frac{h(u)}{u^{1/\ell}} \downarrow 0$ quand $u \downarrow 0$.

(6.4) LEMME. <u>Si</u> $k - 2 - \ell = 0$, <u>alors</u> $N = \mathbb{R}^3$ <u>ou</u> $\mathbb{H}_1$.

DEMONSTRATION. $k = \sum_{i=1}^{\ell} i \dim(m_i)$ et $k - 2 - \ell = 0$ implique immédiatement $k \leq 4$. Si $\ell = 1$, on a forcément $N = \mathbb{R}^3$. Si $\ell = 2$, on a $\dim(m_1) = 2$ et $\dim(m_2) = 1$. Soient alors x et $y \in N$ et $\bar{x}$ et $\bar{y}$ leur projection sur m_1 . La formule de Campbell-Hausdorff donne : $xy = x + y + \frac{1}{2}[\bar{x},\bar{y}]$ et $(\bar{x},\bar{y}) \longmapsto [\bar{x},\bar{y}]$ est une forme bilinéaire alternée sur $\mathbb{R}^2$, donc proportionnelle à la forme déterminant on a donc en coordonnées :

$$x.y = (x_1 + y_1 \ , \ x_2 + y_2 \ , \ x_3 + y_3 + \lambda(x_1 \, y_2 - y_1 \, x_2)) \ ,$$

le groupe obtenu est donc isomorphe à $\mathbb{H}_1$ ∎

Pour la démonstration du théorème, on a besoin de savoir évaluer la capacité d'un cylindre d'axe D .

(6.5) PROPOSITION. _Avec les notations du théorème, soit le cylindre d'axe_ D , _de hauteur_ L _et de rayon_ 1 :

$$S_L = \{x = z + ue \ ; \ |z|_1 \leqslant 1 \text{ et } 0 \leqslant u \leqslant L\} \ .$$

Soit $L_o > 0$, _alors il existe des constantes_ $M > 0$ _et_ $N > 0$ _telles que_ :

1) _si_ $k - 2 - \ell > 0$, _on ait_

$$ML \leqslant C(S_L) \leqslant NL \quad \underline{\text{si}} \quad L \geqslant L_o \ .$$

2) _si_ $k - 2 - \ell = 0$, _on ait_

$$M \frac{L}{\log L} \leqslant C(S_L) \leqslant N \frac{L}{\log L} \quad \underline{\text{si}} \quad L \geqslant L_o \ .$$

DEMONSTRATION. a) Montrons déjà les majorations. Notons qu'une translation à gauche par un élément de D est aussi une translation vectorielle puisque $D \subset m_\ell$, donc le cylindre $S_L^a = \{z + ue \ ; \ |z|_1 \leqslant 1$ et $a \leqslant u \leqslant a + L\}$ a même capacité que le cylindre S_L . La fonction $L \longmapsto C(S_L)$ est non décroissante et

$$C(S_{L_1+L_2}) = C(S_{L_1} \cup S_{L_2}^{L_1}) \leqslant C(S_{L_1}) + C(S_{L_2}) \ ,$$

d'après la propriété (4.2) et ce qui précède. On a donc dans tous les cas : $\lim\limits_{L\to+\infty} \frac{C(S_L)}{L}$ existe, donc en particulier la majoration dans 1). Pour la majoration dans 2), soit Π_{S_L} la mesure d'équilibre de S_L . Comme $V\Pi_{S_L} \leqslant 1$, on a :

$$L \geqslant \int_o^L V\Pi_{S_L}(ue) \, du = \int_o^L \left(\int_{S_L} g(x^{-1} + ue) \, \Pi_{S_L}(dx) \right) du$$

$$= \int_{S_L} \Pi_{S_L}(dx) \int_o^L g(-x+ue)\, du .$$

Pour terminer, il suffit de montrer qu'il existe une constante $N > 0$ telle que

$$\int_o^L g(-x+ue)\, du \geqslant \frac{\log L}{N} .$$

Or, d'après l'expression de g pour $\mathbb{R}^3$ et pour $\mathbb{H}_1$ (voir (3.2)), on a, dans les deux cas :

$$\int_o^L g(-x+ue)\, du \geqslant C^{te} \int_o^L \frac{1}{\sqrt{a^2+(u-x_3)^2}}\, du$$

pour $x = (x_1, x_2, x_3) \in S_L$ et où a^2 est une constante > 0 telle que $x_1^2 + x_2^2 \leqslant a^2$ dans le cas de $\mathbb{R}^3$ et $(x_1^2+x_2^2)^2 \leqslant a^2$ dans le cas de $\mathbb{H}_1$. Il est facile de voir qu'on peut encore minorer l'intégrale précédente par $C^{te} \int_o^L \frac{du}{\sqrt{a^2+u^2}} = C^{te} \log\left(\frac{L}{a} + \sqrt{\left(\frac{L}{a}\right)^2 + 1}\right) \geqslant$ $\geqslant C^{te} \log L$ si $L \geqslant L_o$.

b) Pour la minoration, on peut supposer que L est entier, on aura le résultat pour tout L grâce à la croissance de $L \longmapsto C(S_L)$ et à la concavité des fonctions minorantes. Pour tout j entier, on considère le cylindre S_1^{j-1} défini plus haut et $\Pi_j = \Pi_{S_1^{j-1}}$ sa mesure d'équilibre. On a

$$V\Pi_{j+1}(x) = P_x(T_{S_1^j} < +\infty) = P_{(je)^{-1}x}(T_{(je)^{-1}S_1^j} < +\infty)$$

$$= P_{x-je}(T_{S_1^1} < +\infty) = V\Pi_1(x-je) ,$$

car $je \in m_\ell$. Posons alors $\mu_L = \Pi_1 + \ldots + \Pi_L$. μ_L est une mesure portée par S_L et on a :

$$\mu_L(S_L) = C(S_1)\, L .$$

De plus, $V\mu_L(x) = \sum_{j=1}^{L} V\Pi_1(x-je)$.

Or on a $V\Pi_1 \leqslant 1$ et $\lim_{x\to\infty} |x|^{k-2} V\Pi_1(x) = C(S_1) > 0$ d'après la proposition (4.7), la norme homogène $|x|$ étant celle du théorème (3.1). Par l'équivalence des normes homogènes, il existe alors une constante $A > 0$ telle que :

$$V\Pi_1(x) \leq \left(\frac{A}{|x|_\infty}\right)^{k-2} \wedge 1 \qquad (\forall x \in N)$$

$$\leq \left(\frac{A}{|u|^{1/\ell}}\right)^{k-2} \wedge 1$$

où l'on a posé $x = z + ue$ ($z \in P$, $e \in D$, $u \in \mathbb{R}$) et $|x|_\infty = \sup(|z|_1, |u|^{1/\ell})$. On a donc :

$$V\mu_L(x) \leq \sum_{j=1}^{L} \frac{A}{|u-j|^{\frac{k-2}{\ell}}} \wedge 1 .$$

On a alors deux cas :

(i) si $\frac{k-2}{\ell} > 1$ (i. e. : $k - 2 - \ell > 0$), la série précédente est majorée par une constante $M' > 0$, indépendamment de x et de L. On a alors $V\left(\frac{\mu_L}{M'}\right) \leq 1$ et comme $\frac{\mu_L}{M'}$ est une mesure portée par S_L, on a :

$$C(S_L) \geq \mu_L(S_L)/_{M'} = \frac{C(S_1)}{M'}.L$$

d'après la propriété maximale de la capacité (4.4).

(ii) si $\frac{k-2}{\ell} = 1$ (i. e. : $k - 2 - \ell = 0$), on voit par un calcul facile que pour tout $x \in N$, on a :

$$V\mu_L(x) \leq K \log L$$

et on termine comme en (i), d'où la minoration dans le cas $N = \mathbb{R}^3$ ou $\mathbb{H}_1$. ∎

(6.6) DEMONSTRATION DU THEOREME (6.2). On considère ici la norme homogène : $|x|_\infty = |(z,u)|_\infty = \sup(|z|_1, |u|^{1/\ell})$ et la constante $C_1 \geq 1$ telle que $C_1^{-1}|x|_\infty \leq |x| \leq C_1 |x|_\infty$ $(\forall x \in N)$.

Soit une constante $\alpha > \operatorname{Max}(C_o, C_1^2)$. Posons :

$$A_n = \{x = (z,u) \in A_h \ ; \ \alpha^{-n-1} < |(z,u)| \leq \alpha^{-n}\}$$

$$A_n' = \{x = (z,u) \in A_h ; C_1\alpha^{-n-1} < |(z,u)|_\infty \leq \frac{\alpha^{-n}}{C_1}\}$$

$$A_n'' = \{x = (z,u) \in A_h \ ; \ \frac{\alpha^{-n-1}}{C_1} < |(z,u)|_\infty < C_1 \alpha^{-n}\} .$$

On voit tout de suite que $A_n' \subset A_n \subset A_n''$ (la condition $\alpha > C_1^2$ assure que A_n' n'est pas vide). De plus, comme $(z,u) \in A_h$, on a

$|z| \leq h(u) \leq u^{1/\ell}$ pour u assez proche de zéro. Donc pour n assez grand, on a :

$$A'_n = \left\{x = (z,u) \in A_h \ ; \ (C_1 \ \alpha^{-(n+1)})^{\ell} < u \leq \left(\frac{\alpha^{-n}}{C_1}\right)^{\ell}\right\}$$

et

$$A''_n = \left\{x = (z,u) \in A_h \ ; \ \left(\frac{\alpha^{-(n+1)}}{C_1}\right)^{\ell} < u \leq (C_1 \ \alpha^{-n})^{\ell}\right\} .$$

Considérons alors les cylindres :

$$S'_n = \left\{x = (z,u) \in N \ ; \ |z| \leq h(C_1^{\ell} \ \alpha^{-(n+1)\ell}) \quad \text{et} \quad C_1^{\ell} \ \alpha^{-(n+1)\ell} < u \leq \frac{\alpha^{-n\ell}}{C_1^{\ell}}\right\}$$

et

$$S''_n = \left\{x = (z,u) \in N \ ; \ |z| \leq h(C_1^{\ell} \ \alpha^{-n\ell}) \quad \text{et} \quad \frac{\alpha^{-(n+1)\ell}}{C_1^{\ell}} < u \leq C_1^{\ell} \ \alpha^{-n\ell}\right\} .$$

On a clairement :

$$S'_n \subset A'_n \subset A_n \subset A''_n \subset S''_n ,$$

grâce à la décroissance de la fonction $h(u)$ quand $u \downarrow 0$.

Maintenant à l'aide des propositions (4.7) et (6.5), on obtient, dès que n est assez grand :

a) si $k - \ell - 2 \neq 0$

$$C(S'_n) \geq A.\alpha^{-(n+1)\ell}.\left[h(C_1^{\ell}.\alpha^{-(n+1)\ell})\right]^{k-\ell-2}$$

et

$$C(S''_n) \leq B.\alpha^{-n\ell}.\left[h(C_1^{\ell}.\alpha^{-n\ell})\right]^{k-\ell-2}$$

b) si $k - \ell - 2 = 0$

$$C(S'_n) \geq A \ \alpha^{-(n+1)\ell} \left|\log\left[A' \ \frac{\alpha^{-(n+1)\ell}}{\left[h(C_1^{\ell}.\alpha^{-(n+1)\ell})\right]^{\ell}}\right]\right|^{-1}$$

et

$$C(S''_n) \leq B \ \alpha^{-n\ell} \left|\log\left[B' \ \frac{\alpha^{-n\ell}}{\left[h(C_1^{\ell}.\alpha^{-n\ell})\right]^{\ell}}\right]\right|^{-1}$$

où A , B , A' et B' sont des constantes > 0 . En effet, $\delta_r(S'_n)$

est un cylindre de hauteur

$$r^{\ell}\left(\frac{\alpha^{-n\ell}}{C_1^{\ell}} - C_1^{\ell}\ \alpha^{-(n+1)\ell}\right) = r^{\ell}\ \alpha^{-(n+1)\ell}\left(\frac{\alpha^{\ell}-C_1^{2\ell}}{C_1^{\ell}}\right)$$

et de rayon $r\ h(C_1^{\ell}.\alpha^{-(n+1)\ell})$. Si on choisit δ_r tel que ce rayon soit égal à 1 , on a alors

$$C(\delta_r(S_n')) = r^{k-2}\ C(S_n') \geqslant M.\left(\frac{\alpha^{\ell}-C_1^{2\ell}}{C_1^{\ell}}\right)\alpha^{-(n+1)\ell}\ r^{\ell}\ ,$$

d'où la minoration de $C(S_n')$ grâce à (6.5). Pour la majoration de $C(S_n'')$, on procède de la même façon.

Grâce aux minorations et majorations précédentes, on montre alors par un calcul facile que les deux séries $\sum_n \alpha^{-n(k-2)}\ C(S_n')$ et $\sum_n \alpha^{-n(k-2)}\ C(S_n'')$ convergent ou divergent comme l'intégrale :

$$\int_{0^+}\left(\frac{h(u)}{u^{1/\ell}}\right)^{k-\ell-2}\frac{du}{u} \quad \text{si} \quad k-\ell-2>0$$

ou

$$\int_{0^+}\left|\log\frac{h(u)}{u^{1/\ell}}\right|^{-1}\frac{du}{u} \quad \text{si} \quad k-\ell-2=0\ .$$

Le théorème (6.2) résulte alors du test de régularité de Wiener, compte tenu des inclusions $S_n' \subset A_n \subset S_n''$ et de la relation (4.3) sur les capacités. ∎

(6.7) REMARQUES. a) Notre méthode de démonstration, bien qu'inspirée du cas classique, est plus simple que toutes celles que nous connaissons pour $\mathbb{R}^k$ (cf. [18], p. 256 ou [21], p. 68), car on n'a pas besoin d'avoir de renseignements sur la forme géométrique des cellules A_n .

b) Dans le cas où l'axe de l'épine n'est plus inclus dans m_ℓ mais dans un autre m_i par exemple $(i < \ell)$, le test intégral sera vraisemblablement différent et fera intervenir la fonction $u^{1/i}$ comme le montre déjà le cas du cône (voir (5.3) b)). Bien entendu, dans le cas de $\mathbb{R}^k$, il n'y a pas de problème d'axe grâce à l'invariance du mouvement brownien par rotations.

c) Si $k-\ell-2 \neq 0$: (i) 0 est régulier pour A_h si $h(u) = u^{\alpha}$ et $\alpha \leqslant \frac{1}{\ell}$, 0 est irrégulier si $\alpha > \frac{1}{\ell}$,

(ii) O est régulier pour A_h si $h(u) = u^{\frac{1}{\ell}} |\log u|^{-\alpha}$ et si $\alpha < \frac{1}{k-2-\ell}$, irrégulier si $\alpha \geq \frac{1}{k-2-\ell}$.

d) Si $k - \ell - 2 = 0$: (i) O est régulier si $h(u) = u^{\alpha}$, quel que soit $\alpha > 0$.

(ii) O est régulier si $h(u) = e^{-|\log u|^{\varepsilon}}$ pour $\varepsilon \leq 1$, irrégulier pour $\varepsilon > 1$.

(6.8) EXEMPLE. Dans le cas du groupe d'Heisenberg $\mathbb{H}_1$, le mouvement brownien $X(t)$ a des composantes sur la base (e_i) données par :

$$(\beta_1(t)\ ,\ \beta_2(t)\ ,\ 2\int_0^t \beta_1(s)\, d\beta_2(s) - \beta_2(s)\, d\beta_1(s))$$

où $\beta_1(t)$ et $\beta_2(t)$ sont deux mouvements browniens indépendants sur $\mathbb{R}$ (cf. [23]). Le théorème (6.2) donne alors un résultat sur le comportement local quand $t \downarrow 0$ de l'intégrale stochastique $\int_0^t \beta_1\, d\beta_2 - \beta_2\, d\beta_1$; ainsi on a :

$$\sqrt{\beta_1^2(t) + \beta_2^2(t)} \leq h\left(\int_0^t \beta_1(s)\, d\beta_2(s) - \beta_2(s)\, d\beta_1(s)\right)$$

infiniment souvent quand $t \downarrow 0$ avec probabilité 0 ou 1 suivant que l'intégrale $\int_{0^+} \left|\log \frac{h(u)}{u^{1/2}}\right|^{-1} \frac{du}{u}$ converge ou diverge. Par exemple, en prenant $h(u) = u$, on voit que l'aire balayée entre les instants 0 et t par la courbe brownienne plane $(\beta_1(s), \beta_2(s))$ dépasse infiniment souvent le rayon vecteur quand $t \downarrow 0$. Par symétrie, on a également :

$$\int_0^t \beta_1(s)\, d\beta_2(s) - \beta_2(s)\, d\beta_1(s) \leq -\sqrt{\beta_1^2(t) + \beta_2^2(t)}$$

infiniment souvent quand $t \downarrow 0$.

7. Problème des ensembles récurrents

On peut traiter le problème de la récurrence et de la transience à l'aide de la notion de capacité par un test de Wiener qui s'énonce comme suit :

(7.1) THEOREME. Avec les notations du théorème (5.1), soient A un borélien de N, α une constante $> C_o$ et $A_n = \{x \in A ;$ $\alpha^n < |x| \leq \alpha^{n+1}\}$. Alors : $P_o(X(t) \in A$ infiniment souvent quand $t \uparrow +\infty) = 0$ ou 1 suivant que la série $\sum_n \alpha^{-n(k-2)} C(A_n)$ converge ou diverge.

DEMONSTRATION. Par la même méthode que pour le théorème (5.1), on obtient $P_o(X(t) \in A$ infiniment souvent quand $t \uparrow +\infty) = 0$ si et seulement si $\sum_n \alpha^{-n(k-2)} C(A_n)$ converge. Le théorème est alors démontré grâce à la loi du zéro-un (2.6). ∎

(7.2) REMARQUE. Dans le cas du groupe d'Heisenberg, le théorème précédent a été obtenu par Cygan (cf. [4]) par une autre méthode.

(7.3) COROLLAIRE. Les cônes sont des ensembles récurrents.

DEMONSTRATION. Analogue à (5.4).

Grâce à (7.1), on obtient aussi une condition intégrale (analogue à (6.2)) pour qu'une épine soit récurrente :

(7.4) THEOREME. Avec les notations de (6.1), on suppose que $D \subset m_\ell$ et $\frac{h(u)}{u^{1/\ell}} \downarrow 0$ quand $u \uparrow +\infty$. Alors :

1) Si $k - 2 - \ell > 0$, $P_o(X(t) \in A_h$ i. s. quand $t \uparrow +\infty) = 0$ ou 1 suivant que l'intégrale $\int^{+\infty} \left(\frac{h(u)}{u^{1/\ell}}\right)^{k-2-\ell} \frac{du}{u}$ converge ou diverge.

2) Si $k - 2 - \ell = 0$, $P_o(X(t) \in A_h$ i. s. quand $t \uparrow +\infty) = 0$ ou 1 suivant que l'intégrale $\int^{+\infty} \left|\log \frac{h(u)}{u^{1/\ell}}\right|^{-1} \frac{du}{u}$ converge ou diverge.

La démonstration de (7.4) suit pas à pas la démonstration du théorème (6.2) ; nous ne la faisons donc pas figurer ici.

(7.5) EXEMPLE. En appliquant le résultat précédent au cas du groupe d'Heisenberg, on obtient (comme en (6.8)) un résultat sur le comportement quand $t \uparrow +\infty$ de l'intégrale stochastique $\int_o^t \beta_1\, d\beta_2 - \beta_2\, d\beta_1$:

$$\sqrt{\beta_1^2(t) + \beta_2^2(t)} \leq h\left(\int_o^t \beta_1(s)\, d\beta_2(s) - \beta_2(s)\, d\beta_1(s)\right)$$

infiniment souvent quand $t \uparrow +\infty$ avec la probabilité 0 ou 1 suivant que l'intégrale $\int^{+\infty} \left|\log \frac{h(u)}{u^{1/2}}\right|^{-1} \frac{du}{u}$ converge ou diverge.

Ainsi, par exemple, avec $h(u) = u$, on a

$$\sqrt{\beta_1^2(t) + \beta_2^2(t)} \leq \int_0^t \beta_1(s)\, d\beta_2(s) - \beta_2(s)\, d\beta_1(s)$$

infiniment souvent quand $t \uparrow +\infty$ avec probabilité 1 .

(7.6) Autre intérêt des capacités. La notion de capacité sur un groupe de Lie nilpotent N (s. c. gradué), outre l'intérêt qu'elle peut présenter en elle-même, semble en fait être un outil particulièrement bien adapté à l'étude du comportement local et asymptotique du mouvement brownien et de certains processus sur N . Ainsi, par exemple, dans un article à paraître (cf. [7]), nous montrons comment on peut estimer avec l'aide de la capacité des probabilités du type $P_0(|X(t)| \leq r$ pour un $t > T)$ $(r > 0 , T \geq 0)$ et comment on peut en déduire des résultats sur le comportement local de $|X(t)|$ quand $t \downarrow 0$ et sur la vitesse de fuite de $|X(t)|$ quand $t \uparrow +\infty$.

BIBLIOGRAPHIE

[1] AZENCOTT R., Espaces de Poisson des groupes localement compacts. Lecture Notes in Math. 148, Springer, Berlin (1970).

[2] BLUMENTHAL R. M. et GETOOR R. K., Markov Processes and Potential Theory. Academic Press, New-York (1968).

[3] BONY J. M., Principe du maximum, inégalité de Harnack et unicité du problème de Cauchy pour les opérateurs elliptiques dégénérés. Ann. Inst. Fourier, Grenoble 19, 1 (1969), p. 277-304.

[4] CYGAN J., Wiener's Test for the brownian Motion on the Heisenberg Group. Colloquium Math. Vol. 39 (1978), p. 367-373.

[5] FOLLAND G. B., Subelliptic estimates and function spaces on nilpotent Lie groups. Arkiv for Math. n° 13, p. 161-206.

[6] FOLLAND G. B., A Fundamental solution for a subelliptic operator. Bull. of the amer. Math. Soc. 79 (1973), p. 373-376.

[7] GALLARDO L., Processus subordonnés au mouvement brownien sur les groupes de Lie nilpotents (à paraître).

[8] GAVEAU B., Principe de moindre action, propagation de la chaleur et estimées sous-elliptiques sur certains groupes nilpotents. Acta Math. 139 (1977), p. 95-153.

[9] GETOOR R. K., Some Asymptotic Formulas Involving Capacity. Z. Wahrscheinlichkeitstheorie verw. Geb. 4 (1965), p. 248-252.

[10] GOODMAN R. W., Nilpotent Lie Groups : Structure and Applications to Analysis. Lecture Notes in Math. 562, Springer, Berlin (1976).

[11] GUIVARC'H Y., Croissance polynomiale et périodes des fonctions harmoniques. Bull. Soc. Math. France, 101, (1973), p. 333-379.

[12] GUIVARC'H Y., KEANE M. et ROYNETTE B., Marches aléatoires sur les groupes de Lie. Lecture Notes in Math. 624, Springer, Berlin (1977).

[13] HELMS L. L., Introduction to Potential Theory. Wiley-Interscience, New-York (1969).

[14] HEYER H., Probability Measures on Locally Compact Groups. Springer Verlag, Berlin-Heidelberg (1977).

[15] HOCHSCHILD G., La structure des groupes de Lie. Dunod, Paris (1968).

[16] HUNT G. A., Semi-groups of measures on Lie groups. Trans. Amer. Math. Soc. 81 (1956), p. 264-293.

[17] HUNT G. A., Markov Processes and Potentials III. Illinois Journal of Math. 2 (1958), p. 151-213.

[18] ITÔ K. et Mc KEAN H. P., Diffusion Processes and their Sample Paths. Springer Verlag, Berlin, Second printing (1974).

[19] LAMPERTI J., Wiener's test and Markov chains. Journal Math. Anal. Appl. 6 (1963), p. 58-66.

[20] PORT S. C. et STONE C. J., Classical Potential Theory and brownian Motion. 6th Berkeley Symposium (1970), p. 143-176.

[21] PORT S. C. et STONE C. J., Brownian Motion and Classical Potential Theory. Academic Press, New-York (1978).

[22] ROTHSCHILD L. P. et STEIN E. M., Hypoelliptic Differential Operators and Nilpotent Groups. Acta Math. 137 (1977), p. 247-320.

[23] ROYNETTE B., Croissance et mouvements browniens d'un groupe de Lie nilpotent et simplement connexe. Zeit. für Wahrscheinlichkeitstheorie und verw. Geb. 32 (1975), p. 133-138.

[24] SPITZER F., Discussion on Professor Kingman's Paper : Subadditive Ergodic Theory. Annals of Proba., Vol. 1, n° 6 (1973), p. 904-905.

STABLE BANACH SPACES, RANDOM MEASURES AND ORLICZ FUNCTION SPACES

by D.J.H. Garling

1. Introduction.

Suppose that $(X, \| \ \|)$ is a Banach space. What can be said about the infinite-dimensional subspaces of X? At one extreme, if X is a Hilbert space, so are all its infinite-dimensional subspaces. At the other, every separable Banach space is isometrically isomorphic to a linear subspace of C[0,1]. Another result, of great antiquity, is the following: if X is an infinite-dimensional subspace of $\ell_p (1 \le p < \infty)$ or c_o, then given $\varepsilon > 0$ there is a sequence in X which is $(1+\varepsilon)$-equivalent to the unit vector basis of ℓ_p (or c_o) (see [2] Chapter 12). Thus there are severe restrictions on what the subspaces of ℓ_p and c_o can be.

The space $L_1(0,1)$ has a much richer subspace structure. It has subspaces isometrically isomorphic to $L_p(0,1)$ for $1 \le p \le 2$, and subspaces isomorphic to a large class of Orlicz spaces. Kadeč and Pełczyński [10] showed that if X is a non-reflexive subspace of $L_1(0,1)$ and $\varepsilon > 0$ then there exists a sequence (x_i) in X which is $(1+\varepsilon)$-equivalent to the unit vector basis of ℓ_1. What about the reflexive subspaces? Rosenthal conjectured that if X is a reflexive subspace of $L_1(0,1)$ then there exists $1 < p \le 2$ such that ℓ_p is isomorphic to a subspace of X. Aldous [1] showed that this conjecture is true, using random measures and probabilistic methods: then Krivine and Maurey ([13] and [14] gave a more general condition (based on ideas from logic) on a Banach space which ensures that the same holds.

In these notes, I have attempted to give an account of part of these two approaches. Sections 2-8 are an account of the work of Krivine and Maurey, and results related to it. In the last four sections, we see how random measures can be used to deal with Orlicz function spaces. Here the ideas of Aldous play a fundamental rôle: there is however no discussion of the probabilistic arguments that he uses.

I owe many debts of gratitude. I would like to thank those mentioned above, together with Sylvie Guerre and Jean-Thierry Lapresté, for sending me preprints of their work. At the Polish-GDR Winter School in Wisła, Poland in January 1980, Jean Bourgain lent me a copy of [13], and Richard Haydon spent much time explaining the details to me; I would like to thank them, to thank the Mathematical Institute of the Polish Academeny of Sciences for inviting me and to thank Piotr Mankiewicz for his many kindnesses. Finally, I would like to thank Herbert Heyer for asking me to speak on this topic at a Symposium on Stochastic Analysis in Tübingen in July 1981, and for providing me with this opportunity for publishing these notes.

2. The completion of a metric space.

Let us begin with a simple problem which illustrates the sort of ideas which we shall use. Suppose that (X,d) is a metric space: can we construct a completion for (X,d)? In other words, can we construct a complete metric space $(\hat{X},\hat{d})$ and an isometric embedding i from (X,d) onto a dense subspace of $(\hat{X},\hat{d})$? The standard way to do this is to consider equivalence classes of Cauchy sequences in (X,d), but anyone who has taught an elementary course on metric spaces will know that it is a tedious business to verify all the details. There is however a simple functional approach, belonging to folk-lore, which provides an alternative way. If S is any non-empty set, let $\ell_\infty(S)$ denote all bounded real-valued functions on S, equipped with the usual metric:

$$\rho(f,g) = \sup_{s\in S}|f(s)-g(s)|.$$

Then $(\ell_\infty(S),\rho)$ is a complete metric space. Now if (X,d) is a metric space, let x_0 be a fixed point of X, and let us define a mapping j from (X,d) to $(\ell_\infty(X),\rho)$ by setting $j(x)(y) = d(x,y)-d(x_0,y)$.

By the triangle inequality,

$$|j(x)(y)| \le d(x,x_0)$$

so that $j(x) \in \ell_\infty(X)$. Also

$$|j(x_1)(y)-j(x_2)(y)| = |d(x_1,y)-d(x_2,y)| \le d(x_1,x_2)$$

so that $\rho(j(x_1),j(x_2)) \le d(x_1,x_2)$;

on the other hand

$$j(x_1)(x_2)-j(x_2)(x_2) = d(x_1,x_2),$$

so that

$$\rho(j(x_1),j(x_2)) = d(x_1,x_2).$$

Thus j is an isometric embedding of (X,d) into the complete metric space $(\ell_\infty(X),\rho)$: $(\overline{j(X)},\rho)$ then provides a suitable completion for (X,d).

3. Local compactifications of metric spaces.

Suppose that (X,d) is a metric space. Can we construct a compactification of (X,d) which reflects the metric structure of X? At one extreme, there is the one-point compactification, and at the other the Stone-Čech compactification, but in general neither of these is a metrizable space. Thus we have the problem: if (X,d) is a metric space, when does there exist a compact metric space (Y,ρ) and a homeomorphic embedding j from (X,d) onto a dense subspace of (Y,ρ)? Since a compact metric space is always separable, and a subspace of a separable metric space is separable, a necessary condition is that (X,d) should be separable. We shall see that this condition is also sufficient.

In fact, though, we shall consider a slightly different problem. In considering completions, we consider isometries, whereas above we only require j to be a homeomorphism, so that the problem remains the same if we replace d by an equivalent metric. For our purposes, we shall wish to take an intermediate position. Recall that a subset A of a metric space (X,d) is bounded if $\sup_{a_1,a_2 \in A} d(a_1,a_2) < \infty$. If x_o is a fixed point of X, the sets $E_n = \{x: d(x,x_o) \leq n\}$, $n = 1,2,\ldots$, form a fundamental sequence of bounded sets, in the sense that A is bounded if and only if A is contained in some E_n. We shall prove the following result.

Theorem 1 Suppose that (X,d) is a separable metric space. There exists a locally compact metric space (T,ρ) and a homeomorphic embedding t from (X,d) onto a dense subspace of (T,ρ) such that a subset A of X is bounded if and only if j(A) is relatively compact in (T,ρ).

Before proving the theorem, let us make two remarks. First, it is easy to see that, since (X,d) has a fundamental sequence of bounded sets, (T,ρ) has a fundamental sequence of compact sets (that is, (T,ρ) is σ-compact), and so the condition that (X,d) be separable is necessary for the conclusion of the theorem to hold. Secondly, we can always replace the metric on X by an equivalent bounded metric. Applying the theorem to this, we obtain a metric space compactification of (X,d).

We now turn to the proof of the theorem.

Let

$$\lambda_1(X) = \{f: X \longrightarrow R : |f(x)-f(y)| \leq d(x,y)\};$$

that is, $\lambda_1(X)$ is the space of real-valued functions on X which satisfy a Lipschitz condition with constant 1. We give $\lambda_1(X)$ the topology τ_s of simple convergence. Recall that a topological space (Y,τ) is a Polish space if τ can be given by a metric under which Y is complete and separable.

Lemma 1 $(\lambda_1(X),\tau_s)$ is a Polish space which is locally compact and σ-compact. If $x_o \in X$, a fundamental sequence of compact sets is given by

$$\{f : |f(x_o)| \leq n\}_{n=1,2,\ldots} .$$

Proof. Let $D = \{d_1, d_2, \ldots\}$ be a dense sequence in (X,d). Suppose that

$$N(f) = \{g : |g(y_i)-f(y_i)| < \varepsilon,\ 1 \leq i \leq n\}$$

is a basic neighbourhood of f. There exist $d_{j_1}, \ldots, d_{j_n}$ in D such that $d(y_i, d_{j_i}) < \varepsilon/3$ for $1 \leq i \leq n$. Then

$$N(f) \supseteq \{g : |g(d_{j_i})-f(d_{j_i})| < \varepsilon/3,\ 1 \leq i \leq n\}.$$

Thus τ_s is the same as the topology of simple convergence on D. $\lambda_1(X)$, considered as a space of functions on D, is closed, in the topology of simple convergence, in the space of all functions on D. Thus $\lambda_1(X)$ is a Polish space.

Also a subset A of $\lambda_1(X)$ is relatively compact if and only if it is pointwise bounded on D, and, by the Lipschitz condition, this happens if and only if it is bounded at any one fixed point. Thus $\lambda_1(X)$ is σ-compact and locally compact, and the sets $\{f : |f(x_o)| \leq n\}_{n=1,2,\ldots}$ form a fundamental collection of compact sets.

Next we observe that there is a natural map of X into $\lambda_1(X)$. If $y \in X$, let $t_y(x) = d(x,y)$. Then $t_y \in \lambda_1(X)$, by the triangle inequality.

Lemma 2 The map $t : y \longrightarrow t_y$ is a homeomorphism of (X,d) into $(\lambda_1(X), t_s)$.

Proof. If $d(y,z) < \varepsilon$, $|t_y(x)-t_z(x)| < \varepsilon$ for all x in X, so that t is continuous. If $y \in X$ and $\varepsilon > 0$, let

$$N(t_y) = \{g : |g(y)-t_y(y)| < \varepsilon\}$$
$$= \{g : |g(y)| < \varepsilon\}.$$

$N(t_y)$ is a neighbourhood of t_y in $\lambda_1(X)$, and, if $t_z \in N(t_y)$,

$$d(y,z) = t_z(y) < \varepsilon.$$

Thus t is a homeomorphism of (X,d) into $(\lambda_1(X), t_s)$.

To complete the proof of the theorem, we set $T = \overline{t(X)}$, and let ρ be any metric on T defining the topology τ_s.

To conclude this section, we make one further observation: It follows from a theorem of Sierpiński ([18], or [11] §15, 11(6)) that $t(X)$ is a G_δ subset of (T,ρ) if and only if (X,d) is a Polish space. We shall not use this result, but it may well have application to the study of measures on metric spaces.

4 Stable metric spaces.

Let (X,d) be a separable metric space, and let (T,ρ) be the space constructed in Theorem 1. Although the conclusions of Theorem 1 relate only to the topology of X and the collection of bounded sets, the construction of T depends, *via* the Lipschitz condition, exactly on the metric d, and we shall investigate this further. Note also that it is the topology which ρ defines that is important, rather than ρ itself.

We shall call T the *space of types* on (X,d). If $t \in T$, there is a bounded sequence (x_m) in (X,d) such that

$$t(x) = \lim_m d(x,x_m)$$

for each x in X. We shall call such a sequence an *approximating sequence*. A simple diagonal argument shows that any bounded sequence has a subsequence which is an approximating sequence.

Notice that if $x \in X$,

$$d(x,y) = t_x(y)$$

for each y in X. This suggests defining a function $\tilde{d}$ on $T\times X$ by setting

$$\tilde{d}(t,y) = t(y).$$

Can this be extended to a function on $T\times T$ in a sensible way? Suppose that (y_n) is an approximating sequence for s. It would be appropriate to set

$$\tilde{d}(t,s) = \lim_n t(y_n).$$

Does this limit exist? If so, is it independent of the choice of approximating sequence? In order to answer these closely linked questions, we need a definition. We say that a metric space (X,d) is stable if it is separable and whenever (x_m) and (y_n) are bounded sequences in X then

$$\sup_{m>n} d(x_m,y_n) \geq \inf_{m<n} d(x_m,y_n).$$

Observe that a subspace of a stable metric space is stable, as in the completion of a stable metric space.

Theorem 2. A separable metric space (X,d) is stable if and only if whenever (x_m) and (y_n) are approximating sequences the double limits

$$\lim_m \lim_n d(x_m,y_n) \text{ and } \lim_n \lim_m d(x_m,y_n)$$

both exist and are equal.

Proof. If (X,d) is not stable, there exist bounded sequences (x_m) and (y_n) such that

$$\sup_{m>n} d(x_m,y_n) < \inf_{m<n} d(x_m,y_n).$$

If we extract subsequences, this inequality still holds, and by extracting subsequences we may suppose that (x_m) and (y_n) are approximating sequences. Then

$$\lim_n \sup(\lim_m d(x_m,y_n)) < \lim_m \inf(\lim_n d(x_m,y_n))$$

so that if both double limits exists they cannot be equal.

Conversely, suppose that (x_m) and (y_n) are approximating sequences for which

$$\ell = \lim_n \inf(\lim_m d(x_m,y_n)) < \lim_m \sup(\lim_n d(x_m,y_n)) = L.$$

We can extract subsequences (x_{m_i}) and (y_{n_j}) such that

$$\ell = \lim_j \lim_m d(x_m,y_{n_j}) = \lim_j \lim_i d(x_{m_i},y_{n_j})$$

and

$$L = \lim_i \lim_n d(x_{m_i},y_n) = \lim_i \lim_j d(x_{m_i},y_{n_j});$$

thus we need only consider the case where both limits exist and are different:

$$\ell = \lim_n \lim_m d(x_m,y_n) < \lim_m \lim_n d(x_m,y_n) = L.$$

Choose a and b so that $\ell < a < b < L$. There exist m_1 and n_1 such that

$$\lim_n d(x_m,y_n) > b \quad \text{for } m \ge m_1 \tag{1}$$

and

$$\lim_m d(x_m,y_n) < a \quad \text{for } n \ge n_1. \tag{2}$$

Suppose we have found $m_1 <\ldots< m_k$ and $n_1 <\ldots< n_k$ such that

$$d(x_{m_i},y_{n_j}) < a \quad \text{for } 1 \le j < i \le k$$

and

$$d(x_{m_i},y_{n_j}) > b \quad \text{for } 1 \le i < j \le k.$$

Then we can find $m_{k+1} > m_k$ and $n_{k+1} > n_k$ such that

$$d(x_{m_i},y_{n_{k+1}}) > b \text{ for } 1 \le i \le k, \text{ by (1)}$$

and

$$d(x_{m_{k+1}},y_{n_j}) < a \text{ for } 1 \le j \le k, \text{ by (2)}.$$

Then $\sup d(x_{m_i},y_{n_j}) \le a < b \le \inf_{i<j} d(x_{m_i},y_{n_j})$, and so (X,d) is not stable.

Corollary Suppose that (X,d) is a stable metric space, that t and s are types on (X,d) and that (x_m) is an approximating sequence for s. Then $\lim_m t(x_m)$ exists and is independent of the choice of (x_m). The function $\tilde{d}(t,s) = \lim_m t(x_m)$ on $T\times T$ satisfies

$$\tilde{d}(t,s) = \tilde{d}(s,t)$$

$$\tilde{d}(t,s) \le \tilde{d}(s,u) + \tilde{d}(u,t).$$

Proof. Let (y_n) be an approximating sequence for t. Then

$$t(x_m) = \lim_n d(x_m,y_n)$$

so that

$$\lim_m t(x_m) = \lim_m \lim_n d(x_m,y_n) = \lim_n \lim_m d(x_m,y_n)$$
$$= \lim_n d(y_n).$$

This shows that the limit exists, and is independent of the choice of (x_m), and also shows that $\tilde{d}(t,s) = \tilde{d}(s,t)$. Finally if u is another type, defined by an approximating sequence (z_o), then, for any p,

$$\begin{aligned}\tilde{d}(t,s) &= \lim_m \lim_n d(x_m,y_n) \\ &\le \lim_m \lim_n d(x_m,z_p + d(z_p,y_n)) \\ &= s(z_p) + t(z_p).\end{aligned}$$

Letting $p \longrightarrow \infty$,

$$\tilde{d}(t,s) \le \tilde{d}(s,u) + \tilde{d}(u,t).$$

The function $\tilde{d}$ on $T \times T$ is seldom jointly continuous:

Proposition 1 Suppose that (X,d) is a stable metric space. $\tilde{d}$ is jointly continuous on $(T,\rho)\times(T,\rho)$ if and only if each bounded subset of (X,d) is precompact.

Proof. If the condition is satisfied, the mapping t extends to a homeomorphism of $(\hat{X},\hat{d})$ onto (t,ρ), and $\tilde{d}(t(\hat{x}),t(\hat{y})) = \hat{d}(\hat{x},\hat{y})$. As $\tilde{d}$ is jointly continuous on $(\hat{X},\hat{d})\times(\hat{X},\hat{d})$, $\tilde{d}$ is jointly continuous on $(T,\rho)\times(T,\rho)$.

If the condition is not satisfied, there exists a bounded sequence (x_n) in (X,d) with

$$\inf_{m,n} d(x_m,x_n) = \varepsilon > 0.$$

By extracting a subsequence if necessary, we may suppose that (x_n) is an approximating sequence, defining a type s. Then

$$(t_{x_n},t_{x_n}) \longrightarrow (s,s) \text{ in } (T,\rho)\times(T,\rho)$$

while

$$\tilde{d}(t_{x_n},t_{x_n}) = 0 \text{ for all } n$$

and

$$\tilde{d}(s,s) = \lim_m \lim_n d(x_m,x_n) \ge \varepsilon.$$

Fortunately, we have separate continuity:

Theorem 3. Suppose that (X,d) is a stable metric space. For each s in T, the function $t \longrightarrow \tilde{d}(s,t)$ is continuous on (T,ρ).

Proof. Suppose that $t^{(k)} \longrightarrow t$ in (T,ρ), and that $\tilde{d}(s,t^{(k)})$ does not converge to $\tilde{d}(s,t)$. Let (x_m) be an approximating sequence for s, (y_n) an approximating sequence for t and $(y_n^{(k)})$ an approximating sequence for $t^{(k)}$. If z is a fixed point of X,

$$t(z) = \lim_{k\to\infty} t^{(k)}(z) = \lim_{k\to\infty} \lim_{n\to\infty} d(z, y_n^{(k)}),$$

and so, by extracting subsequences if necessary, we can suppose that the double array $(y_n^{(k)})_{n=1,k=1}^{\infty\ \infty}$ is bounded in X.

As $\tilde{d}(s,t^{(k)})$) is a bounded sequence, by considering a suitable subsequence we can suppose that

$$\tilde{d}(s,t^{(k)}) \longrightarrow b \neq a = \tilde{d}(s,t).$$

Let $\varepsilon = |b-a|/3$. By considering further subsequences we can suppose successively that

$$|\tilde{d}(s,t^{(k)})-b| < \varepsilon \text{ for all } k$$

and that

$$|\lim_m d(x_m, y_n^{(k)})-b| < \varepsilon \text{ for all } n \text{ and } k; \tag{1}$$

we can also suppose that

$$|t(x_m)-a| < \varepsilon/3 \text{ for all } m. \tag{2}$$

Now set $n_1 = k_1 = 1$. By (1) there exists m_1 such that

$$|d(x_{m_1}, y_{n_1}^{(k_1)})-b| < \varepsilon.$$

Suppose that we have found $m_1 <\ldots< m_r$, $n_1 <\ldots< n_r$ and $k_1 <\ldots< k_r$ such that

$$|d(x_{m_j}, y_{n_i}^{(k_i)})-b| < \varepsilon \text{ for } 1 \le i \le j \le r \text{ and} \tag{3}$$

$$|d(x_{m_j}, y_{n_i}^{(k_i)})-a| < \varepsilon \text{ for } 1 \le j < i \le r. \tag{4}$$

Since $t^{(k)} \longrightarrow t$ in (T,ρ), there exists $k_{r+1} > k_r$ such that

$$|t^{(k_{r+1})}(x_{m_j})-t(x_{m_j})| < \varepsilon/3 \text{ for } 1 \le j \le r. \tag{5}$$

From the definition of $(y_n^{(k_{r+1})})$ there exists $n_{r+1} > n_r$ such that

$$|t^{(k_{r+1})}(x_{m_j})-d(x_{m_j}, y_{n_{r+1}}^{(k_{r+1})})| < \varepsilon/3 \text{ for } 1 \le j \le r. \tag{6}$$

Combining (2), (5) and (6), it follows that

$$|d(x_{m_j}, y_{n_{r+1}}^{(k_{r+1})})-a| < \varepsilon \text{ for } 1 \le j \le r.$$

It now follows from (1) that there exists $m_{r+1} > m_r$ such that

$$|d(x_{m_{r+1}}, y_{n_i}^{(k_i)}) - b| < \varepsilon \text{ for } 1 \le i \le r+1.$$

As the sequence $(y_{n_i}^{(k_i)})$ is bounded, there exists a subsequence, (w_h) say, which is an approximating sequence. Then

$$|\lim_m \lim_n d(x_m, w_h) - a| \le \varepsilon, \text{ by (4)}$$

and

$$|\lim_n \lim_m d(x_m, w_h) - b| \le \varepsilon, \text{ by (3)}.$$

This contradicts the fact that the two double limits are equal, and completes the proof.

The fact that $\tilde{d}$ is only separately continuous leads to some difficulties: fortunately there are results from general topology which enable us to deal with them, and which ensure that there are plenty of points of continuity. We shall establish these results here.

Proposition 2. *Suppose that* (X,d), (Y,ρ) *and* (Z,σ) *are metric spaces, and that* $f : X\times Y \longrightarrow Z$ *is separately continuous. Then if* F *is closed in* (Z,σ), $f^{-1}(F)$ *is a* G_δ*-set in* $(X,d)\times(Y,\rho)$.

Proof. For each n, let

$$F_n = \{z : \text{there exists } w \text{ in } F \text{ with } \sigma(z,w) < 1/n\}.$$

F_n, the open neighbourhood of F of radius $1/n$, is an open subset of Z. Since F is closed, $F = \bigcap_{n=1}^{\infty} F_n$.

Now for each n and each $x' \in X$, let

$$A_{n,x'} = \{x : d(x,x') < 1/n\}\times\{y : f(x',y) \in F_n\}.$$

By the separate continuity of f, $A_{n,x'}$ is an open subset of $(X,d)\times(Y,\rho)$; so therefore is

$$B_n = \bigcup_{x'\in X}(A_{n,x'}), \text{ for each } n.$$

Let $C = \bigcap_{n=1}^{\infty} B_n$. Then C is a G_δ-subset of $(X,d)\times(Y,\rho)$, and the proof will be complete if we show that $C = f^{-1}(F)$.

If $(x,y) \in f^{-1}(F)$, $f(x,y) \in F \subseteq F_n$, so that $(x,y) \in A_{n,x}$, for each n. Thus $(x,y) \in \bigcap_{n=1}^{\infty} B_n = C$. Conversely suppose that $(x,y) \in C$. For each n there exists $k \ge 2n$ such that if $d(x,x') < 1/k$ then $\rho(f(x,y), f(x',y)) < 1/2n$. Since $(x,y) \in C$, $(x,y) \in B_k$ and so there exists x' such that $(x,y) \in A_{k,x'}$. Thus $\rho(f(x,y),f(x',y)) < 1/2n$ and $f(x',y) \in F_k \subseteq F_{2n}$. This implies that $f(x,y) \in F_n$; as this holds for each n, $f(x,y) \in F$. This completes the proof.

Proposition 3. Suppose that f is a mapping from a topological space (X_1,τ_1) to a second countable topological space (X_2,τ_2) such that whenever F is closed in (X_2,τ_2) then $f^{-1}(F)$ is a G_δ-subset of (X_1,τ_1). Then the set D of points of discontinuity of f is the first category in X.

Proof. Let $(U_r)_{r=1}^{\infty}$ be a base for the topology τ_2. A point x is in D if and only if, for some r, $x \in f^{-1}(U_r)$ and $x \notin \mathrm{int}(f^{-1}(U_r))$. Thus $D = \bigcup_{r=1}^{\infty} G_r$, where $G_r = f^{-1}(U_r)\setminus(\mathrm{int}\, f^{-1}(U_r))$. Each G_r is an F_σ-set without interior points, so that each G_r is of the first category in X: so, therefore is D.

Combining Propositions 2 and 3 and the Baire category theorem, we have the following corollary.

Corollary Suppose that (X,d), (X_1,d_1) and (X_2,d_2) are metric spaces, that (X,d) is complete, that (g_n) is a sequence of continuous mappings of (X,d) into $(X_1,d_1)\times(X_2,d_2)$ and that (f_n) is a sequence of separately continuous real-valued functions on $(X_1,d_1)\times(X_2,d_2)$. Then there exists a point of (X,d) at which each of the functions $g_n \circ f_n$ is continuous.

5. Stable Banach spaces and typical norms.

We now turn our attention to the case where $(X,\|\ \|)$ is a separable real Banach space, with its natural translation invariant metric

$$d(x,y) = \| x-y \| .$$

It turns out that, when $(X,\|\ \|)$ is stable, the linear structure of X is reflected in properties of the space T of types on X.

First let us consider scalar multiplication. Suppose that t is a type, defined by an approximating sequence (y_n), and that α is a non-zero scalar. Then

$$\| x-\alpha y_n \| = |\alpha| \, \| \frac{x}{\alpha} - y_n \|$$

so that $\lim_n \| x-\alpha y_n \|$ exists, and equals $|\alpha| t(x/\alpha)$. Thus (αy_n) is an approximating sequence: we denote the type which it defines, which clearly does not depend on the choice of approximating sequence, by

$D_\alpha t$. We also define $D_0 t(x)$ to be the <u>trivial type</u> $t_0(x) = \|x\|$; thus, whatever α is, $D_\alpha t(x) = \lim_n \|x - \alpha y_n\|$.

It is clear that if $\alpha \neq 0$, then D_α is a homeomorphism of T (with inverse $D_{1/\alpha}$) which extends scalar multiplication on X: that is to say that $D_\alpha j(x) = j(\alpha x)$, for all x in E.

We can also define translation. If $t \in T$ and $x \in E$, let $\tau_x(t)(y) = t(x+y)$. It is easy to see that $\tau_x(t) \in T$, and that τ_x is a homeomorphism of T.

In order to deal with addition more fully, we need to suppose that $(X, \|\ \|)$ is stable - in other words, that the metric space (X,d) is stable. Suppose that s and t are types on X, defined by approximating sequences (x_m) and (y_n) respectively. Then $(x-x_m)$ is also an approximating sequence for each x in X. By Theorem 2,

$$\lim_m \lim_n \|(x-x_m)-y_n\| \quad \text{and} \quad \lim_n \lim_m \|(x-x_m)-y_n\|$$

both exist, and are equal. By the Corollary to Theorem 2, the resulting quantity does not depend upon the choice of approximating sequences. Further, considered as a function of x, it is clearly a type on X, which we denote by $s \star t$. Notice that $s \star t = t \star s$, and that this operation is associative.

Let ϕ denote the vector space of all scalar sequences with only finitely many non-zero terms. We denote the nth unit vector in ϕ by e_n. We shall show that if t is a type on a stable Banach space $(X, \|\ \|)$ then t defines a seminorm on $X \oplus \phi$.

<u>Theorem 4</u>. *Suppose that* t *is a type on a stable Banach space* $(X, \|\ \|)$. *If* $x+\alpha = x+\sum_{i=1}^k \alpha_i e_i \in X \oplus \phi$, *let*

$$\|x+\alpha\|_t = D_{\alpha_1}(t) \star \ldots \star D_{\alpha_k}(t)(x).$$

If there exists y *in* X *such that* $t(x) = \|x-y\|$, *then*

$$\|x+\alpha\|_t = \|x - (\textstyle\sum_{i=1}^k d_i) y\|.$$

Otherwise, $\|\ \|_t$ *is a norm on* $X \oplus \phi$, *and*

$$\|x + \textstyle\sum_{i=1}^k \alpha_i e_i\|_t = \|x + \sum_{i=1}^k \alpha_{\sigma(i)} e_i\|_t,$$

for every permutation σ *of* $\{1,\ldots,k\}$.

Proof. Let (y_n) be an approximating sequence for t. Then

$$\|x+\alpha\|_t = \lim_{n_1\to\infty} \dots \lim_{n_k\to\infty} \|x-\alpha_1 y_{n_1} - \dots -\alpha_k y_{n_k}\|,$$

from which it follows readily that $\| \ \|_t$ is a seminorm on $X \oplus \phi$, which has the given form if $t(x) = \|x-y\|$. If (y_n) has a subsequence (y_{n_k}), convergent to y say, than $t(y) = \lim_n \|y-y_n\| = 0$, so that y_n converges to y and $t(x) = \|x-y\|$. If (y_n) has no convergent sunsequence, there exist n_o and $\delta > 0$ such that $\|y_k-y_\ell\| > \delta$ for $k > \ell > n_o$. Suppose that $z = x+\alpha_1 e_1+\dots+\alpha_k e_k \neq 0$ and that $\|z\|_t = 0$. Since $\|x\| = \|x\|_t$, not all α_i are zero: let j be the largest index for which $\alpha_j \neq 0$. There exists $n_o \leq n_1 < n_2 < \dots n_j$ such that

$$\|x-\alpha_1 y_{n_1}-\dots-\alpha_{j-1}y_{j-1}-\alpha_j y_r\| < |\alpha_j|\delta/2$$

for $r \geq n_j$, and so by the triangle inequality, $\|y_{n_j}-y_{n_j+1}\| < \delta$, giving a contradiction.

Finally

$$\|x + \textstyle\sum_{i=1}^k \alpha_i e_i\|_t = \|x + \textstyle\sum_{i=1}^k \alpha_{\sigma(i)} e_i\|_t,$$

since

$$D_{\alpha_1}(t) \star \dots \star D_{\alpha_k}(t) = D_{\alpha_{\sigma(1)}}(t) \star \dots \star D_{\alpha_{\sigma(k)}}(t).$$

We shall call a norm defined on $X \oplus \phi$ by a type, as in Theorem 4, a <u>typical norm</u>.

Let us give one application of this theorem. Recall that a sequence (x_n) of vectors in a Banach space $(E, \| \ \|)$ is a <u>basic sequence</u> if each element x of the closed linear space $[x_n]$ of the sequence (x_n) can be written uniquely in the form

$$x = \textstyle\sum_{n=1}^\infty \alpha_n x_n.$$

A basic sequence (x_n) is <u>symmetric</u> if $\sum_{n=1}^\infty \alpha_n x_{\sigma(n)}$ converges whenever $\sum_{n=1}^\infty \alpha_n x_n$ converges and σ is a permutation of the positive integers. A basic sequence (x_n) is <u>subsymmetric</u> if $\sum_{k=1}^\infty \varepsilon_k \alpha_k x_{n_k}$ converges whenever $\sum_{k=1}^\infty \alpha_k x_k$ converges, $\varepsilon_k = \pm 1$ and $n_1 < n_2 < \dots$. A symmetric basic sequence is subsymmetric, but the converse is not always true (see Chapter 3 of [16] for details).

<u>Theorem 5</u>. <u>If</u> $(X, \|\ \|)$ <u>is a stable Banach space, every subsymmetric basic sequence</u> (x_n) <u>in</u> X <u>is symmetric</u>.

<u>Proof</u>. There exist $0 < m \le M < \infty$ such that

$$m\|\sum_{i=1}^{n}\alpha_i x_i\| \le \|\sum_{i=1}^{n}\alpha_i x_{m_i}\| \le M\|\sum_{i=1}^{n}\alpha_i x_i\| \qquad (*)$$

for all $n, \alpha_1, \ldots, \alpha_n$ and $m_1 < \ldots < m_n$. In particular, (x_n) is a bounded sequence, and $\inf\{\|x_m - x_n\| : m \ne n\} > 0$. Thus (x_n) has an approximating subsequence $(y_i) = (x_{m_i})$, and the type t which (y_i) determines defines a typical norm on $X \oplus^1 \phi$. It follows from (*) that

$$m\|\sum_{i=1}^{n}\alpha_i x_i\| \le \|\sum_{i=1}^{n}\alpha_i e_i\|_t \le M\|\sum_{i=1}^{n}\alpha_i x_i\|$$

for all $\alpha = \sum_{i=1}^{n}\alpha_i e_i$ in ϕ, and as

$$\|\sum_{i=1}^{n}\alpha_i e_i\|_t = \|\sum_{i=1}^{n}\alpha_{\sigma(i)} e_i\|_t$$

it follows that

$$m\|\sum_{i=1}^{n}\alpha_{\sigma(i)} x_i\| \le M\|\sum_{i=1}^{n}\alpha_i x_i\| ;$$

this ensures that (x_i) is a symmetric basic sequence.

This result shows that stable Banach spaces form a restricted class of Banach spaces. We shall see that L_p spaces are stable: the result of Theorem 5 in this case was proved by Dacunha-Castelle and Krivine [6]; their arguments provide an interesting precursor both of Aldous' ideas and of those of Krivine and Maurey.

6. Some examples.

In this section we give some examples. These are important not only because they show that some well-known spaces are stable and others are not, but also because they lead to some ideas which we shall use later on.

<u>Example 1</u> Let $E = c_o$, with the usual supremum norm. Let e_n denote the $n^{\underline{th}}$ unit vector, and let $f_m = e_1 + \ldots + e_m$. Then

$$\begin{aligned} d(e_n, -f_m) = \|e_n + f_m\| &= 2 \text{ if } m \ge n \\ &= 1 \text{ if } m < n. \end{aligned}$$

Thus $\sup_{m<n} d(e_n, -f_m) < \inf_{m>n} d(e_n, -f_m)$, and so c_o is not stable.

Example 2. Let $E = \ell_p$ $(1 \le p < \infty)$, with its usual norm. Let us suppose that t is a type on ℓ_p, defined by an approximating sequence $(y^{(n)})$. As $(y^{(n)})$ is bounded, it is coordinate-wise bounded, and, by extracting a subsequence if necessary, we can suppose that $y^{(n)}$ converges coordinate-wise to an element $\bar{y}$ of ℓ_p. Let $z^{(n)} = y^{(n)} - \bar{y}$, so that

$$t(x) = \lim_n \| x-\bar{y}-z^{(n)} \|$$

and $z^{(n)} \longrightarrow 0$ coordinatewise. By extracting a further sequence if necessary, we can suppose that $\| z^{(n)} \| \longrightarrow \mu$. Then an easy calculation shows that

$$t(x) = (\| x-\bar{y}\|^p + \mu^p)^{1/p}.$$

From this it follows that $\bar{y}$ and μ are uniquely determined by t, so that $y^{(n)}$ must converge to $\bar{y}$, and $\| y^{(n)}-\bar{y}\|$ must converge to μ.

If now

$$s(x) = (\| x-\bar{x}\|^p + \lambda^p)^{1/p}$$

is another type, defined by an approximating sequence $(x^{(m)})$,

$$\begin{aligned}\lim_m \lim_n \| x^{(m)}-y^{(n)} \| &= \lim_m (\| x^{(m)}-\bar{y}\|^p + \mu^p)^{1/p} \\ &= (\| \bar{x}-\bar{y}\|^p + \lambda^p + \mu^p)^{1/p} \\ &= \lim_n \lim_m \| x^{(m)}-y^{(n)} \|,\end{aligned}$$

so that ℓ_p is stable, by Theorem 2.

In the case where $y^{(n)} \longrightarrow 0$ coordinatewise, the type t has a particularly simple form:

$$t(x) = (\| x\|^p + \mu^p)^{1/p};$$

the corresponding typical norm is also very simple:

$$\| x+\alpha\|_t = (\| x\|^p + \mu^p \textstyle\sum_{i=1}^n |\alpha_i|^p)^{1/p}.$$

This leads us to make the following definitions.

If $(E,\| \; \|)$ is a stable Banach space, a type t on X is <u>symmetric</u> if $t(x) = t(-x)$ for each x in E; in terms of the corresponding typical norm, t is symmetric if and only if $\| x+\alpha\|_t = \| x-\alpha\|_t$ for each $x+\alpha$ in $X \oplus \phi$. A type t is an <u>ℓ_p-type</u> if

$$D_\alpha t \star D_\beta t = D_{\gamma(\alpha,\beta)}(t)$$

for all α and β, where $\gamma(\alpha,\beta) = (|\alpha|^p + |\beta|^p)^{1/p}$; in terms of the corresponding typical norm, an easy calculation shows that t is an ℓ_p-type if and only if

$$\| x+\alpha \|_t = \| x + \|\alpha\|_p e_1 \|_t$$

for each $x+\alpha$ in $X \oplus \phi$.

Thus any symmetric type on ℓ_p is an ℓ_p-type.

If $(E, \| \ \|)$ is an infinite-dimensional stable Banach space, there are plenty of non-trivial symmetric types on E. There exists a bounded sequence with no convergent subsequence, and consequently there exists an approximating sequence (x_m) with the same property. Let t be the type which it defines: then $t \star D_{-1}t$ is symmetric, and

$$(t \star D_{-1}t)(0) = \lim_{m,n\to\infty} \| x_m - x_n \| \neq 0$$

so that $t \star D_{-1}t$ is not the trivial type.

The next theorem shows the importance of ℓ_p-types:

<u>Theorem 6</u>. <u>Suppose that</u> t <u>is a</u> ℓ_p-<u>type defined by an approximating sequence</u> (y_n) <u>and that</u> $\varepsilon > 0$. <u>Then there exists a subsequence</u> (y_{n_k}) <u>which is</u> $1+\varepsilon$ <u>equivalent to the unit vector basis of</u> ℓ_p.

<u>Proof</u>. Let $M = \sup_n \| y_n \|$. For each positive integer k let

$$D_k = \{n\varepsilon/(k+1)2^{k+2}M : n \in Z,\ |n\varepsilon| \le (k+1)2^{k+2}M\}.$$

Inductively we can find $n_1 < m_1 < n_2 < m_2 < \ldots$ such that

$$\Big| \, \| \textstyle\sum_{j<k} \beta_j y_{n_j} + (|\beta_k|^p + |\beta_{k+1}|^p)^{1/p} y_{n_k} \| -$$

$$- \| \textstyle\sum_{j<k} \beta_j y_{n_j} + (|\beta_k|^p + |\beta_{k+1}|^p)^{1/p} e_1 \|_t \, \Big| < \varepsilon/2^{k+2} \qquad \text{(A)}$$

and

$$\Big| \, \| \textstyle\sum_{j=1}^{k} \beta_j y_{n_j} + \beta_{k+1} y_n \| - \| \textstyle\sum_{j<k} \beta_j y_{n_j} + \beta_k e_1 + \beta_{k+1} e_2 \|_t \, \Big| < \varepsilon/2^{k+2} \qquad \text{(B)}$$

for $n \ge m_k$, $\beta_1, \ldots, \beta_{k+1}$ in D_k, for $k = 1,2,\ldots$.

[Suppose that n_i and m_i have been found for $i < k$: we can find $n_k > m_{k-1}$ such that (A) is satisfied, and

$$\left|\lim_{n\to\infty} \| \textstyle\sum_{j<k}\beta_j y_{n_j} + \beta_k y_{n_k} + \beta_{k+1} y_n \| - \| \textstyle\sum_{j<k}\beta_j y_{n_j} + \beta_k e_1 + \beta_{k+1} e_2 \|_t \right| < \varepsilon/2^{k+2};$$

we can then find $m_k > n_k$ so that (B) is satisfied.]
Then

$$\left| \| \textstyle\sum_{j=1}^{k+1}\beta_j y_{n_j} \| - \| \textstyle\sum_{j<k}\beta_j y_{n_j} + \beta_k e_1 + \beta_{k+1} e_2 \|_t \right| < \varepsilon/2^{k+2},$$

so that

$$\left| \| \textstyle\sum_{j=1}^{k+1}\beta_j y_{n_j} \| - \| \textstyle\sum_{j<k}\beta_j y_{n_j} + (|\beta_k|^p)^{1/p} y_{n_k} \| \right| < \varepsilon/2^{k+1}$$

for $\beta_1,\ldots,\beta_{k+1}$ in D_k, for $k = 1,2,\ldots$.

Now suppose that $\sum_{j=1}^{k+1}|\alpha_j|^p = 1$. There exist $\beta_1,\ldots,\beta_{k+1}$ in D_k such that $|\alpha_j-\beta_j| < \varepsilon/(k+1)2^{k+2}M$, and so

$$\left| \| \textstyle\sum_{j=1}^{k+1}\alpha_j y_{n_j} \| - \| \textstyle\sum_{j=1}^{k+1}\beta_j y_{n_j} \| \right| < \varepsilon/2^{k+2}$$

and

$$\left| \| \textstyle\sum_{j<k}\alpha_j y_{n_j} + (|\alpha_k|^p + |\alpha_{k+1}|^p)^{1/p} y_{n_k} \| - \| \textstyle\sum_{j<k}\beta_j y_{n_j} + (|\beta_k|^p + |\beta_{k+1}|^p)^{1/p} y_{n_k} \| \right| < \varepsilon/2^{k+2}.$$

Consequently

$$\left| \| \textstyle\sum_{j=1}^{k+1}\alpha_j y_{n_j} \| - \| \textstyle\sum_{j<k}\alpha_j y_{n_j} + (|\alpha_k|^p + |\alpha_{k+1}|^p)^{1/p} y_{n_k} \| \right| < \varepsilon/2^k.$$

Repeating this inequality k times,

$$\left| \| \textstyle\sum_{j=1}^{k+1}\alpha_j y_{n_j} \| - (\textstyle\sum_{j=1}^{k+1}|\alpha_j|^p)^{1/p} \| y_{n_1} \| \right| < \varepsilon.$$

<u>Corollary</u> (Banach) Suppose that X is an infinite dimensional closed subspace of ℓ_p and that $\varepsilon > 0$. Then X has a subspace which is $(1+\varepsilon)$-isomorphic to ℓ_p.

<u>Proof</u>. Let (x_n) be a sequence in X which converges coordinatewise to 0, but which has no norm convergent subsequence. Then (x_n) has a subsequence which is an approximating sequence in ℓ_p, and which therefore defines an ℓ_p-type. The result now follows from the theorem.

This is an old result: moreover, the arguments which have been used to prove it reduce, for the most part, to the argument given in Banach's book [2].

Example 3. The quotient of a stable Banach space need not be stable. For any separable Banach space is isometrically iosmorphic to a quotient of ℓ_1.

Example 4. Stability is not preserved under distortion of the norm, even by small amounts. For example, define a new norm on ℓ_p by setting

$$|||x||| = \|x\|_p + \varepsilon \max\{\|x_{2k}+x_{2\ell-1}\| : k < \ell\}.$$

Then

$$\lim_{\ell} \lim_{k} |||e_{2k}+e_{2\ell-1}||| = 2^{1/p} + 2\varepsilon$$

while

$$\lim_{k} \lim_{\ell} |||e_{2k}+e_{2\ell-1}||| = 2^{1/p} + 3\varepsilon.$$

This example was shown to me by Bela Bollobás.

On the other hand, if $(E,\|\ \|)$ is not stable, it remains unstable under small perturbations of the norm.

This means that if $(E,\|\ \|)$ is a stable Banach space, there is no non-trivial type t on E which satisfies

$$D_\alpha t * D_\beta t = D_{\max(\alpha,\beta)} t;$$

that is, there is no c_o-type on E. For if there were, arguing as in Theorem 6, for each $\varepsilon > 0$, there would be a subspace of E which is $(1+\varepsilon)$-isomorphic to c_o. As c_o is not stable, this is not possible.

7. The existence of ℓ_p-types.

This section will be devoted to proving the following fundamental theorem.

Theorem 7. *If $(E,\|\ \|)$ is an infinite-dimensional stable Banach space, there exists an ℓ_p-type t on E, for some $1 \leq p < \infty$.*

Combining this with Theorem 6, we have the following corollary:

Corollary. *If $(E,\|\ \|)$ is a stable Banach space, there exist p (with $1 \leq p < \infty$) such that for each $\varepsilon > 0$ there is a subspace of E which is $(1+\varepsilon)$-isomorphic to ℓ_p.*

We shall give the proof of Krivine and Maurey. We shall break the proof into several parts. We shall need a result of Bohnenblust [4]. Recall that a Banach lattice F is a vector lattice with a norm $\| \ \|$ such that $(F,\| \ \|)$ is a Banach space, and such that

$$\|x\| \le \|y\| \text{ if } |x| \le |y|.$$

Bohnenblust's result is the following:

Theorem 8. Suppose that $(F,\| \ \|)$ is a Banach lattice of dimension at least 3 with the following property: if $|x_1|\wedge|x_2| = 0 = |y_1|\wedge|y_2|$ and $\|x_1\| = \|y_1\|$, $\|x_2\| = \|y_2\|$, then $\|x_1+x_2\| = \|y_1+y_2\|$. Then either there exists $1 \le p < \infty$ such that if $|x_1|\wedge|x_2| = 0$, then $\|x_1+x_2\| = (\|x_1\|^p + \|x_2\|^p)^{1/p}$, or if $|x_1|\wedge|x_2| = 0$, then $\|x_1+x_2\| = \max(\|x_1\|, \|x_2\|)$.

We shall not prove this: the proof is given in [15] pp137-138 and [17] p18.

Lemma 3. Suppose that t is a non-trivial symmetric type on E. t is an ℓ_p-type, for some $1 \le p < \infty$, if for each $\alpha \ge 0$ there exists $\beta \ge 0$ such that

$$t \star D_\alpha t = D_\beta t.$$

Proof. Let $R^3 = \text{span}(e_1,e_2,e_3)$ in $E \oplus \phi$, with its natural ordering. R^3 is a Banach lattice under the norm $\| \ \|_t$ of dimension 3. We shall verify that the hypotheses of Theorem 8 are satisfied. By homogeneity, and because

$$\|\alpha_1e_1+\alpha_2e_2+\alpha_3e_3\|_t = \|\alpha_{\sigma(1)}e_1+\alpha_{\sigma(2)}e_2+\alpha_{\sigma(3)}e_3\|_t$$

for each σ in $\sum_3$, it is sufficient to consider the case where $x_1 = e_1+\alpha e_2$, $y_1 = \beta e_1$, $x_2 = y_2 = \gamma e_3$. But then if (z_n) is an approximating sequence for t,

$$\begin{aligned}\|x_1+x_2\|_t &= \lim_p \lim_m \lim_n \|\gamma z_p+z_m+\alpha z_n\| \\ &= \lim_p (t \star D_\alpha t)(z_p) \\ &= \lim_p D_\beta t(z_p) \\ &= \lim_p \lim_m \|\gamma z_p+\beta z_m\| = \|y_1+y_2\|_t .\end{aligned}$$

Thus either there exists $1 \le p < \infty$ such that

$$\beta = (1+\alpha^p)^{1/p}$$

in which case t is an ℓ_p-type, or $\beta = \max(1,\alpha)$. But we saw at the end of the last section that the latter case is not possible.

Lemma 3 clearly helps us to recognise ℓ_p-types. In order to see this in another way, we need a definition. Let T denote the space of types on E, given the topology of pointwise convergence on E (as in §3). Let S denote the set of symmetric types on E. S is clearly closed in T. A subset C of S will be called a <u>cone</u> if

(a) C is closed in S

(b) $C \neq \{t_o\}$

(c) if $t \in C$, then $D_\alpha t \in C$ for all real α

and (d) if t_1 and t_2 are in C, then $t_1 \star t_2 \in C$.

It now follows from Lemma 3 that if t is a non-trivial symmetric type, then $\{D_\alpha t : \alpha \in \mathbb{R}\}$ is a cone if and only if t is an ℓ_p-type. In our search for ℓ_p-types, it is therefore worth considering small cones.

If C is a cone, let $C_1 = \{t \in C : t(0) = 1\}$. C_1 is a compact subset of T, and C_1 is non-empty, since if t is a non-trivial type in C, and $\alpha = t(0)^{-1}$, $D_\alpha t \in C_1$. For the same reason,

$$C = \{D_\lambda t : t \in C_1, \lambda \in \mathbb{R}\}$$

and so C_1 determines C.

<u>Lemma 4</u>. <u>Every cone</u> C <u>contains a minimal cone</u>.

<u>Proof</u>. Let $\mathcal{B}$ be a set of cones contained in C which is totally ordered by inclusion. By compactness,

$$A_1 = \bigcap\{B_1 : B \in \mathcal{B}\}$$

is non-empty, and so $A = \bigcap\{B : B \in \mathcal{B}\}$ is non-empty. A is clearly a cone contained in each B of $\mathcal{B}$, and so the lemma follows by applying Zorn's lemma.

Let C be a minimal cone. We shall show that every non-trivial type in C is an ℓ_p-type. We need the following elementary result from spectral theory.

<u>Proposition 4</u>. <u>Suppose that</u> λ <u>is a boundary point of a continuous linear operator</u> T <u>on a Banach space</u> F. <u>There exists a sequence</u> (x_n) of unit vectors in F such that $Tx_n - \lambda x_n \longrightarrow 0$.

Proof. By considering $T-\lambda I$, we can suppose that $\lambda = 0$. Let (λ_n) be a sequence of points such that $\lambda_n \longrightarrow 0$ and such that $T-\lambda_n I$ is invertible. Since

$$T = T-\lambda_n I+\lambda_n I = (T-\lambda_n I)(I+\lambda_n (T-\lambda_n I)^{-1})$$

and T is not invertible, $\| (T-\lambda_n I)^{-1}\| \geq |\lambda_n|^{-1}$. Thus there exists y_n with

$$\| y_n\| < 2|\lambda_n| , \| (T-\lambda_n I)^{-1}y_n\| = 1.$$

Let $x_n = (T-\lambda_n I)^{-1}y_n$. Then $Tx_n = \lambda_n x_n+y_n \longrightarrow 0$.

Lemma 5. Suppose that $\alpha \geq 0$. There exists $1 \leq \beta \leq 1+\alpha$ such that for each n there exists σ_n in C_1 such that

$$|(\sigma_n \star \alpha\sigma_n)(x)-\beta\sigma_n)(x)| \leq 1/n \qquad (*)$$

for each x in E.

Proof. Let $t \in C$. Let F be the completion of the normed space $(\phi, \| \ \|_t)$. If $x = \sum_{i=1}^n \xi_i e_i \in \phi$, let $Tx = \sum_{i=1}^n \xi_i (e_{2i}+\alpha e_{2i+1})$. Then

$$\| x\| \leq \| Tx\| \leq (1+\alpha)\| x\| ,$$

and T extends by continuity to a continuous linear operator on F, again denoted by T, with the same property. As $e_1 \notin T(F)$, 0 is in the spectrum of T. This implies that there is a real non-negative β in the boundary of the spectrum of T. By Proposition 4, for each n there exists γ_n in F with $\| \gamma_n\|_t = 1$ such that $\| T\gamma_n-\beta\gamma_n\|_t < 1/n$. By making a small perturbation if necessary, we can suppose that $\gamma_n \in \phi$. Note that $1 \leq \beta \leq 1+\alpha$, since $1 \leq \| T\gamma_n\| \leq 1+\alpha$. Now if

$$\gamma_n = \sum_{j=1}^{k_n} g_{j,n}e_j \text{ and } \delta_n = T\gamma_n = \sum_{j=1}^{\ell_n} d_{j,n}e_j,$$

let

$$\sigma_n = D_{g_{1,n}}(t) \star \ldots \star D_{g_{k_n,n}}(t)$$

and

$$\tau_n = D_{d_{1,n}}(t) \star \ldots \star D_{d_{\ell_n,n}}(t).$$

Then σ_n is in C_1 and τ_n is in C, and

$$(D_\beta\sigma_n)(x) = \| x+\beta\gamma_n\|_t , \quad \tau_n(x) = \| x+\delta_n\|_t$$

for each x in E. Bearing in mind that $s \star u = u \star s$, $\tau_n = \sigma_n \star \alpha\sigma_n$, and so

$$|(\sigma_n \bigstar \alpha\sigma_n)(x) - (D_\beta\sigma_n)(x)| = \left| \|x+\beta\gamma_n\|_t - \|x+T\gamma_n\|_t \right|$$

$$\le \|\beta\gamma_n - T\gamma_n\|_t < 1/n.$$

This proves the lemma.

This lemma shows that there are types in C_1 which are nearly ℓ_p-types. Unfortunately, Proposition 1 shows that we cannot use a simple approximation argument: if (σ_{n_k}) is a subsequence which converges to σ, say there is no immediate reason to suppose that $\sigma_{n_k} \bigstar \alpha\sigma_{n_k}$ converges to $\sigma \bigstar \alpha\sigma$. Instead we shall need to appeal to the Corollary to Proposition 3.

A sequence (σ_n) in C for which (*) holds will be called an (α,β) <u>approximating sequence</u>.

<u>Lemma 6</u>. <u>Suppose that</u> $\alpha \ge 0$. <u>There exists</u> $1 \le \beta \le 1+\alpha$ <u>such that each</u> t <u>in</u> C <u>is the limit of an</u> (α,β) <u>approximating sequence</u>.

<u>Proof</u>. Let β be as in Lemma 5, and let D be the set of those t in C which are the limit of an (α,β) approximating sequence. Lemma 5 and the compactness of C_1 show that D does not reduce to the trivial type. We shall show that D is a cone: the minimality of C then implies that C = D.

D <u>is closed</u>. If $t \in \bar{D}$, there is a sequence (t_m) in D such that $t_m \longrightarrow t$. For each m there is an (α,β) approximating sequence $(\sigma_{m,n})$ such that $\sigma_{m,n} \longrightarrow t_m$. There exists a sequence (n_m), with $n_m \ge m$ for each m, such that $\sigma_{m,n_m} \longrightarrow t$. If $x \in E$

$$|(\sigma_{m,n_m} \bigstar \alpha\sigma_{m,n_m})(x) - (D_\beta\sigma_{m,n_m})(x)| \le 1/n_m \le 1/m,$$

so that $(\sigma_{m,n_m})_{m=1}^{\infty}$ is an (α,β) approximating sequence, and $t \in D$.

<u>If</u> $t \in D$ <u>and</u> $\lambda \in \mathbb{R}$, <u>then</u> $D_\lambda t \in D$. If $\lambda = 0$ $D_\lambda t = t_0 \in D$. If $\lambda \ne 0$, choose an integer k such that $k \ge |\lambda|$. Then if (σ_m) is an (α,β) approximating sequence such that $\sigma_m \longrightarrow t$, and $x \in E$

$$|(D_\lambda\sigma_{km} \bigstar D_\alpha(D_\lambda\sigma_{km}))(x) - (D_\beta D_\lambda\sigma_{km})(x)|$$

$$\doteq |\lambda| \left| (\sigma_{km} \bigstar D_\alpha\sigma_{km})(\tfrac{x}{x}) - D_\beta\sigma_{km})(\tfrac{x}{x}) \right| \le \frac{|\lambda|}{km} \le \frac{1}{m},$$

so that $(D_\lambda\sigma_{km})_{m=1}^{\infty}$ is an (α,β) approximating sequence. As

$D_\lambda \sigma_{km} \longrightarrow D_\lambda t$, $D_\lambda t \in D$.

<u>If</u> t_1 <u>and</u> t_2 <u>are in</u> D, <u>then</u> $t_1 \bigstar t_2 \in D$. Suppose that (σ_m) is an (α,β) approximating sequence which converges to t_1, (τ_n) an (α,β) approximating sequence which converges to t_2. By separate continuity,

$$\sigma_m \bigstar t_2 \longrightarrow t_1 \bigstar t_2, \text{ and } \sigma_m \bigstar \tau_n \longrightarrow \sigma_m \bigstar t_2$$

for each m. Thus there exists an increasing sequence (n_m) such that $\sigma_m \bigstar \tau_{n_m} \longrightarrow t_1 \bigstar t_2$.

Now if s is any type, defined by an approximating sequence (x_n), and $x \in E$,

$$(\sigma_m \bigstar D_\alpha \sigma_m \bigstar s)(x) = \lim_n (\sigma_m \bigstar D_\alpha \sigma_m)(x-x_n)$$

and

$$(D_\beta \sigma_m \bigstar s)(x) = \lim_n (D_\beta \sigma_m)(x-x_m)$$

so that

$$|(\sigma_m \bigstar D_\alpha \sigma_m \bigstar s)(x) - (D_\beta \sigma_m \bigstar s)(x)| \le 1/m,$$

and similar inequalities hold for (τ_n). Thus

$$\begin{aligned} &|(\sigma_m \bigstar \tau_{n_m} \bigstar D_\alpha(\sigma_m \bigstar \tau_{n_m}))(x) - D_\beta(\sigma_m \bigstar \tau_{n_m})(x)| \le \\ &|(\sigma_m \bigstar D_\alpha \sigma_m) \bigstar (\tau_{n_m} \bigstar D_\alpha \tau_{n_m})(x) - (D_\beta \sigma_m) \bigstar (\tau_{n_m} \bigstar D_\alpha \tau_{n_m})(x)| \\ &+ |(\tau_{n_m} \bigstar D_\alpha \tau_{n_m}) \bigstar D_\beta \sigma_m(x) - (D_\beta \tau_{n_m} \bigstar D_\beta \sigma_m)(x) \\ &\le \frac{1}{m} + \frac{1}{n_m} \le \frac{2}{m}. \end{aligned}$$

Consequently $(\sigma_{2m} \bigstar \tau_{n_{2m}})$ is an (α,β) approximating sequence, and so $t_1 \bigstar t_2 \in D$.

Bearing in mind that if $\sigma_n \longrightarrow t$, $\sigma_n(0) \longrightarrow t(0)$, the following corollary follows easily:

<u>Corollary</u>. <u>Suppose that</u> $\alpha \ge 0$. <u>There exists</u> $\beta \ge 1$ <u>such that each</u> t <u>in</u> C_1 <u>is the limit of an</u> (α,β) <u>approximating sequence in</u> C_1.

We are now in a position to prove Theorem 7. Let (x_n) be a dense sequence in E. For each non-negative rational q and each t in C_1, let $g_q(t) = (t, D_q t)$; g_q is a continuous map from C_1 into T×T. If $(s,t) \in T \times T$, let $f_n(s,t) = (s \bigstar t)(x_n)$. By Theorem 3, each f_n is a separately continuous function on T×T. For each (n,q), let

$$h_{n,q}(t) = f_n g_q(t) = (t \bigstar D_q t)(x_n).$$

By the Corollary to Proposition 3, there exists $\tilde{t}$ in C_1 such that each $h_{n,q}$ is continuous at $\tilde{t}$. Now for each q there exists $1 \le \beta_q \le 1+q$ such that $\tilde{t}$ is the limit of a (q,β_q) approximating sequence $(\sigma_{m,q})_{m=1}^{\infty}$ in C_1. Then

$$\begin{aligned}(\tilde{t} \bigstar D_q\tilde{t})(x_n) &= h_{n,q}(\tilde{t}) = \lim_m h_{n,q}(\sigma_{m,q}) \\ &= \lim_m(\sigma_{m,q} \bigstar D_q\sigma_{m,q})(x_n) \\ &= \lim_m(D_{\beta_q}\sigma_{m,q})(x_n) = (D_{\beta_q}\tilde{t})(x_n)\end{aligned}$$

for each n. By continuity,

$$(\tilde{t} \bigstar D_q\tilde{t})(x) = (D_{\beta_q}\tilde{t})(x)$$

for each x in E: thus $\tilde{t} \bigstar D_q\tilde{t} = D_{\beta_q}\tilde{t}$.

Now if $\alpha \ge 0$, there exists a sequence of rationals (q_n) such that $0 \le q_n \le \alpha$ and $q_n \longrightarrow \alpha$. As $1 \le \beta_{q_n} \le 1+q_n \le 1+\alpha$, we can suppose, by extracting a subsequence if necessary, that β_{q_n} converges, to β say. Then $\tilde{t} \bigstar D_{q_n}\tilde{t} \longrightarrow \tilde{t} \bigstar D_\alpha\tilde{t}$, and $D_{\beta_{q_n}}\tilde{t} \longrightarrow D_\beta\tilde{t}$, and so $\tilde{t} \bigstar D_\alpha\tilde{t} = D_\beta\tilde{t}$. Theorem 7 therefore follows from Lemma 3.

8 Weak sequential completeness of stable Banach spaces.

In this section we shall establish some topological properties of stable Banach spaces, due to Guerre and Lapresté ([8],[9]).

Theorem 9. *A stable Banach space is weakly sequentially complete.*

Proof. Suppose that $(E,\|\ \|)$ is a stable Banach space and that (x_n) is a weak Cauchy sequence which does not converge weakly. By extracting a subsequence if necessary, we can suppose that (x_n) is a basic sequence ([16] page 5); we can also suppose that (x_n) and $(x_n - x_{n+1})$ are approximating sequences. Let $y_n = x_n - x_{n+1}$; note that $y_n \longrightarrow 0$ weakly. Let t be the type, and $\|\ \|_t$ the typical norm, that (x_n) defines, and let $(F,\|\ \|_t)$ be the completion of $(\phi,\|\ \|_t)$. If L is the basis constant of (x_n), if $\alpha = \sum_{i=1}^{j}\alpha_i e_i \in \phi$ and if $k < j$

$$\begin{aligned}\|\alpha_1e_1+\dots+\alpha_ke_k\|_t &= \lim_{n_1}\dots\lim_{n_k}\|\alpha_1x_{n_1}+\dots+\alpha_kx_{n_k}\| \\ &\le L\lim_{n_1}\dots\lim_{n_j}\|\alpha_1x_{n_1}+\dots+\alpha_jx_{n_j}\| \\ &= L\|\alpha_1e_1+\dots+\alpha_je_j\|_t\end{aligned}$$

so that (e_n) is a basis for $(F, \| \ \|_t)$. Since

$$\| \sum_{i=1}^{j} \alpha_i e_i \|_t = \| \sum_{i=1}^{j} \alpha_{\sigma(i)} e_i \|_t$$

for any permutation σ, (e_n) is a symmetric basis and is therefore 1-unconditional ([16] p.113).

Since (x_n) is weakly Cauchy and does not converge weakly to 0, there exists ϕ in E^* such that $\phi(x_n) \longrightarrow 1$. Suppose that $\beta \in \ell_\infty$. If $\alpha = \sum_{i=1}^{j} \alpha_i e_i \in \phi$, let

$$\begin{aligned}\psi_\beta(\alpha) &= \lim_{n_1\to\infty} \dots \lim_{n_j\to\infty} \phi(\alpha_1\beta_1 x_{n_1} + \dots + \alpha_j \beta_j x_{n_j}) \\ &= \alpha_1\beta_1 + \dots + \alpha_j\beta_j .\end{aligned}$$

Then
$$\begin{aligned}|\psi_\beta(\alpha)| &\le \|\phi\| \lim_{n_1\to\infty} \dots \lim_{n_j\to\infty} \| \alpha_1\beta_1 x_{n_1} + \dots + \alpha_j\beta_j x_{n_j} \| \\ &= \|\phi\| \, \| \sum_{i=1}^{j} \alpha_i \beta_i e_i \|_t \le \|\phi\| \, \|\beta\| \, \| \sum_{i=1}^{j} \alpha_i e_i \|_t ,\end{aligned}$$

by the unconditionality of (e_i). Consequently ψ maps ℓ_∞ isomorphically into F^*. On the other hand, ψ is clearly onto, so that (e_n) is equivalent to the unit vector basis of ℓ_1.

Now let $s = t \bigstar D_{-1} t$. s is symmetric, and if $x \in E$

$$s(x) = \lim_n \| x - y_n \| = \lim_n \| x - x_n + x_{n-1} \|$$

so that

$$\| x + \sum_{i=1}^{j} \alpha_i e_i \|_s = \| x + \sum_{i=1}^{j} \alpha_i (e_{2i-1} - e_{2i}) \|_t .$$

This means that there exists $K > 0$ such that

$$\frac{1}{K}\left(\sum_{i=1}^{j} |\alpha_j|\right) \le \| \sum_{i=1}^{j} \alpha_i e_i \|_s \le K \sum_{i=1}^{j} |\alpha_i|$$

for all $\alpha = \sum_{i=1}^{j} \alpha_i e_i$ in ϕ.

Now let D denote all types v of the form

$$v = D_{\alpha_1}(s) \bigstar \dots \bigstar D_{\alpha_j}(s) \qquad (\alpha_j \ge 0)$$

and let $C = \bar{D}$. C is a cone. By the arguments of §7, there exists an ℓ_p-type u in C_1. We shall show that u is an ℓ_1-type. Suppose that (v_n) is a sequence in $C_1 \cap D$ such that $v_n \longrightarrow u$, and suppose that

$$v_n = D_{\alpha_{1,n}}(s) \bigstar \dots \bigstar D_{\alpha_{j_n,n}}(s).$$

Then $1 = v_n(0) = \| \sum_{i=1}^{j_n} \alpha_{i,n} e_i \|_s$, so that

$$1/K \le \sum_{i=1}^{j_n} \alpha_{i,n} \le K, \text{ for each } n.$$

Also,

$$(v_{n_1} \star \dots \star v_{n_r})(0) = \| \sum_{q=1}^{r} (\sum_{i=1}^{j_{n_q}} \alpha_{i,n_q} e_{N_q+i}) \|_s$$

(where $N_q = \sum_{p<q} j_{n_p}$), so that

$$r/K^2 \le v_{n_1} \star \dots \star v_{n_r}(0) \le rK^2$$

Since $\| e_1 + \dots + e_r \|_v = \lim_{n_1} \dots \lim_{n_r} v_{n_1} \star \dots \star v_{n_r}(0)$

it follows that

$$r/K^2 \le \| e_1 + \dots + e_r \|_u \le rK^2$$

and that u is an ℓ_p-type.

Now, if $x \in E$,

$$v_n(x) = \lim_{m_1} \dots \lim_{m_{j_n}} \| x - \alpha_{1,n} y_{m_1} - \dots - \alpha_{j_n,n} y_{m_{j_n}} \|,$$

and so by an inductive argument (using a countable dense subset of E), there exists $m_1 < m_2 < \dots$ such that, setting $z_i = y_{m_i}$,

$$u(x) = \lim_{n\to\infty} \| x - \alpha_{1,n} z_{N_n+1} - \dots - \alpha_{j_n,n} z_{N_n+j_n} \|$$

for each x in E. Thus if

$$w_n = \sum_{i=1}^{j_n} \alpha_{i,n} z_{N_n+i}$$

(w_n) is an approximating sequence for u. By Theorem 6, there is a subsequence equivalent to the unit vector basis for ℓ_1, so that w_n does not converge to 0 weakly. But if $\theta \in E^*$,

$$|\theta(w_n)| \le \sum_{i=1}^{j_n} \alpha_{i,n} |\theta(z_{N_n+i})|$$

$$\le K \max_{m>n} |\theta(y_m)|$$

and this last quantity tends to 0, since $y_m \longrightarrow 0$ weakly. This gives a contradiction, and completes the proof of the theorem.

<u>Theorem 10</u>. <u>Suppose that</u> $(E, \| \ \|)$ <u>is a stable Banach space. The following are equivalent</u>:

(i) E <u>is reflexive</u>;

(ii) ℓ_1 <u>is not isomorphic to any subspace of</u> E;

(iii) <u>if</u> t <u>is an</u> ℓ_p<u>-type on</u> E, <u>then</u> $p > 1$.

<u>Proof</u>. Clearly (i) implies (ii), and (ii) implies (iii), by Theorem 6. Suppose that F is a subspace of E which is isomorphic to ℓ_1. Let T be an isomorphism of ℓ_1 onto F, and let $T(e_i) = x_i$. By extracting subsequences if necessary, we can suppose that (x_i) and $(x_{2i}-x_{2i})$ are approximating sequences in E. Let $y_i = x_{2i-1}-x_{2i}$. Then $[y_i]$ is isomorphic to ℓ_1, and if s is the type defined by (y_i), s is symmetric. If $(G, \| \ \|_s)$ is the completion of $(\phi, \| \ \|_s)$, it is easy to see that $(G, \| \ \|_s)$ is isomorphic to ℓ_1 (with (e_i) equivalent to the unit vector basis). Arguing as in Theorem 9, if u is an ℓ_p-type in the cone generated by s, then u is an ℓ_1-type. Consequently (iii) implies (ii).

Finally, suppose that ℓ_1 is not isomorphic to any subspace of E; then every element of E** is the $\sigma(E^{**},E^*)$-limit of a sequence in E ([16] Theorem 2.e.7) and so, as E is weakly sequentially complete, E must be reflexive.

9. Random measures.

We now turn to the problem of the stability of function spaces. Although Maurey and Krivine show that $L_p(E)$ is stable if E is stable and $1 \le p < \infty$ fairly directly, we shall use the approach which Aldous took, involving random measures. In many ways, random measures form a concrete setting for some of the more abstract ideas of the earlier sections.

We begin by recalling some results about non-random measures. In what follows, (T,ρ) will be a complete σ-compact locally compact metric space, with a fundamental increasing sequence of compact sets $(K_n)_{n=1}^{\infty}$, with $K_n \subseteq \operatorname{int} K_{n+1}$, for each n. We denote by $C_b(T)$ the Banach space of bounded real continuous functions on T, with the supremum norm. For each n, there exists f_n in $C_b(T)$ with $0 \le f_n \le 1$,

$$\begin{aligned} f_n(t) &= 0 \text{ if } t \in K_n \\ &= 1 \text{ if } t \notin K_{n+1}. \end{aligned}$$

We denote by $M(T)$ the space of bounded signed Borel measures on T; by $M_+(T)$ the subset consisting of non-negative Borel measures on T, and by $P(T)$ the subset of Borel probability measures on T.

We give $M(T)$ the narrow topology - that is, the weak topology $\sigma(M(T),C_b(T))$. Let us recall some fundamental results: for details see Bourbaki [5], Chapitre IX. Every element of $M(T)$ is regular ([5], p49). A subset L of $M(T)$ is narrowly relatively compact if and only if

(i) $\sup\{\|\mu\| : \mu \in L\} < \infty$

and (ii) for each $\varepsilon > 0$, there exists n such that

$$|\mu|(T\backslash K_n) < \varepsilon, \text{ for } \mu \in L$$

(Prokhorov's theorem ([5] pp63-64)). Although $M(T)$ is not metrizable in the narrow topology, $M_+(T)$ is a Polish space (that is, its topology is given by a metric under which it is complete and separable), and so therefore is $P(T)$. Further, $\mu_n \longrightarrow \mu$ in $P(T)$ if (and only if) $\int f\, d\mu_n \longrightarrow \int f\, d\mu$ for each uniformly continuous f in $C_b(T)$ ([3] Theorem 2.1).

We now turn to random measures. The definition of a random measure is deceptively simple. Suppose that (Ω,Σ,P) is a probability space. Then a random measure ξ is simply an equivalence class of measurable mappings from Ω into $P(T)$, with the narrow topology (as usual, identifying two mappings if they are equal almost everywhere). In fact, there are several ways of looking at random measures. In order to do so, we shall assume that (Ω,Σ,P) is a complete probability space, and we shall also assume that it is the completion of a countably generated space (Ω,Σ_0,P) with Σ_0 generated by a countable ring of sets $A_1,A_2,\ldots$.

Let us give three examples: they all play a fundamental role in Aldous' work, but in fact only the first will really concern us.

i) If $X \in L_0(\Omega;T) = L_0(T)$ (the space of all T measurable functions on Ω), let

$$\xi(\omega) = \delta_{x(\omega)}$$

where x is a representative of X, and δ_y denotes the point mass at y. Up to equality almost everywhere, this does not depend on the choice of x. It is quite easy to see that ξ is measurable. Such a random measure will be called a degenerate random measure. We therefore have a mapping i from $L_o(T)$ into $\pi(T)$, the collection of all T-valued random measures on Ω.

ii) Let $\mu \in P(T)$. Let $\xi(\omega) = \mu$ for all ω: ξ is a constant random measure.

iii) The third example concerns stable probability measures. If $0 < q \le 2$ and $a \in R$ there exists a measure $\sigma(q,a)$ on R, with characteristic function $e^{-|at|^q}$: $\sigma(q,a)$ is called a stable (symmetric) measure with index q and scale factor $|a|$. Note that $\sigma(2,a)$ is simply a centred Gaussian distribution with variance a^2. Note also that

$$\sigma(q,a) \star \sigma(q,b) = \sigma(q,(|a|^q)^{1/q}):$$

this follows immediately by considering characteristic functions.

Now suppose that $\alpha \ge 0$ is a real-valued random variable on Ω and that $0 < q \le 2$. We define

$$\sigma(q,\alpha)(\omega) = \sigma(q,\alpha(\omega)).$$

Again it is easy (but instructive) to verify that $\sigma(q,\alpha)$ is a measurable map of Ω into $P(R)$:

$\sigma(q,\alpha)$ is called a q-stable random measure.

We now consider how to topologise the space $\pi(T)$ of random measures in T. This can be done in many ways: Aldous considers two topologies on $\pi(T)$, but we shall restrict attention to just one, which he calls the ωm-topology.

In order to define this topology, we consider random measures as operators: if $\xi \in \pi(T)$ and $f \in C_b(T)$, let

$$T_\xi(f)(\omega) = \int_T f(t)\xi(\omega,dt).$$

T_ξ is a norm-decreasing linear mapping of $C_b(T)$ into $L_\infty(\Omega)$. We give $L_\infty(\Omega)$ the weak topology $\sigma(L_\infty,L_1)$: the resulting simple topology defines the ωm-topology on $\pi(T)$: a sub-base of neighbourhoods of ξ is given by

$$N_{(f,g)}(\xi) = \xi\eta: |\int_\Omega g(\omega)|(T_\eta f)(\omega)-(T_\xi f)(\omega)P(d\omega)| \le 1\},$$

where $f \in C_b(T)$ and $g \in L_1(\Omega)$.

If $f \in C_b(T)$, the set $\{T_\xi(f): \xi \in \pi(T)\}$ is bounded in L_∞, and is therefore relatively compact in $\sigma(L_\infty, L_1)$. In order to define the ωm-topology it is therefore sufficient to let g run through a separating subset of L_1. We may for example consider a sub-base of neighbourhoods of the form

$$N_{(f,E)}(\xi) = \{\eta: \left|\int_E (T_\eta f)(\omega) - (T_\xi f)(\omega) P(d\omega)\right| \leq 1\}$$

where $f \in C_b(T)$ and $E \in \Sigma$.

We have represented random measures as linear operators: it is also helpful to consider the transposed operators. $C_b(T)$ is isometrically isomorphic to $C(\beta T)$, the space of continuous functions on the Stone-Čech compactification βT of T. We can therefore identify the dual $(C_b(T))^*$ with the space $M(\beta T)$ of regular Borel measures on βT. Suppose that $\xi \in \pi(T)$. Since T_ξ is norm-decreasing, T_ξ^* maps $L_1(\Omega)$ into $M(\beta T)$. In fact we can say more.

Theorem 11. If $g \in L_1(\Omega)$, $T_\xi^*(g) \in M(T)$.

Proof. Since $M(T)$ is norm-closed in $M(\beta T)$, it is enough to show that $T_\xi^*(\chi_A) \in M(T)$ for each $A \in \Sigma$.

Suppose that $\varepsilon > 0$. Let $L(\xi)$ denote the law of ξ on $P(T)$. Since $L(\xi)$ is a Borel measure on the Polish space $P(T)$, $L(\xi)$ is regular: there therefore exists a compact subset K of $P(T)$ such that $L(\xi)(K) > 1 - \varepsilon/2$; that is, $P(\xi \in K) > 1 - \varepsilon/2$. Since K is compact, by Prokhorov's theorem there exists n_o such that

$$\int f_n(t)\mu(dt) < \varepsilon/2$$

for $\mu \in K$ and $n \geq n_o$. Then if $n \geq n_o$

$$\begin{aligned}\int f_n d(T_\xi^*(\chi_A)) &= \int_A T_\xi(f_n)dP \\ &= \int_{A\cap(\xi\in K)} \left(\int f_n(t)\xi(\omega,dt)\right)P(d\omega) + \\ &\quad + \int_{A\cap(\xi\notin K)} \left(\int f_n(t)\xi(\omega,dt)\right)P(d\omega) \\ &\leq (\varepsilon/2)P(A\cap(\xi\in K)) + P(A\cap(\xi\notin K)) \\ &\leq \varepsilon.\end{aligned}$$

Since $T_\xi^*(\chi_A)$ is a non-negative measure, this ensures that $T_\xi^*(\chi_A)$ is in $M(T)$.

If $g \in L_1(\Omega)$, we shall write ξ_g for $T_\xi^*(g)$ and if $A \in \sum$, we shall write ξ_A for $T_\xi^*(\chi_A)$. If $A \in \sum$ and $P(A) > 0$, we shall write $E(\xi|A)$ for $\xi_A/P(A)$: note that $E(\xi|A) \in P(T)$.

Bearing in mind the remark following the definition of the ωm-topology, $\xi_\nu \longrightarrow \xi$ in the ωm-topology if and only if $E(\xi_\nu|A_i) \longrightarrow E(\xi|A_i)$ for each i. This shows that $\pi(T)$ is separable and metrizable in the ωm-topology.

__Theorem 12__. __The mapping__ $i: L_o(T) \longrightarrow i(L_o(T))$ __is a homeomorphism of__ $L_o(T)$ __into__ $\pi(T)$.

__Proof__. Suppose that $X_n \longrightarrow X$ in $L_o(T)$, and let $(Y_k) = (X_{n_k})$ be a subsequence. There exists a further subsequence $(Z_r) = (Y_{k_r})$ such that $Z_r \longrightarrow X$ almost surely. Now if $f \in C_b(T)$ and $g \in L_1(\Omega)$

$$T_{i(Z_r)}(f) = f(Z_r) \longrightarrow f(X) = T_{i(X)}(f)$$

almost surely, so that by dominated convergence

$$\int g\, T_{i(Z_r)}(f)dP \longrightarrow \int g\, T_{i(X)}(f)dP.$$

Thus $i(Z_r) \longrightarrow i(X)$ in $\pi(T)$, and i is continuous.

Now suppose that $X \in L_o(T)$ and that $0 < \varepsilon < 1/2$. There exists a measurable partition $\Omega = A_1 \cup \ldots \cup A_n \cup B$ of Ω and elements $t_1,\ldots,t_n$ of T such that $P(B) < \varepsilon/2$ and $\rho(X(\omega),t_i) < \varepsilon^2/8$ on A_i. Set

$$h_i(t) = \frac{8}{\varepsilon^2}(\rho(t,t_i) \wedge 1),$$

$$g_i = \chi_{A_i}/(P(A_i)),$$

and let $N = \bigcap_{i=1}^n N_{(h_i,g_i)}(i(X))$. N is a ωm-neighbourhood of $i(X)$. Suppose that $i(Y) \in N$. Then

$$\frac{1}{P(A_i)} \left|\int_{A_i} (\rho(Y,t_i)\wedge 1)-(\rho(X,t_i)\wedge 1)dP\right| \le \varepsilon^2/8$$

for $1 \le i \le n$. But $\rho(X,t_i) \le \varepsilon^2/8$ on A_i, so that

$$\frac{1}{P(A_i)} \int_{A_i} (\rho(X,t_i)\wedge 1)dP \le \varepsilon^2/8$$

and consequently

$$\frac{1}{P(A_i)} \int_{A_i} (\rho(Y,t_i) \wedge 1)\, dP \le \varepsilon^2/4$$

for $1 \le i \le n$. Thus

$$P((\rho(Y,t_i) > \varepsilon/2) \cap A_i) < \varepsilon P(A_i)/2$$

for $1 \le i \le n$, and so

$$P((\rho(X,Y) > \varepsilon) \cap A_i) < \varepsilon P(A_i)/2.$$

Adding up,

$$P(\rho(X,Y) > \varepsilon) < \varepsilon.$$

Thus i is a homeomorphism of $L_o(T)$ into $\pi(T)$.

In order to consider the space of random measures further, and in particular to consider compactness properties, it is convenient to represent random measures as linear functionals on a suitable Banach space.

The setting which we consider is related to the general problem: if X is a Banach space, what is the dual of the space $L_1(X)$ of Bochner integrable X-valued functions on a probability space Ω? If X* has the Radon-Nikodym property, it is well known that the dual can be identified with $L_\infty(X^*)$ (cf. [5]): in the case that concerns us, however, this does not happen.

Let us recall that if X is a Banach space and (Ω, Σ, P) is a probability space, an X-<u>valued vector measure</u> F is a finitely additive set function on Σ taking values in X. We set

$$\| F\|_\infty = \inf\{k > 0: \| F(E)\| \le kP(E) \text{ for } E \in \Sigma\}$$

and set

$$V_\infty(X) = \{F: \| F\|_\infty < \infty\}.$$

It is straightforward to verify that $(V_\infty, \| \ \|_\infty)$ is a Banach space under the natural algebraic operations and that if $F \in V_\infty(X)$ then F is σ-additive.

<u>Theorem 13</u>. <u>The dual of</u> $L_1(X)$ <u>is naturally isometrically isomorphic to</u> $V_\infty(X^*)$.

<u>Proof</u>. If $F \in V_\infty(X^*)$ and $f = \sum_{i=1}^n x_i \chi_{E_i}$ is a simple function in $L_1(X)$, let

$$J(F)(f) = \sum_{i=1}^n F(E_i) x_i.$$

Then

$$|J(F)f| \le \sum_{i=1}^{n} \| F(E_i) \| \, \| x_i \|$$
$$\le \| F \|_\infty \sum_{i=1}^{n} \| x_i \| P(E_i)$$
$$= \| F \|_\infty \| f \|_1$$

so that $J(F)$ can be extended by continuity to an element of $(L_1(X))^*$, which we again denote by $J(F)$. The inequality implies that $\| J(F) \| \le \| F \|_\infty$, so that J is a norm-decreasing linear map of $V_\infty(X^*)$ into $(L_1(X))^*$.

Conversely, if $\phi \in (L_1(X))^*$, let

$$K(\phi)(E)(x) = \phi(x\chi_E)$$

Then

$$|K(\phi)(E)(x)| \le \| \phi \| . \| x \| .P(E)$$

so that

$$K(\phi)(E) \in X^*, \text{ and}$$
$$\| K(\phi)(E) \| \le P(E) \| \phi \| .$$

Clearly $K(\phi): \Sigma \longrightarrow X^*$ is finitely additive, and

$$\| K(\phi) \|_\infty \le \| \phi \|$$

so that

$$K(\phi) \in V_\infty(X^*).$$

Finally $JK(\phi) = \phi$ and $KJ(F) = F$, from which we get the result.

Notice that this result is very elementary: we need only "integrate" simple functions. In order to obtain a satisfactory representation of $(L_1(X))^*$, however, we need to have a more detailed representation of $V_\infty(X^*)$.

In order to do this, we need to have a clear understanding of scalar measurability. Suppose that (Ω, Σ, P) is a probability space, that (E,G) is a separating dual pair of vector spaces, that we give E the weak topology $\sigma(E,G)$ and that C is a closed convex subset of E. Suppose that f maps Ω into C. We say that f is <u>scalarly measurable</u> if the map $\omega \longrightarrow \langle f(\omega), g\rangle$ is measurable for each g in G. We say that f is <u>weakly Borel measurable</u> if $f^{-1}(B) \in \Sigma$ for each $\sigma(E,G)$ Borel set B of C. We say that f is a <u>weak random variable</u> if f is weakly Borel measurable and $f(P)$ is a <u>regular</u> Borel measure on C.

In general, each definition is stronger than its predecessor. If however C is a Polish space under $\sigma(E,G)$ (and this will be the case later when we take $C = P(T)$, $E = M(T)$ and $G = C_b(T)$) then any scalarly measurable function is a weak random variable.

In general, too, we need to be careful with statements involving 'almost everywhere'. Let us restrict our attention to the case where G is a Banach space X and $C = E = X^*$, with the w*-topology $\sigma(X^*,X)$. Let

$$\mathcal{L}_\infty^*(X^*) = \{f: \Omega \longrightarrow X^*: f \text{ is scalarly measurable and } \|f\|_\infty^* = \sup_\omega \|f(\omega)\| < \infty\}$$

$\mathcal{L}_\infty^*(X^*)$ is a Banach space of functions. Let

$$N = \{f \in \mathcal{L}_\infty^*(X^*): \text{the map } \omega \longrightarrow f(\omega)(x) \text{ vanishes almost everywhere, for each } x \text{ in } X\}.$$

Of course, the fact that X need not be separable means that we cannot conclude that if $f \in N$, then $f = 0$ almost everywhere. N is however a closed linear subspace of $\mathcal{L}_\infty^*(X^*)$: we set

$$L_\infty^*(X^*) = \mathcal{L}_\infty^*(X^*)/N,$$

and give $L_\infty^*(X^*)$ the quotient norm, which we again denote by $\|\ \|_\infty^*$.

<u>Theorem 14</u>. <u>If</u> (Ω,Σ,P) <u>is a complete probability space</u> $(L_\infty^*(X^*),\|\ \|_\infty^*)$ <u>is naturally isometrically isomorphic to</u> $(V_\infty(X^*),\|\ \|)$.

<u>Proof</u>. Suppose first that $f \in L_\infty^*(X^*)$. Let $x \in X$, and let ϕ be a representative for f. Then $\phi(\cdot)(x) \in \mathcal{L}_\infty$, and

$$\|\phi(\cdot)(x)\|_\infty < \|\phi\|_\infty^* \|x\| .$$

Let $(Af)(x) = \phi(\cdot)(x) \in L_\infty$. This clearly does not depend on the choice of ϕ, and

$$\|(Af)(x)\|_\infty \le \|f\|_\infty^* \|x\| .$$

Let $(Jf)(E)(x) = \int_E (Af)(x)dP.$

Then $(Jf)(E) \in X^*$, and $\|J(F)(E)\| \le \|f\|_\infty^* P(E)$. Jf is clearly finitely additive, and $\|Jf\|_\infty \le \|f\|_\infty^*$.

In order to go in the opposite direction we need to use the linear lifting theorem [19]. Recall that, if (Ω,Σ,P) is a complete probability space, there exists a linear lifting: that is, a linear map ρ from $L_\infty(\Omega,\sum,P)$ to $\mathcal{L}_\infty(\Omega,\sum,P)$ such that

(i) ρ is an isometry

(ii) ρ is positive and $\rho(\tilde{1}) = 1$

and (iii) $\widetilde{\rho(x)} = x$ for all $x \in L_\infty$.

Now if $F \in V_\infty(X^*)$ and $x \in X$, the map $E \longrightarrow F(E)(x)$ is a measure $\mu(F,x)$ on $\sum$. $\mu(F,x)$ is absolutely continuous with respect to P, and

$$|\mu(F,x)(E)| \le \|F\|_\infty \|x\| P(E),$$

so that $\mu(F,x)$ has a Radon-Nikodym derivative $f(F,x)$ in L_∞, with $\|f(F,x)\|_\infty \le \|F\|_\infty \|x\|$. Let us set

$$K(F)(\omega)(x) = \rho(f(F,x))(\omega)$$

Then

$$|K(F)(\omega)(x)| \le \|F\|_\infty \|x\|$$

and $K(F)(\omega)(x)$ is linear in X, so that $K(F)(\omega) \in X^*$ and $\|K(F)(\omega)\| \le \|F\|_\infty$. Thus

$$K(F) \in \mathcal{L}^*_\infty(X^*), \text{ and } \|K(F)\|^*_\infty \le \|\ \|_\infty.$$

Finally, let us set $\tilde{K}(F) = K(F) \in L^*_\infty(X^*)$. It is readily verified that $\tilde{K}J(f) = f$ and $J\tilde{K}(F) = F$, and so the proof is complete.

Combining Theorems 13 and 14, we see that the dual of $L_1(X)$ is naturally isometrically isomorphic to $L^*_\infty(X)$. With the usual conventions of identifying functions almost everywhere, if $f = \sum_{i=1}^n x_i \chi_{E_i}$ is a <u>simple</u> function in $L_1(X)$ and $g \in L^*_\infty(X^*)$, the pairing given by

$$\langle f,g\rangle = \sum_{i=1}^n \int_{E_i} g(w)(x_i)P(dw).$$

We can however extend the duality to the whole of $L_1(X)$. Suppose that $f \in L_1(X)$ and that $g \in L^*_\infty(X^*)$. For each n there are a simple f_n in $L_1(X)$ and A_n in $\sum$ such that

$$\|f_n(\omega) - f(\omega)\| < \frac{1}{2^n} \text{ for } \omega \notin A_n$$

$$f_n(\omega) = 0 \text{ on } A_n \text{ and } P(A_n) < 1/2^n.$$

Then $f_n \longrightarrow f$ in $L_1(X)$, so that $\langle f_n,g\rangle \longrightarrow \langle f,g\rangle$. But $\langle f_n,g\rangle = \int_\Omega g(\omega)f_n(\omega)P(d\omega)$ and

$$|g(\omega)f_n(\omega)| \le (\|f(\omega)\| + 1)\|g\|^*_\infty$$

so that as $g(\omega)f_n(\omega) \longrightarrow g(\omega)f(\omega)$ almost surely, $g(\omega)f(\omega)$ is measurable, and by dominated convergence

$$\int_\Omega g(\omega)f_n(\omega)P(d\omega) \longrightarrow \int_\Omega g(\omega)f(\omega)P(d\omega)$$

Consequently

$$\langle f,g\rangle = \int_\Omega g(\omega)f(\omega)P(d\omega).$$

In particular, considering $P(T)$ as a subset of $M(\beta T)$, we see that we can identify the set $\pi(T)$ with a subset of $(L_1(C_b(T)))^*$, and that the ωm-topology is that induced by the weak*-topology

$$\sigma(L_1(C_b(T)))^*, L_1(C_b(T))).$$

We are now in a position to characterise compactness in the ωm-topology.

<u>Definition</u>. A subset A of $\pi(T)$ is <u>uniformly tight</u> if $\lim_{n\to\infty} E_\omega(\xi(\omega)\ K_n) = 1$ uniformly on A.

Thus A is uniformly tight if and only if $\{\xi_\Omega : \xi \in A\}$ is uniformly tight in $P(R)$.

<u>Theorem 15</u>. <u>A subset</u> A <u>of</u> $\pi(T)$ <u>is</u> ωm-<u>relatively compact if and only if it is uniformly tight.</u>

<u>Proof</u>. The map $E: \xi \longrightarrow \xi_\Omega$ from $\pi(T),\omega m)$ to $(P(T)$, narrow) is continuous, so that the condition is necessary.

Conversely, suppose that A is uniformly tight. $\bar{A}$, the closure in $L_\infty^*(M(\beta T))$, is $\sigma(L_\infty^*(M(\beta T)), L_1(C_b(T)))$ compact. Let $\theta \in \bar{A}$. As $\|\xi\|_\infty^* = 1$ for all $\xi \in A$, $\|\theta\|_\infty^* \le 1$. Also, if $\xi \in A$,

$$E_\omega \int 1\xi(\omega)dt) = 1$$

so that

$$E_\omega \int 1\theta(\omega,dt) = 1, \text{ and } \theta(\omega) \in P(\beta T), \text{ for almost all } \omega.$$

Given $\varepsilon > 0$, there exists n_0 such that

$$E_\omega \int f_n(t)\xi(\omega,dt) < \varepsilon^2 \text{ for } n \ge n_0,\ \xi \in A.$$

Thus

$$E_\omega \int f_n(t)\theta(\omega,dt) < \varepsilon^2 \text{ for } n \ge n_0$$

and so

$$P(\theta(\omega,K_{n+1}) < 1-\varepsilon) \leq P(\int f_n(t)\theta(\omega,dt) > \varepsilon)$$
$$< \varepsilon.$$

Thus $\theta(\omega) \in P(T)$ for almost all ω, and $\theta \in \pi(T)$.

Corollary 1. If B is a subset of $L_o(T)$, $i(B)$ is ωm-relatively compact if and only if

$$P(X \in K_n) \longrightarrow 1 \text{ uniformly on } B, \text{ as } n \longrightarrow \infty.$$

Now suppose that (X,d) is a complete separable metric space. Let (A_n) be a fundamental sequence of bounded sets of (X,d). Let $L_o(X)$ denote the space of Borel measurable mappings of $(\Omega,\sum,P)$ into X. We say that a subset B of $L_o(X)$ is bounded if

$$P(h \in A_n) \longrightarrow 1 \text{ uniformly on } B, \text{ as } n \longrightarrow \infty.$$

Let (T,ρ) be the space of types on X, equipped with a suitable metric ρ. The mapping j induces a homeomorphic mapping $\tilde{j}$ from $L_o(X)$ into $L_o(T)$. Combining Corollary 1, Theorem 12 and Theorem 1, we have the following.

Corollary 2. The mapping $h = i\tilde{j}$ is a homeomorphic mapping of $L_o(X)$ into $(\pi(T),\omega m)$. If B is a subset of $L_o(X)$, $h(B)$ is ωm-relatively compact if and only if B is bounded in $L_o(X)$.

10. Operations on measures and random measures.

Suppose now that (X,d) is a separable Banach space, and that (T,ρ) is the space of types on (X,d), equipped with a suitable metric. If $\mu \in P(T)$ and α is a non-zero scalar, we define $s_\alpha(\mu)(E)=\mu(D_{\alpha^{-1}}(E))$, for each Borel set E in T. $s_\alpha(\mu)$ is again an element of $P(T)$. If $f \in C_b(T)$,

$$\int f(t)(s_\alpha(\mu))(dt) = \int f(D_\alpha t)\mu(dt)$$

and so s_α is a homeomorphism of $\pi(T)$. In the case $\alpha = 0$, we set $s_o(\mu) = \delta_o$. Similarly if $\xi \in \pi(T)$, we define $s_\alpha(\xi)$ by

$$s_\alpha(\xi)(\omega) = s_\alpha(\xi(\omega))$$

for ω in Ω. Again if $\alpha \neq 0$, s_α is a homeomorphism of $\pi(T)$ with the ωm-topology. If $s_{-1}(\xi) = \xi$, we shall say that ξ is a symmetric measure.

We can also translate measures and random measures. Suppose that $f \in C_b(T)$, that $\mu \in P(T)$ and that $x \in X$. We define

$$\tau_x(f)(t) = f(\tau_x(t))$$
$$\tau_x(\mu)(E) = \mu(\tau_{-x}(E)).$$

Then τ_x: $C_b(T) \longrightarrow C_b(T)$ is an isometry, $\tau_x(\mu) \in P(T)$ and

$$\int f(t)(\tau_x(\mu))(dt) = \int (\tau_x f)(t)\mu(dt).$$

τ_x: $P(T) \longrightarrow P(T)$ is a homeomorphism.

Proposition 5. The map $(x,\mu) \longrightarrow \tau_x(\mu)$: $X \times P(T) \longrightarrow P(T)$ is jointly continuous.

Proof. Suppose that $x_n \longrightarrow x_o$ in X and that $\mu_n \longrightarrow \mu_o$ in P(T). Let f be a bounded uniformly continuous function on T. Then $\tau_{x_n} f \longrightarrow \tau_{x_o} f$ uniformly on T, and so

$$\int (\tau_{x_n} f)d\mu_n - \int (\tau_{x_o} f)d\mu_n \longrightarrow 0 \text{ as } n \longrightarrow \infty.$$

But $\int (\tau_{x_o} f)d\mu_n \longrightarrow \int (\tau_{x_o} f)d\mu_o$, so that

$$\int (\tau_{x_n} f)d\mu_n \longrightarrow \int (\tau_{x_o} f)d\mu_o:$$

that is, $\tau_{x_n}\mu_n \longrightarrow \tau_{x_o}\mu_o$.

If now $\xi \in \pi(T)$ and $x \in L_o(X)$, we define $\tau_x(\xi)(\omega) = \tau_{x(\omega)}(\xi(\omega))$. By expressing x and ξ as pointwise limits of simple functions, and using Proposition 5, it follows that $\tau_x(\xi) \in \pi(T)$.

Proposition 6. The map $(x,\xi) \longrightarrow \tau_x(\xi)$: $L_o(X) \times \pi(T) \longrightarrow \pi(T)$ is jointly continuous.

Proof. Suppose that $x_n \longrightarrow x_o$ in $L_o(X)$ and $\xi_n \longrightarrow \xi_o$ in $\pi(T)$. Since it is enough to show that a subsequence of $(\tau_{x_n}(\xi_n))$ converges to (x_o,ξ_o) we can suppose that $x_n \longrightarrow x_o$ almost surely. Suppose that f is a bounded uniformly continuous function on (T,ρ), with $\|f\| \le 1$, that $E \in \Sigma$ and that $\varepsilon > 0$. By Egoroff's theorem, there exists $A \subseteq E$ with $P(A) < \varepsilon/2$ such that $x_n(\omega) \longrightarrow x_o(\omega)$ uniformly on $E \backslash A$. Since f is uniformly continuous,

$$f(\tau_{x_n(\omega)}(t)) \longrightarrow f(\tau_{x_o(\omega)}(t))$$

uniformly on $T\times E\backslash A$. Thus

$$\int_T f\tau_{x_n(\omega)}(t)\xi_n(\omega,dt) - \int_T f_{\tau_o(\omega)}(t)\xi_n(\omega,dt) \longrightarrow 0$$

uniformly on $E\backslash A$. But $f_{\tau_o(\omega)}\chi_{E\backslash A}(\omega) \in L_1(C_b(T))$, and so

$$\int_{E\backslash A}\left(\int_T f_{\tau_o(\omega)}(t)\xi_n(\omega,dt)\right)P(d\omega) \longrightarrow \int_{E\backslash A}\left(\int_T f_{\tau_o(\omega)}(t)\xi_o(\omega,dt)P(d\omega)\right)$$

Consequently

$$\int_{E\backslash A}\left(\int_T f(t)\tau_{x_n(\omega)}\xi_n(\omega,dt)\right)P(d\omega) \longrightarrow$$
$$\int_{E\backslash A}\left(\int_T f(t)\tau_{x_o(\omega)}\xi_o(\omega,dt)\right)P(d\omega).$$

Since
$$\left|\int_A\left(\int_T f(t)\tau_{n(\omega)}\xi_n(\omega,dt)\right)P(d\omega) - \int_A\left(\int_T f(t)\tau_{x_o(\omega)}\xi_o(\omega,dt)P(d\omega)\right|\right. < \varepsilon$$

the result is established.

In order to consider convolution, we must suppose that X is stable, so that convolution is defined in T.

First suppose that μ and ν are in $P(T)$. If $f \in C_b(T)$, the function $(s,t) \longrightarrow f(s \star t)$ is measurable on $T\times T$ (by Proposition 2, for example): let

$$\phi(f) = \int f(s \star t)\mu(ds)\nu(dt).$$

Then ϕ is a positive linear functional on $C_b(T)$ and $\phi(1) = 1$, so that there exists a regular Borel probability measure π on the Stone-Čech compactification βT of T such that $\phi(f) = \int_{\beta T}\tilde{f}(u)\pi(du)$, where $\tilde{f}$ is the continuous extension of f to βT.

Now μ and ν are regular, and so given $0 < \varepsilon < 1$ there exists n such that, setting $K_n = \{t: t(0) \le n\}$,

$$\mu(K_n) \ge 1-\varepsilon \text{ and } \nu(K_n) \ge 1-\varepsilon.$$

Now $(s \star t)(0) \le s(0)+t(0)$, so that $K_n \star K_n \subseteq K_{2n}$, and so

$$\pi(K_{2n}) \ge (1-\varepsilon)^2.$$

This implies that $\pi(T) = 1$, and so we can consider π as an element of $P(T)$. We denote π by $\mu \star \nu$.

Proposition 7. *The mapping* $(\mu,\nu) \longrightarrow \mu \star \nu\colon P(T)\times P(T) \longrightarrow P(T)$ *is separately continuous. It is jointly continuous if and only if* $(X,\|\ \|)$ *is finite-dimensional.*

Proof. Since the mapping $(s,t) \longrightarrow s \star t$ is separately continuous, if $f \in C_b(T)$ and $t \in T$, the function F_t defined by

$$F_t(s) = f(s \star t)$$

is in $C_b(T)$. Further if $t_n \longrightarrow t$, $F_{t_n}(s) \longrightarrow F_t(s)$ for each s. Then since $\|F_t\| \leq \|f\|$, for each t,

$$\int f(s \star t_n)\mu(ds) \longrightarrow \int f(s \star t)d\mu(s)$$

for each $\mu \in P(T)$, by the theorem of bounded convergence. Thus if we set $G_\mu(t) = \int f(s \star t)\mu(ds)$, $G_\mu \in C_b(T)$. Consequently, if $\nu_n \longrightarrow \nu$ in $P(T)$

$$\int fd(\mu \quad \nu_n) = \int G_\mu(t)\nu_n(dt) \longrightarrow \int G_\mu(t)\nu(dt)$$

$$= \int fd(\mu \quad \nu).$$

Thus $\mu \star \nu_n \longrightarrow \mu \star \nu$. Since $\mu \star \nu = \nu \star \mu$, convolution is separately continuous.

If $(X,\|\ \|)$ is infinite dimensional, the mapping $(s,t) \longrightarrow s \star t$ is not jointly continuous. Since the mapping $s \longrightarrow \delta_s$ is a homeomorphism of T into $P(T)$, and since $\delta_{s \star t} = \delta_s \star \delta_t$, the mapping $(\mu,\nu) \longrightarrow \mu \star \nu$ cannot be jointly continuous. On the other hand, if $(X,\|\ \|)$ is finite-dimensional, $X = T$, and it is well-known that convolution of measures is jointly continuous.

Next suppose that ξ and η are in $\pi(T)$. We define $\xi \star \eta$ in the obvious way:

$$(\xi \star \eta)(\omega) = \xi(\omega) \star \eta(\omega).$$

Proposition 8. *The mapping* $(\xi,\eta) \longrightarrow \xi \star \eta\colon \pi(T)\times\pi(T) \longrightarrow \pi(T)$ *is separately continuous.*

Proof. Suppose that $f \in C_b(T)$, and that $\xi \in \pi(T)$. If we set

$$G_\xi(\omega)(t) = \int_T f(s \star t)\xi(\omega,ds)$$

then as in Proposition 5, $G_\xi(\omega) \in C_b(T)$, and $\|G_\xi(\omega)\| \le \|f\|$. Consequently $G_\xi \in L_\infty(C_b(T))$.

Now suppose that $\eta_n \longrightarrow \eta$ in $\Pi(T)$. If $g \in L_1(\Omega)$, the function $G_\xi g$ is in $L_1(C_b(T))$. Identifying $\pi(T)$ with a subset of $L_1(C_b(T))^*$, as in §9,

$$\int_\Omega g(\omega)\left(\int_T G_\xi(\omega)(t)\eta_n(\omega,dt)\right)P(d\omega) \longrightarrow \int_\Omega g(\omega)\left(\int_T G_\xi(\omega)(t)\eta(\omega,dt)P(d\omega)\right).$$

But this simply states that

$$\int_\Omega g(\omega)(T_{\xi\star\eta_n}(f))(\omega)P(d\omega) \longrightarrow \int_\Omega g(\omega)(T_{\xi\star\eta}(f)(\omega)P(d\omega).$$

Since this holds for each f in $C_b(T)$ and each g in $L_1(\Omega)$,

$$\xi \star \eta_n \longrightarrow \xi \star \eta.$$

Even in the case where $X = T = R$, the real line, convolution of random measures is not jointly continuous. For example, let (X_j) be a sequence of non-trivial symmetric independent identically distributed random variables in $(\Omega,\sum,P)$, with law μ. Then if $f \in C_b(R)$,

$$T_{i(X_j)}(f) = f(X_j)$$

so that if $g \in L_1(\Omega)$,

$$\int_\Omega T_{i(X_j)}(f)g\,dP \longrightarrow E(f(X_1))E(g)$$

On the other hand, $T_\mu(f) = E(f(X_1)).1$, so that

$$\int_\Omega T_\mu(f)g\,dP = E(f(X_1))E(g).$$

Thus $i(X_j) \longrightarrow \mu$ in the wm-topology. By symmetry, $i(-X_j) \longrightarrow \mu$. Now $i(X_j) \star i(-X_j) = \delta_o$, and $\mu \star \mu \neq \delta_o$, since the sequence is non-trivial, and so convolution is not jointly continuous.

11. The stability of Orlicz function spaces.

We now consider the stability of Orlicz function spaces. We begin by recalling some definitions. An Orlicz function ϕ is a non-negative convex function on the positive real axis which vanishes only at the origin. We shall suppose that ϕ satisfies the Δ_2-condition:

there exists a constant $k > 1$ such that

$$\phi(2t) \le k\phi(t) \quad \text{for all } t.$$

Recall that this, and the convexity of ϕ, imply that

$$\phi(a)-\phi(a-b) \le \phi(a+b)-\phi(a) \le (k-1)\frac{b}{a}\,\phi(a) \qquad (*)$$

for $0 < b \le a$.

Now suppose that (Ω,Σ,P) is the completion of a countably generated probability space and that $(X,\|\ \|)$ is a separable Banach space. Then we define the Orlicz space $L_\phi(X)$ to be

$$\{f \in L_0(X): \int \phi(\|f\|)\,dP < \infty\}.$$

If $B_\phi = \{f \in L_0(X): \int \phi(\|f\|)\,dP \le 1\}$,

B_ϕ is an absolutely convex absorbing subset of $L_\phi(X)$, and its gauge is a norm $\|\ \|_\phi$ on $L_\phi(X)$, under which $L_\phi(X)$ is a Banach space, with closed unit ball $B_\phi(X)$.

It is well-known ([12] Theorem 9.4) that, since ϕ satisfies the Δ_2-condition, $x_n \longrightarrow x$ in $L_\phi(X)$ if and only if $E(\phi\|x_n-x\|)) \longrightarrow 0$. The following result is of fundamental importance.

<u>Theorem 16</u>. $x_n \longrightarrow x$ <u>in</u> $L_\phi(X)$ <u>if and only if</u> $x_n \longrightarrow x$ <u>in probability</u> <u>and</u> $E(\phi(\|x_n\|)) \longrightarrow E(\phi(\|x\|)) \longrightarrow E(\phi(\|x\|))$.

<u>Proof</u>. First let us introduce some terminology that we shall use throughout this section. For each j let us set

$$\phi_j(t) = \phi(t) \text{ for } 0 \le t \le j;$$

$$\phi_j(t) = (n+1-t)\phi(t) \text{ for } j \le t \le j+1;$$

$$\phi_j(t) = 0 \text{ for } t \ge j+1.$$

Suppose that $x_n \longrightarrow x$ in $L_\phi(X)$. Then certainly $x_n \longrightarrow x$ in probability. Suppose that $\varepsilon > 0$. Since $\phi(\|x\|)$ is integrable, there exists j such that

$$\int_{\|x\| \ge j} \phi(\|x\|)\,dP < \varepsilon$$

and so

$$E(\phi(\|x\|) - \phi_{2j}(\|x\|) < \varepsilon.$$

Now $E(\phi_{2j}(\|x_n\|)) \longrightarrow E(\phi_{2j}(\|x\|))$ and $E(\phi(\|x_n-x\|)) \longrightarrow 0$ as $n \longrightarrow \infty$, and so there exists n_0 such that

$$|E(\phi(\|x\|)-\phi_{2j}(\|x_n\|))| < \varepsilon \text{ and } E(\phi(\|x_n-x\|)) < \varepsilon$$

for $n \geq n_o$.

Now if $\|x_n\| \geq 2\|x\|$, $\|x_n\| \leq 2\|x_n-x\|$, so that $\phi(\|x_n\|) \leq k\phi(\|x_n-x\|)$, while if $2j \leq \|x_n\| \leq 2\|x\|$, $\|x\| \geq j$ and $\phi(\|x_n\|) \leq k\phi\ \|x\|)$. Thus

$$0 \leq E(\phi(\|x_n\|)-\phi_{2j}(\|x_n\|)) \leq \int_{\|x_n\| \geq 2j} \phi\ \|x_n\|)\,dP$$

$$\leq kE(\phi\ \|x_n-x\|)) + k\int_{\|x\| \geq j} \phi(\|x\|)\,dP \leq 2k\varepsilon$$

for $n \geq n_o$, and so $|E\phi(\|x\| - E\phi(\|x_n\|)| < (2k+1)\varepsilon$ for $n \geq n_o$.

Conversely suppose that $x_n \longrightarrow x$ in probability and that $E\phi(\|x_n\|) \longrightarrow E\phi(\|x\|)$. In order to show that $x_n \longrightarrow x$ in $L_\phi(X)$ it is sufficient to prove that there is a subsequence which converges to x, and so we may suppose that $x_n \longrightarrow x$ almost surely. Suppose that $\varepsilon > 0$. Since $\phi(\|x\|)$ is uniformly integrable, there exists $\delta > 0$ such that if $P(A) < \delta$ then $\int_A \phi(\|x\|)\,dP < \varepsilon$. By Egoroff's theorem, there exists A, with $P(A) < \delta$, such that

$$\phi(\|x_n-x\|) \longrightarrow 0 \text{ and } \phi(\|x_n\|) \longrightarrow \phi(\|x\|)$$

uniformly on $\Omega\backslash A$. In particular,

$$\int_{\Omega\backslash A} \phi(\|x_n\|)\,dP \longrightarrow \int_{\Omega\backslash A} \phi(\|x\|)\,dP,$$

and so

$$\int_A \phi(\|x_n\|)\,dP \longrightarrow \int_A \phi(\|x\|)\,dP.$$

Thus there exists n_o such that

$$\int_A \phi(\|x_n\|)\,dP < \varepsilon \text{ and } |\phi(\|x_n(\omega)\|) - \phi(\|x(\omega)\|)| < \varepsilon$$

for ω in $\Omega\backslash A$, for $n \geq n_o$. As

$$\int_A \phi(\|x_n-x\|)\,dP \leq k\int_A (\phi\ \|x_n\|) + \phi(\|x\|))\,dP,$$

$E(\phi(\|x_n-x\|)) < (2k+1)\varepsilon$ for $n \geq n_o$, and so $x_n \longrightarrow x$.

Now let T be the space of types on $(X, \|\ \|)$ and let $\pi(T)$ denote the space of random measures on T. If $\xi \in \pi(T)$, $x \in L_o(X)$ and ψ is a continuous non-negative function on $[0,\infty)$, we set

$$M_\psi^{(x)}(\xi)(\omega) = \int_T \psi(t(x(\omega)))\xi(\omega,dt).$$

As $M_\psi^{(x)}(\xi)(\omega) = \lim M_{\psi\wedge n}^{(x)}(\xi)(\omega)$, $M_\psi^{(x)}(\xi)$ is a measurable $[0,\infty]$ valued function on Ω. We set $I_\psi^{(x)}(\xi) = E(M_\psi^{(x)}(\xi))$. In the case where $x = 0$, we write $M_\psi\ (\xi)$ for $M_\psi^{(x)}(\xi)$, and $I_\psi(\xi)$ for $I_\psi^{(x)}(\xi)$. We set $\pi_\psi(T) = \{\xi\colon I_\psi(\xi) < \infty\}$. Note that if $f \in L_0(X)$ then $I_\psi(h(f)) = E(\psi(\|f\|)$.

Now let us return to our Orlicz function ϕ. If $\xi \in \pi_\phi(T)$, $I_\phi(\xi) = \lim_n I_{\phi_n}(\xi)$. We shall say that a subset A of $\pi_\phi(T)$ is ϕ-uniformly integrable if $I_{\phi_n}(\xi) \to I_\phi(\xi)$ uniformly on A. Suppose that $C \subseteq L_\phi$. Then $h(C)$ is ϕ-uniformly integrable if and only if $\{\phi(\|f\|) : f \in C\}$ is a uniformly integrable subset of L_1.

Proposition 9. If $\xi_k \in \pi_\phi(T)$ for $k = 1,2,\ldots$ and $\xi_k \to \xi$ in the wm-topology, then

$$I_\phi(\xi) \le \liminf I_\phi(\xi_k).$$

The sequence (ξ_k) is ϕ-uniformly integrable if and only if $I_\phi(\xi_k) \to I_\phi(\xi)$.

Proof. By monotone convergence and ω^m-convergence,

$$I_\phi(\xi) = \lim_n I_{\phi_n}(\xi) = \lim_n \lim_k I_{\phi_n}(\xi_k).$$

As $I_{\phi_n}(\xi_k) \le I_\phi(\xi)$,

$$I_\phi(\xi) \le \liminf I_\phi(\xi_k).$$

The assertion about ϕ-uniform integrability follows from the definition and the equation above.

Note that in general if $I_\phi(\xi_k) \to \ell$ then

$$\ell - I_\phi(\xi) = \lim_n \lim_k (I_\phi(\xi_k) - I_{\phi_n}(\xi_k)).$$

In order to deal with stability, we need two results concerning the truncation of functions in $L_\phi(X)$.

Proposition 10. Suppose that (y_j) is a bounded sequence in $L_\phi(X)$, that $N_j \to \infty$, and that

$$\int_{\|y_j\| \geq N_j} \phi(\|y_j\|) \longrightarrow \beta \text{ as } j \longrightarrow \infty.$$

Let $z_j = y_j \chi \|y_j\| < N_j$. If $x \in L_\phi(X)$ and $E(\phi \|x-y_j\|)) \longrightarrow \alpha$, then

$$E(\phi \|x-z_j\|)) \longrightarrow \alpha-\beta.$$

Proof $\quad |E(\phi \|x-y_j\|)) - E(\phi \|x-z_j\|)) -\beta| =$

$$= |\int_{\|y_j\| \geq N_j} (\phi(\|x-y_j\| -\phi(\|x\|)) dP-\beta|$$

$$\leq |\int_{\|y_j\| \geq N_j} (\phi(\|x-y_j\| -\phi(\|y_j\|)) dP| + \int_{\|y_j\| \geq N_j} \phi(\|x\|) dP +$$

$$+ |\int_{\|y_j\| \geq N_j} \phi \|y_j\| dP-\beta|.$$

Now

$$\phi(N_j)P(\|y_j\| \geq N_j) \leq \sup_k E(\phi(\|y_k\|)) = L, \text{ say,}$$

so that $P(\|y_j\| \geq N_j) \longrightarrow 0$ and $\int_{\|y_j\| \geq N_j} \phi(\|x\|) dP \longrightarrow 0$.

Thus it is sufficient to show that

$$\int_{\|y_j\| \geq N_j} (\phi(\|x-y_j\| -\phi(\|y_j\|)) dP \longrightarrow 0.$$

Suppose that $\varepsilon > 0$. Choose r so that $2^{-r}(k-1)L < \varepsilon$ (where k is the constant of the Δ_2-condition). If $\|y_j\| \leq 2^r\|x\|$,

$$\phi(\|x-y_j\| \leq k^{r+1}\phi(\|x\|))$$

and $\quad \phi(\|y_j\|) \leq k^r\phi)\|x\|)$

whiile if $\|y_j\| > 2^r\|x\|$

$$|\phi(\|x-y_j\|)-\phi(\|y_j\|) \leq 2^{-r}(k-1)\phi(\|y_j\|)$$

by the inequality(*) at the beginning of the section. Consequently,

$$|\int_{\|y_j\| \geq N_j} (\phi(\|x-y_j\|) -\phi(\|y_j\|)) dP \leq$$

$$(k^{r+1}+k^r)\int_{\|y_j\| \geq N_j} \phi(\|x\|) dP + 2^{-r}(k-1)L < \varepsilon$$

for large enough j.

Proposition 11. Suppose that (X_n) is a bounded sequence in $L_\phi(X)$. There exists a subsequence $(y_j) = (x_{n_j})$ such that, setting $z_j = y_j \chi_{\|y_j\| < j}$, the sequence $(\phi(\|z_j\|))$ is uniformly integrable.

Proof. We can find a subsequence (y_j), such that for each positive integer m there exists β_m such that

$$\int_{\|y_j\| \geq m} \phi \; \|y_j\|)\, dP \longrightarrow \beta_m,$$

and by extracting a further subsequence if necessary, we can suppose that

$$\left|\int_{\|y_j\| \geq m} \phi(\|y_j\|)\, dP - \beta_m\right| < \frac{1}{m}$$

for $j \geq m$. The sequence (β_m) is decreasing, so that $\beta_m - \beta_j \longrightarrow 0$ as $m, j \longrightarrow \infty$. Now if $j > m$,

$$\begin{aligned}\int_{\|z_j\| \geq m} \phi(\|z_j\|)\, dP &= \int_{\|y_j\| \geq m} (\phi\|y_j\|) dP - \int_{\|y_j\| \geq j} (\phi \; \|y_j\|) dP \\ &\leq \left|\int_{\|y_j\| \geq m} (\phi(\|y_j\|) dP - \beta_m\right| + (\beta_m - \beta_j) + \\ &\quad + \left|\beta_j - \int_{\|y_j\| \geq j} (\phi(\|y_j\|)\, dP\right| \\ &\leq \frac{1}{m} + \frac{1}{j} + (\beta_m - \beta_j)\end{aligned}$$

while $\displaystyle\int_{\|z_j\| \geq m} \phi(\|z_j\|)\, dP = 0,$

so that

$$\sup_j \int_{\|z_j\| \geq m} \phi(\|z_j\|) \longrightarrow 0 \text{ as } m \longrightarrow \infty.$$

We shall also need the following lemma:

Lemma 7. Suppose that A and B are ϕ-uniformly integrable subsets of $\pi(T)$, and that α is a scalar. Then $A \star B = \{\xi \star \eta : \xi \in A, \eta \in B\}$ and $s_\alpha(A) = \{s_\alpha(\xi) : \xi \in A\}$ are also ϕ-uniformly integrable.

Proof. Suppose that $\xi \in A$, $\eta \in B$ and $K > 0$. Then, bearing in mind that $(s \star t)(0) \leq s(0) + t(0)$,

$$\int_\Omega \left(\int_{t(0) \geq K} \phi(t(0))\,(\xi \star \eta)(\omega, dt)\right) P(d\omega) =$$

$$= \int_\Omega \left(\int_{(s \star t)(0) \geq K} \phi((s \star t)(0)\,\xi(\omega, ds)\,\eta(\omega, dt)\right) P(d\omega) \leq$$

$$\leq \int_\Omega \Big(\int_{\substack{t(O)\leq s(O)\\ s(O)\geq K/2}} \phi(2s(O))\xi(\omega,ds)\eta(\omega,dt)\Big)P(d\omega) \; +$$

$$+ \int_\Omega \Big(\int_{\substack{s(O)\leq t(O)\\ t(O)\geq K(2)}} \phi(2t(O))\xi(\omega,ds)\eta(\omega,dt)\Big)P(d\omega) \;\leq$$

$$\leq k\int_\Omega \Big(\int_{s(O)\geq K/2} \phi(s(O))\xi(\omega,ds)\Big)P(d\omega) \; +$$

$$+ k\int_\Omega \Big(\int_{t(O)\geq k/2} \phi(t(O)\eta(\omega,dt)\Big)P(d\omega)$$

which implies that $A \star B$ is ϕ-uniformly integrable. A similar but easier argument shows that $s_\alpha(A)$ is ϕ-uniformly integrable.

We can now establish stability.

<u>Theorem 16</u>. <u>Suppose that</u> ϕ <u>is an Orlicz function which satisfies the</u> Δ_2<u>-condition, that</u> $(\Omega,\sum,P)$ <u>is a separable probability space, and that</u> X <u>is a stable Banach space. Then</u> $L_\phi(X)$ <u>is stable</u>.

<u>Proof</u>. We can clearly suppose that $(\Omega,\sum,P)$ is complete. If $L_\phi(X)$ is not stable, there exist approximating sequences (x_m) and (y_n) such that $\lim_m \lim_n \|x_m+y_n\|_\phi$ and $\lim_n \lim_m \|x_m+y_n\|_\phi$ exist and

$$\lim_m \lim_n \|x_m+y_n\|_\phi < 1 < \lim_n \lim_m \|x_m+y_n\|.$$

By extracting subsequences if necessary, we can suppose that the double limits

$$\lim_m \lim_n E(\phi(\|x_m+y_n\|)) \text{ and } \lim_n \lim_m E(\phi(\|x_m+y_n\|))$$

both exist and

$$\lim_m \lim_n E(\phi(\|x_m+y_n\|)) < 1 < \lim_n \lim_m E(\phi(\|x_m+y_n\|))$$

and we can also suppose that, setting

$$x_m' = x_m\chi_{(\|x_m\|<m)},\; y_n' = y_n\chi_{(\|y_n\|<n)},$$

the sequences $(h(x_m'))$ and $(h(y_n'))$ are ϕ-uniformly integrable.

Using Corollary 2 to Theorem 15, we can find subsequences (x_{m_j}) and (y_{n_k}), constants β and γ, and ξ and η in $\pi_\phi(T)$ such that

$$\int_{\|xm_j\|\geq m_j} \phi(\|x_{m_j}\|)\,dP \longrightarrow \beta,\; \int_{\|y_{n_k}\|\geq n_k} \phi(\|y_{n_k}\|)\,dP \longrightarrow \gamma$$

and $h(x'_{m_j}) \longrightarrow \xi$, $h(y'_{n_k}) \longrightarrow \eta$ in the ωm-topology. Let $A = \{h(x'_{m_j}): j = 1,2,\ldots\} \cup \{\xi\}$, $B = \{h(y'_{n_k}): k = 1,2,\ldots\} \cup \{\eta\}$. Then by Proposition 9 and Lemma 7, $A \bigstar B$ is ϕ-uniformly integrable. Applying Propositions 9 and 10,

$$\begin{aligned}\lim_j \lim_k E(\phi(\| x'_{m_j} + y'_{n_k} \|) &= \lim_j \lim_k I_\phi(h(x'_{m_j} + y'_{n_k}) \\ &= \lim_j \lim_k I_\phi(h(x'_{m_j}) \bigstar h(y'_{n_k})) \\ &= \lim_j I_\phi(h(x'_{m_j}) \bigstar \eta) = I_\phi(\xi \bigstar \eta)\end{aligned}$$

and similarly

$$\lim_k \lim_j E(\phi(\| x'_{m_j} + y'_{n_k} \|) = I_\phi(\xi \bigstar \eta):$$

thus we obtain a contradiction.

12. The representation of types on Orlicz function spaces.

The proof of Theorem 16 suggests that if X is any separable Banach space, then a type on $L_\phi(X)$ can be represented in terms of a suitable random measure in $\pi_\phi(T)$, but that also something else is needed, to deal with the fact that if $\xi_j \longrightarrow \xi$ is the ωm-topology then $I_\phi(\xi) \le \liminf I_\phi(\xi_n)$. We shall see this is indeed so: in the process we shall see that we obtain another local-compacification for $L_\phi(X)$.

Suppose that $\xi \in \pi_\phi(T)$. If $\lambda \ge 0$, we set

$$\phi_\xi(\lambda) = I_\phi(s_\lambda(\xi)).$$

Then it is immediate that ϕ_ξ is a convex non-negative function on R^+ which satisfies the Δ_2-condition, for the same constant k as ϕ. Let us therefore set

$$\Omega_k = \{f: R^+ \longrightarrow R^+: f(0) = 0, f \text{ is convex and } f(2\lambda) \le kf(\lambda) \text{ for all } \lambda \ge 0\}$$

On Ω_k, the topology of simple convergence and the topology of uniform convergence on compact sets coincide. Under this topology Ω_k is a locally compact metrizable space, and the sets,

$$\{f: f(1) \le n\}_{n=1,2,\ldots}$$

form a fundamental system of compact sets.

We now set $S_\phi(X) = \pi_\phi(T)\times\Omega_k$. If $(\xi,\alpha) \in S_\phi(X)$, let $\theta(\xi,\alpha) = (\xi,\phi_\xi+\alpha)$. θ is a one-one mapping of $S_\phi(X)$ into $\pi(T)\times\Omega_k$. We give $S_\phi(X)$ the topology defined by θ and the product topology on $\pi(T)\times\Omega_k$. Thus $(\xi_n,\alpha_n) \longrightarrow (\xi,\alpha)$ if and only if $\xi_n \longrightarrow \xi$ is the ωm-topology and $\phi_{\xi_n}+\alpha_n \longrightarrow \phi_\xi+\alpha$ in Ω_k. If $\xi \in \pi_\phi(T)$, let $c(\xi) = (\xi,0)$.

Theorem 17. The map $j = ch: L_\phi(X) \longrightarrow S_\phi(X)$ is a homeomorphism of $L_\phi(X)$ into $S_\phi(X)$.

Proof. If $x_n \longrightarrow x$, $h(x_n) \longrightarrow h(x)$ in the ωm-topology, by Corollary 2 of Theorem 15. Also if $\lambda \geq 0$, $E(\phi(\lambda x_n)) \longrightarrow E\phi(\lambda x)$ by Theorem 16, and so $\phi_{h(xn)}(\lambda) \longrightarrow \phi_{h(x)}(\lambda)$. Thus j is continuous.

Conversely if $j(x_n) \longrightarrow j(x)$, $x_n \longrightarrow x$ in probability, again by Corollary 2 of Theorem 15, and $E(\phi(x_n)) \longrightarrow E(\phi(x))$ so that $x_n \longrightarrow x$, by Theorem 16.

Theorem 18. $S_\phi(X)$ is a locally compact metrizable space. A subset A of $S_\phi(X)$ is relatively compact if and only if

$$\operatorname{Sup}\{\phi_\xi(1)+\alpha(1): (\xi,\alpha \in A\} < \infty.$$

Proof. First we show that $\theta(S_\phi(X))$ is closed in $\pi(T)\times\Omega_k$. Suppose that $\theta((\xi_n,\alpha_n)) \longrightarrow (\xi,\beta)$, so that $\xi_n \longrightarrow \xi$ in the ωm-topology and $\phi_{\xi_n}+\alpha_n \longrightarrow \beta$. Then (ϕ_{ξ_n}) and (α_n) are relatively compact in Ω_k, and (ξ_n) is relatively compact in $\pi(T)$. There exists a subsequence $((\xi_{n_j},\alpha_{n_j})) = ((\eta_j,\delta_j))$ and elements γ and δ of Ω_k such that $\phi_{n_j} \longrightarrow \gamma$ and $\delta_j \longrightarrow \delta$, and such that

$$\lim_{j\to\infty} E\left(\int_{t(0)\leq n} \phi(\lambda t(0))\eta_j(\omega,dt)\right)$$

exists for each n and each $\lambda \geq 0$. Then arguing as in Proposition 9,

$$\gamma - \phi_\xi = \lim_n \lim_k E\int_{t(0)>n} \phi(\lambda t(0))\eta_j(\omega,dt)$$

from which it follows that $\gamma - \phi_\xi \in \Omega_k$. Consequently $\beta - \phi_\xi = \delta+(\gamma-\phi_\xi) \in \Omega_k$, and

$$(\xi_n,\alpha_n) \longrightarrow (\xi,\beta-\phi_k) \in \theta(S_\phi(X)).$$

This implies that a subset A of $S_\phi(X)$ is relatively compact if and only if $\{\xi: (\xi,\alpha) \in A\}$ is relatively compact in $\pi(T)$ and $\{\phi_\xi+\alpha: (\xi,\alpha) \in A\}$ is relatively compact in Ω_k. The second of these conditions implies the first, and is equivalent to

$$\sup\{\phi_\xi(1)+\alpha(1): (\xi,\alpha) \in A\} < \infty.$$

As a result,

$$\{(\eta,\beta): |\phi_\eta(1)+\beta(1)-\phi_\xi(1)-\alpha(1)| \le 1\}$$

is a compact meighbourhood of (ξ,α), so that $S_\phi(X)$ is locally compact. Finally $S_\phi(X)$ is metrizable, since $\pi(T)$ and Ω_k are.

Next we show how we can extend the norm $\|\ \|_\phi$ to $S_\phi(X)$. If $f \in \Omega_k$, f is either identically zero or f is a strictly increasing unbounded continuous function. If $f = 0$, we set $n(f) = 0$; otherwise we set $n(f) = 1/f^{-1}(1)$. Elementary calculations show that n is a continuous function on Ω_k.

If now $(\xi,\alpha) \in S_\phi(X)$, we set $N_\phi((\xi,\alpha)) = n(\phi_\xi+\alpha)$. N_ϕ is a continuous function on $S_\phi(X)$, and if $x \in L_\phi(X)$, $N_\phi(j(x)) = \|x\|_\phi$.

In order to prove our main result, we need one further technical result.

<u>Theorem 19</u>. <u>Suppose that</u> (ξ_j) <u>is a sequence in</u> $\pi_\phi(T)$ <u>such that</u> $\xi_j \longrightarrow \xi$ <u>in the</u> ωm<u>-topology and</u>

$$I_\phi(\xi_j) \longrightarrow I_\phi(\xi) + \gamma.$$

<u>Then if</u> $x \in L_\infty(X)$, $I_\phi^{(x)}(\xi_j) \longrightarrow I_\phi^{(x)}(\xi) + \gamma.$

<u>Proof</u>. It is sufficient to show that the conclusion holds for some subsequence of (ξ_k), and we may therefore freely extract subsequences. We may therefore suppose that

$$I_\phi^{(x)}(\xi_j) \longrightarrow I_\phi^{(x)}(\xi) + \delta.$$

Since $\tau_x(\xi_j) \longrightarrow \tau_x(\xi)$ in the ωm-topology (Proposition 6) $\delta \ge 0$, by Proposition 9. Also

$$I_\phi^{(x)}(\xi) = \lim_n \lim_j I_{\phi_n}^{(x)}(\xi_j)$$

and $$\gamma = \lim_n \lim_j (I_\phi(\xi_j) - I_{\phi_n}(\xi_j))$$

so that we can suppose that there is an increasing sequence of integers (u_n), with $u_{n+1} \ge u_n+\|x\|_\infty+1$ for each n, such that, setting $\psi_n = \phi_{u_n}$,

$$|I_\phi^{(x)}(\xi) - I_{\psi_n}^{(x)}(\xi_j)| \le 2^{-n} \qquad (1)$$

and $$|\gamma - I_\phi(\xi_j) - I_{\psi_n}(\xi_j)| \le 2^{-n} \qquad (2)$$

for $j \ge n$.

Suppose that $\varepsilon > 0$. First choose $\alpha > 0$ such that $\alpha\gamma < \varepsilon/2$. Next choose n so that

$$\alpha u_n \ge (k-1)\|x\|_\infty$$

and $$(1+\alpha)2^{-n} < \varepsilon/2 - \alpha\gamma.$$

The reason for these choices should become clear as the calculations proceed. Notice that $2^{-n} < \varepsilon/2$ and that

$$(1-\alpha)(\gamma-2^{-n}) > \gamma-\varepsilon/2 \text{ and } (1+\alpha)(\gamma+2^{-n}) < \gamma+\varepsilon/2.$$

First we show that $\delta \ge \gamma$. If $j > n$,

$$\begin{aligned} I_\phi^{(x)}(\xi_j)-I_\phi^{(x)}(\xi) &= (I_{\psi_n}^{(x)}(\xi_j)-I_\phi^{(x)}(\xi))+(I_\phi^{(x)}(\xi_j)-I_{\psi_n}^{(x)}(\xi_j)) \\ &\ge I_\phi^{(x)}(\xi_j)-I_{\psi_n}^{(x)}(\xi_j)-\varepsilon/2 \quad \text{(by (1))} \\ &\ge E_\omega\Big(\int_{t(x)\ge u_n+1} \phi(t(x(\omega)\xi_j(\omega,dt))-\varepsilon/2, \end{aligned}$$

since $\phi \ge \psi_n$, and $\psi_n(u) = 0$ for $u \ge u_n+1$.

Now $|t(x(\omega))-t(0)| \le \|x\|_\infty$, so that if $t(0) \ge u_{n+1}$, $t(x(\omega)) \ge u_n+1$ and

$$\begin{aligned} \phi(t(x(\omega)) &\ge \Big(1 - \frac{(k-1)\|\ \|_\infty}{t(0)}\Big)\phi(t(0)) \\ &\ge (1-\alpha)\phi(t(0)) \end{aligned}$$

by inequality (*) at the beginning of §11. Thus

$$\begin{aligned} I_\phi^{(x)}(\xi_j)-I_\phi^{(x)}(\xi) &\ge (1-\alpha)E_\omega\Big(\int_{t(0)\ge u_{n+1}} \phi(t(0))\xi_j(\omega,dt)\Big)-\varepsilon/2 \\ &\ge (1-\alpha)(I_\phi(\xi_j)-I_{\psi_{n+1}}(\xi_j))-\varepsilon/2 \\ &\ge (1-\alpha)(\gamma-2^{-n-1})-\varepsilon/2 > \gamma-\varepsilon \end{aligned}$$

Letting $j \to \infty$, $\delta \ge \gamma-\varepsilon$, and so $\delta \ge \gamma$.

Next we show that $\delta \le \gamma$. As before, if $j > n$

$$\begin{aligned} I_\phi^{(x)}(\xi_j)-I_\phi^{(x)}(\xi) &\le I_\phi^{(x)}(\xi_j)-I_{\psi_{n+1}}^{(x)}(\xi_j) + \varepsilon/2 \\ &\le E_\omega\Big(\int_{t(x(\omega))\ge u_{n+1}} \phi(t(x(\omega)))\xi_j(\omega,dt)\Big)+\varepsilon/2 \end{aligned}$$

since $\phi - \psi_{n+1} \le \phi\chi_{[u_{n+1},\infty)}$.

Now if $t(x(\omega)) \ge u_{n+1}$, $t(0) \ge u_n+1$, and

$$\phi(t(x(\omega))) \le (1+\alpha)\phi(t(0))$$

by inequality (*). Thus

$$\begin{aligned} I_\phi^{(x)}(\xi_j)-I_\phi^{(x)}(\xi) &\le (1+\alpha)E_\omega\Big(\int_{t(0)\ge u_n+1}\phi(t(0))\xi_j(\omega,dt)\Big)+\varepsilon/2 \\ &\le (1+\alpha)(I_\phi(\xi_j)-I_{\psi_n}(\xi))+\varepsilon/2 \\ &\le (1+\alpha)(\gamma+2^{-n})+\varepsilon/2 \le \gamma+\varepsilon. \end{aligned}$$

Letting $j \longrightarrow \infty$, $\delta \le \gamma+\varepsilon$, and so $\delta \le \gamma$. This completes the proof.

If $x \in L_\phi(X)$ and $(\xi,\alpha) \in S_\phi(X)$, let us set

$$\tau_x((\xi,\alpha)) = (\tau_x(\xi),\alpha).$$

<u>Corollary</u> <u>If</u> $x \in L_\infty(X)$, τ_x <u>is a homeomorphism of</u> $S_\phi(X)$.

Let us denote the closure of $j(L_\phi(X))$ in $S_\phi(X)$ by $J_\phi(X)$. Note that if $x \in L_\infty(X)$, τ_x is a homeomorphism of $J_\phi(X)$.

We are now in a position to give a concrete representation of the types on $L_\phi(X)$.

<u>Theorem 20</u>. <u>If</u> $x \in L_\phi(X)$ <u>and</u> $(\xi,\alpha) \in J_\phi(X)$, <u>let</u>

$$Q((\xi,\alpha))(x) = N_\phi(\tau_{-x}((\xi,\alpha))).$$

<u>Then</u> $Q((\xi,\alpha))$ <u>is a type on</u> $L_\phi(X)$, <u>and</u> Q <u>maps</u> $J_\phi(X)$ <u>continuously onto the space of types on</u> $L_\phi(X)$.

<u>Proof</u>. Suppose first that $x \in L_\infty(X)$ and that (y_n) is a sequence in $L_\phi(X)$ such that $j(y_n) \longrightarrow (\xi,\alpha)$. Then by the Corollary to Theorem 19,

$$j(y_n-x) = \tau_{-x}(j(y_n)) \longrightarrow \tau_{-x}((\xi,\alpha))$$

and so

$$\|y_n-x\|_\phi = N_\phi(j(y_n-x)) \longrightarrow N_\phi(\tau_{-x}((\xi,\alpha)) = Q((\xi,\alpha))(x).$$

Since this holds for each x in $L_\infty(X)$ (which is dense in $L_\phi(X)$), (y_n) is an approximating sequence, and if τ is the type on $L_\phi(X)$ which it defines, $\tau(x) = Q((\xi,\alpha))(x)$.

Now suppose that $x \in L_\phi(X)$. Let $x_k = x\chi_{\|x\| \le k}$. Then $(h(x_k))$ is a ϕ-uniformly integrable sequence, and $x_k \longrightarrow x$ in $L_\phi(X)$. As $x_k \longrightarrow x$ in probability $\tau_{-x_k}(\xi)$ in the ωm-topology. Also for each $\lambda \ge 0$, $(s_\lambda(\tau_{-x_k}(\xi)))$ is a ϕ-uniformly integrable sequence, by Lemma 7, and so

$$\phi_{\tau_{-x_k}(\xi)} \longrightarrow \phi_{\tau_{-x}(\xi)} \text{ in } \Omega_k.$$

Consequently $Q((x,\alpha))(x_k) \longrightarrow Q((x,\alpha))(x)$. As $\tau(x_k) \longrightarrow \tau(x)$, we have that $\tau(x) = Q((x,\alpha)(x)$ for all x in $L_\phi(X)$.

Suppose that $(\xi_n,\alpha_n) \longrightarrow (\xi,\alpha)$ in $J_\phi(X)$. Then if $x \in L_\infty(X)$, $\tau_{-x}((\xi_n,\alpha_n)) \longrightarrow \xi_{-x}((\xi,\alpha))$, by the Corollary to Theorem 19, and so $Q((\xi_n,\alpha_n))(x) \longrightarrow Q((\xi,\alpha))(x)$. Thus Q is continuous.

Finally, suppose that τ is a type on $L_\phi(X)$. Let (y_n) be an approximating sequence in $L_\phi(X)$ which defines τ. Then $(j(y_n))$ is relatively compact in $J_\phi(X)$; by extracting a subsequence if necessary, we can suppose that $j(y_n) \longrightarrow (\xi,\alpha)$ in $J_\phi(X)$. Then if $x \in L_\infty(X)$

$$\begin{aligned} \tau(x) &= \lim_n \|y_n - x\|_\phi = \\ &= \lim_n N_\phi(\tau_{-x}(y_n)) \\ &= N_\phi(\tau_{-x}(\xi,\alpha)) = Q((\xi,\alpha))(x). \end{aligned}$$

Thus $\tau = Q((\xi,\alpha))$, and Q is onto.

We conclude with a few remarks. First, it can be shown that Theorem 19 holds for x in $L_\phi(X)$. This leads to a simpler proof of Theorem 20.

Secondly, if X is finite-dimensional, and in particular if $X = \mathbb{R}$ or $\mathbb{C}$, we can identify X with T: this leads to obvious simplifications.

Thirdly, if $\phi(t) = t^p$, for some $1 \le p < \infty$, we obtain the space $L_p(X)$. In this case, if $(\xi,\alpha) \in J_p(X)$, α has the form $\alpha(\lambda) = \alpha_o\lambda^p$, and any type τ on $L_p(X)$ has the form

$$\tau(x) = (E_\omega(\int (t(x))^p \xi(\omega,dt)) + \alpha_o^p)^{1/p},$$

so that we can identify $J_p(X)$ with a subspace of

$$\pi_p(T)\times[0,\infty).$$

If $(\xi,\alpha) \in \pi_p(T)\times[0,\infty)$, let

$$\theta'((\xi,\alpha)) = I_p(\xi) + \alpha^p;$$

θ' is a one-one mapping of $\pi_p(T)\times[0,\infty)$ into $\pi(T)\times[0,\infty)$; we give $\pi_p(T)\times[0,\infty)$ the topology defined by θ' and the product topology on $\pi(T)\times[0,\infty)$. Then $J_p(X)$ is homeomorphic to a subspace of $\pi_p(T)\times[0,\infty)$. In fact if (Ω,Σ,P) is atom-free, it is not hard to see that $J_p(X)$ is homeomorphic to $\pi_p(T)\times[0,\infty)$: if $(\xi,\alpha) \in \pi_p(T)\times[0,\infty)$ and $x \in L_p(X)$, let

$$Q'((\xi,\alpha))(x) = (E_\omega(\int (t(x)^p\xi(\omega,dt))+\alpha^p)^{1/p};$$

Q' maps $\pi_p(T)\times[0,\infty)$ continuously onto the space of types on $L_p(X)$.

Even with this simplification, Q (and Q') are not one-one. For example if (x_n) is any independent symmetric identically distributed sequence of random variables with mean 0 and variance 1, (x_n) is an approximating sequence in L_2 which defines the type $t(x) = (\|x\|^2+1)^{1/2}$, while $j(x_n) \longrightarrow (\xi,0)$ where $\xi(\omega) = \mu$, the law of x_1.

References

1. D.J. Aldous. Subspaces of L_1 via random measures (preprint).

2. S. Banach. Opérations linéaires (Chelsea 1955).

3. P. Billingsley. Convergence of probability measures (Wiley 1968).

4. H.F. Bohnenblust. An axiomatic characterization of L_p-spaces. Duke Math. J. 6 (1940) 627-640).

5. N. Bourbaki. Intégration, Chapitre IX (Hermann 1969).

6. D. Dacunha-Castelle. Indiscernability and exchangeability in L_p-spaces. Aarhus University Mathematical Institute Various Publications Series 24 (1974) 50-56.

7. J. Diestel and J.J. Uhl, Jnr. Vector measures A.M.S. Mathematical surveys 15 (1977).

8. S. Guerre and J.-T. Lapresté. Quelques propriétés des espaces de Banach stables. CRAS Paris 290 (1980) 645-647.

9. S. Guerre and J.-T. Lapresté. Quelques propriétés des modèles étalés sur les espaces de Banach. Publications mathématiques de l'Université Paris VI, 1980.

10. M. I. Kadeč and A. Pełczyński. Bases, lacumary sequences and complemented spaces in the spaces L_p. Studia Math. 21 (1961-2) 161-176.

11. G. Köthe. Topological vector spaces I (Springer-Verlag 1969).

12. M.A. Krasnosel'skii and Ya.B. Rutickii. Convex functions and Orlicz spaces (Noordhoff 1961).

13. J.L. Krivine and B. Maurey. Espaces de Banach stables. C.R.A.S. Paris 289 (1979) 679-681.

14. J.L. Krivine and B. Maurey. Espaces de Banach stables. Israel J. Math. 39 (1981)

15. H.E. Lacey. The isometric theory of classical Banach spaces (Springer-Verlag 1974).

16. J. Lindenstrauss and L. Tzafriri. Classical Banach spaces I (Springer-Verlag 1977).

17. J. Lindenstrauss and L. Tzafriri. Classical Banach spaces II (Springer-Verlag 1979).

18. W. Sierpiński. Sur les ensembles complets d'un espace (D). Fund. Math. 11 (1928) 203-205.

19. A.I. Tulcea and C. Tulcea. Topics in the theory of lifting (Springer-Verlag 1969).

Theorem 16 (the stability of $L_\phi(X)$) has also been proved by Yves Raynaud (Thèse de 3ème Cycle, 1980), using methods similar to those of Krivine and Maurey.

Department of Pure Mathematics
and Mathematical Statistics,
16 Mill Lane,
Cambridge,
England.

AUTOCORRELATION, EQUIPARTITION OF ENERGY, AND RANDOM EVOLUTIONS

Jerome A. Goldstein

1. PROPAGANDA. The theory of stochastic processes renders many services to the rest of mathematical analysis. One such service is the provision of powerful tools with which to prove theorems. Another is the suggestion of which theorems one ought to try to prove. This paper gives an illustration of the latter assertion. We show how the well-known connection between the Poisson process and the abstract wave equation suggests interesting results in the theory of equipartition of energy.

2. RANDOM EVOLUTIONS. Let A be a self-adjoint operator on a complex Hilbert space H. Then the initial value problem for the *abstract wave equation*

$$d^2u/dt^2 + A^2u = 0 \qquad (-\infty < t < \infty), \tag{1}$$

$$u(0) = f_1 \in \mathrm{Dom}(A^2), \; du(0)/dt = f_2 \in \mathrm{Dom}(A) \tag{2}$$

is well-posed. Associated with equation (1) is the *dissipative abstract wave equation* or *abstract telegraph equation*

$$d^2u/dt^2 + 2a\, du/dt + A^2u = 0 \quad (t \geq 0) \tag{1a}$$

where A is a nonnegative self-adjoint operator. (Note that (1o) reduces to (1).) Let u^a denote the solution of (1a), (2). Then for $a > 0$ we have the representation

$$u^a(t) = E[u^a(\tau_a(t))] \tag{3}$$

where $\{N_a(t) : t \geq 0\}$ is a Poisson process with intensity parameter a, $N_a(0) \equiv 0$, and

$$\tau_a(t) = \int_0^t (-1)^{N_a(s)}\, ds \qquad (t \geq 0).$$

The idea behind this result was published by Mark Kac in a 1956 Magnolia Petroleum Company publication (of all places!) and was influenced by earlier work of G. I. Taylor (1922) and S. Goldstein (1951); see Kac [10]. R. Griego and R. Hersh substantially generalized this result through their notion of a random evolution [6].

The idea is as follows. Let $A(i)$ generate a (C_0) semigroup $\{T(t; A(i)) : t \geq 0\}$ on a Banach space X for $i = 1,\dots,n$. Let the $n \times n$ matrix Q generate a Markov chain $\{j(t) : t \geq 0\}$ with state space $\{1,\dots,n\}$. Then the initial value problem

$$dV(t)/dt = \begin{pmatrix} A(1) & & \bigcirc \\ & \ddots & \\ \bigcirc & & A(n) \end{pmatrix} V(t) + QV(t) \qquad (t \geq 0), \tag{4}$$

$$V(0) = f = \begin{pmatrix} f_1 \\ \vdots \\ f_n \end{pmatrix} \in \mathrm{Dom}(A(1)) \times \ldots \times \mathrm{Dom}(A(n)) \subset X^n \tag{5}$$

is well-posed and is governed by a (C_0) semigroup $\{S(t) : t \geq 0\}$ obtained in the following manner. Let $\tau_0 \equiv 0$ and for $i \geq 1$, let τ_i be the epoch of the ith jump of $j(t)$. Let $N(t)$ be the total number of jumps up to time t; i.e. $N(t)$ is defined by the formula $\tau_{N(t)} \leq t < \tau_{N(t)+1}$. Define the *random evolution* R by the formula

$$R(t) = T(\tau_1; A(j(\tau_0)))T(\tau_2-\tau_1; A(j(\tau_1))) \ldots$$

$$\ldots T(\tau_N-\tau_{N-1}; A(j(\tau_{N-1})))T(t-\tau_N; A(j(\tau_N))).$$

Then the solution of (4), (5) is obtained from R via the formulas

$$V(t) = S(t)f, \; V(t) = \begin{pmatrix} v_1(t) \\ \vdots \\ v_n(t) \end{pmatrix},$$

$$v_i(t) = E_i[R(t)f_{j(t)}], \quad i = 1,\ldots,n,$$

where E_i denotes the expectation under the condition that $j(0) \equiv i$.

When the operators $T(t;A(i))$ commute for all t and i, the last formula reduces to

$$v_i(t) = E_i[\{\prod_{k=1}^{n} T(\gamma_k(t); A(k))\}f_{j(t)}] \tag{6}$$

where $\gamma_k(t)$ is the amount of time in $[0,t]$ in which the chain is in the state k.

Now specialize to the case when $n = 2$, A_0 generates a (C_0) group, $A(1) = -A(2) = A_0$, and $Q = \begin{pmatrix} -a & a \\ a & -a \end{pmatrix}$ where $a > 0$. Then if $u = v_1 - v_2$, straightforward manipulations show that (6) reduces to (3). This gives the unique solution of (1a), (2) when $X = H$ and $A = iA_0$ is self-adjoint.

Nice survey articles on these matters have been published by R. Hersh [7], [8]. These contain additional results and further references, as does the recent paper of A. Janssen and E. Siebert [9]

3. EQUIPARTITION OF ENERGY.

Let u denote the solution of (1), (2). The total *energy*

$$E = \|du(t)/dt\|^2 + \|Au(t)\|^2$$

is (finite and) conserved, i.e., E is independent of t. Let

$$K(t) = \|du(t)/dt\|^2, \quad P(t) = \|Au(t)\|^2$$

denote the *kinetic* and *potential energies* at time t. We proved the following equipartition of energy results more than a decade ago.

THEOREM 1. [2] $$\lim_{t\to\pm\infty} K(t) = \lim_{t\to\pm\infty} P(t) = E/2$$

for all solutions of (1) *if and only if* $e^{itA} \to 0$ *as* $t \to \pm\infty$ *in the weak operator topology.*

THEOREM 2. [3] $$\lim_{T\to\pm\infty} \frac{1}{T}\int_0^T K(t)dt = \lim_{T\to\pm\infty} \frac{1}{T}\int_0^T P(t)dt = E/2$$

for all solutions of (1) *if and only if* A *is one-to-one.*

If we write $A = \int_{-\infty}^{\infty} \lambda dF_\lambda$ by the spectral theorem, then the conclusion of Theorem 1 is equivalent to

$$\int_{-\infty}^{\infty} e^{it\lambda} d(\|F_\lambda h\|^2) \to 0 \quad \text{as} \quad t \to \pm\infty$$

for each $h \in H$. This is the conclusion of the Riemann-Lebesgue Lemma and a self-adjoint operator satisfying this condition is said to be a *Riemann-Lebesgue operator*. If A is *spectrally absolutely continuous* (i.e., $\lambda \mapsto \|F_\lambda h\|^2$ is absolutely continuous for each $h \in H$), then A is a Riemann-Lebesgue operator; and if A is a Riemann-Lebesgue operator, then A is *spectrally continuous* (i.e., $\lambda \mapsto \|F_\lambda h\|^2$ is continuous for each $h \in H$, which is equivalent to A having no eigenvalues). The converse to each of the above two assertions fails to hold; see [3] for counterexamples.

For a survey of recent results on equipartition of energy, see Goldstein and Sandefur [5].

4. AN OBSERVATION. Let u^a be the solution of (1a), (2). Defining kinetic and potential energies as above we have

$$\begin{aligned} P(t) &= \langle Au^a(t), Au^a(t)\rangle \\ &= \langle E[Au^0(\tau_a(t))], E[Au^0(\tau_a(t))]\rangle \qquad \text{by (3)} \\ &= \int_\Omega\int_\Omega \langle Au^0(\tau_a(t,\omega_1)), Au^0(\tau_a(t,\omega_2))\rangle Pr(d\omega_1)Pr(d\omega_2) \end{aligned}$$

where (Ω,Σ,Pr) denotes the underlying probability space which usually lurks in the background.

This observation suggests two things. Firstly, for solutions of (1a), (2) with $a > 0$, even though

$$E = K(t) + P(t) \ (= E(t))$$

decreases to zero exponentially as $t \to \infty$, there still may be some sort of nontrivial equipartition of energy result. Secondly, it is of interest to study

$$\langle Au^0(s), Au^0(r)\rangle, \ \langle du^0(s)/dt, \ du^0(r)/dt\rangle$$

as $s, r \to \infty$, but with s and r not necessarily equal to one another.

5. THE ABSTRACT TELEGRAPH EQUATION. Theorem 1 can be rewritten as follows:

$$\lim_{t\to\pm\infty} K(t)/P(t) = 1$$

holds for all solutions of (1), (2) if and only if A is a Riemann-Lebesgue operator; in particular, this holds if A is spectrally absolutely continuous.

For Equation (1a) an analogous result was established by Goldstein and Rosencrans [4].

THEOREM 3. *Let* $A \geq bI$ *where* A *is spectrally absolutely continuous, and suppose* $0 < a < \sqrt{8}\, b$. *Then there is a constant* C *such that*

$$0 < C \leq \liminf_{t\to\infty} K(t)/P(t) \leq \limsup_{t\to\infty} K(t)/P(t) \leq C^{-1} < \infty$$

holds for all solutions of (1a). *Moreover,* $C \to 1$ *as* $a \to 0$.

6. MEANS AND AUTOCORRELATIONS. For a solution of (1), define the quantities

$$K_r(t) = \langle du(t+r)/dt, \ du(t)/dt\rangle,$$
$$P_r(t) = \langle Au(t+r), \ Au(t)\rangle,$$
$$E_r = K_r(t) + P_r(t)$$

for r real. One readily checks that $dE_r/dt \equiv 0$ so that E_r is conserved. For a special choice of H and A in equation (1), the averages

$$\lim_{T\to\infty} \frac{1}{T}\int_0^T K_r(t)\,dt, \ \lim_{E\to\infty} \frac{1}{T}\int_0^T P_r(t)\,dt$$

(assuming they exist) correspond to weighted correlations of the velocity and strain fields when u denotes the displacement field of an elastic body; see W. A. Day [1]. When $r = 0$, $K_r(t)$ [resp. $P_r(t)$, E_r] reduces to $K(t)$ [resp. $P(t)$, E]. We have the following analogues of Theorems 1 and 2.

THEOREM 4. $$\lim_{t\to\pm\infty} K_r(t) = \lim_{t\to\pm\infty} P_r(t) = E_r/2$$

for all solutions of (1) *and all real* r *if and only if* A *is a Riemann-Lebesgue operator.*

THEOREM 5. $$\lim_{t\to\pm\infty} \frac{1}{\tau}\int_0^\tau K_r(t)dt = \lim_{t\to\pm\infty} \frac{1}{\tau}\int_0^\tau P_r(t)dt = E_r/2$$

for all solutions of (1) *and all real* r *if and only if* A *is one-to-one.*

For a discussion of the significance of these results in a continuum hypothesis context see Day [1]. In particular, Theorem 5 includes Day's main result as a special case.

Proofs. By the spectral theorem we have the d'Alembert formulas

$$Au(t) = \cos(tA)Af_1 + \sin(tA)f_2, \tag{7}$$
$$du(t)/dt = \cos(tA)f_2 - \sin(tA)Af_1$$

for the solution of (1), (2). It follows that

$$Au(t) = e^{itA}g_+ + e^{-itA}g_- ,$$
$$du(t)/dt = e^{itA}h_+ + e^{-itA}h_-$$

where

$$g_\pm = 2^{-1}(Af_1 \mp if_2),\ h_\pm = 2^{-1}(f_2 \pm iAf_1).$$

Note that

$$-ih_+ = g_+ ,\ ih_- = g_- . \tag{8}$$

Now let $s = r+t$. Using (7) we get

$$P_r(t) = \langle e^{isA}g_+ + e^{-isA}g_-,\ e^{itA}g_+ + e^{-itA}g_-\rangle$$
$$= [\langle e^{irA}g_+,g_+\rangle + \langle e^{-irA}g_-,g_-\rangle]$$
$$+ \{2\ \mathrm{Re}\langle e^{2itA}\ e^{irA}g_+,g_-\rangle\} \equiv \alpha_1 + \beta_1.$$

Similarly,

$$K_r(t) = [\langle e^{irA}h_+,h_+\rangle + \langle e^{-irA}h_-,h_-\rangle]$$
$$+ \{2\mathrm{Re}\langle e^{2itA}e^{irA}h_+,h_-\rangle\} \equiv \alpha_2 + \beta_2.$$

Using (8) we see that $\alpha_1 = \alpha_2$, $\beta_1 = -\beta_2$, and so

$$P_r(t) = E_r/2 + \beta_1,\quad K_r(t) = E_r/2 - \beta_1. \tag{9}$$

Moreover, as (f_1,f_2) varies over all of $\mathrm{Dom}(A^2) \times \mathrm{Dom}(A)$, $(e^{irA}g_+,g_-)$ varies over all of $\mathrm{Dom}(A) \times \mathrm{Dom}(A)$, and conversely. Theorem 4 now follows.

The Césaro equipartition of energy assertion of Theorem 5 holds if and only if

$$\lim_{T\to\pm\infty} \frac{1}{\tau}\int_0^\tau \langle e^{2itA}(e^{irA}g_+),g_-\rangle dt = 0 \tag{10}$$

for all f_1, f_2 as in (2) (see (9)). By the proof given in [3,pp396-397], condition (10) is equivalent to the injectivity of A. Q.E.D.

Day [1] also considered the quantity

$$J_r(t) = \langle u(t+r),\ u(t)\rangle$$

(see also [2]) in the case when zero is in the resolvent set of A. Since $v(t) = A^{-1}u(t)$ also satisfies (1) in this case, it follows that

$$J_r(t) \to \frac{1}{2}E' \quad \text{as} \quad t \to \pm\infty$$

for all solutions of (1) if and only if A is a Riemann-Lebesgue operator, and

$$\frac{1}{\tau}\int_0^\tau J_r(t)\,dt \to \frac{1}{2}E' \quad \text{as} \quad \tau \to \pm\infty$$

holds for all solutions of (1); here E' is the total energy corresponding to the initial data $f_1' = A^{-1}f_1$, $f_2' = A^{-1}f_2$.

Finally, we have the following generalization of a result of Day [1, Cor.3, p.254].

COROLLARY 6. *For the solution of* (1), (2) *and each real* r,

$$\frac{1}{\tau}\int_0^\tau P_r(t) - \frac{1}{\tau}\int_0^\tau K_r(t)\,dt \to 4\ \mathrm{Re}\langle Qg_+, Qg_-\rangle$$

as $\tau \to \pm\infty$, *where* Q *is the orthogonal projection onto the null space of* A *and* $g_\pm = 2^{-1}(Af_1 \mp if_2)$.

PROOF. By (9),

$$\frac{1}{\tau}\int_0^\tau P_r(t)\,dt - K_r(t))\,dt = \frac{4}{\tau}\int_0^\tau \mathrm{Re}\langle e^{(2t+r)iA}g_+, g_-\rangle dt\ . \tag{11}$$

Write $g_\pm = Qg + (I - Q)g_\pm$ and expand the inner product on the right hand side of (11), getting four terms. Each term involving $(I-Q)g_\pm$ tends to zero by the proof of Theorem 5 (cf. [3]). The remaining term is

$$\frac{4}{\tau}\int_0^\tau \mathrm{Re}\langle e^{(2t+r)iA}Qg_+, Qg_-\rangle dt.$$

But since $e^{isA}Qg_+ = Qg_+$ for all real s (since $AQg_+ = 0$), the result follows.

Q.E.D.

ACKNOWLEDGEMENTS. It is a pleasure to thank W. Alan Day of Hertford College, Oxford University for a stimulating and helpful discussion on these matters. This paper was presented at the Symposium on Stochastics and Analysis in June, 1981 at Tübingen, where I enjoyed the gracious hospitality of H. Heyer and E. Siebert. Finally the partial support of a U.S. NSF grant is gratefully acknowledged.

REFERENCES

1. Day, W. A.: Means and autocorrelations in elastodynamics. Arch. Rat. Mech. Anal. 73, 243-256 (1980)

2. Goldstein, J.A.: An asymptotic property of solutions of wave equations. Proc. Amer. Math. Soc. 23, 359-363 (1969)

3. Goldstein, J.A.: An asymptotic property of wave equations, II. J. Math. Anal. Appl. 32, 392-399 (1970)

4. Goldstein, J. A., Rosencrans, S. I.: Energy decay and partition for dissipative wave equations. J. Diff. Equations, 36, 66-73 (1980)

5. Goldstein, J. A., Sandefur, Jr., J. T.: Equipartition of energy. Proc. Séminaire Brézis-Lions, Collège de France. To appear

6. Griego, R., Hersh, R.: Theory of random evolutions with applications to partial differential equations. Trans. Amer. Math. Soc. 156, 405-418 (1971)

7. Hersh, R.: Stochastic solutions of hyperbolic problems. In: Partial Differential Equations and Related Topics (ed. by J. A. Goldstein), Lecture Notes in Math. No.446, Berlin, Heidelberg, New York: Springer, 283-300 (1975)

8. Hersh, R.: Random evolutions: A survey of results and problems. Rocky Mtn. J. Math. 4, 443-477 (1974)

9. Janssen, A., Siebert, E.: Convolution semigroups and generalized telegraph equations. Math. Z. 177, 519-532 (1981).

10. Kac. M.: A stochastic model related to the telegrapher's equation. In: Some stochastic problems in physics and mathematics. Magnolia Petroleum Co., Lectures in Pure and Applied Science, No.2, 1956. Reprinted in: Rocky Mtn. J. Math. 4, 497-504 (1974)

TULANE UNIVERSITY

WESTFIELD COLLEGE (UNIVERSITY OF LONDON)

Stable Probabilities on Locally Compact Groups

Wilfried Hazod

1. Introduction

The aim of this paper is to develop a concept of stability for locally compact groups which is a natural generalization of the "classical" concepts of stable and operator stable distributions on $\mathbb{R}$ and more generally on vector spaces. On the other hand we choose such a generalization which is applicable to the "central limit problems".

We start with the classical Lévy Hinčin formula for stable probabilities on the real line (s.e.g.[2]): A stable probability measure is embeddable into a continuous convolution semigroup (c.c.s.) $(\mu_t)_{t\geq 0} \subseteq M^1(\mathbb{R})$ with the characteristic functions

$$\hat{\mu}_t(y) = \exp t\{i\gamma y - c|y|^\alpha[1 - \beta \operatorname{sign}(y)\, w(y,\alpha)]\},$$

where $c > 0$ (w.l.o.g. $c = 1$ in the sequel),

$$\gamma \in \mathbb{R}, \quad |\beta| \leq 1, \alpha \in (0, 2],$$

and

$$w(y,\alpha) := \tan(\alpha \tfrac{\pi}{2}) \qquad \boxed{\alpha \neq 1}$$

$$:= -\frac{2}{\pi} \log |y| \qquad \boxed{\alpha = 1}$$

Let $A \in \mathcal{D}'(\mathbb{R})$ be the generating distribution of (μ_t), i.e. the distribution with Fourier transform $\hat{A} = \log \hat{\mu}_1$. We use the notation which is convenient in the case of probabilities on groups

$$\mu_t =: \mathcal{E}xp\,(t A), \quad t \geq 0$$

Let $\delta_t, t > 0$ be the homothetical transformation $\delta_t : \mathbb{R} \ni x \mapsto tx \in \mathbb{R}$.

Every continuous transformation $\tau: \mathbb{R} \longrightarrow \mathbb{R}$ induces a linear transformation of the function spaces $C^b(\mathbb{R})$, $\mathcal{D}(\mathbb{R})$

$$\tau : f \mapsto \tau(f) := f \circ \tau .$$

For measures and distributtions we define

$$\langle \tau(\mu), f \rangle := \langle \mu, \tau(f) \rangle, \ f \in C^b(\mathbb{R}), \ \mu \in M^b(\mathbb{R})$$
$$\langle \tau(F), f \rangle := \langle F, \tau(f) \rangle, \ f \in \mathcal{D}(\mathbb{R}), \ F \in \mathcal{D}'(\mathbb{R})$$

From the Lévy-Hinčin formula for stable distributions we see:

(1.1) $\boxed{\alpha \neq 1}$ $\delta_{s^{1/\alpha}}(\mu_t) = \mu_{st} * \varepsilon_{\gamma(s^{1/\alpha}-s)}$ $s, t > 0$

resp.

(1.2) $\boxed{\alpha = 1}$ $\delta_s(\mu_t) = \mu_{st} * \varepsilon_{\frac{2\beta}{\pi} s \log s}$ $s, t > 0$

i.e. progress in the "time variable t" is nothing but a suitable transformation of the "space variable x". This property is obviously a property of the corresponding generating distribution:

(1.1') $\boxed{\alpha \neq 1}$ a) $\delta_{s^{1/\alpha}}(A) = sA + s^{1/\alpha}\gamma(\frac{1}{s} - 1)\frac{d}{dx}$

b) Especially for $\gamma = 0$: $\delta_{s^{1/\alpha}}(A) = sA$

(1.2') $\boxed{\alpha = 1}$ a) $\delta_s(A) = A + \frac{2\beta}{\pi} s \log s \frac{d}{dx}$

b) Especially for $\beta = 0$: $\delta_s(A) = sA$

For $\alpha \in (0,2]$ fixed we denote $\tau_t := \delta_{t^{1/\alpha}}$ $t > 0$

Then $\tau_t \tau_s = \tau_{st}$, $t,s > 0$. Put $\varphi(s) := (s^{1/\alpha}(s^{-1} - 1))$ $\boxed{\alpha \neq 1}$

resp. $\varphi(s) := \frac{2\beta}{\pi} s \log s$ $\boxed{\alpha = 1}$ and finally $\frac{d}{dx} =: D$.

Then we get the simple form $\tau_s(A) = sA + \varphi(s) D$.

Therefore we suggest the following definition

1.3 Definition: Let G be a locally compact group with Lie algebra $\mathfrak{g}$. $\mathfrak{g}$ is identified with the set of primitive generating distributions (s.[73,78,79]). Let $\tau_t: G \to G$, $t > 0$, be a family of continuous transformations, s.t.

(i) $\tau_t \tau_s = \tau_{ts}$ $t,s > 0$ (ii) $\tau_s(e) = e$, $s > 0$,

(iii) $\tau_s \mathcal{D}(G) \subseteq \mathcal{D}(G)$, $s > 0$. ($\mathcal{D}(G)$ is the space of test functions, s.[73] $\tau_s(f) := f \circ \tau_s$).

(iv) $(s, f) \mapsto \tau_s(f)$ is continuous $(0,\infty) \times \mathcal{D}(G) \to \mathcal{D}(G)$.

Let A be a generating distribution, then A resp. the generated c.c.s. ($\mu_t =: \mathcal{E}xp(tA)$) is called

(1.3a) stable (in the generalized sense) w.r.t. (τ_t) if there exist

$X_s \in \mathcal{G}$ s.t. $\tau_s(A) = s A + X_s, s > 0$

(1.3b) stable (in the strict sense) w.r.t. (τ_t) if

$\tau_s(A) = s A, \quad s > 0.$

The concept of "operator stable probabilities on $\mathbb{R}$" was introduced by M. Sharpe [37]; The investigations were extended in the sequel by different authors (especially [23, 38,36,39,25,41...], see also part D, E, F of the literature). It turns out, that the class of operator stable measures in the sense of Sharpe is exactly the class of stable measure (-semigroups) defined in (1.3), where

$\tau_t = t^B$ (= exp(log t)B), $t > 0$, with $B \in \mathrm{Aut}(\mathbb{R}^n)$, and $\tau_t x \longrightarrow 0, x \in \mathbb{R}^n$.

A different concept was introduced by Z.J.Jurek [20,21,22] :Here the transformations $\tau_t: \mathbb{R} \longrightarrow \mathbb{R}$ are nonlinear and not onto. Finally I want to mention the definition of stability for discrete measures (on $\mathbb{Z}_+$), where the defining transformations τ_t operate as (linear) transformations on the probabilities (or equivalently on the Laplace transforms) but not on the space variables (s. e.g. Berg,Forst [19]).

In the case of topological groups which are not vector groups only few attempts had been done to define stable measures: On the one side there exist profound investigations on limit behaviour of identically distributed group valued random variables (s. e.g.[69,54,55,58, 62, 63]), on the other hand ([56, 57, 59, 61]) there is a natural definition of symmetric stable laws as the fundamental solutions of fractional powers of the Laplacians.

Summarizing the papers on this subject which appeared in the last years we see that it is worth to unify different approaches to a generalization of stability,and we see that the definition (1.3) is a quite natural one.

2. General considerations

Let ($\mu_t = \mathcal{E}xp(tA)$, $\mu_0 = \varepsilon_e$) be a c.c.s. For details on c.c.s. and their generating distributions we refer to [72, 73, 78, 79]. $\mathfrak{g}$ is the Lie algebra of the locally compact group G. $\mathfrak{g}$ is identified with the (vector space of) primitive forms on the testfunctions $\mathcal{D}(G)$. Further fix $t > 0$ and let $T: C^b(G) \to C^b(G)$ be a linear transformation, such that

(i) $T \geq 0$, (ii) $\| T \| = 1$, (iii) $T\mathcal{D}(G) \subseteq \mathcal{D}(G)$

Let $\mathcal{M}(G) \subseteq \mathcal{D}'(G)$ be the cone of generating distributions (s. [72]). For $F \in \mathcal{D}'(G)$ define $\langle T(F), f \rangle := \langle F, T(f) \rangle$, $f \in \mathcal{D}(G)$.

(iv) $T\,\mathcal{M}(G) \subseteq \mathcal{M}(G)$.

(v) T is $\sigma(\mathcal{D}', \mathcal{D})$- continuous.

We use the notation (s. [72]) : Let (μ_t) be a c.c.s. with generating distribution A, $\mathcal{E}xp(tA) := \mu_t$, $t \geq 0$.

2.1 Proposition

The following conditions (a), (b), (c) are equivalent:

(a) There exist $x(s) \in G$ such that

$$\langle T(\mu_s), f \rangle = \langle \mu_{ts} * \varepsilon_{x(s)}, f \rangle + o(s), \quad f \in \mathcal{D}(G), \ s \to 0$$

(b) There exists a primitive form $X \in \mathfrak{g}$ such that $T(A) = tA + X$

(c) $\mathcal{E}xp(sT(A)) = \lim_{n\to\infty} [\mu_{s/n} * \varepsilon_{\exp(\frac{s}{n}X)}]^n$

(Here "exp" denotes the exponential mapping $\mathfrak{g} \to G$ and convergence of measures is understood as $\sigma(M^b, C^b)$ - convergence)

If $X = 0$ we get

(c') $\mu_{st} = \mathcal{E}xp(sT(A)) + o(s) = T(\mu_s) + o(s)$,

if X commutes with A (i.e. $\mu_s * \varepsilon_{\exp(rX)} = \varepsilon_{\exp(rX)} * \mu_s$, $r,s > 0$) then we get

(c'') $\mathcal{E}xp(sT(A)) = \mu_{st} * \varepsilon_{\exp(sX)}$, $s,t > 0$

<u>Proof</u> " (a)$\Leftrightarrow$ (b)": $s \to \mu_{st}$, $s \to \mu_{st} * \varepsilon_{\exp(sX)}$ and $s \mapsto \mathcal{E}xp(sT(A))$ ($\mathbb{R}_+ \to \mathcal{D}'(G)$) are differentiable at $s = 0$ with differential quotient tA, tA+D, T(A) respectively. So we obtain the equivalence (a)$\Leftrightarrow$(b). "(b)$\Leftrightarrow$ (c)" is a simple consequence of the Lie Trotter formula for c.c.s., s.[72] .

To see (c') we use that

$$T(\mu_s) = \mathcal{E}xp(s\,T(A)) + o(s) \overset{(a)}{=} \mu_{st} * \varepsilon_{esp(sX)} + o(s)$$

and put X = 0. In the same manner (c'') is proved. □

2.2 Corollary

Let T, A, X be as in 2.1. Assume further that $T: M^b(G) \to M^b(G)$ is a homomorphism with respect to convolutions. Then

$(T(\mu_s))_{s \geq 0}$ is a c.c.s., and 2.1 b) implies

(c''') $$T(\mu_s) = \lim_{n\to\infty} \left[\mu_{st/n} * \varepsilon_{\exp(\frac{s}{n}X)}\right]^n$$

Especially if X = 0

$$T(\mu_s) = \mu_{st} \quad , s \geq 0.$$

If X and A commute

$$T(\mu_s) = \mu_{st} * \varepsilon_{\exp(sX)} \quad , s > 0$$

Proof: $T(A) = \frac{d^+}{ds} T(\mu_s)\Big|_{s=0}$ determines uniquely the c.c.s.

$(T(\mu_s))_{s\geq 0}$. Therefore $o(s) \equiv 0$ in 2.1.a. □

2.3 Proposition

a) For T as in 2.1 the sets

$S(T,t) := \{A \in \mathcal{M}(G) : T(A) = tA + X_t$ for some $X_t \in \mathfrak{g}\}$

and $S_o(T,t) := \{A \in S(T,t) : X_t = 0\}$

are $\sigma(\mathcal{D}', \mathcal{D})$ - closed convex cones in $\mathcal{M}(G)$.

b) If $T(\tilde{\mu}) = T(\mu)^\sim$, $\mu \in M^b(G)$, then for any $A \in S(T,t)$ we have $\tilde{A} \in S(T,t)$ and $A + \tilde{A} \in S_o(T,t)$.

Proof: Assume that $(A_\alpha)_{\alpha\in I}$ is a net in

$S(T,t)$, $A_\alpha \to A \in \mathcal{M}(G)$ in $\sigma(\mathcal{D}', \mathcal{D})$.

$T(A_\alpha) = tA_\alpha + X_t^\alpha$. As $T(A_\alpha) \to T(A)$, $A_\alpha \to A$,

we have $\lim_\alpha X_t^\alpha =: X_t$ exist and $T(A) = tA + X_t$.

b) is obvious. □

Now we assume that $\tau_t: G \to G$ is a group of transformations with the properties (i) - (iv) defined in (1.3). Then for fixed $t > 0$ $Tf := f \circ \tau_t$ has the properties (i) - (v) above.

2.4 Proposition

Assume moreover that $\tau_t \in \mathrm{Hom}(G,G)$. Then τ_t induces a convolution-homomorphism on $M^b(G)$. Moreover for $(\mu_t = \mathcal{E}xp(tA))_{t\geq 0}$

the following conditions are equivalent:

a) there exist $x(s) \in G$ s.t. $\tau_s(\mu_1) = \mu_s * \varepsilon_{x(s)} + o(s)$
(i.e. A is stable see 2.2)

b) there exist $x(t,s) \in G$ s.t. $\tau_t(\mu_s) = \mu_{st} * \varepsilon_{x(s,t)} + o(s)$

Proof: For $t,s > o$ we have

$$\tau_s(\tau_t(\mu_1)) = \tau_s(\mu_t) * \varepsilon_{\tau_s(x(t))} + \tau_s(o(t)).$$
$$\|$$
$$\tau_{st}(\mu_1) = \mu_{st} * \varepsilon_{x(st)} + o(s), \quad s \to o$$

Therefore $\tau_s(\mu_t) = \mu_{st} * \varepsilon_{x(s,t)} + o(s), \; s \to o$

with $x(s,t) := x(st) \; \tau_s(x(t))^{-1}$ □

2.5 Corollary

Let A be stable (in the generalized sense), $\tau_t(A) = tA + X_t$, $X_t \in \mathcal{G}$. Further for ($\mu_t = \mathcal{E}xp(tA)$) $\tau_s(\mu_t) = \mu_{st} * \varepsilon_{x(s,t)} + o(s)$.

Then we have

$$\tau_s(X_t) = X_{st} - X_s, \quad s,t > 0$$

resp. $\tau_s(x(t)) = x(st)^{-1} \, x(s,t)$, $s,t > 0, s \to 0$.

Now we want to sketch the connection of the definition of stability given in (1.3) and the central limit problem.

2.6 Definition

Let $(\tau_t)_{t > 0}$ be defined as before, let $(\mu_t = \mathcal{E}xp(tA))_{t \geq 0}$ be a c.c.s. A measure $\mu \in M^1(G)$ lies in the domain of attraction in the strict sense of (μ_t) w.r.t. the norming group $(\tau_t)_{t > 0}$ if

(2.6.1) $$[\tau_{1/n}(\mu)]^{[nt]} \xrightarrow[n \to \infty]{} \mathcal{E}xp(tA), \quad t > 0 .$$

μ lies in the general domain of attraction of (μ_t) if there exists a sequence $(x(n))_{n \in \mathbb{N}}$ such that

(i) $x(n) \to e$

(2.6.2) (ii) $\tau_{1/n}(\mu) * \varepsilon_{x(n)} \to \varepsilon_e$

(iii) $[\tau_{1/n}(\mu) * \varepsilon_{x(n)}]^{[nt]} \xrightarrow[n \to \infty]{} \mathcal{E}xp(tA), \quad t > 0.$

(This definition is restrictive, as we suppose in (2.6.1) resp. (2.6.2) convergence for every $t > 0$. This is due to the fact that for c.c.s. on groups the semigroup (μ_t) is in general not uniquely determined by μ_1.)

In the following theorem we use the notation of resolvent-convergence (generalized convergence) (s. [7o]): A sequence A_n of generating distributions converges to A in the generalized sense, if the resolvents of the corresponding convolution operators converge, i.e.

$$A_n \rightsquigarrow A \quad \text{if} \quad (A_n - \lambda \varepsilon_e)^{-1} \to (A - \lambda \varepsilon_e)^{-1}, \quad \lambda > 0$$

or equivalently

$$\text{if} \quad \mathcal{E}xp(tA_n) \longrightarrow \mathcal{E}xp(tA) \qquad t \geq 0 \quad .$$

2.7 Theorem

Let $\mu \in M^1(G)$ be in the domain of attraction of $(\mathcal{E}xp\,(tA))_{t\geq 0}$ in the strict (resp. general) sense. Then

$$\tau_t\,(A) = tA \qquad , t > 0$$

$$(\text{resp. } \tau_t\,(A) = tA + X_t \qquad , t > 0, \text{ for some } X_t \in \mathcal{G}).$$

That means: The domain of attraction of $(\mathcal{E}xp(tA))_{t\geq 0}$ is non empty iff

A is stable (in the strict resp. generalized sense).

Proof: A simple application of the approximation of continuous operator semigroups by discrete semigroups ([7o]O. § 2) shows:

If $A_n := n\left[\tau_{1/n}(\mu) - \varepsilon_e\right]\left(\text{resp. } n\left[\tau_{1/n}(\mu) * \varepsilon_{x(n)} - \varepsilon_e\right]\right)$

then $A_n \rightsquigarrow A$ iff

$$\left[\tau_{1/n}(\mu)\right]^{[nt]} \to \mathcal{E}xp(tA) \left(\text{resp. } \left[\tau_{1/n}(\mu) * \varepsilon_{x(n)}\right]^{[nt]} \to \mathcal{E}xp(tA)\right) \; t \geq 0.$$

For rational $t = k/m$, $k, m \in \mathbb{N}$ we get

$$\tau_{k/m}\,(A_{nk}) = \tau_{k/m}\,(nk\,(\tau_{1/nk}(\mu) * \varepsilon_{x(nk)} - \varepsilon_e)) \rightsquigarrow \tau_{k/m}\,(A)$$

On the other hand

$$\tau_{k/m}(A_{nk}) = \frac{k}{m}\, nm\, (\tau_{1/nm}(\mu) * \varepsilon_{\tau_{k/m}(x(nk))} - \varepsilon_e) =$$

$$= \frac{k}{m}\, mn \left[\tau_{1/mn}(\mu) * \varepsilon_{x(nm)}\ {}_{x(nm)^{-1}\ \tau_{k/m}(x(nk))} - \varepsilon_e \right] =$$

$$= \frac{k}{m} \left[A_{nm} * \varepsilon_{y(n)} + mn\ (\varepsilon_{y(n)} - \varepsilon_e) \right]$$

As $y(n) := x(nm)^{-1}\, \tau_{\frac{k}{m}}\ (x\ (nk)) \longrightarrow e$

and $\tau_{k/m}(A_{nk}) \rightsquigarrow \tau_{k/m}(A)$, the limit

$X_{k/m} := \lim_{n\to\infty} mn\ (\varepsilon_{y(n)} - \varepsilon_e)$ exists (and is a primitive form)

and we have

$$\tau_{k/m}(A) = \frac{k}{m} A + X_{k/m}$$

So the theorem is proved for rational t. The continuity of $t \to \tau_t(A)$ implies the proof for real $t > 0$. □

Several authors used Bochner's representation of symmetric stable laws on $\mathbb{R}$ via fractional powers of Laplacians to define stable laws on groups, s.e.g. [56, 57, 59, 61].

Let $(\sigma_t^{(\alpha)})_{t>0}$ be the c.c.s. on $\mathbb{R}_+$ with Laplace transform $\hat{\sigma}_t^{(\alpha)}(y) = e^{-ty^\alpha}$, $\operatorname{Re} y > 0$, $0 < \alpha \leq 1$. For any generating distribution $A \in \mathcal{MD}(G)$ $M^1(\mathbb{R}_+) \ni \lambda \longmapsto \int_{\mathbb{R}_+} \mathcal{E}xp\,(tA)\, d\lambda(t) \in M^1(G)$ is a continuous convolution homomorphism. Therefore

$$(\mu_t^{(\alpha)} := \int_{\mathbb{R}_+} \mathcal{E}xp\,(sA)\, d\sigma_t^{(\alpha)}(s))_{t \geq 0} \quad \text{is a c.c.s.,}$$

the generating distribution is denoted by $A^{(\alpha)}$ (fractional power). We shall show, that stability of A implies stability of $A^{(\alpha)}$, $0 < \alpha \leq 1$.

<u>2.8 Lemma</u> $\delta_{t^{1/\alpha}}(\sigma_r^{(\alpha)}) = \sigma_{rt}^{(\alpha)}$

(where $\delta_t := \mathbb{R}_+ \ni x \to tx \in \mathbb{R}_+$)

2.9 Lemma For $\lambda \in M^1(\mathbb{R}_+)$, $t > 0$, A strictly stable w.r.t. (τ_t),

we have $\tau_t\left(\int_{\mathbb{R}_+} \mathcal{E}xp(sA)\, d\lambda(s)\right) = \int_{\mathbb{R}_+} \mathcal{E}xp(sA)\, d\delta_t(\lambda)(s)$

2.10 Proposition

Let (τ_t) be defined as before. Let A be strictly stable w.r.t. (τ_t). Then for any $\alpha \in (0,1]$ $A^{(\alpha)}$ is strictly stable w.r.t. $(\tau_t^{(\alpha)} := \tau_{t^{1/\alpha}})_{t>0}$.

Proof:

$$\tau_{t^{1/\alpha}}\left(\int_{\mathbb{R}_+} \mathcal{E}xp(sA)\, d\sigma_r^{(\alpha)}(s)\right) = \int_{\mathbb{R}_+} \mathcal{E}xp(sA)\, d\delta_{t^{1/\alpha}}\sigma_r^{(\alpha)}(s) =$$

$$= \int_{\mathbb{R}_+} \mathcal{E}xp(sA)\, d\sigma_{rt}^{(\alpha)}(s), \text{ i.e.}$$

$$\tau_{t^{1/\alpha}}\left(\mathcal{E}xp(rA^{(\alpha)})\right) = \mathcal{E}xp(rtA^{(\alpha)}), \qquad t, r > 0. \quad \square$$

Remark: If we replace A by a generating distribution which is stable in the generalized sense, then $A^{(\alpha)}$ need not to be stable. On the other hand, if we replace $\sigma_t^{(\alpha)}$ by $\sigma_t^{(\alpha)} * \varepsilon_{\gamma t}$ (a shifted one-sided stable distribution), then for strictly stable $A \in \mathcal{M}(G)$ $\int_{\mathbb{R}_+} \mathcal{E}xp(sA)\, d(\sigma_t^{(\alpha)} * \varepsilon_{\gamma t})$ need not to be stable in the generalized sense.

3. An example: Stability on nilpotent simply-connected Lie groups.

Let G be a connected Lie group with Lie algebra $\mathfrak{g}$. Let $\tilde{G}$ be the simply-connected covering group and $\pi: \tilde{G} \longrightarrow G$ the canonical homomorphism. Let further φ: Aut (G) $\longrightarrow$ Aut $(\tilde{G})$, ψ: Aut $(\tilde{G}) \longrightarrow$ Aut $(\mathfrak{g})$ be the canonical homomorphisms. φ is injective and ψ is bijective. Let $\mathbb{R}_+ \ni t \longmapsto \tau_t \in$ Aut (G) be a continuous homomorphism, such that $\tau_t \tau_s = \tau_{ts}$, $t, s > 0, \tau_0 = id$. (τ_t) is called contracting, if for any compact $K \subseteq G$, for any neighbourhoood $V \in \mathfrak{U}(e)$ there exists $t_0 > 0$, such that $\tau_t(K) \subseteq V$, $t < t_0$. Obviously we have:

3.1. Lemma

For any $\mu \in M^1(G)$ and any fixed contracting (τ_t) the map $t \longmapsto \tau_t(\mu) \in M^1(G)$ is continuous with $\lim_{t \downarrow 0} \tau_t(\mu) = \varepsilon_e$.

3.2. Corollary

Assume $\frac{d^+}{dt} \tau_t(\mu)\Big|_{t=0} =: A \in \mathscr{D}'(G)$ exists. Then A is generating distribution of a c.c.s. $(\mathcal{E}xp(tA))_{t \geq 0}$ s.t.

$$\mathcal{E}xp(tA) = \lim_{n \to \infty} \left[\tau_{t/n}(\mu)\right]^n, \qquad t > 0.$$

[[This is a simple application of the Lie-Trotter-product formula, s. [70]]]

Therefore μ is embeddable into a stable c.c.s. w.r.t. (τ_t) if

$$\tau_t(\mu) * \tau_s(\mu) = \tau_{t+s}(\mu), \quad t, s > 0.$$

It is natural to raise the question, for which groups G contracting automorphisms(τ_t) exist. If G admits contracting automorphisms(τ_t), then $(\varphi(\tau_t))_{t>0}$ is a group of contracting automorphisms in Aut $(\tilde{G})$. On the other hand the covering map $\pi: \tilde{G} \longrightarrow G$ induces homomorphisms $M^1(\tilde{G}) \to M^1(G)$ and $\mathcal{M}(\tilde{G}) \longrightarrow \mathcal{M}(G)$. So it is quite natural to restrict the considerations to the case of simply-connected Lie groups.

3.3 Lemma

If $(\tau_t)_{t>0}$ is a contracting group $\subseteq$ Aut (G) then $\psi(\tau_t)_{t>0}$ is a contracting group $\subseteq$ Aut $(\mathfrak{g})$ and vice versa.

3.4 Lemma

(s.R. Goodman [65, 66, 67]). Let $\mathfrak{g}$ be a real finite dimensional Lie algebra. Assume that $\mathfrak{g}$ admits a graduation, i.e. a representation

$$\mathfrak{g} = \sum_{i=1}^{r} \oplus \mathfrak{m}_i, \quad [\mathfrak{m}_i, \mathfrak{m}_j] \subseteq \mathfrak{m}_{i+j}, \quad i, j = 1, \ldots r.$$

Then $\mathfrak{g}$ is nilpotent and $\mathfrak{g}$ admits a contracting group of automorphisms.

More generally: Assume $\mathfrak{g}$ admits a positive graduation ([66]), $\sum_{\lambda>0}\oplus\mathfrak{m}_\lambda$ then $\mathfrak{g}$ is nilpotent and admits a contracting group of automorphisms. (A positive graduation is a representation $\mathfrak{g} = \sum_{r>0}\oplus\mathfrak{m}_r$, $\mathfrak{m}_r \neq \{0\}$ for a finite set of r's, s.t. $[\mathfrak{m}_r, \mathfrak{m}_s] = \mathfrak{m}_{r+s}$, r, s > 0).

The following inversion of 3.4 seems to be well known, but I couldn't find a reference. So I give a sketch of a proof:

3.5. Lemma

Let $\mathfrak{g}$ be a real finite dimensional Lie algebra which admits a contracting group $(\tau_t) \subseteq Aut(\mathfrak{g})$. Then $\mathfrak{g}$ admits a positive graduation (and is therefore nilpotent).

The proof goes along the following lines:

Let $\mathfrak{g}_c := \mathfrak{g} \oplus i\mathfrak{g}$ be the complexification of $\mathfrak{g}$. The group of automorphisms is represented as $\tau_t = t^B$ where $B: \mathfrak{g} \to \mathfrak{g}$ is a derivation. (τ_t) is contracting iff the spectrum of B is contained in the half-plane $\{\lambda : \operatorname{Re}\lambda > 0\}$. Let $\{\lambda_i := a_i + i b_i\}$ be the eigenvalues of B.

We represent B in Jordan-form:

$$\mathfrak{g}_c = \sum_{j=1}^{r}\oplus V_j, \qquad B\big|_{V_j} =: B_j = \lambda_j \operatorname{id}_{V_j} + N_j, \qquad N_j \quad \text{nilpotent.}$$

$\pi_j : \mathfrak{g}_c \longrightarrow V_j$ denotes the canonical projection. For $X \in V_i$, $Y \in V_j$ we get $\pi_m t^B[X, Y] = \pi_m[t^B X, t^B Y]$

i.e. (with s: = log t)

$$e^{s\lambda_m}\Big(\operatorname{id}_{V_m} + \sum_{k\geq 1}\frac{s^k}{k!}N_m^k\Big)\pi_m[X, Y] =$$

$$= e^{(\lambda_i+\lambda_j)s}\,\pi_m\Big\{[X, Y] + \sum_{k+l\geq 1}\frac{s^{k+l}}{k!\,l!}\,[N_i^k X, N_j^l Y]\Big\}$$

As N_i, N_j, N_m are nilpotent, these are polynomials. Applying N_m^p p=1,2,.. to both sides of the equation we get easily that either

$\lambda_i + \lambda_j = \lambda_m$ (and therefore $a_i + a_j = a_m$)

or $\pi_m[V_i, V_j] = \{0\}$.

Now define $V_1, \dots V_r \subseteq \mathfrak{g}_c$ as before, $V_{r+1} := \{0\}$.

Define for $1 \leq i, j \leq r$:

$$m(i,j) := m \quad \text{if } \pi_m[V_i, V_j] \neq \{0\}$$

$$:= r+1 \text{ if } [V_i, V_j] = \{0\}$$

So we get: $[V_i, V_j] \subseteq V_{m(i,j)}$, $1 \leq i, j, \leq r$

Finally let $0 < c_1 < \dots < c_k$ be the real parts of the eigenvalues of B in strict monotone order. Define

$$\tilde{V}_s := \sum_{Re\lambda_j = s} \oplus V_j \qquad \text{if } s \in \{c_1, \dots, c_k\}, \; V_s := \{0\} \text{else.}$$

Then $\mathfrak{g}_c = \sum_{\beta > 0} \oplus \tilde{V}_\beta$, $[\tilde{V}_\beta, \tilde{V}_\gamma] \subseteq \tilde{V}_{\beta+\gamma}$, $\beta, \gamma > 0$.

The eigenvalues of B are either real or complex conjugate, therefore the decomposition $\mathfrak{g}_c = \sum \oplus \tilde{V}_\beta$ induces a decomposition of $\mathfrak{g}$

$\mathfrak{g} = \sum \oplus W_\beta$, i.e. a positive graduation of $\mathfrak{g}$. □

Remark: If $\mathfrak{g} = \sum_{\beta > 0} \oplus W_\beta$ is a positive graduation of $\mathfrak{g}$ with $W_{c_1}, \dots, W_{c_k} \neq \{0\}$. Then there exists a "natural" group of contracting automorphisms $(\tau_t := t^B)$: Define $B := \sum_{j=1}^{k} \oplus \; c_j \; \mathrm{id}_{W_j}$,

then B is a derivation and (t^B) is contracting. There exist profound investigations on homogeneous distributions on simply-connected nilpotent Lie groups, especially for stratified groups, where homogenity means homogenity w.r.t. the "natural" group of automorphisms, i.e. $t^B(A) = t^\alpha A$, $\forall t > 0$, some $\alpha > 0$.

If we define $\tilde{\tau}_t := t^{(1/\alpha)B}$, then the homogenity condition reads $\tilde{\tau}_t(A) = tA$. I.e. stable generating distributions (in the strict sense) are special cases of homogenous distributions. (Details and hints may be found in [75, 65]).

The most interesting case are the homogeneous Gauß-generators, i.e. homogeneous sub-Laplacians and fractional powers of these operators. The corresponding fundamental solutions are just the Gauß-semigroups, which are stable (in the strict sense) w.r.t. $(\tilde{\tau}_t)$.

See especially [56, 58, 59, 60, 61].

These papers give information on the existence of absolute-continuous stable Gauß-semigroups, the behaviour of the densities, limit theorems etc.. Beyond this for special groups (e.g. Heisenberg groups) much is known about hypoellipticity and analytic-hypoellipticity of stable Gauß generators (s. e.g. [58, 80, 81]). In the sequel we shall show, that for nilpotent simply-connected positively graduated Lie groups the structure of the generators of stable semigroups (via Lévy Hinčin representation) is completely known, whereas the information about the structure of the corresponding semigroups is not so complete.

Let G be a nilpotent simply-connected Lie group of dimension r, let $\mathfrak{g}$ be the Lie algebra and $\mathfrak{g} = \sum_{\beta} \oplus V_{\beta}$, $[V_{\beta}, V_{\beta'}] \subseteq V_{\beta+\beta'}$ be a positive graduation .

(i) $\exp : \mathfrak{g} \longrightarrow G$ and $\log = \exp^{-1} : G \longrightarrow \mathfrak{g}$ are one-one, onto and C^{∞}.

(ii) ($\mathfrak{g}$,+) is a commutative Lie group, simply connected of dimension r. The group structures of ($\mathfrak{g}$, +) and (G, .) are related by the Campbell-Hausdorff formula

$$\log(\exp(X)\exp(Y)) = X + Y + \tfrac{1}{2}[X, Y] + \tfrac{1}{12}[X, [X,Y]] + \ldots$$

which is a polynomial. So G may be represented as (the topological space) G $= \mathbb{R}^r$, where the multiplication $(x, y) \to x \cdot y$ is a polynomial in x, y .

(iii) In this representation the Haar measure on $\mathfrak{g}$ (= Lebesque measure on $\mathbb{R}^r$) and the Haar measure on G coincide: Denote by $\omega_{\mathfrak{g}}$ resp. ω_G the Haar measures on ($\mathfrak{g}$,+) resp. (G,·), then $\exp(\omega_{\mathfrak{g}}) = \omega_G$ $\log(\omega_G) = \omega_{\mathfrak{g}}$.

(iv) Via exp and log the function-, measure- and distribution-spaces $C^b(G)$, $C^b(\mathfrak{g})$; $\mathcal{D}(G)$, $\mathcal{D}(\mathfrak{g})$; $M^b(G)$, $M^b(\mathfrak{g})$; $\mathcal{D}'(G)$, $\mathcal{D}'(\mathfrak{g})$ coincide. Especially for the cones of generating distributions of c.c.s. we have: $\exp \mathcal{M}(\mathfrak{g}) = \mathcal{M}(G)$, $\log \mathcal{M}(G) = \mathcal{M}(\mathfrak{g})$.

(v) Let $(\tau_t) \subseteq \mathrm{Aut}(G)$ be a group of contracting automorphisms and let $(\delta_t) \subseteq \mathrm{Aut}(\mathfrak{g})$ be the induced group of automorphisms of $\mathfrak{g}$. ($\delta_t = t^B$ are Lie algebra-automorphisms and therefore automorphisms of the group ($\mathfrak{g}$,+)). Then there is a one-one-correspondence between (τ_t) and (δ_t) via $\tau_t = \exp\circ\,\delta_t\circ\log$.

Now we turn back to the stable distributions:

3.6 Theorem

The stable generating distributions $A \in \mathcal{MD}(G)$ w.r.t. (τ_t) coincide via log with the stable generating distributions on the (vector space) $\mathfrak{g}$ ($= \mathbb{R}^r$).

Attention: exp resp. log induce isomorphisms of the vector spaces $M^b(G)$ and $M^b(\mathfrak{g})$, $\mathcal{D}'(G)$ and $\mathcal{D}'(\mathfrak{g})$ (especially log: $\mathcal{MD}(G) \to \mathcal{MD}(\mathfrak{g})$ is an isomorphism w.r.t. the cone-structure), but log: $M^b(G) \longrightarrow M^b(\mathfrak{g})$ is - if G is non Abelian - not a convolution homomorphism. So the theorem reads:

For $A \in \mathcal{MD}(G)$ s. t. $\tau_t(A) = tA + X_t$, $X_t \in \mathfrak{g}$

put $\mathring{A} := \log A$, then $\delta_t(\mathring{A}) = t\mathring{A} + X_t$ and vice versa.

(We identify $\mathfrak{g}$ with the invariant 1st-order differential operators on $\mathfrak{g}$ and G.)

This does of course not imply

$$\log(\mathcal{E}xp_G(tA)) = \mathcal{E}xp_{\mathfrak{g}}(t\mathring{A}), \quad t \geq 0.$$

The proof is obvious: $\langle \delta_t(\mathring{A}), \mathring{f} \rangle = \langle \mathring{A}, \mathring{f} \circ \delta_t \rangle =$

$$= \langle A, \mathring{f} \circ \delta_t \circ \log \rangle = \langle A, (f \circ \log) \circ \tau_t \rangle = \langle \tau_t(A), f \rangle =$$

$$(\text{with } \mathring{f} \circ \log =: f) \quad = t \langle A, f \rangle + \langle X_t, f \rangle =$$

$$= t \langle \mathring{A}, \mathring{f} \rangle + \langle X_t, \mathring{f} \rangle . \qquad \square$$

3.7 Proposition

If $A \in \mathcal{MD}(G)$, $\mathring{A} = \log A \in \mathcal{MD}(\mathfrak{g}) = \mathcal{MD}(\mathbb{R}^n)$.

Then $\varphi: \mathbb{R}_+ \ni t \longmapsto \exp(\mathcal{E}xp(t\mathring{A})) \in M^b(G)$ is differentiable at $t = 0$, $\frac{d^+}{dt}\varphi(t)\Big|_{t=0} = A$ and we obtain the following relation between the c.c.s. generated by $\mathring{A}$ (in $M^1(\mathfrak{g})$) and by A (in $M^1(G)$):

$$\mathcal{E}xp_G(tA) = \lim_{n\to\infty}\left[\exp \mathcal{E}xp_{\mathfrak{g}}\left(\frac{t}{n}\mathring{A}\right)\right]^n .$$

Proof: Application of the Lie-Trotter-product formula (s. [70]) : Obviously $A = \frac{d^+}{dt}\varphi(t)\Big|_{t=0}$. Therefore $\mathcal{E}xp_G(tA) = \lim_n \left[\varphi\left(\frac{t}{n}\right)\right]^n$. $\square$

3.8 Corollary

If $A \in \mathcal{M}(G)$ is stable in the strict sense w.r.t. (τ_t), $\mathring{A}$ stable w.r.t. (δ_t), then

$$\mathcal{E}xp_G(t\,A) = \lim_n \left[\tau_{1/n}(\exp \mathcal{E}xp_{\mathfrak{g}}(t\mathring{A})) \right]^n$$

An analogous formula yields for stable distributions in the generalized sense.

Proof: We use the abbreviations $\mu_t := \mathcal{E}xp_G(tA)$,

$$\nu_t := \exp\left(\mathcal{E}xp_{\mathfrak{g}}(t\mathring{A})\right).$$

Then (μ_t) is a c.c.s. $\subseteq M^1(G)$, ν_t is differentiable with $\frac{d^+}{dt}\nu_t\Big|_{t=0} = \frac{d^+}{dt}\mu_t\Big|_{t=0} = A.$

On the other hand $\delta_s\left(\mathcal{E}xp_{\mathfrak{g}}(t\mathring{A})\right) = \mathcal{E}xp_{\mathfrak{g}}(ts\,A)$, $t, s > 0$

Therefore $t\mathring{A} = \lim_n n\left[\delta_{1/n}(\mathcal{E}xp_{\mathfrak{g}}(t\mathring{A}) - \varepsilon_0)\right]$

resp. $tA = \lim_n n\left[\tau_{1/n}(\nu_t - \varepsilon_e)\right]$.

This yields

$$\mu_t = \mathcal{E}xp_G(tA) = \lim_n \left[\tau_{1/n}(\nu_t)\right]^n$$

(s. [70]0.§2, I.§2). □

3.9 Corollary

Let $A \in \mathcal{M}(G)$ be stable in the strict sense (= homogeneous of degree 1) w.r.t. (τ_t). Then

$$\tau_t(\mathcal{E}xp_G(sA)) = \mathcal{E}xp_G(t\,s\,A)$$

$$\delta_t(\mathcal{E}xp_{\mathfrak{g}}(s\mathring{A})) = \mathcal{E}xp_{\mathfrak{g}}(t\,s\,\mathring{A}), \quad t, s > 0$$

3.10 Corollary

Let G, $\mathfrak{g}$ be as before. If $(\tau_t) \subseteq \mathrm{Aut}(G)$ is a contracting group and A is stable, w.r.t. (τ_t), then A is the generating distribution of a c.c.s. of (not necessarily full) operator-stable (in the sense of M. Sharpe) measures on the vector space $\mathfrak{g} = \mathbb{R}^r$. On the other hand let $\mathring{A}$ be the generating distribution of an operator stable measure $\mu_1 = \mathcal{E}xp_{\mathfrak{g}}\mathring{A} \in M^1(\mathfrak{g})$ (- the measure defines uniquely the semigroup and therefore $\mathring{A}$ -). Assume that there exists a contracting group of automorphisms ($\delta_t = t^B) \subseteq$ Aut ($\mathbb{R}^r$), such that (i) the exponent B is a derivation of $\mathfrak{g}$ and

(ii) $\delta_t\ (\mathring{A}) = t\mathring{A} + X_t$, (for some primitive X_t, $t > 0$).

(Equivalently: $\delta_t\, \mathcal{E}xp_{\mathfrak{g}}\,(s\mathring{A}) = \mathcal{E}xp_{\mathfrak{g}}\,(st A) * \mathcal{E}_{x(s,t)}$
for some $x\,(s,\,t) \in \mathfrak{g}$, $s,t > 0$,

Then $A := \exp\,(\mathring{A})$ is stable w.r.t. (τ_t) (defined via $\tau_t = \exp\circ\delta_t\circ\log$).

<u>3.11 Remarks</u>

1. If ($\mathbb{R}$,+) is regarded as a commutative Lie algebra, identified with the corresponding Lie group, then the stable generating distributions w.r.t. some (t^B) are just the generating distributions of operator stable probabilities in the sence of M. Sharpe. So via the Lévy-Hinčin-representation ([37, 38,25,29 32]) the structure is completely known: There exists a vectorspace-decomposition
$\mathfrak{g} = \mathbb{R}^r = V_1 \oplus V_2$ (V_1 or V_2 may be $\{0\}$),
$A = A_1 \oplus A_2$, such that A_1 is Gaussian (on V_1) and A_2 is completely non Gaussian. A_2 admits a desintegration $A_2 = \int_\Gamma B_\gamma\, d\nu(\gamma)$, where the B_γ ' s are generating distributions concentrated on t^B-orbits and the mixing measure ν is concentrated on an analytic set, which is intersected by each t^B-orbit in exactly one point.

2. The measure μ_1 determines uniquely the c.c.s. and therefore the generating distribution, but the group (t^B) is not uniquely determined (s. e.g. [24, 26]).

3. If we identify the Lie algebra $\mathfrak{g}$ - as vectorspace - with $\mathbb{R}^r$, then the structure of the stable generating distributions w.r.t. $(\delta_t = t^B)$ is completely known (s. 1.)
<u>Attention:</u> the vectorspace-decomposition $\mathfrak{g} = V_1 \oplus V$ is in general not compatible with the Lie algebra structure. This is easily shown if we look at the simplest nilpotent, non-commutative algebras, e.g. the Heisenberg algebra. Therefore, if $A = A_1 \oplus A_2$ is the decomposition of A in a Gaussian and a non Gaussian part, then $\mathcal{E}xp_G(tA)$ is in general not representable as $\mathcal{E}xp_G\,(tA_1) \otimes \otimes\, \mathcal{E}xp_G(tA_2)$. (Of course via the Lie Trotter formula we have

$$\mathcal{E}xp_G\,(tA) = \lim_n \left[\mathcal{E}xp_G\,(\tfrac{t}{n}A_1) * \mathcal{E}xp_G\,(\tfrac{t}{n}A_2)\right]^n.)$$

4. Changing the point of view we may look at the problem in the following way: Fix $r \geqslant 1$. On $\mathbb{R}^r$ we have different brackets $[\ ,\ ]_\alpha$ such that $\mathfrak{g}_\alpha = (\mathbb{R}^r, [\ ,\ ]_\alpha)$ becomes a nilpotent Lie algebra. W.l.o.g. we suppose (see 3.5) that these algebras admit positive graduations. Let $\mathfrak{g}_0$ be the commutative Lie algebra $\mathbb{R}^r$. For any α let G_α be the corresponding simply-connected (nilpotent) Lie group.
Define: $\mathcal{O}_p(r) := \{A \in \mathcal{MD}(\mathbb{R}^r): A \text{ is operator stable}\}$

For any α let D_α be the set of devivations B of $\mathfrak{g}_\alpha$, s.t. (t^B) is contracting. Then we define for fixed α, $B \in D_\alpha$:

$$\text{Stab}(\alpha, B) := \{A \in \mathcal{MD}(\mathbb{R}^r): A \text{ is stable w.r.t. } (t^B)\},$$

and $\text{Stab}(\alpha) := \bigcup_B \text{Stab}(\alpha, B)$ the set of stable generating distributions on $\mathfrak{g}_\alpha$.
Then $\mathcal{O}_p(r) = \bigcup_\alpha \text{Stab}(\alpha) = \text{Stab}(o)$.
As any simply-connected nilpotent Lie group G which admits a contracting group of automorphisms is isomorphic to one of the G_α's, we get a description of the stable generating distributions for any dimension $\dim(G) = r$.

5. Our knowledge of the structure of the generating distribution includes only partial answers to important questions. We shall explain this in the following discussion:
<u>Question</u>: For which stable distributions A the c.c.s. $(\mathcal{E}xp_G(tA))$ on the group G consists of absolutly continuous measures. We know (s. [74] Korollar 5): If the absolutly continuous part of A is infinite, then $(\mathcal{E}xp_G(tA))$ consists of absolutly continuous measures. As the Haar measures on simply-connected nilpotent Lie groups of fixed dimension r are equivalent via $\exp_G$ to the Lebesque measure on $\mathbb{R}^r$, one might suppose: If $\mathcal{E}xp_G(tA)$ is absolutly continuous, then for any group structure G' on $\mathbb{R}^r$ $\mathcal{E}xp_{G'}(tA)$ is absolutly continuous. But it is easily shown, that this is not true, for example if we examine stable Gauss-c.c.s. on $\mathbb{R}^{2n+1}$ and on the Heisenberg group $\mathbb{H}_n$.

6. In his thesis [77] E. Siebert introduced Gauß-measures on arbitrary locally compact groups: measures which are embeddable into a c.c.s. , the generating distribution of which is of local character. It turned out, that this generalization of the classical

situation is the "correct" one, as it covers several possible generalizations as subcases. (For a discussion the reader is referred to [73,77]). But there is no connection to stability in the general situation. In the situation studied in this section (simply-connected nilpotent Lie groups) it is possible to describe the intersection of the stable (w.r.t. contracting groups of automorphisms) and the Gaussian generating distributions: First we have to remark, that there exist nilpotent Lie algebras, admitting only unipotent automorphisms. So even in the nilpotent case there do exist Gaussian c.c.s. which are not stable (in the sense defined above). Assume that G admits a contracting group $(\tau_t) \subseteq \mathrm{Aut}(G)$, assume further, that A generates a Gauß semigroup which is stable w.r.t. (τ_t). Then $\mathring{A} := \log(A)$ is Gaussian too and stable w.r.t. (δ_t) (and vice versa). So it is sufficient to describe the set of Gaussian generating distributions on (the vector space) $\mathfrak{g}$, which are operator stable w.r.t. (δ_t). We identify $\mathfrak{g}$ with the first order invariant differential operators. Let $\mathfrak{g} = \sum_{\gamma>0}^{\oplus} V_\gamma$ be a positive graduation with $\delta_t V_\gamma = V_\gamma, t, \gamma > 0$. Let further $D_{\gamma 1}, \ldots D_{\gamma k_\gamma}$ be a basis of V_γ $(\neq \{0\})$ and we represent $\mathring{A}$ in the form

$$\mathring{A} = \sum_{\gamma,\gamma',i,j} a_{\gamma\gamma' ij} \; D_{\gamma i} \; D_{\gamma' j} + \sum_{\gamma,i} b_{\gamma i} \; D_{\gamma i} .$$

Now $\delta_t = t^B$ with $B = \sum \oplus B_\gamma$

and the stability condition $\delta_t \mathring{A} = t\mathring{A} + X_t$ for Gaussian generating distributions implies that on the subspace of $\mathfrak{g}$, on which $\mathring{A}$ is non degenerate the exponent B has no nilpotent part (s. [38, 24]). This gives a system of equations determining the possible $(a_{\gamma\gamma' ij}, b_{\gamma i})$ which correspond to (δ_t)-stable Gaussian distributions. These equations become less complicated if we restrict to the case of canonical automorphisms, i.e. to $B_\gamma = r(\gamma)\, \mathrm{id}_{V_\gamma}$, $r(\gamma) > 0$

with $r(\gamma) + r(\gamma') = r(\gamma + \gamma')$ if $[V_\gamma, V_{\gamma'}] \neq \{0\}$

Especially for the Heisenberg groups we get a simple description of the stable Gauß-generators. The details are left to the reader.

7. Stable Gauß semigroups play an important role in the central limit theorems, s.e.g. [55,62,63 also 69,58]. There the possible limits of $\{(\tau_{t_n}(\mu) * \varepsilon_{x(n)})^n\}$ are considered. The papers are organized as follows: In the first step the "norming constants" τ_t are assumed to be automorphisms, the limits are therefore stable (Gaussian) measures. In the second step (in the nilpotent case) the τ_t's are

assumed to be "almost automorphisms" (s. [65,66]), i.e. automorphisms of a graduated Lie group which corresponds in a canonical manner to the given one. These "almost automorphisms" are one of the reasons why in the definition of stability (1.3) we allowed the norming group (τ_t) to belong to a rather general class of transformations.

7a. Conjecture: In the case of nilpotent (not necessarily positive graduated) Lie groups the stable measures w.r.t. a group of "almost automorphisms" behave like the stable measures described above.

7b. Conjecture: It is possible to describe the domain of attraction of stable measures w.r.t. (τ_t) via moment condition similar as in [68, 55, 62, 63]. (The moments are defined w.r.t. a (τ_t) homogeneous norm, i.e. an invariant metric $\|\cdot\|$ on G, s.t. $\|\tau_t(\cdot)\| = t\|\cdot\|$.)

8. In connection with the central limit theorem it is necessary to mention the important paper [76] : Siebert gives necessary conditions for the convergence of triangular systems of probabilities in terms of the (operator valued) Fourier transforms. Let $(\tau_t)_{t\geq 0} \subseteq \mathrm{Aut}(G)$ be as above and $\mu \in M^1(G)$. Then $\{\mu_{nk} = \tau_{t_n}(\mu),\ 1 \leq k \leq n\}$ is a triangular system, the row products are just the measures $[\tau_{t_n}(\mu)]^n$ which appeared in (2.6, 2.7).

The automorphisms τ_t operate in a canonical manner on $\hat{G}$ (the equivalence classes of irreducible representations of G . Therefore $\tau_t(\mu)^\wedge(\gamma) = \hat{\mu}(\tau_t(\gamma))$, $t > 0$ $\gamma \in \hat{G}$.
so it is easy to formulate Siebert's condition in this case.

There exists an interesting class of groups containing the Heisenberg groups, for which
(i) there exist canonical automorphisms (τ_t) and
(ii) a set $\Gamma \subsetneq \hat{G}$ - in the case of the Heisenberg group $\mathbb{H}_1$ a two-point set - such that the complement of $\{\tau_t(\gamma), t > 0, \gamma \in \hat{G}\}$ is a zero-set w.r.t. Plancherel measure. So in this case the values of $\hat{\mu}(\gamma), \gamma \in \Gamma$ decide wether μ is in the domain of attraction of a stable law.

There are still some open problems and interesting questions concerning stability on nilpotent Lie groups. Details will appear in a subsequent paper.

4. Concluding remarks

4.1 In 3. we restricted to the case of nilpotent groups, as we assumed the existence of a contracting group of automorphisms. (τ_t) is contracting iff for any $\mu \in M^1(G)$ $\lim_{t \downarrow 0} \tau_t(\mu) = \varepsilon_e$.

If we consider c.c.s. (μ_t) with non trivial idempotent factors ω_K - the Haar measure on a compact subgroup $K \subseteq G$ - the contraction condition may be replaced by the following: (τ_t) is K-contracting if for $V \in \mathcal{U}(e)$, M compact $\subseteq G$ $\tau_t(M) \subseteq K + V$ for small t.

In his talk at the last conference on probabilities on groups in Oberwolfach P. Baldi [54] gave a definition of stable measures on the group of motions of $\mathbb{R}^d$. It turns out that the measures $\mu \in M^1(\mathbb{R}^d \circledS SO(d))$ which are stable in the sense of Baldi (Definition 2) are embeddable into a c.c.s. μ_t, with idempotent factor $\omega_{SO(d)}$, such that $\tau_t(\mu_s) = \mu_{st}$, t,s > 0 . There (τ_t) is a group of SO(d)-contracting automorphisms $(x, k) \longrightarrow (t x, k_t^{-1} k k_t)$, $t > 0$. $k_t \in SO(d)$.

Baldi's results encourage to the following conjecture: For a wide class of compact extensions of nilpotent Lie groups with positive graduation (K compact, N simply-connected, nilpotent, G extension of N by K) there exist K-contracting automorphisms (τ_t), and the stable c.c.s. on G with nontrivial idempotent factor may be represented as stable c.c.s. in the sence of (1.3) on N, which are invariant under some compact group of automorphisms of N.

4.2 Let G be a connected Lie group und let $(\tau_t)_{t>0}$ be a group of automorphisms. Then (τ_t) is in general neither contracting nor K-contracting. If we consider measures $\mu \in M^1(G)$ which are concentrated on a part of G, where the τ_t's operate contracting, then the investigations of 4. may be useful to gain limit theorems on $(\tau_{t_n}(\mu))^n$ resp. on stability on non-nilpotent Lie groups.

4.3 There is no problem now to define (and examine) more general classes of limit measures, e.g. semistable, selfdecomposable, mixed - stable measures. Especially for the case of simply connected nilpotent groups with contracting (τ_t) one can use the Lévy-Hinčin representations of the generating distributions in the vector space case (see part B,C,F of the literature) and translate to the group case along the lines in 3.

4.4 The most general concept of stability was introduced by A. Tortrat (s.e.g. [64]). There stability is defined via the transformations $\tau_n : x \longrightarrow x^n, n \in \mathbb{N}$. In the non Abelian case this concept seems to be quite different from the stability concept developed in this paper. But it is still an interesting question to characterize measures, which are stable in the above and in Tortrat's sense.

4.5 There is some danger that the concept of stability becomes to wide, if we relax the conditions on (τ_t). We want to show this in the following example. (Examples of this type were invented by H.Carnal at the 1st conference on Probabilities on groups in Oberwolfach 1970, s. [70, 71, 73]). There exist (non-continuous) convolution semigroups with strange properties, which satisfy a stability condition w.r.t. a rather general group of transformation (τ_t):

Let $\mathbb{Q}$ be the rational numbers endowed with the discrete topology. Let T be the torus, $K := T^{\mathbb{Q}}$, $K_s := T^{\mathbb{Q} \cap (-\infty, s]}$. Define $\beta_t \in \mathrm{Aut}(K)$

$$(\beta_t f)(u) := f(u+t), \quad t, u \in \mathbb{Q}, \quad f \in T^{\mathbb{Q}}.$$

Via β_t the semidirect product $G = K \circledS \mathbb{Q}$ is defined.

Put $\mu_t := \omega_{(K_t, 0)} * \varepsilon_{(0,t)}$. Then (μ_t) is a non-continuous c.s.

Define $\gamma_t : K \ni f \longmapsto \gamma_t f$, $(\gamma_t f)(u) = f(tu)$, $u, t \in \mathbb{Q}$,

and $\tau_t : (f, r) \longmapsto (\gamma_t f, r)$.

Then we get

$$\tau_t(\mu_s) = \mu_{st} * \varepsilon_{y(s,t)} \quad \text{for some } y(s,t) \in G.$$

The discussion in [70, 71, 73] shows, that these measure-semigroups should perhaps be excluded. This is a good motivation to suppose that the norming group (τ_t) depends continuously on $t \in (0, \infty)$.

There is a huge literature on stability and related questions. The following hints are only a sample concentrated on the papers on stability on vector spaces which seem to be important for the case of non abelian groups. The great interest in this field encouraged me to suggest a new definition of stability.

Literature

A) Stability on vector spaces

1. DETTWEILER, E.:"Stabile Maße auf Badrikianschen Räumen". Math.Z. 146 (1976) 149-166.
2. HALL, P.: "A comedy of errors: The canonical form for a stable characteristic function". Bull.London Math.Soc.13 (1981) 23-27.
3. JUREK Z.J., URBANIK K.: "Remarks on stable measures on Banach spaces". Coll. Math. 38 (1978) 269-276.

B) Generalizations (Semistable and self decomposable measures

4. FORST G.: "A characterization of self decomposable probabilities on the half line". Z. Wahrscheinlichkeitstheorie verw.Geb.49 (1979) 349-352.
5. KRAPAVICKAITE D.: "Certain classes of probability distributions". Lith. Math. J.2o (1981) 298-3o4.
6. KRAPAVICKAITE D.: " Generalized semistable probability distributions in Hilbert space". Lith.Math.J. 2o (198o) 111-118.
7. KRUGLOV V.M.: "On an extension of the class of stable distribution functions". Theory Prob.Appl.17 (1972) 723-732.
8. LAHA R.G., ROHATGI V.K.: "Stable and semistable probability measures on a Hilbert space". Essays in Statistics and probability . North Holland, Pub.Co (1981).
9. PILLAI R.N.: "Semistable laws as limit distributions". Ann.Math. Stat. 42 (1971) 78o-783.
1o. SATO K.: "Class L of multivariate distributions". J.Mult.Anal.1o (198o) 2o7-232.
11. SHIMIZU R.: "Characteristic functions satisfying a functional equation I, II". Ann.Inst.Stat.Math. 2o (1968) 187-2o9 and 21 (1969) 391-4o5.
12. URBANIK K.: " Remarks on B-stable probability distributions". Bull Acad.Pol.Sci. (1976) 783-787.
13. URBANIK K.: "A representation of self decomposable distributions". Bull. Acad. Pol.Sci. 16 (1968) 2o9-214.
14. URBANIK K.: "Limit laws for sequences of normed sums satisfying some stability conditions". Multivariate Analysis III (1973) 225-237, Academic Press, New York.
15. URBANIK K.: "Self decomposable probability measures on $\mathbb{R}^m$". Zastosowania Mat. 1o (1969) 91-97.

16. VAN THU, NGUYEN: "Multiply self decomposable probability measures in Banach spaces". Studia Math. 66 (1979/8o) 161-175.

17. VAN THU, NGUYEN: "A characterization of mixed stable laws". Bull. Acad. Pol.Sci. 27 (1979) 629-63o.

18. WOLFE S.J.: "A characterization of Lévy probability distribution functions on Euclidean spaces". J.Mult. Analysis 1o (198o)379-384.

C) Further generalizations

19. BERG CH., FORST G.: "Multiply self decomposable probability measures on $\mathbb{R}$ and Z_+". Preprint (1981) Nr.6, Kopenhavns Univ. Matematisk Inst.

2o. JUREK Z.J.: "Some characterization of the class of s-self-decomposable distributions". Bull.Acad.Pol.Sci.26 (1978) 719-725.

21. JUREK Z.J.: "Properties of s-stable distribution function". Bull. Acad.Pol.Sci. 27 (1979) 135-141.

22. JUREK Z.J.: "Limit distributions for shrunken random variables". Diss.Math. 85 (1981) 1-46.

D) Operator stable measures

23. HUDSON W.N.: "Operator stable distributions and stable marginals". J. Mult. Analysis 10 (198o) 26-37.

24. HUDSON W.N., MASON J.D.: "Exponents of operator stable laws." Probabilities in Banach Spaces 3. Proceedings . Lecture Notes in Math. 86o (1981) 291-298.

25. HUDSON W.N., MASON J.D.: "Operator stable laws". J.Mult.Analysis 11 (1981) 434-447.

26. HUDSON W.N., MASON J.D.: "Operator stable measures on $\mathbb{R}^2$ with multiple exponents". Ann.Prob. 9 (1981) 482-489.

27. JUREK Z.J.: "On stability of probability measures in Euclidean spaces." Probability theory on Banach spaces II.Proceedings . Lecture Notes Math. 828 (198o) 129-145.

28. JUREK Z.J., SMALARA J.: "On integrability with respect to infinitely divisible measures". Bull.Acad.Pol.Sci. (198o).

29. KRAKOWIAK W.: "Operator stable probability measures on Banach Spaces". Coll. Math. 41 (1979) 313-326.

3o. KUCHARCZAK J.: "On operator stable probability measures". Bull. Acad. Pol. Sci.23 (1975) 571-576.

31. KUCHARCZAK J.: "Remarks on operator stable measures". Coll. Math. 34 (1976) 1o9-119.

32. KUCHARCZAK J., URBANIK K.: "Operator stable probability measures on some Banach spaces". Bull.Acad.Pol.Sci.25 (1977) 585-588.

33. Michalicek J.: " Die Randverteilungen der operatorstabilen Maße im zweidimensionalen Raum". Z.Wahrscheinlichkeitstheorie verw.Geb.21 (1972) 135-146.

34. MINCER B., URBANIK K.:"Completely stable measures on Hilbert spaces". Coll. Math. 42 (1979) 3o1-3o7.

35. PARTHASARATHY K.R.:"Every completely stable distribution is normal." Sankhyā 35 (1973) Ser. A, 35-38

36. PARTHASARATHY K.R., SCHMIDT K.: "Stable positive definite functions". Trans. Amer. Math.Soc. 2o3 (1975) 161-174.

37. SHARPE M.: "Operator stable probability measures on vector groups" Trans. Amer.Math.Soc. 136 (1969) 51-65.

38. SCHMIDT K.: "Stable probability measures on $\mathbb{R}^n$ ". Z. Wahrscheinlichkeitstheory verw. Geb. 33 (1975) 19-31.

39. SEMORSKI S.V.:"Operator stable laws of distributions". Soviet Math. Doklady 2o (1979) 139-142

E) Domains of attraction for operator stable measures

4o. JUREK Z.J.: "Central limit theorem in Euclidean spaces". Bull.Acad. Pol.Sci. 28 (198o) 81-86.

41. HAHN M.G., KLASS M.J.: "Matrix normalization of sums of random vectors in the domain of attraction of the multi variate normal". Ann.Prob. 8 (198o) 262-28o.

42. HAHN M.G., KLASS M.J.: "A survey of generalized domains of attraction and operator norming methods". Probabilities on Banach Spaces III, Proceedings . Lecture Notes Math. 86o (1981), 187-218.

43. HAHN M.G., KLASS M.J.: "The generalized domain of attraction of spherically symmetric stable laws on $\mathbb{R}^d$. Probability on Vector spaces II, Proceedings". Lecture Notes Math. 828 (198o) 52-81.

44. HAHN M.G.: "The generalized domain of attraction of a gaussian law in a Hilbert space." Probabilities on Banach spaces 2 .Oberwolfach Lecture Notes in Math. 7o9 (1978) 125-144.

45. SEMORSKII V.M.: "The central limit theorem for sums of random vectors normalised by linear operators". Sovjet Math.Doklady 2o (1979) 356-359.

F) Generalization of operator-stability (semistable and self-decomposable measures)

46. JAJTE R.: "Semistable probability measures on $\mathbb{R}^N$ ". Studia Math.61 (1977) 29-39.

47. JAJTE R.: "S-decomposable measures on a Hilbert space". Bull.Acad. Pol.Sci. 27 (1979) 625-628.

48. JAJTE R.: "V-decomposable measures on a Hilbert space." Probability theory on vector spaces II, Proceedings . (Blazejewko 1979) Lecture Notes in Math. 828 (198o) 1o8-127.

49. JUREK Z.J.: "Limit distributions and one parameter groups of linear operators on Banach spaces". Report 81o6 (1981) Math.Inst.Kath. Univ. Nijmegen.

5o. JUREK Z.J.: "Convergence of types, self decomposibility and stability of measures on linear spaces." Probability on Banach spaces III. Proceedings . Lecture Notes in Math. 86o (1981) 257-284.

51. KRAKOWIAK W.: "Operator semistable probability measures in Banach spaces". Coll. Math. 53 (198o) 351-363.

52. URBANIK K.: "Lévy's probability measures on Banach Spaces". Studia Math. 63 (1978) 283-3o8.

53. URBANIK K.: "Lévy's probability measures on Euclidean Spaces". Studia Math. 44 (1972) 119-148.

G) Stability on groups

54. BALDI P.: "Lois stables sur les déplacements de $\mathbb{R}^d$". Probability measures on groups. Proceedings .Lecture Notes in Math.706 (1979) 1-9.

55. CREPEL P., RAUGI A.: "Théorem central limite sur les groups nilpotents". Ann.Inst. H. Poinceré 14 (1978) 145-162.

56. GALLARDO L.: "Processus subordonnés au mouvement brownien sur les groupes de Lie nilpotents". Compt.Rend.Acad.Sc.Paris 292 (1981) 413-416.

57. HAZOD W.: "Subordination von Faltungs- und Operatorhalbgruppen." Probability measures on groups. Proceedings . Lecture Notes Math. 706 (1979).

58. HULANICKI A.: "The distribution of energy in the Brownien motion in the gaussian field and analytic hyppoellipticity of certain subelliptic operators on the Heisenberg group". Studia Math.16 (1976) 165-173.

59. HULANICKI A.: "A Tauberien property of the convolution semigroup generated by $X^2 - |Y|^\gamma$ on the Heisenberg group." Proceedings . Symposia Pure Math. (AMS) 35 (1979) 403-405.

60. HULANICKI A.: "Commutative subalgebras of L^1 (G) associated with a subelliptic operator on a Lie group G". Bull.Amer.Math.Soc.81 (1975) 121-124.

61. HULANICKI A.: " A class of convolution semi-groups of measures on a Lie group." Probability measures on vector spaces II.Proceedings . Lecture Notes Math. 828 (1980) 82-101.

62. RAUGI A.: "Théoreme de la limite centrale sur les groups nilpotents". Z. Wahrscheinlichkeitstheorie verw.Geb. 43 (1978)149-172.

63. RAUGI A.: "Théoreme de limite centrale pour un produit semidirect d'un groupe de Lie résoluble simplement connexe de type rigide par un groupe compact." Probability measures on groups.Proceedings Lecture Notes Math. 706 (1979) 257-324.

64. TORTRAT A.: "Lois stables dans un groupe". Ann.Inst.H.Poincaré Sect.B17 (1981) 51-61.

H) Convolution semigroups, generating distributions, structure of groups

65. GOODMAN R.: "Nilpotent Lie groups: Structure and applications to analysis". Lecture Notes Math. 562 (1976).

66. GOODMAN R.: "Filtration and asymptotic automorphisms on nilpotent Lie groups". J. Diff. Geometry 12 (1977) 183-196.

67. GOODMAN R.: "Filtrations and Canonical Coordinates on Nilpotent Lie groups". Trans. Amer.Math.Soc. 237 (1978) 189-204.

68. GUIVARC'H Y.: "Croissance polynomiale et pédiodes des fonctions harmoniques". Bull.Soc.Math.de France 101 (1973) 333-379.

69. GUIVARC'H Y.: "Sur la loi des grands nombres et le rayon spectral d'un marche aléatoire". Asterisque 74 (1980) 47-98.

70. HAZOD W.: "Stetige Halbgruppen von Wahrscheinlichkeitsmaßen und erzeugende Distributionen". Lecture Notes Math.Vol.595 (1977).

71. HAZOD W.: "Über Faltungshalbgruppen von Wahrscheinlichkeitsmaßen". Österreich Acad.Wiss.Math.-Natur.Kl.S.-B.II, 181, 29-47 (1973).

72. HAZOD W.: "Über Faltungshalbgruppen von Wahrscheinlichkeitsmaßen II: Einige Klassen topologischer Gruppen". Monatsh. für Math. 79, 25-45 (1975)

73. HEYER H.: "Probability measures on locally compact groups." Ergebnisse der Math. Berlin-Heidelberg-New York, Springer 1977

74. JANSSEN A.: "Charakterisierung stetiger Faltungshalbgruppen durch das Lévy-Maß".Math.Ann. 246, 233-240 (1980).

75. NAGEL A., STEIN E.M.: "Lecture on pseudodifferential operators". Princeton Univ. Press 1979.

76. SIEBERT E.: "Fourier Analysis and Limit theorems for convolution semigroups on a locally compact group". Adv.Math.39(1981)111-154.

77. SIEBERT E.: "Wahrscheinlichkeitsmaße auf lokalkompakten maximal fastperiodischen Gruppen". Dissertation, Tübingen 1972.

78. SIEBERT E.: "Über die Erzeugung von Faltungshalbgruppen auf beliebigen lokalkompakten Gruppen". Math.Z.131, 313-333 (1973).

79. SIEBERT E.: "On the generation of convolution semigroups on arbitrary locally compact groups II". Arch.Math.(Basel) 28, 139-148 (1977).

80. METIVIER, G.: "Hypoellipticité analytique des groupes nilpotents de rang 2" Duke M.J.47 (1980) 195-221.

81. HELFFER, B.: "Hypoellipticité analytique sur des groupes nilpotents de rang 2" (d'apres G.Metivier). Seminaire Goulaouic-Schwartz 1979/80 Nr. 1

Wilfried Hazod
Universität Dortmund
Abteilung Mathematik
Postfach 500 500
D-4600 Dortmund 50

Supplement to the preceding paper

Stable Probabilities on locally compact groups.

(On a theorem of E. Lukacs)

1. E. Lukacs [1] proved the following theorem:

Theorem 1 Assume that X_1, X_2 are independent, identically distributed real random variables whose distribution is symmetric stable with exponent α. Assume further, that Z_1, Z_2 are nonnegative real random variables.
Then the distribution of

$$W := \left(\frac{Z_1}{Z_1+Z_2}\right)^{\frac{1}{\alpha}} \cdot X_1 + \left(\frac{Z_2}{Z_1+Z_2}\right)^{\frac{1}{\alpha}} \cdot X_2$$

is a symmetric stable distribution with exponent α.

In the preceding paper we introduced a new concept of stability which extends the class of operator stable laws on $\mathbb{R}^n$. So it is quite natural to ask, if a suitable version of this theorem holds in the general context.
We put $f := Z_1 / (Z_1+Z_2)$, then $0 \leq f \leq 1$ a.s. For fixed $\alpha \in (0,2]$ and $t > 0$ we define $\tau_t^{(\alpha)} : x \to t^{1/\alpha} x$. Finally let $(\mu_t)_{t \geq 0}$ be the uniquely determined semigroup (of symmetric stable measures) such that the distribution of X_i is μ_1, $i = 1,2$.
Then theorem 1 reads as follows:
The distribution of the random variable

$$(1.1) \quad W(\cdot) := \tau_{f(\cdot)}^{(\alpha)} (X_1(\cdot)) + \tau_{1-f(\cdot)}^{(\alpha)} (X_2(\cdot)) \text{ is } \mu_1.$$

If F is the distribution of f, $x \in (0,1)$, then we have for any bounded Borel function h

$$(1.2) \quad E(h(W) \mid f = x) = \int h \, d(\tau_x^{(\alpha)}(\mu_1) * \tau_{1-x}^{(\alpha)}(\mu_1))$$

and therefore

$$(1.3) \quad \mu_1 = \int_{[0,1]} (\tau_x^{(\alpha)}(\mu_1)) * (\tau_{1-x}^{(\alpha)}(\mu_1)) \, dF(x).$$

(We define $\tau_0^{(\alpha)}(\mu_1) := \mu_0 = \varepsilon_0$.)

2. It turns out, that (1.3) holds for strictly stable probabilities in the sense of the definition of the preceding paper:
Let G be a locally compact group. Assume that there exists a group of transformations $\tau_t : G \to G$, $t > 0$, s.t. $\tau_t \tau_s = \tau_{ts}$, $t,s, > 0$. Assume further that the continuous convolution semigroup $(\mu_t)_{t \geq 0}$ is strictly stable (in the sense of the definition of the preceding paper) w.r.t. (τ_t),
(2.1) i.e. $\tau_t(\mu_s) = \mu_{st}$, $s,t \geq 0$. (We define again $\tau_o(\mu_s) := \mu_o$)

<u>Theorem 2</u> Let X_1, X_2 be independent identically distributed G-valued random variables with distribution μ_1. Then the distribution of

$$(1.1') \qquad W(\cdot) := \tau_{f(\cdot)}(X_1) \cdot \tau_{1-f(\cdot)}(X_2)$$

is again μ_1 for any real random variable f $(0 \leq f(\cdot) \leq 1)$.

Equivalently:

$$\int_\Omega (\tau_{f(u)}(\mu_1)) * (\tau_{1-f(u)}(\mu_1)) \, d P(u) =$$

$$(1.3') \qquad \int_{[0,1]} \tau_x(\mu_1) * \tau_{1-x}(\mu_1) \, dF(x) = \mu_1 .$$

(The integrals converge in $\sigma(M^b(G), C_o(G))$.)

Proof: (2.1) implies that $\tau_t(\mu_1) = \mu_t$, $t \geq 0$

and therefore $\tau_x(\mu_1) * \tau_{1-x}(\mu_1) = \mu_x * \mu_{1-x} = \mu_1$ for any $x \in [0,1]$.

Therefore (1.3') holds trivially. On the other hand we have for any $h \in C_o(G)$:

$$E(h(W)) = \int_{[0,1]} E(h(W) \mid f = x) \, dF(x) =$$

$$= \int_{[0,1]} \int_G h(z) d(\tau_x(\mu_1) * \tau_{1-x}(\mu_1)) (z) \, dF(x),$$

i.e. the distribution of W is $\int_{[0,1]} \tau_x(\mu_1) * \tau_{1-x}(\mu_1) \, dF(x)$.

In the preceding paper we pointed out, that there is a large class of convolution semigroups (μ_t) and of transformations (τ_t) for which theorem 2 holds.

3. Assume that X_1, X_2 are independent G-valued random variables with distribution μ, such that (1.3') holds. If we consider constant random

variables $f \equiv x$, $x \in (0,1)$, we obtain the following relation:

$$(3.1) \quad \tau_x(\mu) * \tau_{1-x}(\mu) = \mu, \quad x \in (0,1).$$

The proof of theorem 2 shows that the distributions μ which fulfill the weak stability conditions (3.1) are just the distributions for which E.Lukacs' theorem holds.

Literature:

[1] E. Lukacs, On some properties of symmetric stable distributions. Analytic function methods in probability theory. Coll.Math.Soc. J.Bolyai 21, North Holland, Amsterdam, 1980

ZEITGEORDNETE MOMENTE DES WEISSEN KLASSISCHEN

UND DES WEISSEN QUANTENRAUSCHENS

P.D.F. ION[+)]
Mathematical Reviews
University of Michigan
611 Church Street,
Ann Arbor, Michigan 48109

W. v. WALDENFELS[+)]
Institut für Angewandte Mathematik
Universität Heidelberg
Im Neuenheimer Feld 294
6900 Heidelberg

ZUSAMMENFASSUNG

Da wir uns nur für die Momente interessieren, werden die Rauschprozesse als lineare Funktionale auf Tensoralgebren dargestellt. Die zeitgeordneten Momente des weißen Rauschens werden nach dem Stratonowitsch-Verfahren durch Approximation mit farbigem Rauschen berechnet. Um die zeitgeordneten Momente algebraisch als Funktional einer Algebra verstehen zu können, muß die Algebra erweitert werden mit Hilfe des shuffle- oder Hurwitz-Produkts. Diese Überlegungen eröffnen die Möglichkeit, die stochastische Differentialgleichung zu diskutieren, die das Verhalten eines Zwei-Niveau-Atoms im Wärmebad von Photonen beschreibt.

§ 1. EINLEITUNG

Das klassische weiße Rauschen ist im einfachsten Fall ein reellwertiger (verallgemeinerter) stochastischer Prozeß, der Gaußsch und δ-korreliert ist. Seine Momente sind

$$E\, X(t)X(s) = \sigma^2 \delta(t-s)$$

$$E\, X(t_1)X(t_2)X(t_3)X(t_4) = \sigma^4 \left[\delta(t_1-t_2)\delta(t_3-t_4) + \delta(t_1-t_3)\delta(t_2-t_4) + \delta(t_1-t_4)\delta(t_2-t_3) \right]$$

Die höheren geraden Momente werden in ähnlicher Weise berechnet, die ungeraden Momente verschwinden alle. Wir interessieren uns hier für

+) Die Arbeit wurde im Rahmen des Sonderforschungsbereichs 123 "Stochastische Mathematische Modelle" mit Unterstützung der Deutschen Forschungsgemeinschaft angefertigt.

Ausdrücke der Form

$$E \iint\limits_{0 \le t_1 \le t_2 \le t} dt_1\, dt_2\; X(t_2)\, X(t_1)$$

$$E \iiiint\limits_{0 \le t_1 \le t_2 \le t_3 \le t_4} dt_1\, dt_2\, dt_3\, dt_4\; X(t_4)\, X(t_3)\, X(t_2) X(t_1)$$

die wir zeitgeordnete Momente nennen können. Diese sind durch die oben angeführten gewöhnlichen Momente nicht eindeutig definiert. Es ist z.B. formal

$$E \iint\limits_{0 \le t_1 \le t_2 \le t} X(t_2)\, X(t_1)\, dt_1\, dt_2 = \sigma^2 \iint\limits_{0 \le t_1 \le t_2 \le t} \delta(t_2 - t_1)\, dt_1\, dt_2$$

$$= \sigma^2 \int_0^t dt_2 \int_0^{t_2} dt_1\, \delta(t_1)$$

Das Integral $\int_0^{t_2} \delta(t_1)\, dt_1$ ist aber nicht eindeutig definiert und hängt davon ab, wie die δ -Funktion approximiert wird.

Im Rahmen der stochastischen Integration hat man zwei Möglichkeiten. Entweder man faßt $\int_0^t X(t_1) dt_1 = B(t)$ als Brownsche Bewegung auf und schreibt das Integral

$$\iint\limits_{0 \le t_1 \le t_2 \le t} X(t_2)\, X(t_1)\, dt_1\, dt_2 = \int_0^t B(t_2)\, dB(t_2)$$

als Ito-Integral. Dann ist der Erwartungswert 0. Oder man interpretiert das Integral als Stratonowitsch-Integral. Dann muß man das weiße Rauschen durch farbiges Rauschen approximieren, d.h. man nimmt an, daß

$$E\, X(t)\, X(s) = k(t-s)$$

eine stetige Funktion ist, und geht dann mit k zur δ -Funktion über.

Da k symmetrisch ist, ergibt sich im Limes stets

$$E \iint_{0 \le t_1 \le t_2 \le t} X(t_2) X(t_1)\, dt_1\, dt_2 = \frac{\sigma^2}{2} t$$

Allgemein erhält man

$$E \int_{t_1 \le \dots \le t_{2m}} \cdots \int X(t_{2m}) X(t_{2m-1}) \cdots X(t_1)\, \Phi(t_1, t_2, .., t_{2m})\, dt_1\, dt_2 \dots dt_{2m}$$

$$= \left(\frac{\sigma^2}{2}\right)^m \int_{s_1 \le \dots \le s_m} \cdots \int \Phi(s_1, s_1, s_2, s_2, \dots, s_m, s_m)\, ds_1 \dots ds_m$$

Während der Erwartungswert eines Monoms vom Grade 2m eine Summe über alle Paarzerlegungen von $\{1, \dots, 2m\}$ ist, also $(2m)!/(2^m m!)$ Terme enthält, tritt beim zeitgeordneten Moment nur ein Term auf.

Das komplexe weiße Rauschen Z(t) ist ein verallgemeinerter komplexer stochastischer Prozeß mit den Korrelationen

$$E\, Z(t)\, Z(s) = E\, \bar{Z}(t)\, \bar{Z}(s) = 0$$

$$E\, Z(t)\, \bar{Z}(s) = \sigma^2 \delta(t-s)$$

Berechnet man zeitgeordnete Momente 2. Ordnung nach dem Stratonowitsch-Verfahren durch Approximation durch farbiges Rauschen, so findet man

$$E \iint_{0 \le t_1 \le t_2 \le t} Z(t_2)\, \bar{Z}(t_1)\, dt_1\, dt_2 = \varkappa t$$

$$E \iint_{0 \le t_1 \le t_2 \le t} Z(t_1)\, \bar{Z}(t_2)\, dt_1\, dt_2 = \bar{\varkappa} t$$

wo $2\,\mathrm{Re}\,\varkappa = \sigma^2$ ist. Hier kann also eine rein imaginäre Konstante hinzutreten, deren Wert nicht durch σ^2, also durch die Korrelationsfunktion gegeben ist. Höhere zeitgeordnete Momente können ähnlich wie beim reellen weißen Rauschen berechnet werden. Das <u>weiße Quantenrauschen</u> ist ein verallgemeinerter stochastischer nichtkommutativer Prozeß $F(t)$, den wir formal definieren wollen, um nicht in große analytische Schwierigkeiten hineingezogen zu werden. Er ist gegeben durch die Korrelationsfunktionen

$$E\, F(t)\, F(s) = E\, F^*(t)\, F^*(s) = 0$$

$$E\, F(t)\, F^*(s) = (1+\vartheta)\sigma^2 \delta(t-s)$$

$$E\, F^*(t)\, F(s) = \vartheta\, \sigma^2\, \delta(t-s)$$

mit $\vartheta \geqslant 0$. Durch Approximation durch farbiges Rauschen erhält man die Ausdrücke

$$E \iint_{0 \le t_1 \le t_2 \le t} F(t_2)\, F^*(t_1)\, dt_1\, dt_2 = (1+\vartheta)\, \varkappa\, t$$

$$E \iint_{0 \le t_1 \le t_2 \le t} F(t_1)\, F^*(t_2)\, dt_1\, dt_2 = (1+\vartheta)\, \bar{\varkappa}\, t$$

$$E \iint_{0 \le t_1 \le t_2 \le t} F^*(t_2)\, F(t_1)\, dt_1\, dt_2 = \vartheta\, \bar{\varkappa}\, t$$

$$E \iint_{0 \le t_1 \le t_2 \le t} F^*(t_1)\, F(t_2)\, dt_1\, dt_2 = \vartheta\, \varkappa\, t$$

mit $\sigma^2 = 2\,\mathcal{R}\,\varkappa$.

Bei zeitgeordneten Momenten hat man also auf die Zeitreihenfolge und die Reihenfolge der Faktoren zu achten. Daher wird der Ausdruck für höhere zeitgeordnete Momente schwieriger, er besteht zwar nur aus einem Term, wir müssen aber eine allgemeine Zeitreihenfolge $t_{\pi(1)} \le \dots \le t_{\pi(2m)}$ berücksichtigen, wo π eine Permutation der Zahlen von 1 bis 2m ist.

Da uns nur Momente interessieren, und da wir, wie oben angedeutet, so analytisch Schwierigkeiten vermeiden können, werden wir das weiße

Rauschen algebraisch fassen als Funktionale auf Tensoralgebren. Für die zeitgeordneten Momente benötigen wir etwas mehr: Tensoralgebren mit zusätzlichem Hurwitz- oder shuffle-Produkt (vgl. [1], [2]).

Bei der Definition des Quantenrauschens ohne Zeitordnung ist es möglich, von der Tensoralgebra zur Weyl-Algebra überzugehen, in der die $F(t)$, $F^*(t)$ der zusätzlichen Relation

$$F(t)\,F^*(s) - F^*(s)\,F(t) = \sigma^2\,\delta(t-s)$$

genügen [3]. Diese algebraische Operation scheint bei zeitgeordneten Momenten und bei Tensoralgebren mit Hurwitz-Produkt weit umständlicher zu sein und wurde daher weggelassen.

§ 2. GAUSS'SCHE FUNKTIONALE

Sei V ein komplexer Vektorraum, sei $T(V)$ die Tensoralgebra auf V, und sei β eine Bilinearform $V \times V \to \mathbb{C}$. Ein Gaußsches Funktional γ auf $T(V)$ bezüglich β ist ein lineares Funktional auf $T(V)$ mit den Eigenschaften $(x_i \in V)$

$$\gamma(1) = 1$$
$$\gamma(x_1 \otimes x_2) = \beta(x_1, x_2)$$
$$\gamma(x_1 \otimes x_2 \otimes x_3 \otimes x_4) = \beta(x_1, x_2)\,\beta(x_3, x_4) + \beta(x_1, x_3)\,\beta(x_2, x_4) + \beta(x_1, x_4)\,\beta(x_2, x_3)$$

Allgemeiner

$$\gamma(x_1 \otimes \cdots \otimes x_{2m+1}) = 0$$

$$\gamma(x_1 \otimes \cdots \otimes x_{2m}) = \sum_{\{S_1, \ldots, S_m\}} \beta(x_{S_1}) \cdots \beta(x_{S_m})$$

Dabei läuft $\{S_1, \ldots, S_m\}$ durch alle Paarzerlegungen von $\{1, \ldots, 2m\}$. Eine Paarzerlegung ist eine Zerlegung, in der die zur Zerlegung gehörenden Teilmengen genau zwei Elemente haben. Sei $S = \{i,j\}, i<j$ solch ein Paar, dann setzen wir $\beta(x_S) = \beta(x_i, x_j)$

Das klassische reelle weiße Rauschen kann wie folgt algebraisch definiert werden. Wir setzen $V = L^2(\mathbb{R})$ und betrachten $T(V)$, die algebraische Tensoralgebra. Das weiße Rauschen wird definiert als das Gaußsche Funktional, das zur Bilinearform

$$f,g \in L^2(\mathbb{R}) \longmapsto \beta(f,g) = \sigma^2 \int fg = \sigma^2 \int f(t) g(t)\, dt$$

gehört. In der Tat identifizieren wir f mit

$$X(f) = \int X(t) f(t)\, dt$$

so ist

$$E\, X(f) X(g) = \sigma^2 \int fg$$

und z. B.

$$\begin{aligned} & E\, X(f_1) X(f_2) X(f_3) X(f_4) \\ & = \sigma^4 \Big[\int f_1 f_2 \cdot \int f_3 f_4 + \int f_1 f_3 \cdot \int f_2 f_4 \\ & \qquad + \int f_1 f_4 \cdot \int f_2 f_3 \Big] \\ & = \gamma(f_1 \otimes f_2 \otimes f_3 \otimes f_4). \end{aligned}$$

Das farbige Rauschen wird durch das Gaußsche Funktional, das zur Bilinearform

$$\beta(f,g) = \iint f(s)\, k(s-t)\, g(t)\, ds\, dt$$

gehört, definiert. Dabei ist k eine stetige positiv definite reelle symmetrische Funktion.

Für das komplexe Rauschen setzen wir $V = L^2(\mathbb{R}) \otimes \mathbb{C}^2$. Die Elemente können wir also schreiben $x = \binom{f}{g}$, $f, g \in L^2(\mathbb{R})$ und identifizieren x mit $\begin{pmatrix} Z(f) \\ \bar{Z}(g) \end{pmatrix}$ wo

$$Z(f) = \int Z(t) f(t)\, dt$$

$$\bar{Z}(g) = \int \bar{Z}(t) g(t)\, dt$$

bedeutet. Das weiße Rauschen wird dann definiert als das Gaußsche Funktional auf T(V), das zur Bilinearform

$$\beta\left(\binom{f_1}{g_1}, \binom{f_2}{g_2}\right) = \sigma^2\left(\int f_1 g_2 + \int g_1 f_2\right)$$

$$= \sigma^2 \int\int ds\, dt \quad (f_1, g_1)(s) \begin{pmatrix} 0 & \delta(t-s) \\ \delta(t-s) & 0 \end{pmatrix} \binom{f_2}{g_2}(t)$$

gehört.

Das komplexe farbige Rauschen wird definiert als das Gaußsche Funktional auf T(V), das zur Bilinearform

$$\beta\left(\binom{f_1}{g_1}, \binom{f_2}{g_2}\right) = \int\int ds\, dt \quad (f_1, g_1)(s) \begin{pmatrix} 0 & k(s-t) \\ \bar{k}(s-t) & 0 \end{pmatrix} \binom{f_2}{g_2}(t)$$

gehört. Dabei ist k(t) positiv definit und stetig. In der ursprünglichen Schreibweise $Z(t)$, $\bar{Z}(t)$ bedeuten diese Definitionen, daß das weiße Rauschen die Kovarianzfunktionen

$$E\, Z(s)\, \bar{Z}(t) = \sigma^2\, \delta(s-t)$$

$$E\, Z(s)\, Z(t) = E\, \bar{Z}(s)\, \bar{Z}(t) = 0$$

und das farbige Rauschen die Kovarianzfunktionen

$$E\, Z(s)\, \bar{Z}(t) = k(s-t)$$

$$E\, Z(s)\, Z(t) = E\, \bar{Z}(s)\, \bar{Z}(t) = 0$$

hat.

Derselbe Vektorraum $V = L^2(\mathbb{R}) \otimes \mathbb{C}^2$ dient zur Definition des Quantenrauschens. Wir identifizieren $x = \binom{f}{g}$ mit $\begin{pmatrix} F(f) \\ F^*(g) \end{pmatrix}$ mit

$$F(f) = \int F(t) f(t)\, dt$$

$$F^*(f) = \int F^*(t) f(t)\, dt$$

Während beim weißen komplexen und reellen Rauschen die Größen Z(t) bzw. X(t) eine unabhängige Bedeutung als verallgemeinerter stochastischer Prozeß besaßen, schreiben wir F(t), $F^*(t)$ eine solche unabhängige Bedeutung nicht zu. Wir verstehen sie nur als symbolischen Integranden für F(f) und dies nur als Ausdruck für $\binom{f}{0}$. Es gibt analytische Definitionen für F(t), $F^*(t)$ (s. [4]), doch wollen wir hier auf diese nicht eingehen.

Das weiße Quantenrauschen ist definiert durch das Gaußsche Funktional, das zur Bilinearform

$$\beta\left(\binom{f_1}{g_1}, \binom{f_2}{g_2}\right) = \iint ds\, dt\ (f_1, g_1)(s) \begin{pmatrix} 0 & (1+\vartheta)\sigma^2 \delta(s-t) \\ \vartheta \sigma^2 \delta(s-t) & 0 \end{pmatrix} \binom{f_2}{g_2}(t)$$

$$= (1+\vartheta)\sigma^2 \int f_1 g_2 + \vartheta \sigma^2 \int f_2 g_1$$

gehört. Man hat also die symbolischen Relationen

$$E\ F(s) F^*(t) = (1+\vartheta)\sigma^2 \delta(s-t)$$
$$E\ F^*(s) F(t) = \vartheta \sigma^2 \delta(s-t)$$
$$E\ F(s) F(s) = E\ F^*(s) F^*(t) = 0$$

Das farbige Quantenrauschen wird definiert durch die Bilinearform

$$\beta\left(\binom{f_1}{g_1}, \binom{f_2}{g_2}\right) = \iint ds\, dt\ (f_1, g_1)(s) \begin{pmatrix} 0 & (1+\vartheta) k(s-t) \\ \vartheta \bar{k}(s-t) & 0 \end{pmatrix} \binom{f_2}{g_2}(t)$$

wo k positiv definit und stetig ist. Dazu gehören die symbolischen Relationen

$$E\,F(s)\,F^*(t) = (1+\vartheta)\,k(s-t)$$
$$E\,F^*(s)\,F(t) = \vartheta\,\bar{k}(s-t)$$
$$E\,F(s)\,F(t) = E\,F^*(s)\,F^*(t) = 0$$

Eine Weyl-Algebra zum Vektorraum V und zur schiefsymmetrischen Bilinearform h ist die von V erzeugte assoziative komplexe Algebra mit 1 mit den definierenden Relationen

$$xy - yx = h(x,y)$$

für $x, y \in V$. Sie entsteht aus $T(V)$ durch Division durch das beidseitige Ideal, das von

$$x \otimes y - y \otimes x - h(x,y)$$

erzeugt wird.

Sei γ ein Gaußsches Funktional auf $T(V)$ zur Bilinearform β, so verschwindet γ auf dem beidseitigen Ideal, das von

$$x \otimes y - y \otimes x - \beta(x,y) + \beta(y,x)$$

erzeugt wird. Division durch dieses Ideal liefert die Weyl-Algebra zur Bilinearform

$$h(x,y) = \beta(x,y) - \beta(y,x)$$

Man kann also sehr wohl γ auf dieser Weyl-Algebra definieren (vgl. [3]).

Ist $h = 0$, so ist die zugehörige Weyl-Algebra die symmetrische Tensoralgebra.

Wir wenden diese Überlegungen auf die vorher definierten Rauschprozesse an. Beim reellen und komplexen Rauschen ist β symmetrisch, also $h = 0$, und somit können wir ohne Schwierigkeiten zur symmetrischen Tensoralgebra übergehen, oder, anders formuliert, wir vertauschen

$$X(s)\,X(t) = X(t)\,X(s)$$

$$Z(s)\,Z(t) = Z(t)\,Z(s)$$

$$Z(s)\,\bar{Z}(t) = \bar{Z}(t)\,Z(s) \qquad \text{usw.}$$

Beim Quantenrauschen ist β nicht symmetrisch. Hier erhalten wir nach Division durch das genannte Ideal die Vertauschungsrelation, die wir symbolisch schreiben

a) weißes Quantenrauschen

$$[F(s), F(t)] = [F^*(s), F^*(t)] = 0$$

$$[F(s), F^*(t)] = \sigma^2 \delta(s-t)$$

b) farbiges Quantenrauschen

$$[F(s), F(t)] = [F^*(s), F^*(t)] = 0$$

$$[F(s), F^*(t)] = k(s-t)$$

Wie in der Einleitung erwähnt, wollen wir bei den zeitgeordneten Momenten wegen algebraischer Schwierigkeiten diesen Gedanken nicht weiter verfolgen.

§ 3. BERECHNUNG DER ZEITGEORDNETEN MOMENTE

Wir berechnen die zeitgeordneten Momente nach dem Stratonowitsch-Verfahren durch Approximation durch farbiges Rauschen. Wir diskutieren und beweisen nur den kompliziertesten Fall, den des Quantenrauschens, und geben nur die Resultate in den anderen Fällen, die sich ebenso behandeln lassen.

In der symbolischen Schreibweise interessieren wir uns für Ausdrücke der Form

$$\text{(1)} \quad E \int_{t_{\pi(1)} \leq \cdots \leq t_{\pi(2m)}} \cdots \int F_{2m}(t_{2m}) \cdots F_1(t_1)\, f_1(t_1) \cdots f_{2m}(t_{2m})\, dt_1 \cdots dt_{2m},$$

wo F_i gleich F oder F^* und $f_i \in L^2(\mathbb{R})$ und π eine Permutation von 1, ..., 2m ist. Die ungeraden Momente interessieren nicht, da sie ohnehin verschwinden.

Wir stellen zunächst fest, daß das Integral überhaupt nicht definiert ist. Solche Integrale hatten wir nur symbolisch definiert, und dieses Integral fiel nicht unter die symbolischen Definitionen. Wir können jedoch beim farbigen Rauschen

$$E\, F_{2m}(t_{2m}) \cdots F_1(t_1) = R(t_{2m}, \ldots, t_1)$$

definieren und dann das Integral

$$(2) \qquad \int_{B_\pi} \cdots \int R(t_{2m}, \ldots, t_1) f_1(t_1) \cdots f_{2m}(t_{2m})\, dt_1 \cdots dt_{2m}$$

definieren, falls dieses existiert. Dabei haben wir zur Abkürzung

$$B_\pi = \{(t_1, \ldots, t_{2m}) : t_{\pi(1)} \leq \cdots \leq t_{\pi(2m)}\}$$

gesetzt. Nach den Regeln für Gaußsche Funktionale ist der Kandidat für R

$$(3) \qquad R = \sum_{\{S_1, \ldots, S_m\}} \rho_{S_1}(\Delta t_{S_1}) \cdots \rho_{S_m}(\Delta t_{S_m})$$

wo für $S = \{i, j\},\ i < j$

$$\rho_S(\Delta t_S) = \rho_{ji}(t_j - t_i) = E\, F_j(t_j) F_i(t_i)\,; \quad \Delta t_S = t_j - t_i$$

gesetzt wird. Die Werte für ρ_{ji} ergeben sich aus der Tabelle

F_j	F_i	$\rho_{ji}(t_j - t_i)$
F	F	0
F^*	F^*	0
F	F^*	$(1+\vartheta)k(t_j - t_i)$
F^*	F	$\vartheta \bar{k}(t_j - t_i)$

Wir betrachten nun (1) als formalen Ausdruck und setzen ihn gleich (2), wo R durch (3) definiert wird.

Wir ersetzen nun k durch k_ε, $k_\varepsilon(t) = \frac{1}{\varepsilon} k\left(\frac{t}{\varepsilon}\right)$ und beweisen

<u>Satz 1</u>. Sei $\int |k(t)|\, dt$ und $\int |t|\, |k(t)|\, dt$ endlich und $f_1, \ldots, f_{2m} \in L^2(\mathbb{R})$. Dann konvergiert für $\varepsilon \downarrow 0$

$$E \int_{t_{\pi(1)} \leq \cdots \leq t_{\pi(2m)}} \cdots \int dt_1 \cdots dt_{2m}\, F_{2m}(t_{2m}) \cdots F_1(t_1)\, f_{2m}(t_{2m}) \cdots f_1(t_1)$$

gegen

$$c_{\pi(1),\pi(2)}\, c_{\pi(3),\pi(4)} \cdots c_{\pi(2m-1),\pi(2m)}$$

$$\int_{s_1 \leq \cdots \leq s_m} \cdots \int ds_1 \cdots ds_m\, f_{\pi(1)}(s_1)\, f_{\pi(2)}(s_1)\, f_{\pi(3)}(s_2)\, f_{\pi(4)}(s_2) \cdots f_{\pi(2m-1)}(s_m)\, f_{\pi(2m)}(s_m)$$

Dabei sind die $c_{i,j}$ gegeben durch die Tabelle

$F_{\max(i,j)}$	$F_{\min(i,j)}$	$i \gtrless j$	$c_{i,j}$
F	F	$\gtrless$	0
F^*	F^*	$\gtrless$	0
F	F^*	$<$	$(1+\vartheta)\varkappa$
F	F^*	$>$	$(1+\vartheta)\bar{\varkappa}$
F^*	F	$<$	$\vartheta\bar{\varkappa}$
F^*	F	$>$	$\vartheta\varkappa$

Die Konstante $\varkappa = \int_0^\infty k(t)\, dt$.

Wenn wir den entsprechenden Satz für das komplexe und das reelle weiße Rauschen aufstellen wollen, genügt es, die Standard-Reihenfolge $t_1 \leq \ldots \leq t_{2m}$ zu betrachten, da die Faktoren vertauschbar sind oder, genauer, das Gaußsche Funktional nicht von der Reihenfolge der Faktoren abhängt. Wir ersetzen wieder beim farbigen Rauschen k durch $k_\varepsilon(t) = \frac{1}{\varepsilon} k\left(\frac{t}{\varepsilon}\right)$ und gehen mit $\varepsilon \downarrow 0$.

<u>Satz 2</u>. Sei $\int |k(t)|\, dt$ und $\int |t|\,|k(t)|\, dt$ endlich und $f_1, \ldots, f_{2m}$ quadratintegrabel. Dann konvergiert für $\varepsilon \downarrow 0$

$$E \int_{t_1 \leq \ldots \leq t_{2m}} \cdots \int dt_1 \ldots dt_{2m}\, Z_{2m}(t_{2m}) \ldots Z_1(t_1) f_{2m}(t_{2m}) \ldots f_1(t_1)$$

gegen

$$c_1 c_2 \ldots c_m \int_{s_1 \leq \ldots \leq s_m} \cdots \int ds_1 \ldots ds_m\, f_1(s_1) f_2(s_1) \ldots f_{2m-1}(s_m) f_{2m}(s_m)$$

Dabei sind die c_k gegeben durch die Tabelle

Z_{2k}	Z_{2k-1}	c_k
Z	Z	0
$\bar{Z}$	$\bar{Z}$	0
Z	$\bar{Z}$	$\varkappa$
$\bar{Z}$	Z	$\bar{\varkappa}$

und $\varkappa = \int_0^\infty k(t)\,dt.$

Für reelles Rauschen wird der Satz noch einfacher. Sei $X(t)$ ein farbiges reelles Rauschen mit der Kovarianzfunktion $k(t-s)$ $= E\,X(t)\,X(s)$ und $k_\varepsilon(t) = \frac{1}{\varepsilon}\,k\left(\frac{t}{\varepsilon}\right)$

Satz 3. Sei $\int |k(t)|\,dt$ und $\int |t\,k(t)|\,dt$ endlich und $f_1, \ldots$ $\ldots, f_{2m}$ quadratintegrabel. Dann konvergiert für $\varepsilon \downarrow 0$

$$E \int\limits_{t_1 \le \cdots \le t_{2m}} \cdots \int X(t_{2m}) \cdots X(t_1)\, f_{2m}(t_{2m}) \cdots f_1(t_1)\, dt_1 \cdots dt_{2m}$$

gegen

$$\left(\frac{\sigma^2}{2}\right)^m \int\limits_{s_1 \le \cdots \le s_m} \cdots \int f_1(s_1)\, f_2(s_1) \cdots f_{2m-1}(s_m)\, f_{2m}(s_m)\, ds_1 \cdots ds_m$$

mit $\sigma^2 = \int k(t)\,dt.$

Wir beweisen, wie angekündigt, nur Satz 1.

Hilfsbehauptung 1. Der Satz 1 gilt für $m = 1$.

Beweis. Die Fälle, da beide, F_2 und F_1, entweder gleich F oder F^* sind, ergeben Null. Wir diskutieren einen nichttrivialen Fall

$$E \iint\limits_{t_1 \le t_2} F(t_2)\, F^*(t_1)\, f_1(t_1)\, f_2(t_2)\, dt_1\, dt_2$$

$$= (1+\vartheta) \iint\limits_{t_1 \le t_2} k_\varepsilon(t_2 - t_1)\, f_1(t_1)\, f_2(t_2)\, dt_1\, dt_2$$

$$= (1+\vartheta) \int_0^\infty du\, \frac{1}{\varepsilon}\, k\left(\frac{u}{\varepsilon}\right) \int_{-\infty}^{+\infty} ds\, f_1(s)\, f_2(s+u)$$

$$= (1+\vartheta) \int_0^\infty du\, k(u) \int_{-\infty}^{+\infty} ds\, f_1(s)\, f_2(s+\varepsilon u)$$

$$\longrightarrow (1+\vartheta)\, \varkappa \int ds\, f_1(s)\, f_2(s).$$

Das Integral über ϑ ist nämlich in u eine stetige beschränkte Funktion, die im Unendlichen gegen Null geht.

Zur Abkürzung setzen wir

$$I_\varepsilon = E \int_{B_\pi} \cdots \int F_{2m}(t_{2m}) \cdots F_1(t_1)\, f_{2m}(t_{2m}) \cdots f_1(t_1)\, dt_1 \cdots dt_{2m}$$

$$= \int_{B_\pi} \cdots \int R_\varepsilon(t_{2m}, \ldots, t_1)\, f_{2m}(t_{2m}) \cdots f_1(t_1)\, dt_1 \cdots dt_{2m}$$

$$= \sum_{\{S_1, \ldots, S_m\}} I_\varepsilon(S_1, \ldots, S_m)$$

mit

$$I_\varepsilon(S_1, \ldots, S_m) = \int_{B_\pi} \cdots \int \rho_{S_1,\varepsilon}(\Delta t_{S_1}) \cdots \rho_{S_m,\varepsilon}(\Delta t_{S_m})\, f_{2m}(t_{2m}) \cdots f_1(t_1)\, dt_1 \cdots dt_{2m}$$

wo der Index ε bedeutet, daß $k(t)$ durch $k_\varepsilon(t) = \frac{1}{\varepsilon} k\left(\frac{t}{\varepsilon}\right)$ zu ersetzen ist.

<u>Hilfsbehauptung 2</u>. Für alle $\varepsilon > 0$ gilt

$$|I_\varepsilon| \leq \sum_{\{S_1, \ldots, S_m\}} C(S_1) \cdots C(S_m)$$

wo für $S = \{i, j\},\ i < j$ gilt

$$C(S) = C(\{i,j\}) = \int |\rho_{j,i}|(t)\, dt\ \|f_i\|_{L^2}\ \|f_j\|_{L^2}$$

<u>Beweis.</u> Wir ersetzen in jedem $I_\varepsilon(S_1, \ldots, S_m)$ die Funktionen durch ihre Absolutbeträge und integrieren über den ganzen Raum anstelle über B_π. Diese obere Abschätzung für $I_\varepsilon(S_1, \ldots, S_m)$ zerfällt in das Produkt von Ausdrücken der Form

$$\int\int |\rho_{j,i,\varepsilon}(t_j - t_i)|\, |f_j(t_j)|\, |f_i(t_i)|\, dt_i\, dt_j$$

$$\leq \int |\rho_{j,i,\varepsilon}(t)|\,dt\ \|f_j\|_{L^2}\ \|f_i\|_{L^2}$$

$$= \int |\rho_{j,i}(t)|\,dt\ \|f_j\|_{L^2}\ \|f_i\|_{L^2}$$

unabhängig von ε.

Folgerung aus Hilfsbehauptung 2. Es genügt, den Satz für $f_1, \ldots, f_{2m}$ stetig mit kompaktem Träger zu beweisen.

Hilfsbehauptung 3. Sei $\{S_1, \ldots, S_m\} = \{\{\pi(1), \pi(2)\}, \ldots$ $\ldots, \{\pi(2m-1), \pi(2m)\}\}$, so konvergiert $I_\varepsilon(S_1, \ldots, S_m)$ gegen den im Satz 1 genannten Ausdruck.

Beweis. Wir setzen für $k = 1, \ldots, m$

$$t_{\pi(2k-1)} = s_k$$

$$t_{\pi(2k)} = s_k + u_k$$

und setzen

$$S_k = \{\pi(2k-1), \pi(2k)\}$$

Dann ist

$$\rho_{S_k,\varepsilon}(\Delta t_{S_k}) = \rho_{k,\varepsilon}(u_k)$$

mit

$$\rho_{k,\varepsilon}(u) = \begin{cases} \rho_{\pi(2k),\pi(2k-1),\varepsilon}(u) & \text{falls } \pi(2k-1) < \pi(2k) \\ \rho_{\pi(2k-1),\pi(2k),\varepsilon}(-u) = \bar{\rho}_{\pi(2k-1),\pi(2k),\varepsilon}(u) & \\ \text{falls } \pi(2k-1) > \pi(2k) & \end{cases}$$

Außerdem ist

$$f_1(t_1)\cdots f_{2m}(t_{2m}) = \prod_{j=1}^{2m} f_j(t_j) = \prod_{j=1}^{2m} f_{\pi(j)}(t_{\pi(j)}) =$$

$$= f_{\pi(1)}(s_1)\,f_{\pi(2)}(s_1+u_1)\cdots f_{\pi(2m-1)}(s_m)\,f_{\pi(2m)}(s_m+u_m)$$

Dann wird

$$I_\varepsilon(S_1,..,S_m) = \int_0^\infty \rho_{1,\varepsilon}(u_1)\,du_1 \cdots \int_0^\infty \rho_{m,\varepsilon}(u_m)\,du_m$$

$$\underset{s_1 \le s_1+u_1 \le s_2 \le s_2+u_2 \le \cdots \le s_m \le s_m+u_m}{\iint \cdots \cdots \int} ds_1 \cdots ds_m \; f_{\pi(1)}(s_1)$$

$$f_{\pi(2)}(s_1+u_1)\, f_{\pi(3)}(s_2)\, f_{\pi(4)}(s_2+u_2) \cdots f_{\pi(2m-1)}(s_m)\, f_{\pi(2m)}(s_m+u_m)$$

Da wir ohne Einschränkung der Allgemeinheit annehmen können, daß die f_i stetig mit kompaktem Träger sind, ist das Integral über $s_1, \ldots, s_m$ und $(u_1, \ldots, u_m)$ stetig, beschränkt und verschwindet im Unendlichen. Also konvergiert der Ausdruck für $\varepsilon \downarrow 0$ gegen

$$\int_0^\infty \rho_1(u)\,du \cdots \int_0^\infty \rho_m(u)\,du \underset{s_1 \le \cdots \le s_m}{\int \cdots \int} ds_1 \cdots ds_m$$

$$f_{\pi(1)}(s_1)\, f_{\pi(2)}(s_1) \cdots f_{\pi(2m-1)}(s_m)\, f_{\pi(2m)}(s_m)$$

und, wie man sich leicht überzeugt, ist

$$\int_0^\infty \rho_k(u)\,du = c_{\pi(2k-1),\,\pi(2k)}\,.$$

<u>Hilfsbehauptung 4</u>. Sei $\{S_1, \ldots, S_m\}$ verschieden von $\{\{\pi(1), \pi(2)\}, \ldots, \{\pi(2m-1), \pi(2m)\}\}$. Dann konvergiert $I_\varepsilon(S_1, \ldots, S_m)$ gegen Null.

<u>Beweis</u>. Es gibt ein Paar $\{\pi(2k-1), \pi(2k)\}$, das nicht in der Zerlegung $\{S_1, \ldots, S_m\}$ vorkommt. Also ist $\pi(2k-1)$ mit einem dritten Element $\pi(p)$ und $\pi(2k)$ mit einem vierten Element $\pi(q)$ in dieser Zerlegung verbunden. Wir können ohne Einschränkung der Allgemeinheit annehmen:

$$S_1 = \{\{\pi(2k-1), \pi(p)\}$$
$$S_2 = \{\pi(2k), \pi(q)\}$$

Die Zahl p ist entweder $> 2k$ oder $< 2k-1$. Wir behandeln nur den

Fall $p < 2k$. Der andere Fall geht ebenso. Wir ersetzen in $I_\varepsilon(S_1,\dots,S_m)$ alle Funktionen durch ihre Absolutbeträge und integrieren anstelle über B_π über dem größeren Bereich $B' = \{t_{\pi(2k-1)} \le t_{\pi(2k)} \le t_{\pi(p)}\}$. In dieser oberen Abschätzung können wir über die in $S_3, \dots, S_m$ enthaltenen Variablen integrieren und finden mit Hilfe der Abschätzungen von Hilfsbehauptung 2

$$|I_\varepsilon(S_1,\dots,S_m)| \le C(S_3)\dots C(S_m) \int\int\int\int_{t_{\pi(2k-1)} \le t_{\pi(2k)} \le t_{\pi(p)}}$$

$$dt_{\pi(2k-1)}\, dt_{\pi(2k)}\, dt_{\pi(p)}\, dt_{\pi(q)}\, |\rho_{S_1,\varepsilon}|(t_{\pi(p)} - t_{\pi(2k-1)})$$

$$|\rho_{S_2,\varepsilon}|(t_{\pi(2k)} - t_{\pi(q)})\, |f_{\pi(2k-1)}(t_{\pi(2k-1)})|$$

$$|f_{\pi(2k)}(t_{\pi(2k)})|\, |f_{\pi(p)}(t_{\pi(p)})|\, |f_{\pi(q)}(t_{\pi(q)})|$$

Dabei wurde ausgenutzt, daß es auf die Reihenfolge der $\pi(2k-1), \pi(2k), \dots$ nicht ankommt, da die $\rho_{S,\varepsilon}$ symmetrische Funktionen sind. Wir können wieder ohne Einschränkung der Allgemeinheit annehmen, daß die f_i stetig sind mit kompaktem Träger.

Wir schätzen $|f_{\pi(2k)}|$, $|f_{\pi(p)}|$ und $|f_{\pi(q)}|$ durch ihre Maxima ab. Dann führen wir das Integral über $t_{\pi(q)}$ aus und erhalten das Integral über $|\rho_{S_2,\varepsilon}|$, das von ε unabhängig gleich dem Integral über $|\rho_{S_2}|$ ist. Das Integral über $t_{\pi(2k)}$ liefert einen Beitrag $t_{\pi(p)} - t_{\pi(2k-1)}$. Somit

$$|I_\varepsilon(S_1, \dots, S_m)|$$
$$\le \text{const} \int\int_{s \le t} ds\,dt\,(t-s)\, |\rho_{S_1,\varepsilon}|(t-s)\, |f_{\pi(2k-1)}(s)|$$
$$\le \text{const} \int_0^\infty du\, u\, |\rho_{S_1,\varepsilon}|(u)$$
$$= \text{const}\; \varepsilon \int_0^\infty du\, u\, |\rho_{S_1}|(u)$$
$$\to 0$$

da nach Voraussetzung $\int |t|\,|k(t)|\,dt < \infty$ ist.

§ 4. ALGEBRAISCHE BESCHREIBUNG DES ZEITGEORDNETEN WEISSEN RAUSCHENS. ANWENDUNG AUF DAS ZWEI-NIVEAU-ATOM

Aus den §§ 2 und 3 wird klar, daß sich die zeitgeordneten Momente des weißen Quantenrauschens nicht durch ein Funktional auf der Tensoralgebra $T(V)$ beschreiben lassen. Wir müssen diese erweitern. Wir erinnern uns. In § 2 schrieben wir das formale Integral

$$\int \dots \int F_1(t_1) \dots F_n(t_n) f_1(t_1) \dots f_n(t_n) dt_1 \dots dt_n$$
$$= F_1(f_1) \dots F_n(f_n)$$
$$= f_1 \xi_1 \otimes \dots \otimes f_n \xi_n$$

mit

$$f_k \xi_k \in L^2(\mathbb{R}) \otimes \mathbb{C}^2$$
$$f_k \in L^2(\mathbb{R}),\ \xi_k \in \mathbb{C}^2$$

wo ξ_k gleich $\binom{1}{0}$ oder gleich $\binom{0}{1}$ zu setzen ist, je nachdem F_k gleich F oder gleich F^* ist.

Für das formale Integral

$$\int_{t_{\pi(1)} \leq \dots \leq t_{\pi(n)}} \dots \int F_1(t_1) \dots F_n(t_n) f_1(t_1) \dots f_n(t_n) dt_1 \dots dt_n$$

wo π eine Permutation von $1, \dots, n$ ist, schreiben wir die Abkürzung

$$(f_1 \xi_1 \otimes \dots \otimes f_n \xi_n)_\pi .$$

Eine andere Schreibweise als 4-zeilige Matrix ist ebenfalls günstig

$$\begin{pmatrix} f_1 & \dots\dots\dots & f_n \\ \xi_1 & \dots\dots\dots & \xi_n \\ 1 & \dots\dots\dots & n \\ \pi^{-1}(1) & \dots\dots\dots & \pi^{-1}(n) \end{pmatrix}$$

In der vierten Zeile steht der Platz in der Zeitreihenfolge: Die Koordinate t_1 von f_1 kommt an der $\pi^{-1}(1)$-ten Stelle vor. In der dritten Zeile halten wir die Reihenfolge der Faktoren fest. Wir verabreden, daß eine Permutation der Spalten das Symbol ungeändert läßt. Die Matrix ist also gleich

$$\begin{pmatrix} f_{\sigma(1)} & \cdots & f_{\sigma(n)} \\ \xi_{\sigma(1)} & \cdots & \xi_{\sigma(n)} \\ \sigma(1) & \cdots & \sigma(n) \\ \sigma\pi^{-1}(1) & \cdots & \sigma\pi^{-1}(n) \end{pmatrix}$$

falls σ eine Permutation ist und insbesondere auch gleich

$$\begin{pmatrix} f_{\pi(1)} & \cdots & f_{\pi(n)} \\ \xi_{\pi(1)} & \cdots & \xi_{\pi(n)} \\ \pi(1) & \cdots & \pi(n) \\ 1 & \cdots & n \end{pmatrix}$$

Die dritte Zeile bedeutet dann hier, daß $f_{\pi(1)}\xi_{\pi(1)}$ an der $\pi(1)$-ten Stelle als Faktor vorkommt.

Beim Grenzübergang vom farbigen zum weißen Rauschen ergibt sich nach Satz 1 von § 3, daß der Erwartungswert des zeitgeordneten formalen Integrals für $n = 2$ gegen

$$\tilde{c}_{\pi(1),\pi(2)} \cdots \tilde{c}_{\pi(2k-1),\pi(2k)} \int_{s_1 \leq \ldots \leq s_m} \cdots \int f_{\pi(1)}(s_1) f_{\pi(2)}(s_1) \ldots f_{\pi(2m-1)}(s_m) f_{\pi(2m)}(s_m) ds_1, \ldots ds_m$$

konvergiert, wobei die $\tilde{c}_{i,j}$ sich aus den $\tilde{c}_{i,j}$ des § 3 ergeben, wenn man berücksichtigt, daß die Reihenfolge der Faktoren umgedreht wurde. Wir definieren ein Funktional λ auf den $(f_1\xi_1 \otimes \ldots \otimes f_n\xi_n)_\pi$ durch diesen Ausdruck. Also

$$\lambda \begin{pmatrix} f_1 \cdots f_n \\ \xi_1 \cdots \xi_n \\ 1 \cdots n \\ \pi^{-1}(1) \quad \pi^{-1}(n) \end{pmatrix} = \mu \begin{pmatrix} \xi_1 \cdots \xi_n \\ 1 \cdots n \\ \pi^{-1}(1) \cdots \pi^{-1}(n) \end{pmatrix} \nu \begin{pmatrix} f_1 \cdots f_n \\ \pi^{-1}(1) \cdots \pi^{-1}(n) \end{pmatrix}$$

wobei wir bei der Definition von μ und ν die irrelevanten Zeilen der Matrix weggelassen haben.

Dabei ist

$$\nu\begin{pmatrix} f_1 & \cdots & f_n \\ \pi^{-1}(1) & \cdots & \pi^{-1}(n)\end{pmatrix} = \int\limits_{s_1\leq\cdots\leq s_m}\!\!\cdots\int ds_1\cdots ds_m\, f_{\pi(1)}(s_1)$$

$$f_{\pi(2)}(s_1)\cdots f_{\pi(2m-1)}(s_m)\, f_{\pi(2m)}(s_m)\,.$$

Dieser Ausdruck ist in der Tat invariant gegen die Vertauschung von Spalten, z.B. auch gleich

$$\nu\begin{pmatrix} f_{\pi(1)} & \cdots & f_{\pi(2m)} \\ 1 & \cdots & 2m\end{pmatrix}$$

Der Ausdruck

$$\mu\begin{pmatrix} \xi_1 & \cdots & \xi_{2m} \\ 1 & \cdots & 2m \\ \pi^{-1}(1) & \cdots & \pi^{-1}(2m)\end{pmatrix} = \mu\begin{pmatrix} \xi_{\pi(1)} & \cdots & \xi_{\pi(2m)} \\ \pi(1) & \cdots & \pi(2m) \\ 1 & \cdots & 2m\end{pmatrix}$$

$$= \mu\begin{pmatrix} \xi_{\pi(1)} & \xi_{\pi(2)} \\ \pi(1) & \pi(2) \\ 1 & 2\end{pmatrix}\cdots\mu\begin{pmatrix} \xi_{\pi(2m-1)} & \xi_{\pi(2m)} \\ \pi(2m-1) & \pi(2m) \\ 2m-1 & 2m\end{pmatrix}$$

Dabei müssen wir eine Erweiterung unseres Matrixsymbols durchführen: In den beiden letzten Zeilen können anstelle von den Zahlen von 1, ..., n in einer bestimmten Reihenfolge auch irgendwelche anderen ganzen Zahlen in derselben Reihenfolge stehen. Die kleinste ganze Zahl aus der vorletzten Zeile bedeutet, daß der darüberstehende Teil der erste Faktor ist, in der letzten Zeile bedeutet die kleinste ganze Zahl, daß der entsprechende Term der erste in der Zeitreihenfolge ist. Wir erhalten

$$\mu\begin{pmatrix} \xi_{\pi(2k-1)}, & \xi_{\pi(2k)} \\ \pi(2k-1) & \pi(2k) \\ 2k-1 & 2k\end{pmatrix} = b\left(\xi_{\pi(2k-1)},\, \xi_{\pi(2k)},\, \operatorname{sgn}\left(\pi(2k)-\pi(2k-1)\right)\right)$$

mit

$$b(\xi,\eta,+) = (\xi_1,\xi_2)\begin{pmatrix} 0 & (1+\vartheta)\varkappa \\ \vartheta\bar{\varkappa} & 0 \end{pmatrix}\begin{pmatrix}\eta_1\\ \eta_2\end{pmatrix}$$

$$b(\xi,\eta,-) = (\xi_1,\xi_2)\begin{pmatrix} 0 & \vartheta\varkappa \\ (1+\vartheta)\bar{\varkappa} & 0 \end{pmatrix}\begin{pmatrix}\eta_1\\ \eta_2\end{pmatrix}$$

Wir betten nun die $(x_1\otimes\cdots\otimes x_n)_\pi$, $x_i \in V$ in eine Algebra $T_H(V)$ ein. Wir können dies für einen beliebigen Vektorraum V tun. Wir nennen $T_H(V)$ die zeitgeordnete Tensoralgebra mit Hurwitz- oder shuffle-Produkt (vgl. [1], [2]). Die Algebra $T_H(V)$ ist eine direkte Summe von Teilräumen

$$T_H(V) = \bigoplus_{n=0}^{\infty} \bigoplus_{\pi\in\mathcal{S}_n} (V^{\otimes n})_\pi$$

Die $(V^{\otimes n})_\pi$ sind alle Kopien von $V^{\otimes n}$, sie werden aufgespannt von den $(x_1\otimes \ \ldots \ \otimes x_n)_\pi$, $x_i\in$ V. Das Produkt in $T_H(V)$ wird definiert durch die Festsetzung

$$(x_1\otimes\cdots\otimes x_m)_\rho \otimes (y_1\otimes\cdots\otimes y_n)_\sigma =$$

$$= \sum_{\pi\in H(\rho,\sigma)} (x_1\otimes\cdots\otimes x_m\otimes y_1\otimes\cdots\otimes y_n)_\pi$$

Dabei ist $H(\rho,\sigma)$ eine Menge von Permutationen von m+n Elementen. Fassen wir ρ als Permutation der Menge $\{1, \ldots, m\}$ auf und beschreiben es durch die Folge $\rho_1, \ldots, \rho_m$ und fassen wir σ als Permutation von $\{m+1, \ldots, m+n\}$ auf und beschreiben es durch die Folge $\sigma_{m+1}, \ldots, \sigma_n$, so können die Permutationen von $H(\rho,\sigma)$ durch die Folge beschrieben werden

$\rho_1, \ldots, \rho_m, \sigma_{m+1}, \ldots, \sigma_{m+n}$

$\rho_1, \ldots, \rho_{m-1}, \sigma_{m+1}, \rho_m, \sigma_{m+2}, \ldots, \sigma_{m+n}$

. . . .

$\rho_1, \ldots, \rho_{m-1}, \sigma_{m+1}, \sigma_{m+2}, \ldots, \sigma_{m+n}, \rho_m$

$\rho_1, \ldots, \rho_{m-2}, \sigma_{m+1}, \rho_{m-1}, \rho_m, \sigma_{m+2}, \ldots, \sigma_{m+n}$

. . . .

. . . .

$\sigma_{m+1}, \ldots, \sigma_{m+n}, \rho_1, \ldots, \rho_m.$

$H(\rho, \sigma)$ besteht also aus denjenigen Permutationen, die die relative Ordnung der ρ_i und σ_k, d.h. die Zahlen von 1, ..., m bzw. von m+1, ..., n invariant lassen.

Man zeigt leicht, daß $T_H(V)$ eine assoziative Algebra ist. Die Abbildung λ läßt sich auf $T_H(V)$ eindeutig definieren. Die Abbildung

$$\eta : T(V) \to T_H(V)$$

$$x_1 \otimes \cdots \otimes x_n \longmapsto \sum_{\pi \in \mathcal{S}_n} (x_1 \otimes \cdots \otimes x_n)_\pi$$

ist ein umkehrbar eindeutiger Homomorphismus. Die lineare Abbildung $\lambda \circ \eta$ von T(V) in $\mathbb{C}$ ist gleich dem in § 2 definierten Gaußschen Funktional γ, das das weiße Quantenrauschen beschreibt, mit $\sigma^2 = 2 \operatorname{Re} \varkappa$.

Wir sind nunmehr in der Lage, das Problem zu formulieren, dessentwegen wir diese Untersuchung durchgeführt haben. Wir betrachten ein Zwei-Niveau-Atom in einem Wärmebad von Photonen. Das Atom wird beschrieben durch die vier Operatoren $|i\rangle\langle k| = E_{ik}$, i, k = 1, 2. Dabei ist $|1\rangle$ der Grundzustand, $|2\rangle$ der angeregte Zustand. Die Energiedifferenz sei $\hbar\omega_0$. In einem Wechselwirkungsbild findet man für die vier Operatoren in Näherung die folgende stochastische Differentialgleichung

$$\frac{d}{dt}\begin{pmatrix} E_{11} \\ E_{22} \\ E_{12} \\ E_{21} \end{pmatrix} = \begin{pmatrix} \varkappa & \bar{\varkappa} & -iF & iF \\ -\varkappa & -\bar{\varkappa} & +iF & -iF \\ -iF & +iF & \varkappa & 0 \\ iF & -iF & 0 & \bar{\varkappa} \end{pmatrix} \begin{pmatrix} E_{11} \\ E_{22} \\ E_{12} \\ E_{21} \end{pmatrix}$$

Dabei ist F das in § 2 definierte weiße Quantenrauschen, $\varkappa = \int_0^\infty k(t)dt$ die in Satz 1; § 3 definierte Konstante. Die in der Definition von F vorkommende Konstante $\vartheta = (\exp(\hbar\omega_0/kT) - 1)^{-1}$

wo k die Boltzmannkonstante und T die Temperatur des Wärmebades ist.

Schreiben wir für den Vektor $(E_{ik})_{i,k}$ das Symbol X, so läßt sich die Gleichung schreiben

$$\frac{d}{dt} X(t) = A(t)\, X(t)$$
$$A(t) = A^0 + F(t) A^1 + F^*(t) A^2$$

Die formale Lösung ist

$$X(t) = \mathcal{U}(t,0)\, X(0)$$

mit $(s \leq t)$

$$\mathcal{U}(t,s) = 1 + \int_s^t A(\tau)\, d\tau + \iint\limits_{s \leq \tau_1 \leq \tau_2 \leq t} A(\tau_2) A(\tau_1)\, d\tau_1 d\tau_2$$
$$+ \iiint\limits_{s \leq \tau_1 \leq \tau_2 \leq \tau_3 \leq t} A(\tau_3) A(\tau_2) A(\tau_1)\, d\tau_1\, d\tau_2 d\tau_3$$

In den Integralen kommen nach Ausmultiplikation die Terme vom Typ

$$\int \cdots \int\limits_{s \leq \tau_1 \leq \tau_2 \leq \cdots \leq \tau_n \leq t} F_n(\tau_n) \cdots F_1(\tau_1)\, d\tau_1 \cdots d\tau_n$$

vor. Das formale Integral können wir durch das Element

$$(\xi_n f \otimes \cdots \otimes \xi_1 f)_\varepsilon \quad , f = 1\!\!1_{[s,t]} = \begin{cases} 1 & \text{auf } [s,t] \\ 0 & \text{sonst} \end{cases}$$

aus $T_H(V)$ beschreiben, wo ε die identische Permutation ist. Um den Konvergenzproblemen zu entgehen, führen wir den formalen Parameter z ein, der mit allen anderen Größen vertauschbar ist, und schreiben

$$\frac{d}{dt} X(t) = z\, A(t)\, X(t)$$

Dann kann $\mathcal{U}(t,s)$ als formale Potenzreihe in z definiert werden. Man kann zeigen, daß die $\mathcal{U}(t,s)$ für Intervalle $[s,t]$, die höchstens einen Randpunkt gemeinsam haben, unabhängig sind. Man erhält mit dem oben definierten Funktional λ auf $T_H(V)$

$$\lambda(\mathcal{U}(t,s)) = \exp(t-s)\,\mathcal{A}(z)$$

$$\mathcal{A}(z) = zA^0 + z^2(1+\vartheta)\varkappa A^1A^2 + z^2\vartheta\bar{\varkappa}\,A^2A^1$$

mit

$$\mathcal{A}(1) = \begin{pmatrix} -\vartheta\sigma^2 & (1+\vartheta)\sigma^2 & 0 & 0 \\ \vartheta\sigma^2 & -(1+\vartheta)\sigma^2 & 0 & 0 \\ 0 & 0 & -(2\vartheta+1)\varkappa & 0 \\ 0 & 0 & 0 & -(2\vartheta+1)\bar{\varkappa} \end{pmatrix}$$

Die linke obere Ecke ist die erzeugende Matrix für einen Markovprozeß mit einem zweipunktigen Zustandsraum und dem stationären Zustand

$$\frac{1}{1+e^{-\frac{\hbar\omega_0}{kT}}}\left(1,\ e^{-\frac{\hbar\omega_0}{kT}}\right)$$

Die Erwartungswerte von E_{11} und E_{22} sind ja auch als die Besetzungswahrscheinlichkeiten für die Zustände $|1\rangle$ und $|2\rangle$ zu deuten, und somit ist der stationäre Zustand derjenige, den man nach dem Boltzmannschen Prinzip zu erwarten hatte.

Diese Überlegungen sollen in einer noch zu veröffentlichenden Arbeit genauer ausgeführt werden.

LITERATURVERZEICHNIS

1 G. Baron und P. Kirschenhofer: Operatorenkalkül über freien Monoiden I: Strukturen. Monatshefte für Mathematik, 91, 89-103(1981).

2 S. Eilenberg: Automata, Languages and Machines. New York-London, Academic Press 1974.

3 N. Giri und W. von Waldenfels: An algebraic version of the central limit theorem. Z. Wahrscheinlichkeitstheorie verw. Gebiete, 42, 129-134(1978).

4 R.L. Hudson und R.F. Streater: Ito's formula is the chain rule with Wick ordering. Physics Letters 86A, 277-279(1981).

SOME ZERO-ONE LAWS FOR SEMISTABLE AND SELF-DECOMPOSABLE MEASURES ON LOCALLY CONVEX SPACES

Arnold Janssen

Several zero-one laws for some classes of infinitely divisible probability measures μ on locally convex vector spaces E have been established by various authors, see [5], [6], [10], [11]. These papers contain laws in the sense that some measurable subgroups H of E have either measure 0 or 1. In the preceding paper [4] the author proves a Theorem which deduces these results from the Lévy measure F of μ (see Theorem 4 below). If F is zero or unbounded on the complement H^c of H then either $\mu(H+x)$ is equal to zero or one for fixed $x \in E$.

In this paper we shall give further applications of this result. At first we examine the case $0 < F(H^c) < \infty$ which completely solves the problem in terms of the Lévy measure F. Under these assumptions the relation $\mu(H+x) < 1$ holds for each coset of H (Theorem 5). In chapter 4 and 5 the concept of semistable and self-decomposable measures on locally convex vector spaces E are introduced. These distributions appear in the theory of triangle systems of random variables. They extend the class of stable measures. For these measures zero-one laws (in the sense mentioned above) can be obtained by applying the main result. Since zero-one laws for Gauss measures have been studied carefully we only consider infinitely divisible probability measures without Gaussian component.

1. Preliminaries.

Let E always denote a real locally convex vector space. If $N \subset E$ is a closed linear subspace with finite codimension then $p_N : E \longrightarrow E/_N$ is the canonical projection. For each positive real number a and $b \in E$ the mapping $H_{a,b}$ on E is defined by $H_{a,b}(x) = ax+b$. Let $M_+(E)$ ($M_1(E)$) denote the set of all bounded and tight measures (the subset of probability measures) which are defined on the Borel σ-field of E. Together with the convolution multiplication $*$ the set $M_+(E)$ is a topological semigroup with respect to the weak topology. A net $(\mu_i)_{i \in I}$ in $M_+(E)$ is said to be weakly convergent to μ if $\int f\, d\mu_i \longrightarrow \int f\, d\mu$ for all bounded, continuous, real-valued functions f on E. Let ε_x denote the Dirac measure of $x \in E$ and ν^n the n-th convolution product of $\nu \in M_+(E)$. Put $\nu^0 = \varepsilon_0$. For $\nu \in M_+(E)$ we define $\|\nu\| = \nu(E)$. Then for each $\nu \in M_+(E)$ fulfilling $\nu(\{0\}) = 0$ the Poisson group of ν is by definition

$e(t\nu) = \exp(-t\|\nu\|) \sum_{n=0}^{\infty} t^n \nu^n/_{n!}$. Hence $e(t\nu)$ belongs to $M_1(E)$ for $t \geq 0$ and it is a signed bounded Borel measure for $t < 0$. A family $(\mu_t)_{t>0}$ in $M_1(E)$ is called a continuous convolution semigroup if $\mu_t * \mu_s = \mu_{t+s}$ holds for all $t, s > 0$ and μ_t converges weakly to ε_0 as $t \longrightarrow 0$. Let m be a measure and A be a measurable subset of the measure space. Then 1_A, A^c, $m_{|A}$, $\varphi(m)$ denotes the indicator function of A, the complement of A, the restriction of the measure m on A and the image measure of m with respect to φ if φ is a measurable function on the measure space.

A measure $\mu \in M_1(E)$ is said to be a generalized Poisson measure if there exists an upward directed family $(\nu_i)_{i \in I}$ in $M_+(E)$ with $\nu_i(\{0\}) = 0$ and a net of points $(x_i)_{i \in I}$ in E such that $\lim_{i \in I} e(\nu_i) * \varepsilon_{x_i} = \mu$ weakly. Corollary 1.10 of [2] shows that each generalized Poisson measure is infinitely divisible which means that the n-th root exists for all natural numbers n. On the other hand each infinitely divisible measure $\mu \in M_1(E)$ can be decomposed in the product of a Gauss measure γ and a generalized Poisson measure ν satisfying $\mu = \gamma * \nu$ [2], S.1.9. If μ is a generalized Poisson measure then the abstract measure $F = \sup_{i \in I} \nu_i$ is said to be the Lévy measure of μ which is independent of the special net $e(\nu_i) * \varepsilon_{x_i}$. It is not hard to see that our definition of the Lévy measure coincides with the concept of E. Dettweiler, [2] (see also [13]). If μ admits a Gauss factor γ then the Lévy measure of ν is defined to be the Lévy measure of μ. It is well-known that each infinitely divisible measure $\mu \in M_1(E)$ can be embedded in an uniquely determined continuous convolution semigroup, always denoted by $(\mu_t)_{t>0}$, such that $\mu_1 = \mu$ if E is quasi-complete ([12], Satz 4).

<u>Lemma 1</u>. Suppose that μ and $\nu \in M_1(E)$ are infinitely divisible with Lévy measures F_1 and F_2. Then $\mu * \nu$ has the Lévy measure $F_1 + F_2$.

Proof: Clearly, the assertion is true for Poisson measures. We may assume that μ and ν are generalized Poisson measures $\mu = \lim_{i \in I} e(\rho_i) * \varepsilon_{x_i}$, $\nu = \lim_{j \in J} e(\beta_j) * \varepsilon_{y_j}$ such that $(\rho_i)_{i \in I}$ and $(\beta_j)_{j \in J}$ are (upward directed) measures in $M_+(E)$. Consider the directed family $I \times J$ ($(i_1, j_1) \leq (i_2, j_2)$ iff $i_1 \leq i_2$ and $j_1 \leq j_2$). Then the equality $\mu * \nu = \lim_{(i,j) \in I \times J} e(\rho_i + \beta_j) * \varepsilon_{x_i + y_j}$ follows from the continuity of the convolution. The Lemma now follows from the uniqueness of the Lévy measure. □ In Lemma 2 and 3 let E be quasi-complete.

Lemma 2. Let μ be a generalized Poisson measure with Lévy measure F on E. Suppose that $\rho \in M_+(E)$ fulfils $\rho \leq F$. Then there exists a continuous convolution semigroup $(\nu_t)_{t>0}$ in $M_1(E)$ such that $\mu_t = \nu_t * e(t\rho)$ holds for each t and ν_1 has the Lévy measure $F - \rho$.

Proof: It is known that μ_t has the Lévy measure tF for $t > 0$. (Regard the arguments used in [4], Example 3.) Put $\nu_t = \mu_t * e(-t\rho)$. Then the remark after definition 1.3 of [2] shows that $e(t\rho)$ is a factor of μ_t and $\nu_t \in M_1(E)$ follows. Finally we apply Lemma 1 which proves the assertion. □

Lemma 3. Let μ and μ_n be generalized Poisson measures with Lévy measures F and F_n for each $n \in \mathbb{N}$. Suppose that $(\nu_n)_{n \geq 2}$ is a sequence of generalized Poisson measures such that $\mu = \mu_1 * \ldots * \mu_{n-1} * \nu_n$ and $\lim_{n \to \infty} \nu_n = \varepsilon_0$ holds with respect to the weak topology.
Then $F = \sum_{n=1}^{\infty} F_n$.

Proof: Clearly, $F \geq \sum_{n=1}^{\infty} F_n$ holds. For the converse inequality it suffices to prove (see [2], p. 288) : If $\rho \in M_+(E)$ with $\rho \leq F$ then $\rho \leq \sum_{n=1}^{\infty} F_n =: \nu$ follows.
Assume that there exists a Borel set $A \subset E$ such that $\rho(A) > \nu(A)$ is valid. Then $\nu_{|A}$ is a finite measure on E. Let $\rho = \rho_1 + \rho_2$ denote the Lebesgue decomposition of ρ with respect to $\nu_{|A}$ where ρ_1 fulfils $\rho_1 \ll \nu_{|A}$ and $\rho_2 \perp \nu_{|A}$. Then $\beta = \rho_{2|A} + f \cdot \nu$ belongs to $M_+(E) - \{0\}$ if $f \cdot \nu$ denotes the

measure with the ν-density $f = \left(\frac{d\varrho_1}{d\nu_{|A}} - 1\right) 1_{A \cap \{\frac{d\varrho_1}{d\nu_{|A}} > 1\}}$.

From the definition of β we obtain $F \geq \nu + \beta$. Let α_n denote the Lévy measure of ν_n. Then the equality $F = \sum_{i=1}^{n-1} F_i + \alpha_n$ implies $\alpha_n \geq \beta$ if $n \geq 2$. Now Lemma 2 shows that there is a sequence $(\theta_n)_{n\geq 2}$ in $M_1(E)$ fulfilling $\nu_n = \theta_n * e(\beta)$. Since $e(\beta)$ is invertible and ν_n converges weakly to ε_0 the sequence $(\theta_n)_{n \geq 2}$ converges weakly to a probability measure $\theta \in M_1(E)$. The continuity of the convolution multiplication proves $\varepsilon_0 = \theta * e(\beta)$. But this equality is only possible if $\beta = 0$ holds which yields a contradition. □

2. Infinitely divisible measures.

In this section we will apply the next zero-one law. The proof is given in the preceding paper [4],T. 9, 10.

Theorem 4. Let μ be a generalized Poisson measure on E with Lévy measure F. Suppose that $H \subset E$ is a measurable subgroup. Then the following assertions hold:

(i) $\mu(H+x) = 0$ for each $x \in E$ if $F(H^c) = \infty$.

(ii) If $F(H^c) = 0$ and $x \in E$ then either $\mu(H+x) = 0$ or 1.

Now we shall solve the problem of determining the probability of a coset completely for quasi-complete spaces by dealing with the case $0 < F(H^c) < \infty$. There are upper and lower bounds for the value of $\max\{\mu(H+x) : x \in E\}$. If E is quasi-complete then Theorem 5 holds.

Theorem 5. Let μ and H fulfil the assumption of Theorem 3. Suppose $0 < F(H^c) < \infty$. We put $b = \max\{F(H+x) : x \in E, H+x \neq H\}$ and

$a = \sum_{H+x \neq H} F(H+x)$ where the sum runs over all cosets $H+x \neq H$.

Then either (i) or (ii) is valid for the semigroup $(\mu_t)_{t>0}$ (with $\mu = \mu_1$).

(i) $\max\{\mu_t(H+x) : x \in E\} = 0$ for all $t > 0$.

(ii) $\exp(-tF(H^c)) \leq \max\{\mu_t(H+x) : x \in E\} \leq$

$$\leq \exp(-tF(H^c))\left(\max\{1, tb\} + \sum_{n=2}^{\infty} \frac{t^n}{n!} ba^{n-1}\right) < 1 \text{ for all } t > 0.$$

Proof: We note that $F_{|H^c}$ belongs to $M_+(E)$. Hence Lemma 2 shows that there exists a continuous convolution semigroup $(\nu_t)_{t>0}$ with

(1) $\mu_t = \nu_t * e(tF_{|H^c})$ for each $t>0$

such that ν_1 has the Lévy measure $F - F_{|H^c}$. Now the zero-one law can be applied which proves that either $\nu_t(H+x) = 0$ for each coset H+x or there is exactely one coset with $\nu_t(H+x) = 1$ for fixed $t>o$. If ν_t vanishes on all cosets for one $t>o$ then (i) clearly holds for all $t>o$. On the other hand we conclude

(2) $\max\{\mu_t(H+x) : x \in E\} = \max\{e(tF_{|H^c})(H+x) : x \in E\}$

if ν_t is concentrated on one coset of H. We put

(3) $\varrho = \sum_{H+x \neq H} F_{|H+x}$

where the sum runs over all cosets $H+x \neq H$. Then

(4) $e(t(F_{|H^c} - \varrho))(H+x) = \exp(-t(F(H^c)-a))$ if $x \in H$ and zero otherwise.

Regarding $e(tF_{|H^c}) = e(t\varrho) * e(t(F_{|H^c} - \varrho))$ we obtain

(5) $e(tF_{|H^c})(H+x) = \exp(-t(F(H^c)-a))e(t\varrho)(H+x)$.

Since $\|\varrho\|$ is equal to a the first inequality holds. Suppose now that H+x is one coset of H. Then

(6) $\varrho^n(H+x) \leq a^{n-1}b$ for $n \geq 1$

which is easy to prove. Observe now that $\varrho(H+x) \leq b$ if $x \notin H$ and $\varrho(H+x) = 0$ if $x \in H$. Hence

(7) $e(t\varrho)(H+x) \leq \exp(-ta)(\max\{1,tb\} + \sum_{n=2}^{\infty} \frac{t^n}{n!} ba^{n-1})$.

The statements (2) and (5) imply the result. □

<u>Remark</u>. If $(\mu_t)_{t>o}$ is symmetric the proof of Theorem 5 contains an estimate for the value of $\mu_t(H)$. Then either (i) $\mu_t(H) = 0$ for each $t>o$ or (ii) $\exp(-tF(H^c)) \leq \mu_t(H^c) \leq \exp(-tF(H^c))(1 + \sum_{n=2}^{\infty} \frac{t^n}{n!} ba^{n-1})$ holds for all $t>o$. Observe that ν_t is symmetric and hence $\nu_t(H) = 1$ or each coset has ν_t-measure zero.

We give another application of Theorem 4.

<u>Corollary 6</u>. Suppose that $\mu \in M_1(E)$ is infinitely divisible. Then the assertions (i) and (ii) are equivalent :

(i) $\mu(\{x\}) > 0$ for some $x \in E$,

(ii) $\mu = e(\nu) * \varepsilon_y$ is a shift of a Poisson measure ($y \in E$, $\nu \in M_+(E)$).

Proof: Let $\mu = \gamma * \nu$ denote the decomposition of μ in the product of a Gauss measure γ and a generalized Poisson measure ν with Lévy measure F. Let (i) be satisfied. Then γ must be degenerated which can be seen by regarding the finite dimensional marginal distributions, [2]. Choose $H = \{0\}$. Then $F(H^c) < \infty$ follows. □

3. Semistable measures.

For more than 40 years ago P. Lévy already knew the concept of semi-stable laws on the euclidean space $\mathbb{R}$. Later many other authors examined semistable measures on linear spaces, see for example A. Kumar [7], D. M. Chung, B. Rajput and A. Tortrat [1], D. Louie and B. Rajput [9]. One of the most general works concerning with this subject is the paper [1]. The authors deal with "convexly tight" measures on topological vector spaces. We shall consider the concept of tight semistable probability measures on locally convex vector spaces which is weaker for non-quasi-complete spaces. At first we briefly show that the theory of r-semistable measures on locally convex spaces E can be developed along the same line used by E. Dettweiler [3] who describes stable measures on E. The next definition arises from problems concerning with limit distributions of triangle systems of probability measures which are important in probability theory.

<u>Definition 7</u>. Let r be a real number with $0 < r \leq 1$. A measure $\mu \in M_1(E)$ is called r-semistable if there exists a measure $\nu \in M_1(E)$, sequences $(a_n)_{n\in\mathbb{N}}$ with $a_n > 0$, $(x_n)_{n\in\mathbb{N}}$ in E and an increasing sequence of positive integers $(k_n)_{n\in\mathbb{N}}$ $k_n \uparrow \infty$ such that $\frac{k_n}{k_{n+1}} \longrightarrow r$ for $n \longrightarrow \infty$ and $H_{a_n,x_n}(\nu^{k_n}) \longrightarrow \mu$ weakly as $n \longrightarrow \infty$.

The next statements are well-known for r-semistable measures μ on $E=\mathbb{R}^n$.

(8) μ is stable if $r = 1$.

(9) μ is infinitely divisible.

If μ is no Dirac measure and $r < 1$ then the next statements hold.

(10) μ is r-semistable iff there exists an unique real number $\beta \geq 1/_2$ such that for each $n \in \mathbb{N}$ there is a point $a_n \in E$ with $\mu_{r^n} = H_{r^{n\beta},a_n}(\mu)$. The value $\alpha = \beta^{-1}$ is called the order of μ.

(11) For $\alpha = 2$ μ is a Gauss measure. If $\alpha < 2$ then μ is a generalized Poisson measure.

The following Theorem shows that semistable measures have similar properties as stable measures (compare with [3], S. 2.3).

Theorem 8. Suppose that E is quasi-complete. Then a measure $\mu \in M_1(E)$ is r-semistable iff each marginal distribution $p_N(\mu)$ is r-semistable.

Proof: It suffices to show that μ is r-semistable for $r < 1$ if each marginal distribution has this property. Suppose that μ is no Dirac measure. If $N \subset E$ is a closed linear subspace with finite codimension such that $p_N(\mu)$ is no Dirac measure then there exists an uniquely determined pair $(\beta_N, b_N) \in \{z \in \mathbb{R} : z \geq 1/_2\} \times E$ satisfying

$$p_N(\mu_{r^n}) = H_{r^n \beta_N, b_N}(p_N(\mu)) \qquad \text{for fixed } n \in \mathbb{N}.$$

Following the proof of E. Dettweiler [3], S. 2.3 we conclude that $\beta_N = \beta$ is independent of N. The same arguments used in [3], S. 2.3 prove that $\mu_{r^n} = H_{r^n\beta, b_n}(\mu)$ for some $b_n \in E$.

Observe that μ is infinitely divisible [2], S. 1.9. Now choose $\nu = \mu$, $k_n = [r^{-n}]$ (The Gauss bracket of r^{-n}), $x_n = k_n b_n$ and $a_n = r^{n\beta}$. Then $\mu_{k_n r^n} = H_{a_n, x_n}(\mu^{k_n}) \longrightarrow \mu$ converges weakly which proves the assertion. □

Let μ be r-semistable on an arbitrary space E. Then $p_N(\mu)$ is r-semistable and hence μ is infinitely divisible (stable if $r = 1$, respectively). If μ is no Dirac measure then let us consider all closed subspaces $N \subset E$ with finite codimension such that $p_N(\mu)$ is not degenerated. It is easy to see that the order α of $p_N(\mu)$ is independent of N. We shall call α the order of μ. Hence (8), (9) and (11) carry over for r-semistable measures on E. If E is in addition quasi-complete then Theorem 8 implies that (10) also holds.
Zero-one laws for stable measures have been established in [4], Ex. 3.

Theorem 9. Let $\mu \in M_1(E)$ be r-semistable of order $\alpha < 2$ and $r < 1$. Suppose that $H \subset E$ is a measurable subgroup such that there exists a natural number n with $r^{-n/\alpha} H \subset H$. Then either $\mu(H+x) = 0$ or 1 for $x \in E$.

Proof: Suppose that μ is no Dirac measure. If E is not comlete then $i : E \longrightarrow \hat{E}$ denotes the injection into the completion $\hat{E}$ of E. At first we show that $i(\mu)$ is r-semistable of order α if $\mu \in M_1(E)$ is r-semistable of order α. There is a closed subspace $N \subset E$ with finite codimension such that $p_N(\mu)$ is no Dirac measure. Let $\overline{N}$ be the closure of N in $\hat{E}$. Regard the commuting diagram

$$\begin{array}{ccccc} E & \xrightarrow{\ i\ } & \hat{E} & \xrightarrow{\ p_{\overline{N}}\ } & \hat{E}/_{\overline{N}} \\ {\scriptstyle p_N}\downarrow & & & \nearrow{\scriptstyle j} & \\ E/_N & & & & \end{array}$$

where j is the continuous injection defined by $j(x+N) = x+\overline{N}$. Let $(\nu_t)_{t>0}$ $[(\varrho_t)_{t>0}]$ be the continuous convolution semigroup generated by $p_N(\mu)$ $[i(\mu)]$ on $E/_N$ $[\hat{E}$, respectively$]$. Clearly, $i(\mu)$ is r-semistable. Let α_1 $[\alpha_2]$ denote the order of μ $[i(\mu)]$. By statement (10) there is a point $a \in E/_N$ $[b \in \hat{E}]$ such that $\nu_r = H_{r^{1/\alpha_1},a}(p_N(\mu))$ $[\varrho_r = H_{r^{1/\alpha_2},b}(i(\mu))]$ is valid. Since $p_{\overline{N}} \circ i(\mu) = j \circ p_N(\mu)$ implies $p_{\overline{N}}(\varrho_t) = j(\nu_t)$ for each $t>0$ we obtain

$$H_{r^{1/\alpha_1},j(a)}(j \circ p_N(\mu)) = H_{r^{1/\alpha_2},p_{\overline{N}}(b)}(j \circ p_N(\mu)) .$$

Since j is injective Lemma (2.2) of [3] shows $\alpha_1 = \alpha_2$. Put $\beta = \alpha^{-1}$. We remark that there is a σ-compact subgroup $H' \subset H$ such that $\mu * \varepsilon_{-x}(H - H') = 0$ and $r^{-n\beta}H' \subset H'$ is fulfilled. Choose an increasing sequence of compact symmetric subsets $K_m \subset H$ with $\mu * \varepsilon_{-x}(H - K_m) \longrightarrow 0$ and $0 \in K_1$. For a subset $A \subset E$ we put $A^1 = A$ and $A^m = \{x+y\colon x \in A,\ y \in A^{m-1}\}$. Then $H' = \bigcup_{m \in \mathbb{N}} \left(\bigcup_{i=0}^{m} r^{-in\beta} K_m \right)^m$ has the desired properties. Hence $i(H')$ is σ-compact and it is sufficient to prove the Theorem for $i(\mu)$. Therefore we may assume that E is complete.

Let F be the Lévy measure of μ. Then statement (10) implies

$$(12) \qquad r^m F = H_{r^{m\beta},0}(F) \quad \text{for all } m \in \mathbb{N}$$

since μ_{r^m} has the Lévy measure $r^m F$. The relation $r^{-n\beta}H^c \supset H^c$ proves $r^n F(H^c) = F(r^{-n\beta}H^c) \geq F(H^c)$ and $F(H^c) = 0$ or ∞. □

Remark. Zero-one laws for semistable measures appear in the papers of D. Louie, B. Rajput and A. Tortrat [10] and W. Krakowiak [6]. Our assumptions seem to be a little bit weaker than the assumptions of [10] who assumed that H is a subspace over $\mathbb{Q}(c)$ which is a field containing $\mathbb{Q}$ and $c = r^\beta$. On the other hand the article [10] deals with general measurable vector spaces.

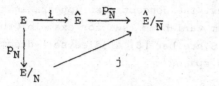

4. Self-decomposable measures.

Another important class of infinitely divisible measures is the set of self-decomposable measures. The concept also appears in the theory of triangle systems of random variables (see for example Urbanik [14]). Later A. Kumar and B. M. Schreiber [8] studied self-decomposable probability measures on Banach spaces.

Definition 10. A measure $\mu \in M_1(E)$ is called self-decomposable if for all $c \in (0,1)$ there exists a measure $\mu^c \in M_1(E)$ such that $\mu = H_{c,0}(\mu) * \mu^c$ is valid.

For self-decomposable measures on $E=\mathbb{R}^n$ the next assertion is well known.

(13) All the measures μ and μ^c are infinitely divisible.

The same result carries over for self-decomposable measures on E which follows from [2], S. 1.9 and the next

Lemma 11. A measure $\mu \in M_1(E)$ is self-decomposable iff each marginal distribution $p_N(\mu)$ is self-decomposable.

Proof: Suppose that $p_N(\mu)$ is self-decomposable for each N. Then there exists an uniquely determined measure μ_N^c on $E/_N$ such that $p_N(\mu) = H_{c,0}(p_N(\mu)) * \mu_N^c$. Hence the assumtions of Lemma 1.7 of [2] are fulfilled. Then there exists a measure $\mu^c \in M_1(E)$ such that $\mu = H_{c,0}(\mu) * \mu^c$ is satisfied (see also A. Tortrat [13], p. 303). □

Theorem 12. Let $\mu \in M_1(E)$ be self-decomposable without Gauss factor. Then $\mu(H+x) = 0$ or 1 for each measurable subgroup $H \subset E$ and $x \in E$.

Proof: Let $i : E \longrightarrow \hat{E}$ be the injection into the completion of E. Then we note that $i(\mu)$ is a self-decomposable generalized Poisson measure. Since there is a σ-compact subgroup $H' \subset H$ with $\mu * \varepsilon_{-x}(H - H') = 0$ we may assume that E is complete (see [4]).

Now it suffices to prove the Theorem for symmetric measures μ and $x = 0$ (see [4]). Let F (F_c) denote the Lévy measure of μ (μ^c). Then Lemma 1 yields

$$F = H_{c,0}(F) + F_c \quad . \tag{14}$$

In the next step we shall use an idea of G. Kallianpur [5], Lemma 5.8-5.10 which has also been needed by W. Krakowiak to prove zero-one laws [6]. Let Π denote the set of all primes and

I_p the set of all integers which are relatively prim to $p \in \Pi$. Then

(15) $$G_p = \bigcup_{q \in I_p} \left(\frac{1}{q} H\right)$$

is a measurable subgroup of E and

(16) $$H = \bigcap_{p \in \Pi} G_p \qquad \text{(see [5])}.$$

The proof is finished if we show $\mu(G_p) = 0$ or 1 for all $p \in \Pi$. For fixed $p \in \Pi$ we choose a natural number $s \neq 1$ which is relatively prim to p. Then we conclude

(17) $$sG_p = G_p .$$

Put $c = s^{-1}$. If $F(G_p^c) = \infty$ the claim is proved (Theorem 4). Otherwise we may assume

(18) $$F(G_p^c) < \infty .$$

Then (14) and (17) imply $F(G_p^c) = F(sG_p^c) + F_c(G_p^c)$ and

(19) $$F_c(G_p^c) = 0 .$$

By induction the next statement holds for $n \in \mathbb{N}$.

(20) $$\mu = H_{c^n,0}(\mu) * H_{c^{n-1},0}(\mu^c) * \ldots * H_{c,0}(\mu^c) * \mu^c$$

For all $j \in \mathbb{N}$ the measure $H_{c^j,0}(\mu^c)$ is infinitely divisible with Lévy measure $H_{c^j,0}(F_c)$. The equalities (17) and (19) now yield

(21) $$H_{c^j,0}(F_c)(G_p^c) = 0 .$$

Since $c < 1$ the sequence $H_{c^n,0}(\mu)$ converges weakly to ε_0 if n tends to infinity. Hence Lemma 3 can be applied which shows

(22) $$F(G_p^c) = \sum_{n=0}^{\infty} H_{c^n,0}(F_c)\,(G_p^c) = 0$$

if we take (20) and (21) into account. Then the zero-one law implies $\mu(G_p) = 0$ or 1. □

Remarks. 1. It should be noted that further applications of Theorem 4 are possible. Let $\mu \in M_1(E)$ be a generalized Poisson measure with Lévy measure F on a quasi-complete space E. Suppose that there is a measurable linear operator $B : E \longrightarrow E$ such that $\mu_t = B(\mu) * \varepsilon_y$ holds for a $t \in (0,1)$ and $y \in E$ (μ is sometimes called operator semistable). Then $tF = B(F)$ follows. Hence a zero-one law can be proved for a measurable subgroup H satisfying $B^{-1}(H) \subset H$ by repeating the arguments used in Theorem 9. This result generalizes Theorem 3 of W. Krakowiak [6] who proved the same result by a different method under the additional assumptions that B is invertible and B(H) = H is fulfilled.

2. Note that Theorem 9 and 12 remain true for non-measurable subgroups H if we consider the inner measure $\mu_*(H+x)$. Observe that there are σ-compact subgroups $H' \subset H$ such that $\mu(H'+x) = \mu_*(H+x)$. The proof of Theorem 9 shows that H' can be chosen such that $r^{-n/\alpha} H' \subset H'$ is satisfied.

REFERENCES

[1] Chung, D. M., Rajput, B. S., Tortrat, A. : Semistable laws on topological vector spaces. Z. Wahrsch. verw. Geb. (to appear).

[2] Dettweiler, E.: Grenzwertsätze für Wahrscheinlichkeitsmaße auf Badrikianschen Räumen. Z. Wahrsch. verw. Geb. 34, 285-311 (1976).

[3] Dettweiler, E.: Stabile Maße auf Badrikianschen Räumen. Math. Z. 146, 149-166 (1976).

[4] Janssen, A. : Zero-one laws for infinitely divisible probability measures on groups. (1981) (to be published).

[5] Kallianpur, G. : Zero-one laws for Gaussian processes. Trans. Amer. Math. Soc. , 149 (1970), 199-211.

[6] Krakowiak, W. : Zero-one laws for stable and semi-stable measures on Banach spaces. Bull. d. l'acad. Polon., ser. sci. math., 27, 1045-1049 (1978).

[7] Kumar, A. : Semistable measures on Hilbert spaces. J. Multiv. Anal., 6 (1976), 309-318.

[8] Kumar, A., Schreiber, B. S. : Self-decomposable measures on Banach spaces. Studia Math., 53 (1975), 55-71.

[9] Louie, D., Rajput B. S. : Support and seminorm integrability theorems for r-semistable probability measures on LCTVS. In : Prob. theory on vector spaces II, Lect. Notes Math., 828 (1979), 179-195. Springer Verlag.

[10] Louie, D., Rajput, B. S., Tortrat, A. : A zero-one dichotomy theorem for r-semi-stable laws on infinite dimensional linear spaces. Sankhya, Ser. A., 42 (1980), 9-18.

[11] Louie, D., Rajput, B. S., Tortrat, A. : A zero-one law for a class of measures on groups. (to appear)

[12] Siebert, E. : Einbettung unendlich teilbarer Wahrscheinlichkeitsmaße auf topologischen Gruppen. Z. Wahrsch. verw. Geb. 28, 227-247 (1974).

[13] Tortrat, A. : Structure de lois indefiniment divisible dans un espace vectoriel topologique. Symposium on probability methods in analysis. Lect. Notes Math. 31, Springer Verlag (1967).

[14] Urbanik, K. : Self-decomposable probability distributions on $\mathbb{R}^m$. Zast. Math. Appl. Math., 10 (1969), 91-97.

Arnold Janssen
Abteilung Mathematik
Universität Dortmund
Postfach 500500

D-4600 Dortmund 50

West-Germany

CONVOLUTION POWERS OF PROBABILITY MEASURES ON LOCALLY COMPACT SEMIGROUPS

FRANZ KINZL

0. Introduction.

The study of the sequence of convolution powers of a probability measure on a locally compact group or more generally on a locally compact semigroup has become very interesting for many years (cf. the monographs from Heyer [3] or Mukherjea and Tserpes [6]). In this connection there is the problem of asymptotic equidistribution of the sequence of the convolution powers of a probability measure and related topics. Many results are known for groups. If one wants to treat these problems on locally compact semigroups there are some difficulties because a semigroup need not have inverse elements and because there is in general no invariant measure.

Let S be a locally compact semigroup. With M(S) we like to designate the set of all bounded regular measures on S, $P(S) \subseteq M(S)$ denotes the subset of all probability measures. M(S) is a Banach Algebra with convolution $*$ as multiplication. The aim of this paper is to study the convolutions of a probability measure λ and its tanslates, that is we study the sequence

(1) $$||\lambda^n - \delta_x * \lambda^n||$$

where $||\cdot||$ denotes the norm or total variation and δ_x is the Dirac-measure in $x \in S$. First we shall see that if S is abelian the limit of the sequence (1) is always 0 or 2 as for groups. In the second part we shall study the problem of asymptotic equidistribution of the power sequence (λ^n).

1. The Zero-Two-Law.

From now on let S be a locally compact semigroup. Let $A \subseteq S$, $x \in S, \lambda \in P(S)$. We shall use the following notations:

$$x^{-1}A = \{z \in S : xz \in A\}, \quad Ax^{-1} = \{z \in S : zx \in A\}$$

$$N(x,\lambda^n) = \|\lambda^n - \delta_x * \lambda^n\| \text{ for } n \geq 1 \text{ and } N(x,\lambda^o) = 2$$

$$L(\lambda) = \{z \in S : \lim N(z,\lambda^n) = 0\}$$

If A is a Borel subset of S then $x^{-1}A$ and Ax^{-1} are also Borel subsets of S. $L(\lambda)$ is a subsemigroup of S and is called the Stam semigroup of λ. Further $\xi(x^{-1}A) = \delta_x * \xi(A)$ for $\xi \in M(S)$.

Theorem 1. Let S be abelian. Then for every element x of S and any probability measure λ we have either

$$\|\lambda^n - \delta_x * \lambda^n\| = 2 \qquad \text{for any } n \in \mathbb{N}$$

or

$$\|\lambda^n - \delta_x * \lambda^n\| = O(\tfrac{1}{\sqrt{n}}).$$

Proof. Assume that $N(x,\lambda^r) < 2$ for some $r \in \mathbb{N}$. This means that λ^r and $\delta_x * \lambda^r$ are not mutually singular. Put $\mu = \lambda^r$ and $\nu = \delta_x * \lambda^r$. Then there exist measures $\alpha, \beta \in M(S)$ so that $\nu = \alpha + \beta$, $\alpha \geq 0$, $\beta \geq 0$, $\|\alpha\| > 0$, $\alpha << \mu$, $\beta \perp \mu$. Since $\beta \perp \mu$ and $\alpha << \mu$ we have also $\alpha \perp \beta$.

Let $f = \frac{d\alpha}{d\mu}$ the Radon-Nikodym derivate of α with respect to μ. Then there is an integer $N > 0$ such that the set $A_N = f^{-1}[0,N]$ has positive α-measure. Now choose a Borel subset E_o of S with $\mu(E_o) = 1$ and $\beta(E_o) = 0$. Now consider the set $A = A_N \cap E_o$ and let us define measures ρ' and σ' as follows:

$$\rho' = (1_A f) \cdot \mu$$

$$\sigma' = \beta + (\alpha - \rho')$$

One sees: $\rho' \geq 0$, $\sigma' \geq 0$, $\nu = \rho' + \sigma'$, $\rho' << \mu$. Furthermore $\rho' \perp \sigma'$ because $||\rho'|| = \alpha(A) = \rho'(A)$ and on the other hand $0 \leq \sigma'(A) =$ $= \beta(A) + \alpha(A) - \rho'(A) \leq \beta(E_o) = 0$. Putting $a = ||\rho'||$, $b = ||\sigma'||$, $\rho = \frac{1}{a}\rho'$, $\sigma = \frac{1}{b}\sigma$,

we obtain

(2) $$\nu = a\rho + b\sigma \qquad (\rho,\sigma \in P(S)).$$

Since $\rho(A) = 1$ and $\sigma(A) = 0$ we have that ρ and σ are mutually singular. Now $\frac{d\rho}{d\mu} = \frac{1}{a} f$ is bounded, therefore we have for the probability measure μ the following form:

(3) $$\mu = c\rho + (1-c)\tau$$

where $\tau \in P(S)$, $0 < c \leq 1$ (cf. [3] (2.6.5)).

Now let us define a measure ω by

$$\omega = (\tfrac{1}{a} 1_{x^{-1}A}) \cdot \mu$$

First ω is a probability measure because $\omega(S) = \frac{1}{a}\mu(x^{-1}A) =$ $= \frac{1}{a}\delta_x * \mu(A) = \frac{1}{a}\nu(A) = \rho(A) = 1$. For every Borel subset E of S we have (see (2))

$$\delta_x * \omega(E) = \omega(x^{-1}E) = \frac{1}{a}\mu(x^{-1}E \cap x^{-1}A) =$$

$$= \frac{1}{a}\delta_x * \mu(E \cap A) =$$

$$= \frac{1}{a} \nu(E \cap A) = \rho(E),$$

this means that $\delta_x * \omega = \rho$. By (3) we get

(4) $\mu = c\ \delta_x * \omega + (1-c)\tau.$

Since $\frac{d\omega}{d\mu} = \frac{1}{a}\ 1_{x^{-1}A} \le \frac{1}{a} < \infty$, the measure μ has the following form as well:

(5) $\mu = d\omega + (1-d)\phi$

where $\phi \in P(S)$ and $0 < d \le 1$. Now add (4) to (5) to obtain

(6) $2\mu = c\ \delta_x * \omega + d\omega + (1-c)\tau + (1-d)\phi$

If $c \le d$ from (6) it follows that

$$\mu = c \cdot \frac{1}{2} (\omega + \delta_x * \omega) + (1-c)\psi \qquad \text{where } \psi \in P(S).$$

Now by [5] p.374 we have $N(x,\omega^n) \le \frac{2}{\sqrt{n+1}}$ and also by [5] Prop. 1 one obtains

$$\| \mu^n - \delta_x * \mu^n \| = \mathcal{O}\left(\frac{1}{\sqrt{n}}\right)$$

Putting $\mu = \lambda^r$ the statement follows. The case $d < c$ is analogous. Now Theorem 1 is proved.

Corollary. $L(\lambda)$ is an F_σ-subset of S.

Proof. Since for any $\mu \in P(S)$ and any continuous function f with compact support the mapping $x \to \delta_x * \mu(f)$ is continuous ([1] (2.1)) and therefore the mapping $x \to \| \mu - \delta_x * \mu \|$ is lower semicontinuous the proof of the corollary is the same as in the group case (cf. [3] (2.6.11)).

Let S be a locally compact abelian semigroup and suppose that there exist absolutely continuous measures on S. A measure $\xi \in M(S)$ is called absolutely continuous if the mappings $x \to \delta_x * \xi$ and $x \to \xi * \delta_x$ are continuous with respect to the norm-topology on M(S) (cf. [4], [6]). With ξ_a we like to designate the absolutely continuous part of ξ. Now we have the following

Proposition 2. Let $\lambda \in P(S)$ such that $\lambda^r{}_a \neq 0$ for some $r \in \mathbb{N}$. Then $L(\lambda)$ is an open and closed subsemigroup of S.

Proof. From the assumption if follows that λ^r has the form

$$\lambda^r = a\mu + b\nu \tag{7}$$

where $\mu, \nu \in P(S)$, $a > 0$, $b \geq 0$, $a + b = 1$ and μ is absolutely continuous.

Let us first prove that $L(\lambda)$ is open. Let $x_o \in L(\lambda)$ and let $\varepsilon > 0$ arbitrary. Then there is a number $n_o \in \mathbb{N}$ so that $N(x_o, \lambda^{rn}) < \varepsilon$ for any $n \geq n_o$. Since μ is absolutely continuous there is a neighbourhood U of x_o such

$$||(\delta_x - \delta_{x_o}) * \mu|| < \frac{\varepsilon}{2} \quad \text{for every } x \in U. \tag{8}$$

Further there is an integer $m \geq n_o$ so that $4\, b^m < \varepsilon$. Now let $n \geq m$ and $x \in U$. By (7) we get

$$||\delta_x * \lambda^{rn} - \delta_{x_o} * \lambda^{rn}|| = ||(\delta_x - \delta_{x_o}) * \sum_{k=0}^{n} \binom{n}{k} a^k b^{n-k} \mu^k * \nu^{n-k}||$$

$$\leq \sum_{k=1}^{n} \binom{n}{k} a^k b^{n-k} ||(\delta_x - \delta_{x_o}) * \mu^k * \nu^{n-k}|| + b^n ||(\delta_x - \delta_{x_o}) * \nu^n||$$

$$\leq (1 - b^n) ||(\delta_x - \delta_{x_o}) * \mu|| + 2\, b^n <$$

$$< \varepsilon \tag{9}$$

Therefore we have for every $n \geq m$ and any $x \in U$:

$$N(x, \lambda^{rn}) \leq N(x_o, \lambda^{rn}) + ||(\delta_x - \delta_{x_o}) * \lambda^{rn}|| < 2\,\varepsilon$$

this means that $U \subseteq L(\lambda)$, i.e. $L(\lambda)$ is open.

To prove the closeness of $L(\lambda)$ let $x_o \in \overline{L(\lambda)}$ and let $\varepsilon > 0$. Let U be a neighbourhood of x_o so that (8) and therefore (9) are satisfied. Since $U \cap L(\lambda) \neq \emptyset$ we can choose an element $x \in U \subset L(\lambda)$. We get

$$N(x_o, \lambda^{rn}) \leq N(x, \lambda^{rn}) + \| (\delta_x - \delta_{x_o}) * \lambda^{rn} \| <$$
$$< N(x, \lambda^{rn}) + \varepsilon .$$

Since $x \in L(\lambda)$ we have also $x_o \in L(\lambda)$ and thus $L(\lambda)$ is closed.

<u>Corollary.</u> Let S be a connected abelian semigroup and let $\lambda \in P(S)$ as in Proposition 2. Then $L(\lambda) = S$ or $L(\lambda) = \emptyset$.

2. Asymptotic Equidistribution.

The statement of the last corollary leads us to the problem of the asymptotic equidistribution of the sequence of the convolution powers of a probability measure. In this section we shall consider semigroups which are not necessary abelian. First we give two definitions:

<u>Definition.</u> The sequence (λ^n) is called asymptotically equidistributed (as.equ.) iff $L(\lambda) = S$.

<u>Definition.</u> A measure $\xi \in M(S)$ is called [weakly] left absolutely continuous if the mapping $x \to \delta_x * \xi$ is continuous with respect to the [weak] norm topology on $M(S)$.

<u>Proposition 3.</u> Let S be an locally compact semigroup which admits

left absolutely continuous measures. If (λ^n) is as.equ. then the sequence of the norms of the left absolutely parts of λ^n tends to 1 as $n \to \infty$.

<u>Proof.</u> Let ξ be a left absolutely continuous probability measure. By Lebesgue's Dominated Convergence Theorem we obtain

$$||\lambda^n - \xi * \lambda^n|| \le \int ||\lambda^n - \delta_x * \lambda^n|| \, d\xi(x) \to 0 \qquad \text{as } n \to \infty$$

Since $\xi * \lambda^n$ is also left absolutely continuous ([4] Satz 2) there is an integer $r' > 0$ so that $\lambda^{r'}$ is of the form $\lambda^{r'} = a'\mu' + (1-a')\nu'$ where $0 < a' \le 1$, $\mu', \nu' \in P(S)$ and μ' is weakly left absolutely continuous ([7] (3.8)). Since $\mu_1 = \mu' * \mu'$ is left absolutely continuous (cf. [7] (5.5)) we get for λ^r $(r = 2r')$:

$$\lambda^r = a\,\mu_1 + b\,\nu$$

where $a = a'^2$, $b = 1 - a$, $\nu \in P(S)$. Since the set of all left absolutely continuous measures is a two-side ideal of $M(S)$ there is a sequence (μ_n) of left absolutely continuous probability measures so that

$$\lambda^{rn} = (1 - b^n)\,\mu_n + b^n\,\nu^n.$$

Therefore Proposition 3 is proved.

<u>Theorem 4.</u> Let S be a locally compact semigroup which has a compact minimal ideal K. Let (λ^n) be an as.equ. convolution sequence. Then there is a probability measure η such that the following two conditions are satisfied:

(i) $\qquad ||\lambda^n - \eta|| \to 0 \qquad$ as $n \to \infty$.

(ii) $\qquad$ Supp (η) contains a compact group G such G is a left ideal of S.

<u>Proof.</u> Let us define $C(S)$ ($C_{oo}(S)$) to be the set of all complex valued bounded continuous functions (with compact support). Let us endow $C(S)^*$ (i.e. the dual of $C(S)$) with the vague or weak $*$-topology and consider $M(S)$ as a subset of $C(S)$.

Any accumulation point of (λ^n) defines a nonnegative linear functional M on $C(S)$ with $M(1) = 1$ and $M({}_xf) = M(f)$ for any $f \in C(S)$ and $x \in S$ (${}_xf$ denotes the function ${}_xf(y) = f(xy)$), because (λ^n) is an as.equ. sequence. From the assumption that S has a compact minimal ideal K it follows that every accumulation point M of the sequence (λ^n) has also the following properties:

(a) $M(h) = 1$ for any $h \in C_{oo}(S)^+$ such $h \leq 1$ and $h(x) = 1$ for every $x \in K$.

(b) $M(u) = 0$ for any $u \in C(S)$ such $u(x) = 0$ for every $x \in K$.

(c) If h is a function as in (a), then $M(f) = M(f.h)$ for every $f \in C(S)$.

To prove (a) let (λ^{n_k}) be a subsequence of (λ^n) which converges vaguely to M. Let $x \in K$. Then $x^{-1}K = S$ and therefore

$$0 \leftarrow \|\lambda^{n_k} - \delta_x * \lambda^{n_k}\| \geq |\lambda^{n_k} - \delta_x * \lambda^{n_k}|(K) \geq$$

$$\geq |\lambda^{n_k}(K) - \lambda^{n_k}(x^{-1}K)| = 1 - \lambda^{n_k}(K)$$

Since $\lambda^n(K) \leq \lambda^{n+1}(K)$ for any $n \in \mathbb{N}$ we have: $\lambda^{n_k}(K) \uparrow 1$. Now let $h \in C_{oo}(S)^+$ such $h \leq 1$ and $h(x) = 1$ for every $x \in K$. Then we get

$$1 \geq M(h) \leftarrow \lambda^{n_k}(h) \geq \lambda^{n_k}(K) \to 1$$

To prove (b) let $u \in C(S)$ so that $u(y) = 0$ for any $y \in K$. Choose $x \in K$. It follows that ${}_xu(z) = u(xz) = 0$ for every $z \in S$ because $xz \in K$. Thus

$$M(u) = M({}_xu) = 0.$$

Since the function $u = f - f.h$ vanishes on K the statement (c) follows from (b).

We have seen that M does not vanish on $C_{oo}(S)$ (property (a)) and therefore the restriction of M to $C_{oo}(S)$ makes sense. Morever $I = M|_{C_{oo}}(S)$ is a left invariant integral on $C_{oo}(S)$, i.e. I is a positive linear functional on $C_{oo}(S)$ so that $I(g) = \bar{I}({}_xg)$ for every $g \in C_{oo}(S)$ and every $x \in S$ ($\bar{I}$ denotes the "canonical extention" of I, (cf. [2] (11.11)). To prove this assertion it is enough to show that $\bar{I}(f) = M(f)$ for every $f \in C(S)$. By property (c) we get for any $f \geq 0$:

$$\bar{I}(f) = \sup\{I(g) : g \in C_{oo}(S),\ g \leq f\} \leq M(f) =$$
$$= M(f.h) = I(f.h) = \bar{I}(f.h) \leq \bar{I}(f).$$

Therefore $\bar{I} = M$ on C(S) and I is a left invariant integral on $C_{oo}(S)$. Now there is a probability measure η which corresponds I and which is left invariant, i.e. $\delta_x * \eta = \eta$ for every $x \in S$ (cf. [4] Satz 6).

Until now we have shown that every vaguely accumulation point of the sequence (λ^n) is a left invariant probability measure. Next we shall show that the sequence (λ^n) converges in the weak$*$-topology, i.e. there is exactly one accumulation point. Let $\lambda^{n_k} \to \eta$ vaguely. Since P(S) endowed with the weak$*$-topology is a topological semigroup we have for every $g \in C_{oo}(S)$ and any $r \in \mathbb{N}$:

$$\eta(g) = \lambda^r * \eta(g) \leftarrow \lambda^r * \lambda^{n_k}(g) = \lambda^{n_k} * \lambda^r(g) \to \eta * \lambda^r(g)$$

This means that

(10) $\eta = \lambda^r * \eta = \eta * \lambda^r$

Now assume that ξ is another accumulation point of (λ^n), e.g. $\lambda^{n_i} \to \xi$ vaguely. Since ξ is also left invariant we get together with (10):

$$\xi = \eta * \xi \leftarrow \eta * \lambda^{n_i} = \eta$$

So the sequence (λ^n) converges vaguely to a left invariant probability measure, say η .

Now to prove the statement (i) of the Theorem use (10) and the Lebesgue Dominated Convergence Theorem to obtain

$$|| \lambda^n - \eta || = || \lambda^n - \eta * \lambda^n || \leq \int || \lambda^n - \delta_x * \lambda^n || d\eta(x) \to 0$$

The statement (ii) can be proved by using similar technics as in [5] (p.380). The details are omitted.

<u>Remark:</u> Also the reversed implication is valid, i.e. if (i) and (ii) are satisfied then (λ^n) is an as.equ. sequence (see [5] for details). In this case it is not necessary to suppose that S has a compact minimal ideal K, this is rather a consequence of (i) and (ii). Therefore it is not possible to weaken the assumption of Theorem 4.

L I T E R A T U R E

1 Anne C. BAKER and J.W. BAKER: Algebras of measures on a locally compact semigroup. J. London Math.Soc. (2) 1, 249 - 259 (1963).

2 E. HEWITT and K.A. ROSS: Abstract harmonic analysis I. Berlin-Göttingen-Heidelberg: Springer 1963.

3 H. HEYER: Probability measures on locally compact groups. Berlin-Heidelberg-New York: Springer 1977.

4 F. KINZL: Absolut stetige Maße auf lokalkompakten Halbgruppen. Mh. Matn. 87, 109 - 121 (1979).

5 F. KINZL: Convolutions of probability measures on semigroups. Semigroup Forum 20, 369 - 383 (1980).

6 A. MUKHERJEA and N.A. TSERPES: Measures on topological semigroups. Lecture Notes in Math.Vol. 547. Berlin-Heidelberg-New York: Springer 1976.

7 G.L.G. SLEIJPEN: Convolution measure algebras on semigroups. Thesis. Nijmegen 1976.

Mathematisches Institut

der Universität Salzburg

Petersbrunnstraße 19

A-5020 Salzburg, Österreich.

THEOREMES LIMITES POUR LES PRODUITS DE MATRICES ALEATOIRES

Emile LE PAGE
Laboratoire de probabilité
Université de Rennes
35042 Rennes Cedex
France

§ 1) Introduction

Soit $(g_n)_{n\geq 1}$ une suite de matrices aléatoires indépendantes de même loi p à valeurs dans le groupe $G = SL(d,\mathbb{R})$ des matrices à coefficients réels d x d de déterminant 1. Désignant par $\| \ \|$ une norme euclidienne sur $\mathbb{R}^d$, nous nous intéressons pour $x \in \mathbb{R}^d-\{0\}$ au comportement asymptotique de la suite de variables aléatoires $(\log\|g_n g_{n-1}\cdots g_1 x\|)_{n\geq 1}$

L'étude des produits de matrices aléatoires a été abordée par divers auteurs.

Une série d'articles concerne l'étude des produits de matrices aléatoires positives : les principaux résultats obtenus sont la loi des grands nombres [9], le théorème de la limite centrale [9], [15], et le théorème de renouvellement [15].

Par ailleurs une situation analogue à celle considérée ici a été étudiée sous l'hypothèse où la probabilité p est étalée : dans ce cadre les principaux résultats obtenus sont la loi des grands nombres [8] le théorème de la limite centrale [20], [21], [22], le théorème de renouvellement [16].

L'objet du présent article est en particulier de retrouver des résultats analogues aux précédents en se passant de cette hypothèse d'étalement.

Pour étudier la suite de variables aléatoires $(\log\|g_n g_{n-1}\cdots g_1 x\|)_{n\geq 1}$ on considère la chaine de Markov à valeurs dans $G \times P(\mathbb{R}^d)$ définie par

$$M_0 = (g,\bar{x}) \qquad M_n = (g_n, g_{n-1}\cdots g_1\bar{x}) \quad n\geq 1$$

où l'on note $g\bar{x}$ l'image sous l'action de $g \in G$ de l'élément $\bar{x}$ de l'espace projectif $P(\mathbb{R}^d)$.

En effet si ρ désigne la fonction de $G \times P(\mathbb{R}^d)$ dans $\mathbb{R}$ définie par

$$\rho(g,\bar{x}) = \operatorname{Log}\|gx\|$$

où $g \in G$ et où x est un vecteur de norme 1, d'image $\bar{x}$ dans $P(\mathbb{R}^d)$, il résulte de la relation

$$\rho(gg',\bar{x}) = \rho(g,g'\bar{x}) + \rho(g',\bar{x}) \qquad g \in G \ , \quad g' \in G \ , \quad \bar{x} \in P(\mathbb{R}^d)$$

que pour tout x tel que $\|x\| = 1$ et tout $n \geq 1$ on a

$$\log\|g_n g_{n-1} \ldots g_1 x\| = \sum_{k=0}^{n} F(M_k)$$

avec $M_0 = (e,\bar{x})$.

La chaine de Markov $(M_n)_{n\geq 0}$ a une probabilité de transition Q définie par

$$Q((g,\bar{x}), A \times B) = p(A)\, P1_B(\bar{x})$$

où $g \in G$, $\bar{x} \in P(\mathbb{R}^d)$, A(resp B) est un borélien de G (resp de $P(\mathbb{R}^d)$) et où P est la probabilité de transition sur $P(\mathbb{R}^d)$) définie par

$$P1_B(\bar{x}) = \int 1_B(g\bar{x})\, p(dg)$$

L'étude de l'opérateur de transition P est menée dans les paragraphes 2 et 3. Si l'on suppose que la probabilité p est étalée l'opérateur de transition P satisfait à la condition de Doeblin [18], ce n'est pas le cas général. Cependant, en faisant sur le support et sur les moments de p des hypothèses convenables il est possible d'établir que P admet sur $P(\mathbb{R}^d)$ une unique probabilité invariante, et qu'il existe un espace de Banach de fonctions Lipschitziennes sur $P(\mathbb{R}^d)$ que nous notons $\mathcal{L}_{\lambda_0}$ sur lequel P est un opérateur de Doeblin-Fortet [19].

Grâce à la théorie de la perturbation d'opérateurs analytiques, on en déduit pour $|\lambda|$ assez petit des propriétés pour les opérateurs $P(\lambda)$ définis par

$$P(\lambda)\, f(\bar{x}) = \int \alpha^{\lambda}(g,\bar{x})\, f(g\bar{x})\, p(dg)$$

$\bar{x} \in P(\mathbb{R}^d)$, $f \in \mathcal{L}_{\lambda_0}$, $\lambda \in \mathbb{C}$

où $\alpha(g,\bar{x}) = \|gx\|$ x étant un vecteur de norme 1 d'image $\bar{x}$ dans $P(\mathbb{R}^d)$.

En particulier pour $|\lambda|$ assez petit on a pour tout $n \geq 1$

$$P^n(\lambda) f = (k(\lambda))^n N_1(\lambda) f + Q^n(\lambda)\, f \qquad f \in \mathcal{L}_{\lambda_0}, \quad \lambda \in \mathbb{C}$$

où $k(\lambda)$ est l'unique valeur propre du plus grand module de $P(\lambda)$.

$N_1(\lambda)$ est le projecteur sur le sous-espace propre de dimension 1 correspondant à $k(\lambda)$.

$Q(\lambda)$ est un opérateur sur $\mathcal{L}_{\lambda_0}$ de rayon spectral strictement inférieur à $|k(\lambda)|$ et tel que $Q(\lambda)\, N_1(\lambda) = 0$.

Comme pour tous $\lambda \in \mathbb{C}$, $x \in \mathbb{R}^d$ de norme 1 on a

$$E(e^{\lambda \log\|g_n g_{n-1} \ldots g_1 x\|}) = P^n(\lambda)\, e(\bar{x})$$

où e désigne la fonction égale à 1 sur $P(\mathbb{R}^d)$, des résultats précédents on peut déduire des propriétés asymptotiques de la suite de variables aléatoires $(\log\|g_n g_{n-1}\dots g_1 x\|)_{n\geq 1}$

Les résultats obtenus établissent que cette suite a un comportement asymptotique analogue à celui d'unesomme de variables aléatoires réelles indépendantes de même loi non centrée.

Au paragraphe 4, on prouve un théorème de la limite centrale, le résultat étant précisé par l'obtention d'une vitesse de convergence. Par ailleurs, nous démontrons le théorème de la limite centrale fonctionnel correspondant.

Au paragraphe 5, à l'aide des résultats obtenus au paragraphe 4 nous démontrons une loi du logarithme itéré.

Le paragraphe 6 est consacré à la preuve d'un théorème limite local.

Dans le paragraphe 7, nous précisons la loi des grands nombres prouvée dans [8] en démontrant un théorème des grands écarts.

§ 2) Hypothèses sur la probabilité p et résultats préliminaires

2-1 Nous notons T_p (resp G_p) le semi-groupe (resp le groupe) fermé engendré par le support de p .

Avant de préciser les hypothèses sur la probabilité p , donnons quelques définitions :

DEFINITION 1 :

Une probabilité p *sur* G *possède des moments exponentiels si toute fonction* f *sous additive de* G *dans* $\mathbb{R}_+$, *c'est-à-dire satisfaisant à :*

$$\forall (g,g') \in G \times G \qquad f(gg') \leq f(g)\, f(g')$$

est intégrable par rapport à p.

Par ailleurs, tout élément y de G s'écrit

$$y = k_1 \exp a\, k_2$$

avec k_1, $k_2 \in SO(d)$ = {matrices orthogonales de déterminant 1}

$$\text{et } a = \exp \begin{bmatrix} x_1 & & (0) \\ & x_2 \cdot \cdot & \\ (0) & & x_d \end{bmatrix} \qquad x_1 \geq x_2 \geq \ldots \geq x_d \qquad \sum_{i=1}^{d} x_i = 0$$

Dans cette décomposition appelée décomposition polaire, l'élément a est déterminé de façon unique.

DEFINITION 2 :

Nous disons qu'une suite $(y_n)_{n\geq 1}$ *d'éléments de* G *s'écrivant dans la décomposition polaire précédente*

$$y_n = k_1(n) \exp \begin{bmatrix} x_1(n) & & (0) \\ & x_2(n) \cdot \cdot & \\ (0) & & \cdot x_d(n) \end{bmatrix} k_2(n)$$

avec $\sum_{i=1}^{d} x_i(n) = 0$ *et* $x_1(n) \geq x_2(n) \geq \ldots \geq x_d(n)$ *est contractante vis-à-vis de l'espace projectif* $P(\mathbb{R}^d)$, *si* $\lim_n x_1(n) - x_2(n) = +\infty$.

Par la suite, nous supposerons que la probabilité p satisfait au groupe d'hypothèses (P) suivant :

(P_1) p admet des moments exponentiels

(P_2) G_p et des sous-groupes d'indice fini ne laissent invariant aucun sous espace propre de $\mathbb{R}^d$

(P_3) T_p contient une suite contractante vis-à-vis de l'espace projectif $P(\mathbb{R}^d)$.

2-2 Notons < > le produit scalaire sur $\mathbb{R}^d$. Sur l'espace projectif $P(\mathbb{R}^d)$, on considère la distance d définie par $d(\bar{x},\bar{y}) = \|x \wedge y\|$ $\bar{x},\bar{y} \in P(\mathbb{R}^d)$ où x,y sont deux vecteurs de norme 1 dans $\mathbb{R}^d$ d'image $\bar{x}$ et $\bar{y}$ dans $P(\mathbb{R}^d)$ et où $\| \ \|$ est la norme associée à la structure euclidienne de $\overset{2}{\wedge}\mathbb{R}^d$ induite par la structure euclidienne de $\mathbb{R}^d$ c'est-à-dire que $\|x \wedge y\|^2 = \|x\|^2 - (\langle x,y\rangle)^2$

Si $\theta(x,y)$ désigne l'angle des vecteurs x et y nous avons

$$d(x,y) = |\sin \theta(x,y)|$$

On a alors le

THEOREME 1

Si la probabilité p *satisfait aux hypothèses* (P) *il existe un réel* λ_0 $0 < \lambda_0 < 1$ *tel que*

$$\lim_n \left\{ \sup_{\substack{\bar{x},\bar{y} \in P(\mathbb{R}^d) \\ \bar{x} \neq \bar{y}}} E\left[\frac{d(g_n g_{n-1} \ldots g_1 \bar{x}, g_n g_{n-1} \ldots g_1 \bar{y})}{d(\bar{x},\bar{y})}\right]^{\lambda_0} \right\}^{1/n} = \rho$$

avec $0 < \rho < 1$

Avant de prouver le théorème 1, précisons quelques notations et donnons quelques résultats préliminaires.

Considérons le G-espace $(P(\mathbb{R}^d) \times P(\mathbb{R}^d) - D)$ où $D = \{(\bar{x},\bar{x})\ ,\ \bar{x} \in P(\mathbb{R}^d)\}$. Ce G-espace peut être compactifié en lui adjoignant l'espace $B_{1,2}$ des drapeaux de dimension 2 de $\mathbb{R}^d$, c'est-à-dire l'espace des couples $(V_1 ; V_2)$ de sous-espaces vectoriels de $\mathbb{R}^d$ tels que $V_1 \subset V_2$ avec $\dim V_i = i \quad i = 1,2$ et en munissant $M_{1,2} = P(\mathbb{R}^d) \times P(\mathbb{R}^d) - D$ de la topologie suivante : $P(\mathbb{R}^d) \times P(\mathbb{R}^d) - D$ est un ouvert de $M_{1,2}$ et une suite $(x_n,y_n)_{n\geq 1} \in P(\mathbb{R}^d) \times P(\mathbb{R}^d) - D$ converge vers $(V_1,V_2) \in B_{1,2}$ si $\lim_n d(x_n,y_n) = 0$ et si $\lim_n (V_1^{(n)}, V_2^{(n)}) = (V_1,V_2)$ où $V_1^{(n)}$ est le sous-espace de $\mathbb{R}^d$ de dimension 1 défini par $\bar{x}_n$ et $V_2^{(n)}$ le sous-espace de $\mathbb{R}^d$ de dimension 2 défini par $\bar{x}_n$ et $\bar{y}_n$ [9].

L'application définie par

$$(1) \qquad \sigma_1(g,(\bar{x},\bar{y})) = \frac{d(g\bar{x},g\bar{y})}{d(\bar{x},\bar{y})} = \frac{\|gx \wedge gy\|}{\|gx\|\,\|gy\|\,\|x\wedge y\|}$$

de $G \times (P(\mathbb{R}^d) \times P(\mathbb{R}^d) - D)$ dans $\mathbb{R}^+$ est un M-cocycle c'est-à-dire vérifie pour $g,h \in G \quad (\bar{x},\bar{y}) \in P(\mathbb{R}^d) \times P(\mathbb{R}^d) - D$

$$\sigma_1(gh,(\bar{x},\bar{y})) = \sigma_1(g,h(\bar{x},\bar{y}))\ \sigma_1(h,(\bar{x},\bar{y}))$$

Ce M-cocycle se prolonge par continuité en un M-cocycle $\tilde{\sigma}_1$ sur $G \times M_{1,2}$ et pour $g \in G$, $(V_1,V_2) \in B_{1,2}$

on a

$$\sigma_1(g,(V_1,V_2)) = \frac{\|gv_2\|}{\|gv_1\|^2}$$

où v_1 est un vecteur quelconque de norme 1 définissant V_1 et v_2 est un bivecteur quelconque de norme 1 définissant V_2.

On a alors la

PROPOSITION 1

Si la probabilité p *satisfait aux hypothèses* (P), *l'expression* $\int_G \int_{M_{1,2}} \log \tilde{\sigma}_1(g,u)\, p(dg)\, \nu_{1,2}(du)$ *est strictement négative et indépendante de la probabilité* p - *invariante* $\nu_{1,2}$ *sur* $M_{1,2}$.

De plus la suite de fonctions $(\frac{1}{n})\, E\{\log \tilde{\sigma}_1(g_n g_{n-1} \dots g_1,.)\}_{n\geq 1}$ *définie sur* $M_{1,2}$ *converge uniformément vers* $\int_G \int_{M_{1,2}} \log \tilde{\sigma}_1(g,u)\, p(dg)\, \nu_{1,2}(du)$.

La démonstration de cette proposition figure dans [13].

Démonstration du théorème 1

D'après la proposition 1 il existe un entier N_0 tel que

$$\beta = \sup_{x \in M_{1,2}} E(\log \tilde{\sigma}_1(g_{N_0} g_{N_0-1} \dots g_1,x)) < 0$$

Posons $f_{N_0}(x,\lambda) = E(\tilde{\sigma}_1^{\lambda}(g_{N_0} g_{N_0-1} \dots g_1,x))$ $\lambda \geq 0$ $x \in M_{1,2}$

et $\delta_1(g) = \sup_{x \in M_{1,2}} |\log \tilde{\sigma}_1(g,x)|$

Par application de la formule de Taylor à l'ordre 2 à la fonction $\lambda \to \tilde{\sigma}_1^{\lambda}(g_{N_0} g_{N_0-1} \dots g_1,x)$ il vient que pour tout $x \in M_{1,2}$, $\lambda > 0$

$$|f_{N_0}(x,\lambda) - 1 - \lambda E(\log(g_{N_0} g_{N_0-1} \dots g_1,x)| \leq \frac{\lambda^2}{2} \int_G (\delta_1(g))^2\, e^{\lambda \delta_1(g)}\, p^{N_0}(dg)$$

l'hypothèse (P_1) assurant que l'intégrale figurant au second membre de cette inégalité est finie.

Il en résulte que : pour tout $\lambda > 0$

$$\sup_{x \in M_{1,2}} f_{N_0}(x,\lambda) \leq 1 + \lambda\beta + \frac{\lambda^2}{2} \int_G \delta_1(g)^2\, e^{\lambda \delta_1(g)}\, p^{N_0}(dg)$$

et donc puisque $\beta < 0$ il existe un réel $0 < \lambda_0 < 1$ tel que

$$(2) \qquad \sup_{x \in M_{1,2}} f_{N_0}(x,\lambda_0) < 1$$

Considérons alors la suite $f_n(\lambda_0) = \sup_{x \in M_{1,2}} (E\, \tilde{\sigma}_1^{\lambda_0}(g_n g_{n-1} \dots g_1,x))$ $n \geq 1$.

Du fait que $\tilde{\sigma}_1$ est un M-cocycle cette suite est sous-multiplicative et par conséquent la suite $(f_n(\lambda_0))^{1/n}$ $n \geq 1$ a une limite égale à :

$$\inf_{n \geq 1} (f_n(\lambda_0))^{1/n} \leq (f_{N_0}(\lambda_0))^{1/N_0}$$

ce qui prouve compte tenu de (2) que

$$\lim_n \sup_{x\in M_{1,2}} \{E(\tilde{\sigma}_1^{\lambda_0}(g_n g_{n-1}\cdots g_1,x))\}^{1/n} = \rho < 1$$

D'après (1) le théorème 1 est ainsi prouvé.

§ 3) Etude d'une famille d'opérateurs

3-1 Commençons par préciser quelques notations :

Soit $\mathcal{C}(P(\mathbb{R}^d))$ l'espace des fonctions continues sur $P(\mathbb{R}^d)$ muni de la norme de la convergence uniforme :

$$|f| = \sup_{\bar{x}\in P(\mathbb{R}^d)} |f(\bar{x})| \qquad f \in \mathcal{C}(P(\mathbb{R}^d))$$

D'autre part, pour $0 < \lambda \leq 1$ et pour toute fonction $f \in \mathcal{C}(P(\mathbb{R}^d))$ on définit

$$m_\lambda(f) = \sup_{\substack{\bar{x},\bar{y}\in P(\mathbb{R}^d)\\ \bar{x}\neq\bar{y}}} \frac{|f(\bar{x}) - f(\bar{y})|}{d^\lambda(\bar{x},\bar{y})}$$

et $\mathcal{L}_\lambda = \{f \in \mathcal{C}(P(\mathbb{R}^d)) / \|f\|_\lambda = |f| + m_\lambda(f) < +\infty\}$

$\mathcal{L}_\lambda$ est une algèbre de Banach unitaire munie de la norme $\| \ \|_\lambda$

De plus, appelons α le M cocycle de $G \times P(\mathbb{R}^d)$ dans $\mathbb{R}^+$ défini par :

$\alpha(g,\bar{x}) = \|gx\|$ où $g \in G$ et x est un vecteur de $\mathbb{R}^d$, de norme 1, d'image $\bar{x}$ dans $P(\mathbb{R}^d)$.

Dans ce paragraphe, nous nous proposons de mettre en évidence quelques propriétés de la famille d'opérateurs $P(\lambda)$ définie par

$$P(\lambda)\, f(\bar{x}) = \int_G \alpha^\lambda(g,\bar{x})\, f(g\bar{x})\, p(dg)$$

$x \in P(\mathbb{R}^d) \qquad f \in \mathcal{C}(P(\mathbb{R}^d)) \qquad \lambda \in \mathbb{C}$

et plus particulièrement celles de sa restriction à $(\mathcal{L}_{\lambda_0}, \| \ \|_{\lambda_0})$ où λ_0 est le réel intervenant dans l'énoncé du théorème 1.

3-2 Nous avons tout d'abord la

PROPOSITION 3

Pour tout $\lambda \in \mathbb{C}$, $P(\lambda)$ est un opérateur continu de $\mathcal{L}_{\lambda_0}$ dans $\mathcal{L}_{\lambda_0}$ et l'application $\lambda \to P(\lambda)$ de $\mathbb{C}$ dans l'espace de Banach $(\mathcal{L}_{\lambda_0}, \mathcal{L}_{\lambda_0})$ des applications linéaires continues de $\mathcal{L}_{\lambda_0}$ dans $\mathcal{L}_{\lambda_0}$ est analytique.

Avant de démontrer cette proposition, énonçons un lemme

LEMME 1

Si $\delta(g) = \sup(\|g\|, \|g^{-1}\|)$

a) Pour tous $\mu \geq 0$ *et* $\nu \in \mathbb{R}$ *il existe une constante* $c(\mu,\nu) < +\infty$ *telle que*

$$\sup_{(\bar{x},\bar{y})\in P(\mathbb{R}^d)\times P(\mathbb{R}^d)-D} \frac{|\alpha^{\mu+i\nu}(g,\bar{x}) - \alpha^{\mu+i\nu}(g,\bar{y})|}{d(\bar{x},\bar{y})} \leq c(\mu,\nu) \ [\delta(g)]^{\mu+4}$$

b) Pour tous $\mu < 0$ *et* $\nu \in \mathbb{R}$ *on a*

$$\sup_{(\bar{x},\bar{y})\in P(\mathbb{R}^d)\times P(\mathbb{R}^d)-D} \frac{|\alpha^{\mu+i\nu}(g,\bar{x}) - \alpha^{\mu+i\nu}(g,\bar{y})|}{d(\bar{x},\bar{y})} \leq c(|\mu|,-\nu) \ [\delta(g)]^{3|\mu|+4}$$

c) Il existe une constante c *telle que*

$$\sup_{(\bar{x},\bar{y})\in P(\mathbb{R}^d)\times P(\mathbb{R}^d)-D} \frac{|\mathrm{Log}\,\alpha(g,\bar{x}) - \mathrm{Log}\,\alpha(g,\bar{y})|}{d(\bar{x},\bar{y})} \leq c[\delta(g)]^6$$

Démonstration du lemme 1

1) Etablissons tout d'abord l'affirmation a).

Soient $\bar{x},\bar{y} \in P(\mathbb{R}^d)$; il existe alors deux éléments $x,y \in S_{d-1}$ d'image $\bar{x}$ et et $\bar{y}$ dans $P(\mathbb{R}^d)$ et tels que $h = x - y$ vérifie l'inégalité $\|h\| \leq \sqrt{2}$.

Soit d'autre part g un élément de $SL(d,\mathbb{R})$. Supposons que $\alpha(g,\bar{x}) \geq \alpha(g,\bar{y})$ (le cas de l'inégalité inverse se traitant de façon analogue).

On a

$$|\alpha^{\mu+i\nu}(g,\bar{x}) - \alpha^{\mu+i\nu}(g,\bar{y})| = \|g\,x\|^{\mu} \left|1 - \left(\frac{\|g\,y\|^2}{\|g\,x\|^2}\right)^{\frac{\mu+i\nu}{2}}\right|$$

$$= \|g\,x\|^{\mu} \left|1 - \left(1 + \frac{\|gh\|^2 + 2\langle gx,gh\rangle}{\|g\,x\|^2}\right)^{\frac{\mu+i\nu}{2}}\right|$$

De plus, on a :

$$\left|\frac{\|g\,h\|^2 + 2\langle gx,gh\rangle}{\|g\,x\|^2}\right| \leq 4\ \delta(g)^4\ \|h\|$$

$\left|1 + (1+u)^{\frac{\mu+i\nu}{2}}\right| \leq K(\mu,\nu)\ |u|$ dès que $|u| \leq \frac{1}{2}$ où $K(\mu,\nu)$ est une constante, et également

$$d(x,y) = \|h\|^2 \left(1 - \frac{\|h\|^2}{4}\right) \geq \frac{\|h\|}{\sqrt{2}}$$

Par conséquent si $\|h\| \leq \dfrac{1}{8\,[\delta(g)]^4}$ on peut conclure que

$$\frac{|\alpha^{\mu+i\nu}(g,\bar{x}) - \alpha^{\mu+i\nu}(g,\bar{y})|}{d(\bar{x},\bar{y})} \leq \sqrt{2}\, K(\mu,\nu)\ [\delta(g)]^{+4}$$

et si $\|h\| > \dfrac{1}{8[\delta(g)]^4}$ on a

$$\frac{|\alpha^{\mu+i\nu}(g,\bar{x}) - \alpha^{\mu+i\nu}(g,\bar{y})|}{d(\bar{x},\bar{y})} \leq \frac{2\,[\delta(g)]^{\mu}\sqrt{2}}{\|h\|} \leq 16\sqrt{2}\,[\delta(g)]^{\mu+4}$$

Les inégalités précédentes suffisent clairement à établir a).

2) Si $\mu < 0$, pour tous $\bar{x} \neq \bar{y} \in P(\mathbb{R}^d)$ et $g \in SL(d,\mathbb{R})$ on a

$$\frac{|\alpha^{\mu+i\nu}(g,\bar{x}) - \alpha^{\mu+i\nu}(g,\bar{y})|}{d(\bar{x},\bar{y})} = \left|\frac{\alpha^{|\mu|-i\nu}(g,\bar{x}) - \alpha^{|\mu|-i\nu}(g,\bar{y})}{d(\bar{x},\bar{y})}\right| \frac{1}{\alpha^{\mu}(g,\bar{x})\,\alpha^{\mu}(g,\bar{y})}$$

En tenant compte de a) et du fait que

$$\sup_{\bar{x}\in P(\mathbb{R}^d)} \frac{1}{\alpha^{\mu}(g,\bar{x})} \leq (\|g^{-1}\|)^{\mu} \leq (\delta(g))^{\mu}$$

on en déduit l'affirmation b).

3) Pour $g \in SL(d,\mathbb{R})$ $\bar{x},\bar{y} \in P(\mathbb{R}^d)$ on a

si $\alpha(g,\bar{x}) \geq \alpha(g,\bar{y})$

$$|\operatorname{Log}\alpha(g,\bar{x}) - \operatorname{Log}\alpha(g,\bar{y})| = \operatorname{Log}\left[\frac{\alpha(g,\bar{x})}{\alpha(g,\bar{y})}\right] \leq \frac{\alpha(g,\bar{x})}{\alpha(g,\bar{y})} - 1$$

On en déduit que pour tout g de $SL(d,\mathbb{R})$

$$\sup_{(\bar{x},\bar{y})\in P(\mathbb{R}^d)\times P(\mathbb{R}^d)-D} \frac{|\operatorname{Log}\alpha(g,\bar{x}) - \operatorname{Log}\alpha(g,\bar{y})|}{d(\bar{x},\bar{y})} \leq \sup_{\substack{\bar{x},\bar{y}\in P(\mathbb{R}^d)\times P(\mathbb{R}^d)\\ \bar{x}\neq\bar{y}}} \frac{|\alpha(g,\bar{x}) - \alpha(g,\bar{y})|}{d(\bar{x},\bar{y})}$$

$$\times \sup_{\bar{x}\in P(\mathbb{R}^d)} \frac{1}{\alpha(g,\bar{x})}$$

L'affirmation c) résulte de l'inégalité précédente et de a).

Pour tout $n \geq 0$ considérons l'opérateur P_n défini par

(3) $$P_n f(\bar{x}) = \int [\mathrm{Log}\ \alpha(g,\bar{x})]^n f(g\bar{x})\ p(dg)$$

$\bar{x} \in P(\mathbb{R}^d)$ $\quad f \in \mathcal{L}_{\lambda_0}$

soit $f \in \mathcal{L}_{\lambda_0}$ on a

(4) $$|P_n f| < \int \delta_2^n(g)\ p(dg)\ |f|$$

où δ_2 est la fonction sous additive sur G définie par $\delta_2(g) = \sup_{\bar{x}\in P(\mathbb{R}^d)} |\mathrm{Log}\ \alpha(g,\bar{x})|$

De plus pour $\bar{x},\bar{y} \in P(\mathbb{R}^d)$, $\bar{x} \neq \bar{y}$ on a

$$\frac{|P_n f(\bar{x}) - P_n f(\bar{y})|}{d^{\lambda_0}(\bar{x},\bar{y})} = \Big|\int (\mathrm{Log}\alpha(g,\bar{x}))^n \frac{f(g\bar{x}) - f(g\bar{y})}{d^{\lambda_0}(\bar{x},\bar{y})}\ p(dg)$$

$$+ \int f(g\bar{x}) \frac{(\mathrm{Log}\alpha(g,\bar{x}))^n - (\mathrm{Log}\alpha(g,\bar{y}))^n}{d^{\lambda_0}(\bar{x},\bar{y})}\ p(dg)\Big|$$

d'où il résulte en tenant compte du lemme 1 que

(5) $$m_{\lambda_0}(P_n f) \leq m_{\lambda_0}(f) \int \delta_2^n(g)\ p(dg)$$

$$+ c|f| \int \delta^6(g) \times n\ \delta_2^{n-1}(g)\ p(dg)$$

D'après (3) et (5) on a

(6) $$\|P_n(f)\|_{\lambda_0} \leq \{\int \delta_2^n(g)\ p(dg) + c\int \delta^6(g) \times n\ \delta_2^{n-1}\ p(dg)\}\ \|f\|_{\lambda_0}$$

Chacune des intégrales $\int e^{\delta_2(g)}\ p(dg)$ et $\int \delta^6(g)\ e^{\delta_2(g)}\ p(dg)$ est convergente, et par conséquent d'après (6) pour tout $\lambda \in \mathbb{C}$ la série $\sum_{n\geq 0} \frac{\lambda^n}{n!} P^n$ converge normalement dans $\mathcal{L}(\mathcal{L}_{\lambda_0},\mathcal{L}_{\lambda_0})$; de plus cette série a pour somme $P(\lambda)$ ce qui établit la proposition 3.

3-3 Désormais, nous notons ν l'unique probabilité p invariante portée par $P(\mathbb{R}^d)$ [1] et nous appelons e la fonction définie par $e(\bar{x}) = 1$, $\bar{x} \in P(\mathbb{R}^d)$. Nous pouvons énoncer la

PROPOSITION 4

Pour toute fonction $f \in \mathcal{L}_{\lambda_0}$ *et pour tout* $n \geq 1$ *on a*

$$P^n(0)f = \nu(f)e + Q^n f$$

où Q *est un opérateur sur* $\mathcal{L}_{\lambda_0}$ *de rayon spectral* $r(Q)$ *strictement inférieur à 1, et tel que* $Qe = 0$.

Enonçons et prouvons deux lemmes utiles à la démonstration de la proposition 4.

LEMME 2

Il existe un entier $n_0 \geq 1$ *et une constante* $r_0 < 1$ *tels que*

$$\forall f \in \mathcal{L}_{\lambda_0} \quad \|P^{n_0}(0)\, f\|_{\lambda_0} \leq r_0 \|f\|_{\lambda_0} + |f|$$

Preuve du lemme 2

Pour $f \in \mathcal{L}_{\lambda_0}$ on a

(7) $\quad |P^n(0)\, f| \leq |f|$

et de plus pour $(\bar{x},\bar{y}) \in P(\mathbb{R}^d) \times P(\mathbb{R}^d) - D$

$$\frac{|P^n(0)f(\bar{x}) - P^n(0)f(\bar{y})|}{d^{\lambda_0}(\bar{x},\bar{y})} = \left|\int \frac{f(g\bar{x}) - f(g\bar{y})}{d^{\lambda_0}(\bar{x},\bar{y})}\, p^n(dg)\right| \leq$$

$$\leq m_{\lambda_0}(f) \sup_{\substack{\bar{x},\bar{y}\in P(\mathbb{R}^d)\\ \bar{x} \neq \bar{y}}} \int \frac{d^{\lambda_0}(g\bar{x},g\bar{y})}{d^{\lambda_0}(\bar{x},\bar{y})}\, p^n(dg)$$

d'où (8) $\quad m_{\lambda_0}(P^n(0)f) \leq \sup_{\substack{\bar{x},\bar{y}\in P(\mathbb{R}^d)\\ \bar{x} \neq \bar{y}}} \int \frac{d^{\lambda_0}(g\bar{x},g\bar{y})}{d^{\lambda_0}(\bar{x},\bar{y})}\, p^n(dy) \times m_{\lambda_0}(f)$.

Les inégalités (7) et (8) et le théorème 1 permettent alors de conclure.

LEMME 3

a) $\forall f \in \mathcal{L}_{\lambda_0} \quad \lim_n \sup_{\bar{x}\in P(\mathbb{R}^d)} |P^n(0)\, f(\bar{x}) - \nu(f)| = 0$

b) 1 est l'unique valeur propre de module 1 de $P(0)$ *et le sous-espace propre correspondant est formé des multiples de* e.

Preuve du lemme 3

On a $\forall n, m \in \mathbb{N}$ et $f \in \mathcal{L}_{\lambda_0}$

$$\sup_{\bar{x}\in P(\mathbb{R}^d)} |P^{n+m}(0)f(\bar{x}) - P^n(0)f(\bar{x})| \leq m_{\lambda_0}(f) \sup_{\bar{x},\bar{y}\in P(\mathbb{R}^d)} d^{\lambda_0}(x,y) \sup_{\substack{\bar{x},\bar{y}\in P(\mathbb{R}^d)\\ \bar{x}\neq\bar{y}}} \int \frac{d^{\lambda_0}(g\bar{x},g\bar{y})}{d^{\lambda_0}(\bar{x},\bar{y})} p^n(dg)$$

A l'aide du théorème 1, on en déduit que pour toute fonction $f \in \mathcal{L}_{\lambda_0}$ la suite $(P^n(0)f)_{n\geq 1}$ converge uniformément sur $P(\mathbb{R}^d)$ et sa limite est $\nu(f)$ car la suite $(\frac{1}{n}\sum_{k=0}^{n-1} P^k(0)f)_{n\geq 1}$ converge uniformément vers $\nu(f)$ puisque ν est l'unique probabilité p invariante portée par $P(\mathbb{R}^d)$.

Démonstration de la proposition 4

Si L est une partie bornée de $(\mathcal{L}_{\lambda_0}, \| \|_{\lambda_0})$, $P^n(0)$ L est une partie bornée et équicontinue de $\mathcal{C}(P(\mathbb{R}^d))$ et donc d'après le théorème d'Ascoli une partie compacte de $(\mathcal{C}(P(\mathbb{R}^d)), | |)$.

En tenant compte du fait que $P(0)$ est une contraction de $(\mathcal{C}(P(\mathbb{R}^d)), | |)$, du lemme 2, et de la remarque précédente on en conclut à l'aide du théorème de Ionescu-Tulcea et Marinescu [19] que l'on peut écrire

$$\forall n \geq 0 \qquad P^n(0) = \sum_{\mu\in S} \mu^n U_\mu + Q^n$$

où S est l'ensemble fini des valeurs propres de module 1 de $P(0)$ et où U_μ $\mu \in S$ et Q sont des opérateurs bornés sur $(\mathcal{L}_{\lambda_0}, \| \|_{\lambda_0})$ tels que $U_\mu^2 = U_\mu$, $U_\mu U_{\mu'} = 0$ si $\mu \neq \mu'$ $U_\mu Q = QU_\mu = 0$ $P_\mu \mathcal{L}_{\lambda_0} = D_\mu$ où $D_\mu \overset{\Delta}{=} \{f\in \mathcal{L}_{\lambda_0} / P(0)f = \mu f\}$ et où Q est de norme spectrale strictement inférieure à 1.

La proposition 4 se déduit alors immédiatement du lemme 4.

Donnons un corollaire de la proposition 4.

Corollaire 1

Pour toute $f \in \mathcal{C}(P(\mathbb{R}^d))$ on a

$$\lim_n \sup_{\bar{x}\in P(\mathbb{R}^d)} |P^n(0) f(x) - \nu(f)| = 0$$

Démonstration du corollaire 1 : La propriété précédente est vraie si $f \in \mathcal{L}_{\lambda_0}$; elle s'obtient aussi pour toute fonction de $\mathcal{C}(P(\mathbb{R}^d))$ car $P(0)$ est une contraction de $\mathcal{C}(P(\mathbb{R}^d))$ et car $\mathcal{L}_{\lambda_0}$ est dense dans $\mathcal{C}(P(\mathbb{R}^d))$ muni de la norme $| |$.

3-4 Les résultats qui précèdent permettent alors d'obtenir la

PROPOSITION 5 :

Il existe un réel $a > 0$ *tel que pour* $\lambda \in \mathbb{C}, |\lambda| < a$ *on ait*

a) $\forall f \in \mathcal{L}_{\lambda_0}$ *et* $n \geq 1$

$P^n(\lambda)f = [k(\lambda)]^n N_1(\lambda)f + Q^n(\lambda)f$

où $k(\lambda)$ *est l'unique valeur propre de plus grand module de* $P(\lambda)$ *et* $|k(\lambda)| > \frac{2+r(Q)}{3}$

$N_1(\lambda)$ *est la projection sur le sous-espace propre* E_λ *de dimension 1, correspondant à* $k(\lambda)$.

$Q(\lambda)$ *est un opérateur de* $\mathcal{L}_{\lambda_0}$ *de rayon spectral*

$r(Q(\lambda)) \leq \frac{1+2r(Q)}{3}$ *et tel que* $Q(\lambda)E_\lambda = 0$

b) Les applications $\lambda \to k(\lambda)$, $\lambda \to N_1(\lambda)$, $\lambda \to Q(\lambda)$ *sont analytiques*

c) (9) $\forall n \geq 1 \quad \|Q^n(\lambda)e\|_{\lambda_0} \leq C_5|\lambda|\rho_1^n$

où C_5 *est une constante et* $0 < \rho_1 < 1$.

Démonstration de la proposition 5

a) Les deux premières affirmations résultent des proposition 3 et 4 et de la théorie générale des perturbations analytiques d'opérateurs [6], [17].

Résumons simplement ici la façon de construire $P_1(\lambda)$ et $Q(\lambda)$. Nous notons $\|T\|_{\lambda_0}$ la norme de tout opérateur T de $\mathcal{L}_{\lambda_0}$ dans $\mathcal{L}_{\lambda_0}$ et nous désignons par N_1 l'opérateur $N_1(f) = \nu(f)e \quad f \in \mathcal{L}_{\lambda_0}$.

Pour $|z| > r(Q)$ et $z \neq 1$ la résolvante de $P(0)$ est

$$R(z) = \frac{1}{z-1} N_1 + \sum_{n=0}^{\infty} \frac{(P(0)-N_1)^{n+1}}{z^{n+1}}$$

Si $\|P(\lambda) - P(0)\|_{\lambda_0} < \frac{1}{\|R(Z)\|_{\lambda_0}}$ la série

$$(10) \qquad \sum_{k=0}^{\infty} R(z) \{(P(\lambda) - P(0) R(z)\}^k$$

converge et détermine la résolvante $R(\lambda,z)$ de $P(\lambda)$.

Considérons alors les cercles I_1 et I_2 de centres 1 et 0 respectivement et de rayon $\rho_1 = \frac{1-r(Q)}{3}$ et $\rho_2 = \frac{1+2r(Q)}{3}$; de plus soit $\delta > 0$ tel que $\delta < \rho_1$ et $r(Q) + \delta < \rho_2$ et $M_\delta = \sup_{z \in \{z/|z|>r(Q)+\delta, |z-1|<\delta\}} \|R(z)\|_{\lambda_0}$

Si $\|P(\lambda) - P(0)\|_{\lambda_0} < \frac{1}{M_\delta}$ les cercles I_1 et I_2 appartiennent à l'ensemble résolvant de $P(\lambda)$. Considérons alors les projections

$$N_1(\lambda) = \frac{1}{2i\pi} \int_{I_1} R(z,\lambda)\, dz$$

$$N_2(\lambda) = \frac{1}{2i\pi} \int_{I_2} R(z,\lambda)\, dz$$

Pour $\|N_1(\lambda) - N\|_{\lambda_0} < 1$, l'image E_λ de $N_1(\lambda)$ est de dimension 1 et on a

$$P(\lambda)\, N_1(\lambda)\, e_\lambda = N_1(\lambda)\, P(\lambda)\, e_\lambda = k(\lambda)\, e_\lambda$$

où $e_\lambda \in \mathcal{L}_{\lambda_0}$ engendre E_λ.

En outre $\forall n \geq 1$ on a

$$P^n(\lambda) = P^n(\lambda)\, N_1(\lambda) + P^n(\lambda)\, N_2(\lambda) = (k(\lambda))^n\, N_1(\lambda) + Q^n(\lambda)$$

où (11) $Q^n(\lambda) = \frac{1}{2i\pi} \int_{I_2} z^n R(z,\lambda)dz$

b) Pour $|\lambda| < a$, on a $R(\lambda,z) = R(z) + \lambda R_1(z,\lambda)$ (cf(10)) et donc d'après (11)

$$Q^n(\lambda)e = \frac{1}{2i\pi} \int_{I_2} z^n R(z)e\, dz + \frac{1}{2i\pi} \int_{I_2} z^n\, R_1(z,\lambda)e\, dz$$

c'est-à-dire

$$Q^n(\lambda)e = \frac{1}{2i\pi} \int_{I_2} z^n\, R_1(z,\lambda)e\, dz$$

d'où $\|Q^n(\lambda)e\|_{\lambda_0} \leq C_5\, |\lambda|\, \rho_2^n$

où $C_5 = \frac{1}{2\pi} \sup_{\substack{|z| = \rho_2 \\ |\lambda| < a}} \|R_1(z,\lambda)\|_{\lambda_0} < +\infty$

ce qui établit l'affirmation c)

3-5 Dans les sections précédentes nous avons mis en évidence des propriétés des opérateurs $P(\lambda)$ $\lambda \in \mathbb{C}$ sur un voisinage de $\lambda = 0$. Complétons cette étude par la

PROPOSITION 6 :

Il existe un réel $0 < b < a$ *tel que pour tout* $\lambda_1 \in \mathbb{R}, |\lambda_1| < b$, $\lambda_2 \in \mathbb{R}$ *l'opérateur* $P(\lambda_1 + i\lambda_2)$ *a un ensemble fini* $G(\lambda_1 + i\lambda_2)$ *de valeurs propres de module* $|k(\lambda_1)|$. *De plus pour chaque* $\mu \in G(\lambda_1 + i\lambda_2)$ *le sous espace propre correspondant* $V_\mu(\lambda_1 + i\lambda_2)$ *est de dimension finie et, il existe des opérateurs linéaires* $U_\mu(\lambda_1 + i\lambda_2)$ $\mu \in G(\lambda_1 + i\lambda_2)$, $Q(\lambda_1 + i\lambda_2)$ *sur* $\mathcal{L}_{\lambda_0}$ *tels que l'on ait :*

$$\forall n \geq 1 \quad P^n(\lambda_1 + i\lambda_2) = \sum_{\mu \in G(\lambda_1 + i\lambda_2)} \mu^n\, U_\mu(\lambda_1 + i\lambda_2) + Q^n(\lambda_1 + i\lambda_2)$$

$$U_\mu(\lambda_1 + i\lambda_2)\, U_{\mu'}(\lambda_1 + i\lambda_2) = 0 \quad \text{si } \mu \neq \mu' \in G(\lambda_1 + i\lambda_2),\quad U^2(\lambda_1 + i\lambda_2) = U_\mu(\lambda_1 + i\lambda_2)$$

$$U_\mu(\lambda_1 + i\lambda_2)\, Q(\lambda_1 + i\lambda_2) = Q(\lambda_1 + i\lambda_2)\, U_\mu(\lambda_1 + i\lambda_2)$$

$$U_\mu(\lambda_1 + i\lambda_2)\mathcal{L}_{\lambda_0} = V_\mu(\lambda_1 + i\lambda_2)$$

En outre, l'opérateur $Q(\lambda_1 + i\lambda_2)$ *est de norme spectrale* $r(Q(\lambda_1 + i\lambda_2))$ *strictement inférieure à* $|k(\lambda_1)|$.

Démonstration de la proposition 6

Elle est analogue à celle de la proposition 4 et repose sur deux lemmes. Notant pour tout opérateur U de $\mathcal{L}_{\lambda_0}$ dans $\mathcal{L}_{\lambda_0}$:

$|U|_{\lambda_0} = \sup_{f \in \mathcal{L}_{\lambda_0}} \frac{|Uf|}{|f|}$ on a tout d'abord

a) LEMME 4

Pour $\lambda_1 \in \mathbb{R}$ $|\lambda_1| < a$, $\lambda_2 \in \mathbb{R}$

$$\sup_{n\geq 0} \frac{|P^n(\lambda_1 + i\lambda_2|_{\lambda_0}}{|k(\lambda_1)|^n} < +\infty$$

Démonstration du lemme 4

Soient $f \in \mathcal{L}_{\lambda_0}$ $\lambda_1 \in \mathbb{R}$ tel que $|\lambda_1| < a$ et $\lambda_2 \in \mathbb{R}$; on a

$$\forall n \geq 1 \quad |P^n(\lambda_1 + i\lambda_2)f| \leq P^n(\lambda_1)\ |f| \leq |f|\ |P^n(\lambda_1)e|$$

or d'après la proposition 5 on a

$$\sup_{n\geq 0} \frac{|P^n(\lambda_1)e|}{|k(\lambda_1)|^n} \leq |N_1(\lambda_1)e| + \sup_{n\geq 0} \frac{\|Q^n(\lambda_1)\|_{\lambda_0}}{|k(\lambda_1)|^n} < +\infty$$

d'où le lemme 4.

b) LEMME 5

Il existe un réel $0 < b < a$, *un réel* $0 < t_0 < 1$ *et un entier* $n_0 \geq 1$ *tels que si* $\lambda_1 \in \mathbb{R}$ $|\lambda_1| < b$, $\lambda_2 \in \mathbb{R}$ *on ait pour toute* $f \in \mathcal{L}_{\lambda_0}$

$$\frac{\|P^{n_0}(\lambda_1 + i\lambda_2)f\|_{\lambda_0}}{|k(\lambda_1)|^{n_0}} \leq t_0 \|f\|_{\lambda_0} + R(n_0,\lambda_1,\lambda_2)\ |f|$$

où $R(n_0,\lambda_1,\lambda_2) < +\infty$.

Démonstration du lemme 5

Pour $(\bar{x},\bar{y}) \in P(\mathbb{R}^d)\times P(\mathbb{R}^d)-D$, $f \in \mathcal{L}_{\lambda_0}$, $n \geq 1$ on a

$$\frac{|P^n(\lambda_1 + i\lambda_2)f(\bar{x}) - P^n(\lambda_1 + i\lambda_2)f(\bar{y})|}{|k(\lambda_1)|^n\ d^{\lambda_0}(\bar{x},\bar{y})} \leq \frac{1}{|k(\lambda_1)|^n} \int \alpha^{\lambda_1}(g,\bar{x})\ \frac{|f(g\bar{x}) - f(g\bar{y})|}{d^{\lambda_0}(\bar{x},\bar{y})}\ p^n(dg)$$

$$+ \frac{|f|}{|k(\lambda_1)|^n} \int \frac{|\alpha^{\lambda_1+i\lambda_2}(g,\bar{x}) - \alpha^{\lambda_1+i\lambda_2}(g,\bar{y})|}{d^{\lambda_0}(\bar{x},\bar{y})}\ p^n(dg)$$

d'où l'on déduit en tenant compte du lemme 1.

$$(12)\quad m_{\lambda_0}\left(\frac{P^n(\lambda_1+i\lambda_2)f}{[k(\lambda_1)]^n}\right) \leq \frac{m_{\lambda_0}(f)}{|k(\lambda_1)|^n} \sup_{\substack{\bar{x},\bar{y}\in P(\mathbb{R}^d)\\ \bar{x}\neq\bar{y}}} \int \alpha^{\lambda_1}(g,\bar{x})\ \frac{d^{\lambda_0}(g\bar{x},g\bar{y})}{d^{\lambda_0}(\bar{x},\bar{y})}\ p^n(dg)$$

$$+ \frac{|f|}{|k(\lambda_1)|^n}\ C_1(n,\lambda_1,\lambda_2)$$

où $C_1(n,\lambda_1,\lambda_2) = c(\lambda_1,\lambda_2) \int \delta^{\lambda_1+4}(g)\, p^n(dg)$ si $\lambda_1 \geq 0$

et où $C_1(n,\lambda_1,\lambda_2) = c(\frac{1}{4}|\lambda_1|, -\lambda_2) \int \delta(g)^{3|\lambda_1|+4}\, p^n(dg)$ si $\lambda_1 < 0$

D'après le théorème 1, il existe un réel $0 < t_0 < 1$ et un entier n_0 tels que

$$\sup_{\substack{\bar{x},\bar{y} \in P(\mathbb{R}^d) \\ \bar{x} \neq \bar{y}}} \int \frac{d^{\lambda_0}(g\bar{x},g\bar{y})}{d^{\lambda_0}(\bar{x},\bar{y})}\, p^{n_0}(dg) < t_0 < 1$$

Par continuité, on en déduit qu'il existe un réel $0 < b < a$ tel que pour $\lambda_1 \in \mathbb{R}$ on ait également

$$(13) \quad \frac{1}{|k(\lambda_1)|^{n_0}} \sup_{\substack{\bar{x},\bar{y} \in P(\mathbb{R}^d) \\ \bar{x} \neq \bar{y}}} \int \alpha^{\lambda_1}(g,\bar{x}) \frac{d^{\lambda_0}(g\bar{x},g\bar{y})}{d^{\lambda_0}(\bar{x},\bar{y})}\, p^{n_0}(dg) < t_0 < 1$$

Le lemme 5 est alors une conséquence immédiate de (12) et (13).

Si L est une partie bornée de $(\mathcal{L}_{\lambda_0}, \|\ \|_{\lambda_0})$, $P^{n_0}(\lambda_1 + i\lambda_2)(L)$ est une partie bornée et équicontinue de $\mathcal{C}(P(\mathbb{R}^d))$ et donc d'après le théorème d'Ascoli une partie compacte de $\mathcal{C}(P(\mathbb{R}^d))$.

Les lemmes 4 et 5 et la remarque précédente permettent alors d'obtenir immédiatement la proposition 5 grâce au théorème de Ionescu Tulcea et Marinescu [19].

§ 4) Théorèmes de la limite centrale

4-1 Etablissons tout d'abord le

THEOREME 2 :

Si les hypothèses (P) *sont vérifiées et si de plus il existe deux entiers* $n_0 \geq 1$ *et* $k_0 \geq 1$ *tels que les supports de* p^{n_0} *et* $p^{n_0+k_0}$ *se rencontrent*

1) $\forall x \in \mathbb{R}^d - \{0\}$

$$\gamma_n(x) = \frac{1}{n} E(\log \| g_n\, g_{n-1} \ldots x\|)$$

converge vers la constante $\gamma = \iint_{G \times P(\mathbb{R}^d)} \operatorname{Log} \alpha(g,\bar{x})\, p(dg)\, \nu(d\bar{x})$

et la convergence est uniforme sur $S_{d-1} = \{x / \|x\| = 1\}$.

2) $\forall x \in \mathbb{R}^d - \{0\}$

$$\sigma_n^2(x) = \frac{1}{n} E(\log \| g_n\, g_{n-1} \ldots g_1 x\| - n\gamma)^2$$

converge vers une constante $\sigma^2 > 0$ *indépendante de* x *et la convergence est uniforme sur* S_{d-1}.

3) $\forall x \in \mathbb{R}^d - \{0\}$ *la suite de variables aléatoires*

$Z_n(x) = \frac{1}{\sigma\sqrt{n}} (\mathrm{Log}\|g_n\, g_{n-1} \cdots g_1 x\| - n\gamma)$ *converge en loi vers une loi normale* $N(0,1)$.

4) *Il existe une constante* $C > 0$

$$(14) \qquad \forall t \in \mathbb{R}, \sup_{\|x\|=1} \left|P(Z_n(x) \leq t) - \frac{1}{\sqrt{2\pi}} \int_{-\infty}^{t} e^{-\frac{u^2}{2}} du\right| \leq \frac{C}{\sqrt{n}}$$

Démonstration du théorème 2

a) Il est immédiat que l'on peut supposer $\|x\| = 1$, ce que nous ferons par la suite

b) Le 1) résulte de l'unicité de la probabilité p invariante portée par $P(\mathbb{R}^d)$

c) $\forall x \in S_{d-1}$ on a pour $\lambda \in \mathbb{R}$ $|\lambda| < a$

$$(15) \quad E(e^{i\lambda \mathrm{Log}\|g_n\, g_{n-1} \cdots g_1 x\|}) = P^n(i\lambda)\, e(\bar{x}) = [k(i\lambda)]^n N_1(i\lambda)\, e(\bar{x}) + (Q(i\lambda))^n e(\bar{x})$$

De plus comme $k(.)$, $N_1(.)$, $Q(.)$ sont analytiques on a pour $|\lambda| < a$

$$(16) \qquad k(i\lambda) = 1 + i\lambda k'(0) - \frac{\lambda^2}{2} k''(0) - i\frac{\lambda^3}{6} k^{(3)}(0) + \lambda^3 \varepsilon_1(\lambda)$$

où $\lim_{\lambda \to 0} \varepsilon_1(\lambda) = 0$

et

$$(17) \quad N_1(i\lambda) = \nu + i\lambda N_1^{(1)} - \frac{\lambda^2}{2} N_1^{(2)} - \lambda^2 N_1^{(3)}(i\lambda)$$

où $N_1^{(1)}, N_1^{(2)}, N_1^{(3)}(i\lambda) \in \mathcal{L}(\mathcal{L}_{\lambda_0}, \mathcal{L}_{\lambda_0})$ et où $\lim_{\lambda \to 0} \|N_1^{(3)}(i\lambda)\|_{\lambda_0} = 0$.

En tenant compte de (12), (13), (14) et (8) on voit que pour $|\lambda| < b$

$$\lim_n E(e^{i\frac{\lambda}{n} \mathrm{Log}\|g_n\, g_{n-1} \cdots g_1 x\|}) = e^{ik'(0)}.$$

Or on sait [8] que p.s. $\lim_n \frac{1}{n} \mathrm{Log}\|g_n g_{n-1} \cdots g_1 x\| = \gamma$ par conséquent (18) $k'(0) = \gamma$

On a alors le

LEMME 6

Il existe un réel $b > 0$ *tel que pour* $|\lambda| < b$ *on ait*

$$E(e^{i\lambda(\mathrm{Log}\|g_n g_{n-1} \cdots g_1 x\| - n\gamma)}) = e^{n[-\frac{\lambda^2}{2}(h''(0) - \gamma^2) + i\lambda^3 A + \lambda^3 \varepsilon_2(\lambda)]}$$

$$\times (1 + i\lambda N_1^{(1)} e(\bar{x}) - \frac{\lambda^2}{2} N_1^{(2)} e(\bar{x}) - \lambda^2 N_1^{(3)}(i\lambda)\, e(\bar{x})) + e^{-i\lambda n\gamma}[Q(i\lambda)]^n e(\bar{x})$$

où $A \in \mathbb{R}$, $\lim_{\lambda\to 0} \varepsilon_2(\lambda) = 0$ $\qquad \lim_{\lambda\to 0} N_1^{(3)}(i\lambda) = 0$

Démonstration du lemme 6

Le lemme 6 est une conséquence immédiate de (15), (16), (17) et (18) et du calcul des développements limités.

d) Calculons $\frac{\partial^2}{\partial\lambda^2} E(e^{i\frac{\lambda}{\sqrt{n}}(\mathrm{Log}\|g_n g_{n-1}\cdots g_1 x\| - n\gamma)}) /_{\lambda=0}$

Remarquons tout d'abord que pour $|\lambda| < a$ $\quad |z| = \rho_2$ nous pouvons développer $R(i\lambda,z)$, sous la forme

$$R(i\lambda,z) = R(z) + i\lambda R^{(1)}(z) - \frac{\lambda^2}{2} R^{(2)}(z) + \lambda^2 R^{(3)}(z,i\lambda)$$

où $R^{(1)}(z)$, $R^{(2)}(z)$, $R^{(3)}(z,i\lambda) \in \mathcal{L}(\mathcal{L}_{\lambda_0}, \mathcal{L}_{\lambda_0})$

On en déduit que

$$(19) \qquad Q^n(\frac{i\lambda}{\sqrt{n}})\, e(\bar{x}) = \frac{1}{2i\pi} \int_{I_2} z^n R(\frac{i\lambda}{\sqrt{n}}, z)\, e(\bar{x})\, dz = \frac{1}{2\pi} \frac{\lambda}{\sqrt{n}} \int_{I_2} z^n R^{(1)}(z)\, e(\bar{x})\, dz$$

$$- \frac{\lambda^2}{2n} \frac{1}{2i\pi} \int_{I_2} z^n R^{(2)}(z)\, e(\bar{x})\, dz - \frac{\lambda^2}{n} \frac{1}{2i\pi} \int_{I_2} z^n R^{(3)}(z,i\lambda)\, e(\bar{x})\, dz$$

Du lemme 6 et de (19) il résulte que

$$\frac{\partial^2}{\partial\lambda^2} E(e^{i\frac{\lambda}{\sqrt{n}}(\mathrm{Log}\|g_n g_{n-1}\cdots g_1 x\| - n\gamma)}) /_{\lambda=0} = -\frac{1}{n} E(\mathrm{Log}\|g_n g_{n-1}\cdots g_1 x\| - n\gamma)^2 =$$

$$= -(k''(0) - \gamma^2) - \frac{1}{n} N_1^{(2)}\, e(\bar{x}) - \frac{1}{2i\pi n} \int_{I_2} z^n R^{(2)}(z)\, e(\bar{x})\, dz$$

$$- \frac{2i\gamma}{2\pi} \int_{I_2} z^n R^{(1)}(z)\, e(\bar{x})\, dz .$$

Cette expression montre puisque $\rho_2 < 1$ qu'il existe une constante $C_6 < +\infty$ telle que

$$(20) \qquad \sup_{\|x\|=1} \left|\frac{1}{n} E(\mathrm{Log}\|g_n g_{n-1}\cdots g_1 x\| - n\gamma)^2 - (k''(0) - \gamma^2)\right| \leq \frac{C_6}{n}$$

Ceci montre la convergence uniforme sur S_{d-1} de la suite $(\sigma_n^2(x))_{n\geq 1}$, vers $\sigma^2 = k''(0) - \gamma^2$.

e) Il reste à prouver que $\sigma^2 > 0$: Cette démonstration se fera par l'absurde en plusieurs étapes ; commençons par préciser quelques notations : pour tout $k \geq 1$ on définit la probabilité de transition $Q_p k$ sur $G_p \times P(\mathbb{R}^d)$ par

$$Q_{p^k} f(g,\bar{x}) = \int f(g',g\bar{x})\, p^k(dg')$$

Q_p est la probabilité de transition de la chaine de Markov $(M'_n)_{n\geq 0}$ à valeurs dans

$G_p \times P(\mathbb{R}^d)$ définie par

$$M'_0 = (g,\bar{x}),\ M'_n = (g_n, g_{n-1} \cdots g_1\ g\bar{x}) \quad n \geq 1$$

De plus, notons

$$\rho(g,\bar{x}) = \mathrm{Log}\ \sigma(g,\bar{x}) \qquad g \in G_p\ ,\ \bar{x} \in P(\mathbb{R}^d)$$

et soit

$$h(g,\bar{x}) = \sum_{k \geq 0} Q_p^k(\rho-\gamma)(g,\bar{x})$$

La convergence uniforme de cette série sur $G_p \times P(\mathbb{R}^d)$ résulte de l'égalité

$$k \geq 1\ Q_p^k(\rho-\gamma)(g,\bar{x}) = Q^{k-1}(\bar{\rho} - \gamma)(g\bar{x})$$

où $\bar{\rho}(\bar{x}) = \int \rho(g,\bar{x}) p(dg)$

et du fait que Q est de norme spectrale strictement inférieure à 1 dans $\mathcal{L}_{\lambda_0}$.

On a alors le

LEMME 7

a) $\sigma^2 = \iint_{G_p \times P(\mathbb{R}^d)} \{h^2(g,\bar{x}) - (Q_p h)^2\ (g,\bar{x})\}\ p(dg)\ \nu(d\bar{x})$

b) si $\sigma^2 = 0$, *pour tout* g *du support de* p *et tout* $\bar{x}$ *du support de* ν *on a* $h(g,\bar{x}) = \int h(g',\bar{x})\ p(dg')$.

Démonstration du lemme 7

La chaine de Markov $(M'_n)_{n \geq 0}$ admet $p \otimes \nu$ pour unique probabilité invariante et on a

$$\sigma^2 = \lim_n \frac{1}{n} E_{p\otimes\nu}\ ((\rho-\gamma)(M'_1) + (\rho-\gamma)(M'_2) + \ldots + (\rho-\gamma)(M'_n))^2$$

Comme $\rho - \gamma = (I-Q_p)h$ on obtient facilement que

$$E_{p\otimes}\ ((\rho-\gamma)(M'_1) + (\rho-\gamma)(M'_2) + \ldots + (\rho-\gamma)(M'_n))^2 =$$
$$n\ p\otimes\nu(h^2) + n\ p\otimes\nu((Q_p h)^2) - 2n\ p\otimes\nu(h\ Q_p\ h) +$$
$$2n\ p\otimes\nu\{(h-Q_p h)(Q_p h)\} - 2\ p\otimes\nu(h-Q_p h)(\sum_{k=1}^{n} Q_p^k\ h)$$

et l'assertion a) du lemme en découle immédiatement.

σ^2 peut encore s'écrire sous la forme

$$\sigma^2 = \iint_{G_p \times P(\mathbb{R}^d)} \{Q_p h^2(g,\bar{x}) - (Q_p h)^2(g,\bar{x})\}\ p(dg)\ \nu(d\bar{x})$$

si $\sigma^2 = 0$ il en résulte que pour $p \otimes \nu$ presque tout $(g,\bar{x}) \in G_p \times P(\mathbb{R}^d)$ on a

$$Q_p(g,\bar{x})\ \{(g',\bar{y})/h(g',\bar{y}) = Q_p h(g,\bar{x})\} = 1$$

c'est-à-dire que

$$p\{\gamma \in G_p\ /\ h(\gamma,g.\bar{x}) = \int h(g',g\bar{x})\ p(dg')\} = 1.$$

On en déduit que pour ν-presque tout $\bar{y} \in P(\mathbb{R}^d)$ on a

$$p\{\gamma \in G_p / h(\gamma,\bar{y}) = \int h(g',\bar{y})\ p(dg')\} = 1$$

et ceci en tenant compte du fait que h est continue établit le b) du lemme.

L'assertion b) du lemme précédent peut être précisée en le

LEMME 8

Si $\sigma^2 = 0$, *on a pour tout* $k \geq 1$, *tout* g *du support de* p^k *et tout* $\overline{x}$ *du support de* ν

$$h(g,\overline{x}) = \int h(g',\overline{x})\, p^k(dg').$$

Démonstration du lemme 8

Pour tout $k \geq 1$ et tout $\overline{x} \in P(\mathbb{R}^d)$

$$\sigma^2 = \lim_n \frac{1}{kn} E(\rho(g_{kn}\, g_{kn-1} \ldots g_1, \overline{x}) - kn\gamma)^2$$

$$= \lim_n \frac{1}{kn} E\{\sum_{j=0}^{n-1} (\rho(g_{jk+k}\, g_{jk+k-1} \ldots g_{jk+1}, g_{jk} \ldots g_1\, \overline{x}) - k\gamma)\}^2$$

En raisonnant comme dans la démonstration du lemme 7, il en résulte que

$$\sigma^2 = \frac{1}{k} \iint_{G_p \times P(\mathbb{R}^d)} h_k^2(g,\overline{x}) - (Q_{p^k}\, h_k(g,\overline{x}))^2\, p^k(dg)\, \nu(d\overline{x})$$

où $h_k(g,\overline{x}) = \sum_{j \geq 0} Q^j_{p^k}(\rho - k\gamma)(g,\overline{x})$

et également si $\sigma^2 = 0$ que pour tout g du support de p^k et $\overline{x}$ du support de ν on a

(21) $\quad h_k(g,\overline{x}) = \int h_k(g',\overline{x})\, p^k(dg')$

Exprimons h_k à l'aide de h ; pour cela remarquons tout d'abord que puisque ρ est un cocycle additif sur $G \times P(\mathbb{R}^d)$ on a

$$Q_{p^k}\, \rho(g,\overline{x}) = \sum_{i=1}^{k} (Q_p)^i\, \rho(g,\overline{x}) \qquad g \in G_p \qquad \overline{x} \in P(\mathbb{R}^d)$$

De plus pour toute fonction F telle que $F(g,\overline{x}) = f(g.\overline{x})$ on a

$$Q_{p^k}\, F(g,\overline{x}) = (Q_p)^k\, F(g,\overline{x})$$

Par conséquent pour tout $j \geq 1$ on a

$$Q^j_{p^k}\, \rho(g,\overline{x}) = (Q_p^{kj-k})[\sum_{i=1}^{k} (Q_p)^i]\, (\rho)(g,\overline{x})$$

Il en résulte que

$$h_k(g,\overline{x}) = \rho(g,\overline{x}) - k\gamma + \sum_{j \geq 1} (Q_p^{kj-1})\, [\sum_{i=1}^{k} (Q_p)^i]\, (\rho - \gamma)(g,\overline{x})$$

ce qui prouve que

$$h_k(g,\overline{x}) = h(g,\overline{x}) - (k-1)\gamma \qquad g \in G_p \qquad \overline{x} \in P(\mathbb{R}^d)$$

De l'égalité (21) on déduit alors que pour tout g du support de p^k et tout

$\bar{x}$ du support de ν on a

$$h(g,\bar{x}) = \int h(g',\bar{x})\, p^k(dg')$$

A l'aide du lemme 8, on peut alors établir la

PROPRIETE 1

Si $\sigma^2 = 0$ *il existe une constante* $c' > 0$ *telle que pour tout* $k \geq 1$ *on ait pour tout* g *dans le support de* p^k *et tout* $\bar{x}$ *dans le support de* ν

$$|\rho(g,\bar{x}) - k\gamma| \leq c'$$

Démonstration de la propriété 1

On peut écrire h sous la forme
$h(g,\bar{x}) = \rho(g,\bar{x}) + f(g\bar{x})$ où f est continue sur $P(\mathbb{R}^d)$.

Il résulte alors du lemme 8 que pour tout g dans le support de p^k $k \geq 1$ et tout $\bar{x}$ dans le support de ν on a

$$(22) \quad |\rho(g,\bar{x}) - \int \rho(g',\bar{x})\, p^k(dg')| \leq 2|f|$$

De plus on a pour $n \geq 1$

$$\frac{1}{n}\int \rho(g',\bar{x})\, p^n(dg') = \frac{1}{n} E \operatorname{Log}\|g_n g_{n-1} \cdots g_1 x\| = \frac{\partial}{\partial\lambda} E(e^{\frac{\lambda}{n}\operatorname{Log}\|g_n g_{n-1}\cdots g_1 x\|})\,/_{\lambda=0} =$$

$$= k'(0) + \frac{1}{n} N_1^{(1)}(0)\, e(\bar{x}) + \frac{1}{2i\pi}\int_{I_2} z^n R^{(1)}(z)\, e(\bar{x})\, dz$$

pour $x \in S_{d-1}$

On en déduit qu'il existe une constante K' telle que pour tout $k \geq 1$, on ait

$$(23) \quad |\int \rho(g',\bar{x})\, p^k(dg') - k\gamma| \leq K' .$$

La propriété 1 est alors une conséquence de (22) et (23).

Cette propriété permet d'établir facilement que $\sigma^2 > 0$. En effet supposons que $\sigma^2 = 0$ et soit alors g un élément commun au support de $p^{n_0+k_0}$ et au support de p^{n_0}. Pour tout $k \geq 1$ et tout $\bar{x}$ du support de ν on a alors

$$|\rho(g^k,\bar{x}) - k(n_0 + k_0)\gamma| \leq c'$$

et

$$|\rho(g^k,\bar{x}) - k\, n_0\, \gamma| \leq c'$$

d'où il résulte que pour tout $k \geq 1$

$$|k\, k_0\, \gamma| \leq 2c'$$

ce qui est impossible puisque $\gamma \neq 0$ et par conséquent $\sigma^2 > 0$.

f) Du lemme 6 et de (8) on déduit immédiatement que pour $|\lambda| < b$

$$\lim_n E(e^{i\frac{\lambda}{\sqrt{n}}(\log\|g_n g_{n-1}\cdots g_1 x\| - n\gamma)}) = e^{-\frac{\lambda^2}{2}(k''(0)-\gamma^2)} = e^{-\frac{\lambda^2}{2}\sigma^2}$$

ce qui prouve l'affirmation 3).

g) Pour prouver le 4) précisons la convergence obtenue dans f)

<u>LEMME 9</u>

Il existe un réel $c > 0$ *tel que pour* $|\lambda| < c\sqrt{n}$ $x \in S_{d-1}$ *et* $n \geq 1$ *on ait*

$$|E(e^{i\lambda Z_n(x)}) - e^{-\frac{\lambda^2}{2}}| \leq e^{-\frac{\lambda^2}{4}}\left(\frac{2|A|}{\sigma^3}\frac{|\lambda|^3}{\sqrt{n}} + c_7\frac{|\lambda|}{\sigma\sqrt{n}}\right) + c_5\frac{|\lambda|}{\sigma\sqrt{n}}\rho_1^n$$

<u>Démonstration du lemme 9</u>

D'après le lemme 6 on a pour $|\lambda| < \sigma b$

$$(24)\quad e^{-i\frac{\lambda\sqrt{n}\,\gamma}{\sigma}}\left(k\left(\frac{i\lambda}{\sigma\sqrt{n}}\right)\right)^n N_1\left(\frac{i\lambda}{\sigma\sqrt{n}}\right)e(\bar{x}) - e^{-\frac{\lambda^2}{2}} = A_n(\lambda) + B_n(\lambda)$$

où

$$A_n(\lambda) = e^{-\frac{\lambda^2}{2}}\left[e^{i\frac{\lambda^3}{\sigma^3\sqrt{n}}A + \frac{\lambda}{\sigma^3\sqrt{n}}\varepsilon_2\left(\frac{\lambda}{\sigma\sqrt{n}}\right)} - 1\right]$$

et

$$B_n(\lambda) = e^{-\frac{\lambda^2}{2}}\, e^{i\frac{\lambda^3}{\sigma^3\sqrt{n}}A + \frac{\lambda^3}{\sigma^3\sqrt{n}}\varepsilon_2\left(\frac{\lambda}{\sigma\sqrt{n}}\right)}\left(\frac{\lambda}{\sigma\sqrt{n}}\right)\Big\{i\,N_1^{(1)}e(\bar{x})$$

$$- \frac{\lambda}{2\sigma\sqrt{n}}N_1^{(2)}e(\bar{x}) - \frac{\lambda}{2\sigma\sqrt{n}}N_1^{(3)}\left(\frac{i\lambda}{\sigma\sqrt{n}}\right)e(\bar{x})\Big\}$$

Il existe un réel $c > 0$ tel que $c < \sigma b$, $\frac{2|A|c}{\sigma^3} < \frac{1}{4}$ et tel que pour $|\lambda| \leq c\sqrt{n}$ on ait

$$\left|\frac{i\lambda^3}{\sigma^3\sqrt{n}}A + \frac{\lambda^3}{\sigma^3\sqrt{n}}\varepsilon_2\left(\frac{\lambda}{\sigma\sqrt{n}}\right)\right| \leq \frac{|\lambda|}{\sqrt{n}}\,\frac{2|A|\lambda^2}{\sigma^3} \leq \frac{1}{4}\lambda^2 \qquad n \geq 1$$

et aussi

$$\sup_{\bar{x}\in P(\mathbb{R}^d)}\left|i\,N_1^{(1)}e(\bar{x}) - \frac{\lambda}{2\sigma\sqrt{n}}N_1^{(2)}e(x) - \frac{\lambda}{2\sigma\sqrt{n}}N_1^{(3)}\left(\frac{i\lambda}{\sigma\sqrt{n}}\right)e(\bar{x})\right| \leq c_7 < +\infty$$

On en déduit en utilisant l'inégalité $|e^z - 1| \leq |z|\,e^{|z|}$ que pour $|\lambda| \leq c\sqrt{n}$ $n \geq 1$ on a

$$(25)\quad |A_n(\lambda)| \leq e^{-\frac{\lambda^2}{2}}\frac{|\lambda|}{\sqrt{n}}\frac{2|A|\lambda^2}{\sigma^3}e^{\frac{|\lambda|}{\sqrt{n}}\frac{2|A|\lambda^2}{\sigma^3}} \leq \frac{|\lambda|}{\sqrt{n}}\frac{2|A|\lambda^2}{\sigma^3}e^{-\frac{\lambda^2}{4}}$$

et aussi

$$(26)\quad |B_n(\lambda)| \le c_7 \frac{|\lambda|}{\sigma\sqrt{n}} e^{-\frac{\lambda^2}{2}} e^{\frac{\lambda^2}{4}} \le c_7 \frac{\lambda}{\sigma\sqrt{n}} e^{-\frac{\lambda^2}{4}}$$

Le lemme 9 se déduit immédiatement de (24), (25), (26) et de (9)

Grâce à l'inégalité de Esseen [7] on a

$\forall T > 0 \quad \forall n \ge 1$

$$\sup_{t\in\mathbb{R}} \left|P(Z_n(x) \le t) - \frac{1}{\sqrt{2\pi}} \int_{-\infty}^{t} e^{-\frac{u^2}{2}} du\right| \le \frac{K}{T} + \frac{1}{\pi} \int_{-T}^{T} \frac{1}{|\lambda|} E\left(e^{i\lambda Z_n(x)} - e^{-\frac{\lambda^2}{2}}\right) d\lambda$$

où $K = \dfrac{24}{\pi\sqrt{2}}$

En posant $T = c\sqrt{n}$ et en tenant compte du lemme 9 il vient

$$\sup_{\substack{x\in S_{d-1}\\ t\in\mathbb{R}}} \left|P(Z_n(x) \le t) - \frac{1}{\sqrt{2\pi}} \int_{-\infty}^{t} e^{-\frac{u^2}{2}} du\right| \le \frac{K}{c\sqrt{n}} + \frac{1}{\sqrt{n}} \int_{-c\sqrt{n}}^{c\sqrt{n}} e^{-\frac{\lambda^2}{4}} (2A\lambda^2 + \frac{c_7}{\sigma}) d\lambda$$

$$(2A\lambda^2 + \frac{c_7}{\sigma})\, d\lambda + \frac{c_5}{\sqrt{n}} \frac{\rho_1^n}{\sigma} \int_{-c\sqrt{n}}^{c\sqrt{n}} d\lambda$$

ce qui compte tenu du fait que $\rho_1 < 1$ et de la convergence de l'intégrale $\int_{-\infty}^{+\infty} e^{-\frac{\lambda^2}{4}} (2A\lambda^2 + \frac{c_7}{\sigma})\, d\lambda$ établit l'assertion 4).

4-2 Précisons les résultats précédents en établissant un théorème de la limite centrale fonctionnel.

Soit $\mathcal{C}[0,1]$ l'espace des fonctions continues sur $[0,1]$ muni de la topologie de la convergence uniforme sur $[0,1]$. Pour $x \in S_{d-1}$ posons

$$S_n^x = \frac{1}{\sigma}(\log\|g_n g_{n-1}\cdots g_1 x\| - n\gamma) \qquad n \ge 1$$

et considérons la fonction aléatoire $X_n^x \in \mathcal{C}[0,1]$ $n \ge 1$ définie par

$$X_n^x(t) = \frac{1}{\sqrt{n}} S^x_{[nt]} + \frac{nt - [nt]}{\sqrt{n}} \log\left(\frac{\|g_{[nt]+1}\, g_{[nt]}\cdots g_1 x\|}{\|g_{[nt]}\, g_{[nt]-1}\cdots g_1 x\|}\right)$$

$x \in S_{d-1} \quad , \quad n \ge 1 \quad , \quad t \in [0,1]$

Notant W la mesure de Wiener sur $\mathcal{C}[0,1]$ [1] on a le

THEOREME 3

Sous les hypothèses du théorème 2, pour tout $x \in S_{d-1}$ la suite de fonctions aléatoires $(X_n^x)_{n>1}$ converge en loi vers W.

Démonstration du théorème 3

Commençons par énoncer et prouver des lemmes utiles à cette démonstration.

a) LEMME 10

Pour tout $x \in S_{d-1}$ *les distributions de dimension finie de* $(X_n^x)_{n\geq 1}$ *convergent vers celles de* W.

Démonstration du lemme 10

Considérons tout d'abord le cas d'un seul instant s. On a

$$(27)\qquad \forall n \geq 1 \quad \left|X_n^x(s) - \frac{1}{\sqrt{n}} S^x_{[ns]}\right| \leq \frac{1}{\sqrt{n}} \log\left(\frac{\|g_{[ns]+1}\, g_{[ns]} \cdots g_1 x\|}{\|g_{[ns]}\, g_{[ns]-1} \cdots g_1 x\|}\right)$$

Lorsque n tend vers l'infini le second membre de cette inégalité converge presque sûrement vers 0, de plus d'après le théorème 2 la suite $(\frac{1}{\sqrt{n}} S^x_{[ns]})_{n\geq 1}$ converge en loi vers B_s, par conséquent la suite $(X_n^x(s))_{n\geq 1}$ converge en loi vers W_s.

Considérons maintenant deux instants s et t avec s < t. Nous allons prouver que

$$(X_n^x(s),\ X_n^x(t))_{n\geq 1} \quad \text{converge en loi vers} \quad (W_s, W_t).$$

Pour celà, il suffit [1] de montrer que $(X_n^x(s),\ X_n^x(t) - X_n^x(s))_{n\geq 1}$ converge en loi vers $(W_s, W_t - W_s)$ ou encore en tenant compte de (27) appliquée aux instants t et s que

$$(28)\qquad \left(\frac{1}{\sqrt{n}} S^x_{[ns]},\ \frac{1}{\sqrt{n}} (S^x_{[nt]} - S^x_{[ns]})\right)_{n\geq 1} \quad \text{converge en loi vers } (W_s, W_t - W_s).$$

Notons $\mathcal{F}_n$ la tribu engendrée par les variables aléatoires $(g_i)_{1\leq i\leq n}$. Soient α, $\beta \in \mathbb{R}$; on a $\forall x \in S_{d-1}$, $n \geq 1$

$$P\left[\frac{S^x_{[ns]}}{\sqrt{n}} \leq \alpha;\ \frac{S^x_{[nt]} - S^x_{[ns]}}{\sqrt{n}} \leq \beta\right] = \int E\left(\frac{S^x_{[nt]} - S^x_{[ns]}}{\sqrt{n}} \leq \beta \,\middle|\, \mathcal{F}_{[ns]}\right) 1_{\left[\frac{S^x_{[ns]}}{\sqrt{n}} \leq \alpha\right]} dP$$

Comme

$$E\left(\frac{S^x_{[nt]} - S^x_{[ns]}}{\sqrt{n}} \leq \beta \,\middle|\, \mathcal{F}_{[ns]}\right) = \int 1_{\left\{\frac{1}{\sigma\sqrt{n}} \log\left(\frac{\|g\, g_{[ns]}\, g_{[ns]-1} \cdots g_1 x\|}{\|g_{[ns]} \cdots g_1 x\|}\right) - ([nt] - [ns])\gamma \leq \beta\right\}} p^{[ns]-[nt]}(dg)$$

il résulte du théorème 2, 4) que $\forall n \geq 1$, $\forall x \in S_{d-1}$

$$\left|E\left(\frac{S^x_{[nt]} - S^x_{[ns]}}{\sqrt{n}} \leq \beta \,\middle|\, \mathcal{F}_{[ns]}\right) - \frac{1}{\sqrt{2\pi}} \int_{-\infty}^{\beta \frac{\sqrt{n}}{\sqrt{[nt]-[ns]}}} e^{-\frac{u^2}{2}} du\right| \leq \frac{C}{\sqrt{[nt]-[ns]}}$$

Par conséquent on a $\forall n \geq 1$ $\forall x \in S_{d-1}$

$$\left| P\left(\frac{S^x_{[ns]}}{\sqrt{n}} \le \alpha , \frac{S^x_{[nt]} - S^x_{[nt]}}{\sqrt{n}} \le \beta\right) - P\left(\frac{S^x_{[ns]}}{\sqrt{n}} \le \alpha\right) \frac{1}{\sqrt{2\pi}} \int_{-\infty}^{\frac{\beta\sqrt{n}}{\sqrt{[nt]-[ns]}}} e^{-\frac{u^2}{2}} du \right| \le \frac{C}{\sqrt{[nt]-[ns]}}$$

d'où l'on déduit facilement que

$$\lim_{n\to+\infty} P\left(\frac{S^x_{[nx]}}{\sqrt{n}} \le \alpha ; \frac{S^x_{[nt]} - S^x_{[ns]}}{\sqrt{n}} \le \beta\right) = \frac{1}{\sqrt{2\pi s}} \int_{-\infty}^{\alpha} e^{-\frac{u^2}{2s}} du . \frac{1}{\sqrt{2\pi(t-s)}} \int_{-\infty}^{\beta} e^{-\frac{u^2}{2(t-s)}} du$$

ce qui établit (28).

Le cas de plus de deux instants se traite de manière analogue, d'où le lemme

b) LEMME 11

Il existe un réel $d_1 > 0$ *tel que* $\forall n \ge 1$, $\forall \varepsilon > 0$, $\forall x \in S_{d-1}$ *on ait*

$$P(\sup_{1\le i\le n} |S^x_i| \ge \varepsilon) \le 4P(|S^x_n| \ge \varepsilon - d_1\sqrt{n})$$

Démonstration du lemme 11

Notons $(S^x_n)^* = \sup_{1\le j\le n} |S^x_j|$

La démonstration du lemme 11 sera obtenue à l'aide des deux lemmes suivants :

α/ LEMME 12

$\forall y, a > 0$, $\forall x \in S_{d-1}$ *et* $\forall n \ge 1$ *on a*

$$P((S^x_n)^* \ge y + a) \le \frac{P(|S^x_n| \ge y)}{\inf_{\substack{z\in S_{d-1}\\ 1\le k\le n-1}} P(|S^z_k| < a)}$$

Démonstration du lemme 12

Notons

$A_1(x) = \{|S^x_1| \ge y + a\}$

$A_k(x) = \{|S^x_k| \ge y + a \; ; \; |S^x_l| < y + a \quad \text{pour} \quad 1 \le l \le k-1\}$

pour $k \ge 2$ $\quad B^n_k(x) = \{|\frac{1}{\sigma} \sum_{j=k+1}^{n} [\log (\frac{\|g_j g_{j-1} \cdots g_1 x\|}{\|g_{j-1} g_{j-2} \cdots g_1 x\|}) - \gamma]| < a\}$

pour $0 \le k \le n-1$ $\qquad C_n(x) = \{|S^x_n| \ge y\}$

On a

$$C_n(x) \supset \bigcup_{k=1}^{n} [A_n(x) \cap B^n_k(x)]$$

et par conséquent $\forall n \ge 1$

$$(29)\quad P(C_n(x)) \geq \sum_{k=1}^{n} P(A_k(x) \cap B_k^n(x))$$

Or on a $\forall n \geq 1 \quad 1 \geq k \geq n$

$$(30)\quad P(B_k^n(x) \cap A_k(x)) = \int 1_{A_k(x)} \; P(B_k^n(x)/\mathcal{F}_k)dP$$

et

$$(31)\quad \text{p.s. } P(B_k^{(n)}(x)/\mathcal{F}_k) = \int 1_{\{g \,/\, \frac{1}{\sigma}\left|\operatorname{Log}\left(\frac{\|g\, g_k\, g_{k-1}\cdots g_1 x\|}{\|g_k\, g_{k-1}\cdots g_1 x\|}\right) - (n-k)\gamma\right| < a\}} \; p^{n-k}(dg)$$

$$\geq \inf_{\substack{z\in S_{d-1}\\ 1\leq k\leq n-1}} P\{\tfrac{1}{\sigma}\left|\operatorname{Log}\|g_{n-k}\cdots g_1 z\| - (n-k)\gamma\right| < a\}$$

soit

$$(32)\quad \text{p.s.}\quad P(B_k^{(n)}(x)\,/\mathcal{F}_k) \geq \inf_{\substack{z\in S_{d-1}\\ 1\leq k\leq n-1}} P(|S_k^z| < a)$$

et par conséquent

$$(33)\quad P(C_n(x)) \geq \left[\sum_{k=1}^{n} P(A_k(x))\right] \inf_{\substack{z\in S_{d-1}\\ 1\leq k\leq n-1}} P(|S_k^z| < a)$$

c'est-à-dire

$$(34)\quad P(C_n(x)) \geq P((S_n^x)^* \geq y + a) \inf_{\substack{z\in S_{d-1}\\ 1\leq k\leq n}} P(|S_k^z| < a)$$

d'où le lemme 12.

β/ LEMME 13

Il existe une constante $d_1 > 0$ *telle que*

$$\forall n \geq 1 \quad \inf_{\substack{z\in S_{d-1}\\ 1\leq k\leq n-1}} P(|S_k^z| < d_1\sqrt{n}) \geq \frac{1}{4}$$

Démonstration du lemme 13

$\forall z\in S_{d-1}$, $\forall n\geq 0$ et $1\leq k\leq n-1$ on a d'après le théorème 2 4)

$$(35)\quad P(|S_k^z| < d_1\sqrt{n}) = P\left(\frac{|S_k^z|}{\sqrt{n}} < d_1\sqrt{\frac{n}{k}}\right) \geq P\left(\frac{|S_k^z|}{\sqrt{k}} < d_1\right) \geq \int_{-d_1}^{d_1} \frac{1}{\sqrt{2\pi}} e^{-\frac{u^2}{2}} du - \frac{2C}{\sqrt{k}}$$

Soit n_0 un entier tel que pour $k \geq n_0$ on ait $\frac{2c}{\sqrt{k}} \leq \frac{1}{4}$; on choisit alors d_1 de sorte que

$$\int_{-d_1}^{d_1} \frac{1}{\sqrt{2\pi}} e^{-\frac{u^2}{2}} du > \frac{1}{2}$$

Dans ces conditions d'après (35), on a

$\forall z\in S_{d-1}$ $\forall n\geq n_0+1$ et $n_0\leq k\leq n-1$

(36) $P(|S_k^z| < d_1\sqrt{n}) \geq \frac{1}{4}$

on peut de plus choisir d_1 suffisamment grand pour que pour $1\leq k\leq n_0$ on ait

(37) $\inf\limits_{z\in S_{d-1}} P(|S_k^z| < d_1) \geq \frac{1}{4}$

d'où le lemme 12.

Le lemme 11 est une conséquence immédiate des lemmes 12 et 13, où l'on a posé $y = \varepsilon - d_1\sqrt{n}$ et $a = d_1\sqrt{n}$

d) LEMME 14

Pour tout $x\in S_{d-1}$ *la suite des distributions des fonctions aléatoires* $(X_n^x)_{n\geq 1}$ *est équitendue.*

Démonstration du lemme 14

Pour établir ce lemme, il suffit d'après [1] de montrer que $\forall\varepsilon>0$ il existe un réel $\lambda>1$ et un entier n_0 tel que pour $n\geq n_0$ on ait :

(38) $P(\sup\limits_{1\leq i\leq n} |S_{k+i}^x - S_k^x| \geq \lambda\sqrt{n}) \leq \frac{\varepsilon}{\lambda^2}$

pour tout k.

Pour $\lambda>0$ on a d'après le lemme 11

(39) $P(\sup\limits_{1\leq i\leq n} |S_{k+i}^x - S_k^x| \geq \lambda\sqrt{n}) \leq 4P(|S_{k+n}^x - S_k^x| \geq \lambda\sqrt{n} - d_1\sqrt{n})$

d'où puisque d'après le théorème 2 4).

$$P(|S_{k+n}^x - S_k^x| \geq \lambda\sqrt{n} - d_1\sqrt{n}) \leq \frac{1}{\sqrt{2\pi}}\int_{\{u|\ |u|\geq\lambda-d_1\}} e^{-\frac{u^2}{2}}\,du + \frac{2C}{\sqrt{n}}$$

Il résulte que

(40) $P(\sup\limits_{1\leq i\leq n} |S_{k+i}^x - S_k^x| \geq \lambda\sqrt{n}) \leq 4\left(\frac{1}{\sqrt{2\pi}}\int_{\{u|\ |u|>\lambda-d_1\}} e^{-\frac{u^2}{2}}\,du + \frac{2C}{\sqrt{n}}\right)$

Par conséquent si $\lambda > 2d_1$ on a $\forall x\in S_{d-1}$, $\forall n\geq 1$ $\forall k\geq 1$

(41) $P(\sup\limits_{1\leq i\leq n} |S_{k+i}^x - S_k^x| \geq \lambda\sqrt{n}) \leq 4\left(\frac{1}{\sqrt{2\pi}}\int_{\{u|\,|u|\geq\frac{1}{2}\lambda\}} e^{-\frac{u^2}{2}}\,du + \frac{2C}{\sqrt{n}}\right)$

De (41) et de l'inégalité

$$\frac{1}{\sqrt{2\pi}}\int_{\{u|\ |u|>\frac{\lambda}{2}\}} e^{-\frac{u^2}{2}}\,du \leq \frac{8}{\lambda^3}\int_{-\infty}^{+\infty} \frac{1}{\sqrt{2\pi}}|u|^3 e^{-\frac{u^2}{2}}\,du = \frac{c_7}{\lambda^3}$$

on déduit que $\varepsilon>0$ étant donné on obtient (38) en choisissant par exemple

$$\lambda = \sup(\frac{2c_7}{\varepsilon}, 4d_1, 2) \quad \text{et} \quad n_0 = [\frac{4\lambda^4 c^2}{\varepsilon^2}] + 1$$

Le lemme 14 est ainsi prouvé.

e) Le théorème 3 est une conséquence immédiate 1 du lemme 10 et du lemme 14.

4-3 Désignons par $\mathcal{N}$ la loi normale centrée réduite sur $\mathbb{R}$; on peut alors énoncer énoncer le

THEOREME 4

Sous les hypothèses du théorème 2, pour tout $x \in \mathbb{R}^d - \{0\}$ *les variables aléatoires* $Z_n(x)$, *et* $g_n \dots g_1\bar{x}$ $n \geq 1$ *à valeurs respectivement dans* $\mathbb{R}$ *et* $P(\mathbb{R}^d)$ *sont asymptotiquement indépendantes et la suite de variables aléatoires* $(Z_n(x), g_n \dots g_1\bar{x})_{n \geq 1}$ *converge en loi vers la loi produit* $\mathcal{N} \otimes \nu$ *sur* $\mathbb{R} \times P(\mathbb{R}^d)$

Démonstration du théorème 4

$\forall x \in S_{d-1}$, $\forall \lambda \in \mathbb{R}$, $|\lambda| < a$ et $\forall f \in \mathcal{L}_{\lambda_0}$ on a facilement à l'aide de (13), (14) et du lemme 6

$$E(e^{i Z_n(x)} f(g_n g_{n-1} \dots g_1 \bar{x})) = P(\frac{i\lambda}{\sigma\sqrt{n}}) f(\bar{x}) = e^{-\frac{\lambda^2}{2} + \frac{i\lambda^3 A}{\sigma\sqrt{n}} + \frac{\lambda^3}{\sigma\sqrt{n}} \varepsilon_2(\frac{\lambda}{\sigma\sqrt{n}})}$$

$$\times [\nu(f) + \frac{i\lambda}{\sigma\sqrt{n}} N_1^{(1)} f(\bar{x}) - \frac{\lambda^2}{2\sigma^2 n} N_1^{(2)} f(\bar{x}) - \frac{\lambda}{\sigma^2 n} N_1^{(3)}(i\frac{\lambda}{\sigma\sqrt{n}}) f(\bar{x})] + e^{\frac{-i\lambda\sqrt{n}}{\sigma}} Q^n(\frac{i\lambda}{\sigma\sqrt{n}}) f(\bar{x})$$

on en déduit immédiatement que

$$\lim_n E(e^{iZ_n(x)} f(g_n g_{n-1} \dots g_1 \bar{x})) = e^{-\frac{\lambda^2}{2}} \nu(f)$$

ce qui suffit à établir le théorème 4.

§ 5 - Une loi du logarithme itéré

THEOREME 5

Sous les hypothèses du théorème 2, pour tout $x \in S_{d-1}$ *l'ensemble des points d'accumulation de la suite* $(\frac{\text{Log}\|g_n g_{n-1} \dots g_1 x\| - n\gamma}{\sigma\sqrt{2n \text{ Log Log } n}})_{n \geq 1}$ *est presque sûrement non aléatoire et égal au segment* $[-1, 1]$.

Démonstration du théorème 5

Elle se fait en deux étapes

a) LEMME 15

$\forall x \in S_{d-1}$, $\forall\varepsilon > 0$

$$P(\overline{\lim_n}\{|S_n^x| \geq (1+\varepsilon)\sqrt{2n \text{ Log Log } n}\}) = 0$$

Démonstration du lemme 15

Soit $d > 1$, posons $n_k = [d^{2k}]$ $k \geq 1$

On a

$$P(\overline{\lim_n}\{|S_n^x| \geq (1+\varepsilon)\sqrt{2n \text{ Log Log } n}\}) \leq P(\overline{\lim_k}\{(S^x_{n_k})^* \geq (1+\varepsilon)\sqrt{2n_{k-1} \text{ Log Log } n_{k-1}}\})$$

Montrons que pour un choix convenable de d

$$(43)\quad \sum_{k=1}^{\infty} P((S^x_{n_k})^* \geq (1+\varepsilon)\sqrt{2n_{k-1} \text{ Log Log } n_{k-1}}) < +\infty)$$

D'après le lemme 11, on a

$$(44)\quad P((S^x_{n_k})^* \geq (1+\varepsilon)\sqrt{2n_{k-1} \text{ Log Log } n_{k-1}}) \leq 4\, P(\frac{|S^x_{n_k}|}{\sqrt{n_k}} \geq a_k - d_1)$$

où $a_k = (1+\varepsilon)\sqrt{2\frac{n_{k-1}}{n_k} \text{ Log Log } n_{h-1}}$

d'où l'on déduit en utilisant le théorème 2 4)

$$(45)\quad P(|S^x_k|^* \geq (1+\varepsilon)\sqrt{2n_{k-1} \text{ Log Log } n_{k-1}}) \leq 8(\int_{a_k-d_1}^{+\infty} \frac{1}{\sqrt{2\pi}}\, e^{-\frac{u^2}{2}} du + \frac{C}{\sqrt{n_k}})$$

Pour tout $t > 0$ on a [15]

$$\int_t^{+\infty} e^{-\frac{u^2}{2}}\, du = \frac{1}{t}\, e^{-\frac{t^2}{2}}\,(1 - \frac{\theta}{t^2}) \qquad 0 < \theta < 1$$

Il en résulte que lorsque k tend vers $+\infty$

$$(46)\quad \frac{1}{\sqrt{2\pi}} \int_{a_k-d_1}^{+\infty} e^{-u^2/2}\, du + \frac{c}{\sqrt{n_k}} \sim \frac{d}{\sqrt{2\pi}\,(1+\varepsilon)\sqrt{2\text{Log } k}} \exp -\frac{1}{2}(a_k - d_1)^2$$

Or on a :

$$\exp -\frac{1}{2}(a_k - d_1)^2 = \exp -\frac{1}{2} d_1^2 \exp -(1+\varepsilon)^2 \frac{n_{k-1}}{n_k} \text{ Log Log } n_{k-1}(1 - \frac{2d_1}{a_k})$$

et pour k assez grand : $\frac{n_{k-1}}{n_k}(1 - \frac{2d_1}{a_k}) \geq \frac{1}{d^2} \frac{(1+\varepsilon/2)^2}{(1+\varepsilon)^2}$

On en déduit que pour k assez grand

$$(47)\quad \exp -\frac{1}{2}(a_k - d_1)^2 \leq \exp -\frac{1}{2} d_1^2 \exp -\frac{1}{d^2}(1+\frac{\varepsilon}{2})^2 \text{ Log Log } n_{k-1}$$

De plus lorsque k tend vers $+\infty$ on a :

(48) $\exp - \frac{1}{d^2}(1+\frac{\varepsilon}{2})^2 \text{ Log Log } n_{k-1} \sim (2 \text{ Log } d)^{-(\frac{1+\varepsilon/2}{d})^2} \; k^{-(\frac{1+\varepsilon/2}{d})^2}$

En choisissant $1 < d < 1 + \frac{\varepsilon}{2}$, on obtient (43) à l'aide de (45), (46), (47), (48).

Le lemme de Borel Cantelli [19] et (42) permettent alors d'obtenir le lemme 15.

b) LEMME 16

$\forall x \in S_{d-1}$, *tout réel* $\alpha \in [-1,1]$ *est presque sûrement point d'accumulation de la suite*

$$\frac{\text{Log}\|g_n g_{n-1} \cdots g_1 x\| - n\gamma}{\sigma \sqrt{2n \text{ Log Log } n}} \qquad n \geq 1$$

Démonstration du lemme 16

Soient $\alpha \in [-1,1]$, $\varepsilon > 0$, $d > 1$. Posons $n_k = [d^{2k}]$ et considérons les ensembles

$$A_k(x) = \{|\frac{S^x_{n_k}}{\sqrt{2n_k \text{ Log Log } n_k}} - \alpha| < \varepsilon\} \qquad k \geq 1$$

et

$$A'_k(x) = \{|\frac{S^x_{n_k} - S^x_{n_{k-1}}}{\sqrt{2n_k \text{ Log Log } n_k}} - \alpha| < \frac{\varepsilon}{2}\} \qquad k \geq 1$$

Commençons par prouver que $P(\overline{\lim_k} A'_k(x)) = 1$ pour un choix convenable de d ; pour cela il suffit d'après le lemme de Borel Cantelli [19] de montrer que la série $\sum_{k=1}^{\infty} P(A'_k(x) / \bigcap_{j=1}^{k-1} C A'_j(x))$ est divergente.

On a pour $k \geq 1$

$$P(A'_k(x) / \bigcap_{j=1}^{k-1} C A'_j(x)) = \int \frac{P(A'_k(x)/\mathcal{F}_{n_{k-1}})}{P(\bigcap_{j=1}^{k-1} CA'_j(x))} 1_{\{\bigcap_{j=1}^{k-1} C A'_j(x)\}} dP$$

et

$$\text{p.s. } P(A'_k(x) / \mathcal{F}_{n_{k-1}}) \geq \inf_{z \in S_{d-1}} P(|\frac{S^z_{n_k - n_{k-1}}}{\sqrt{2n_k \text{ Log Log } n_k}} - \alpha| < \frac{\varepsilon}{2})$$

En tenant compte du théorème 2 4) on en déduit alors que

$$(49) \quad P(A'_k(x) | \bigcap_{j=1}^{k-1} C A'(x)) \geq \frac{1}{\sqrt{2\pi}} \int_{(|\alpha| - \frac{\varepsilon}{2})\sqrt{\frac{2n_k \text{ Log Log } n_k}{n_k - n_{k-1}}}}^{(|\alpha| + \frac{\varepsilon}{2})\sqrt{\frac{2n_k \text{ Log Log } n_k}{n_k - n_{k-1}}}} e^{-\frac{u^2}{2}} du - \frac{2C}{\sqrt{n_k - n_{k-1}}} = f_k(\alpha, \varepsilon)$$

Si $|\alpha| \leq \frac{\varepsilon}{2}$, on déduit de (49) que

$$\lim_k P(A'_k(x) / \bigcap_{j=1}^{k-1} C A'_j(x)) \geq \frac{1}{2}$$

et donc $\sum_{k=1}^{\infty} P(A'_k(x) / \bigcap_{j=1}^{k-1} C A'_j(x)) = +\infty$

Si $|\alpha| > \frac{\varepsilon}{2}$ on voit en raisonnant comme dans la démonstration du lemme 15 que lorsque k tend vers $+\infty$ on a

$$f_k(\alpha,\varepsilon) \sim \frac{1}{\sqrt{2\pi}} \frac{1}{(|\alpha|-\frac{\varepsilon}{2})\sqrt{(1-\frac{1}{d^2})(2 \operatorname{Log} k)}} \exp -\frac{1}{2} \frac{(|\alpha|-\frac{\varepsilon}{2})^2 \, 2n_k \operatorname{Log}\operatorname{Log} n_k}{n_k - n_{k-1}}$$

Le second membre de cette équivalence est le terme général d'une série divergente dès que $\frac{(\alpha-\varepsilon/2)2}{1-\frac{1}{d}} < 1$; donc en choisissant d assez grand pour que cette condition soit réalisée, ce qui est possible car $0 < |\alpha| - \frac{\varepsilon}{2} < 1$ on a également

$$\sum_{k=1}^{\infty} P(A'_k(x) / \bigcap_{j=1}^{k-1} C A'_j(x)) = +\infty$$

Par conséquent pour tout $\alpha \in [-1,1]$ n pour tout $\varepsilon > 0$, il existe un $d_0 > 1$ tel que pour $d > d_0$ on ait

$$(50) \quad P(\overline{\lim_k}\, A'_k(x)) = 1$$

Considérons maintenant les ensembles

$$E_k(x) = \{\frac{|S^x_{n_{k-1}}|}{\sqrt{2n_k \operatorname{Log}\operatorname{Log} n_k}} < \frac{\varepsilon}{2}\} \qquad k \geq 1$$

D'après le lemme 15 il est clair que dès que $\frac{d\varepsilon}{2} > 1$ c'est-à-dire $d > \frac{2}{\varepsilon}$ on a

$P(\overline{\lim_k}\, C(E_k(x)) = 0$ soit encore (51) $P(\underline{\lim_k}\, E_k(x)) = 1$

supposant d assez grand pour que (50) et (51) soient vérifiées on en déduit que

$$(52) \quad P(\overline{\lim_k}\, (A'_k(x) \cap E_k(x)) = 1$$

d'où puisque pour tout $k \geq 1$

$$A'_k(x) \cap E_k(x) \subset A_k(x).$$

Il résulte que

$$(53) \quad P(\overline{\lim_k}\, A_k(x)) = 1$$

ce qui établit le lemme 16.

Le théorème 5 est une conséquence immédiate des deux lemmes précédents.

§ 6 - Un théorème limite local

Dans ce paragraphe, nous nous proposons d'établir un théorème limite local associé au théorème de la limite centrale prouvé au paragraphe 4.

Pour établir ce théorème, il nous faudra outre les hypothèses (P) que la condition suivante (C) soit réalisée

(C) Pour tout $\lambda \in \mathbb{R}$ $\lambda \neq 0$, l'opérateur $P(i\lambda)$ de $\mathcal{L}_{\lambda_0}$ dans $\mathcal{L}_{\lambda_0}$ est de norme spectrale strictement inférieure à 1.

6-1 Nous envisageons deux groupes d'hypothèses (P'_1) et (P'_2) sur le support de p qui assurent la validité de (C).

Avant de donner ces hypothèses précisons quelques notations. Nous noterons S_{p^k} le support de la probabilité p^k $k \geq 1$. De plus nous dirons qu'une matrice $\pi \in SL(d,\mathbb{R})$ est réalisable [17] s'il existe un entier $n \geq 1$ tel que $\pi \in (S_p)^n$ et si de plus π a une valeur propre simple $q(\pi) \in \mathbb{R}$ qui en module excède strictement toutes les autres valeurs propres de π.

Nous considérons alors les hypothèses suivantes :

Hypothèses (P'_1)

1) p satisfait aux hypothèses (P)
2) Il existe deux entiers $n \geq 1$ et $k \geq 1$ tels que
$$S_{p^{n+k}} \cap S_{p^n} \neq \emptyset$$
3) Le groupe engendré par $\Lambda_0 = \{\log |q(\pi)| \ / \ \pi \text{ "réalisable"}\}$ est dense dans $\mathbb{R}$.

Hypothèses (P'_2)

1) p admet des moments exponentiels
2) T_p contient un réseau Γ de $SL(d,\mathbb{R})$

Nous pouvons énoncer la

PROPOSITION 7

Sous l'une ou l'autre des hypothèses (P'_1) *ou* (P'_2) *la condition* (C) *est vérifiée.*

Démonstration de la proposition 7

Pour établir la proposition 7, il suffit d'après la proposition 6, puisque (P) est réalisée sous chacune des hypothèses (P'_1) ou (P'_2) de montrer que $P(i\lambda)$ $\lambda \neq 0$ $\lambda \in \mathbb{R}$ n'admet pas de valeur propre de module égal à 1.

Raisonnons par l'absurde : supposons que pour un réel $\lambda \neq 0$ $P(i\lambda)$ possède une valeur propre μ de module 1 ; soit $f \neq 0$ $f \in \mathcal{L}_{\lambda_0}$ une fonction propre associée ; on a alors : $n \geq 1$ $\bar{x} \in P(\mathbb{R}^d)$.

$$(54) \quad P^n(\lambda)\, f(\bar{x}) = \int \alpha^{i\lambda}(g,\bar{x})\, f(g\bar{x})\, p^n(dg) = \mu^n\, f(\bar{x})$$

Notant S_ν le support de ν on en déduit alors le

LEMME 17

Pour tout $\bar{x} \in S_\nu$, $|f(\bar{x})| = \sup_{y \in P(\mathbb{R}^d)} |f(\bar{y})|$

Démonstration du lemme 17

Soit $\bar{x}_0 \in P(\mathbb{R}^d)$ tel que $|f(\bar{x}_0)| = \sup_{y \in P(\mathbb{R}^d)} |f(\bar{y})|$

de (54) il résulte que

$\forall n \geq 1 \quad |f(x_0)| \leq P^n(0) \quad |f|(\bar{x}_0)$

d'où, puisque d'après le corollaire 1 $\lim P^n(0)\ |f|(x_0) = \nu|f|$

on a

$$\sup_{y \in P(\mathbb{R}^d} |f(\bar{y})| = f(\bar{x}_0) \leq \nu|f|$$

De cette inégalité et de la continuité de f on déduit alors immédiatement puisque ν est une probabilité que

$\forall \bar{x} \in S_\nu \quad f(\bar{x}) = \sup_{y \in P(\mathbb{R}^d)} |f(\bar{y})|$

Le lemme 17 et (54) permettent d'affirmer que

$\forall n \geq 1 \quad , \quad \forall g \in S_{p^n} \quad , \quad \forall \bar{x} \in S$

(55) $\quad \alpha^{i\lambda}(g,\bar{x})\ f(g\bar{x}) = \|gx\|^{i\lambda}\ f(g\bar{x}) = \mu^n\ f(\bar{x})$

où $x \in S_{d-1}$ et a pour image $\bar{x}$ dans $P(\mathbb{R}^d)$ et de plus

$\forall \bar{x} \in S_\nu \quad f(\bar{x}) \neq 0$

a) Envisageons tout d'abord le cas où les hypothèses (P'_1) sont satisfaites.

Soit π une matrice "réalisable" de $SL(d,\mathbb{R})$ et soit $v \neq 0$, un vecteur propre associé à $q(\pi)$. Comme ν ne charge aucune sous variété projective de $P(\mathbb{R}^d)$ [12], il existe un élément $\bar{x}_1$ de S_ν tel que

$$\lim_x \pi^n\ \bar{x}_1 = \bar{v}$$

ce qui, puisque $S_\nu \supset \overline{T_p \bar{x}_1}$ établit que $\bar{v}$ appartient à S_ν.

Supposant que π appartienne à $S_{p^{n_0}}$, on obtient alors en appliquant (55) pour $g = \pi$ et $\bar{x} = \bar{v}$

$$|q(\pi)|^{i\lambda}\ f(\bar{v}) = \mu^{n_0}\ f(\bar{v}) \quad \text{et} \quad f(\bar{v}) \neq 0$$

soit $\quad e^{i\lambda \log |q(\lambda)|} = \mu^{n_0}$

Ceci établit que $\forall \xi \in \Lambda_0$, $e^{i\lambda\xi} \in \{\mu^n ; n \geq 1\}$ et donc aussi puisque le groupe engendré par Λ_0 est dense dans $\mathbb{R}$ que

$\forall t \in \mathbb{R} \quad e^{i\lambda t} \in \{\mu^n ; n \geq 1\}$

ce qui est impossible puisque $\{\mu^n ; n \geq 1\}$ est fini.

La démonstration de la proposition 7 sous l'hypothèse (P'_1) est ainsi achevée.

b) Supposons désormais que les hypothèses (P'_2) sont satisfaites. Sous ces hypothèses pour tout $\bar{x} \in P(\mathbb{R}^d)$ on a :

$$\overline{T_p \bar{x}} \supset \overline{\Gamma \bar{x}} = P(\mathbb{R}^d)$$

et donc $S_\nu = P(\mathbb{R}^d)$

Pour $g \in G_p$ soit $T(i\lambda)(g)$ l'opérateur défini sur $\mathcal{L}_{\lambda_0}$ par

$$T(i\lambda)(g)\,[\phi]\,(\bar{x}) = \alpha^{i\lambda}(g,x)\ \phi(g\bar{x}) \quad f \in \mathcal{L}_{\lambda_0}\ ,\ \bar{x} \in P(\mathbb{R}^d)$$

L'application $g \to T(i\lambda)(g)$ définit une représentation continue de G_p dans $\mathcal{L}_{\lambda_0}$. De (55), et du fait que $S_\nu = P(\mathbb{R}^d)$ il résulte que la droite engendrée par f dans $\mathcal{L}_{\lambda_0}$ est invariante par cette représentation. Il existe donc un caractère χ continu de G_p dans $\mathbb{C}$ tel que pour tous $g \in G_p$, $\bar{x} \in P(\mathbb{R}^d)$ on ait

$$(56)\quad T(i\lambda)(g)\,[f]\,(\bar{x}) = \chi(g)\ f(\bar{x})$$

Distinguons maintenant deux cas

b_1) Supposons $d \geq 3$

Le groupe $\Gamma/_{[\Gamma\ \Gamma]}$ est alors fini [6] et par conséquent $\chi(\Gamma)$ est fini. Il existe alors un entier $k_0 \geq 1$ tel que tout $g \in \Gamma$ on ait $\chi^{k_0}(g) = 1$. Considérons alors la fonction F_{k_0} définie sur $\mathbb{R}^d$ par

$$F_{k_0}(x) = \|x\|^{k_0 i\lambda}\ f^{k_0}(\bar{x}) \qquad x \in \mathbb{R}^d \qquad \bar{x} \in P(\mathbb{R}^d)$$

Cette fonction est continue sur $\mathbb{R}^d$ et non constante puisque pour tout $t \in \mathbb{R}^+$ et tout $x \in \mathbb{R}^d$

$$F_{k_0}(tx) = t^{k_0 i\lambda}\ F_{k_0}(x)$$

D'autre part, F_{k_0} est invariante par Γ. Comme Γ possède une orbite dense dans $\mathbb{R}^d$ [11], il en résulte que F_{k_0} est constante sur $\mathbb{R}^d$. On aboutit ainsi à une contradiction ce qui établit la proposition 7 sous l'hypothèse (P'_2) dans le cas où $d \geq 3$.

b_2) Supposons $d = 2$. Nous envisageons trois cas :

b'_2) Il existe un élément g_0 du support de p tel que $|\text{trace } g_0| \geq 2$

Si $|\text{trace } g_0| > 2$, g_0 admet deux valeurs propres réelles $\lambda_1(g_0)$ et $\lambda_1^{-1}(g_0)$; en écrivant l'égalité (56) en les images dans $P(\mathbb{R}^d)$ des vecteurs propres correspondants il vient :

$$e^{i\lambda \operatorname{Log}|\lambda_1(g_0)|} = \chi(g_0)$$

$$e^{-i\lambda \operatorname{Log}|\lambda_1(g_0)|} = \chi(g_0)$$

d'où $\quad \chi^2\ (g_0) = 1$

Si $|\text{trace } g_0| = 2$, g_0 admet deux valeurs propres égales à +1 ou égales à -1 ; en écrivant l'égalité (56) en l'image dans $P(\mathbb{R}^d)$ d'un vecteur propre correspondant il vient :

$$\chi(g_0) = 1$$

Donc si $|\text{trace } g_0| \geq 2$ on a $\chi^2(g_0) = 1$; or d'après (55) χ est constante sur le support de p ; on a donc également pour tout g du support de p, $\chi^2(g) = 1$, d'où il résulte que :

$$\forall g \in G_p \supset \Gamma \qquad \chi^2(g) = 1$$

Le même raisonnement que celui effectué en b_1) permet de conclure

b''_2) Il existe un élément g_0 du support de p tel que $g_0^{n_0} = I$ pour un entier $n_0 \geq 1$, et tel que $|\text{trace } g_0| < 2$

On a alors $\chi^{n_0}(g_0) = 1$, d'où il résulte en tenant compte du fait que χ est constante sur le support de p que

$$\forall g \in G_p \supset \Gamma \qquad \chi^{n_0}(g) = 1$$

ce qui en raisonnant à nouveau comme dans b_1) permet de conclure.

b'''_2) Tout élément g du support de p est tel que $|\text{trace } g| < 2$ et n'a aucune puissance finie non nulle égale à l'identité.

Alors pour tout g du support de p, il existe une suite $(n_k(g))_{k\geq 1}$ telle que

$$g^{-1} = \lim_k g^{n_k(g)}$$

Il en résulte que $T_p = G_p$

Or nous pouvons énoncer lemme

LEMME 18

Si H *est un sous groupe fermé de* $SL(d,\mathbb{R})$ *contenant un réseau* Γ *, alors on a*

1- ou bien $H = SL(d,\mathbb{R})$

2- ou bien H *est un réseau de* $SL(d,\mathbb{R})$

Démonstration du lemme 18

L'algèbre de Lie $\mathcal{H}_0$ de H est invariante par la représentation adjointe de dans l'algèbre de Lie $\mathcal{G}$ de $SL(d,\mathbb{R})$. Or d'après le théorème de Borel cette représentation est irréductible et on a soit $\mathcal{H}_0 = \{0\}$ soit $\mathcal{H}_0 = \mathcal{G}$, ce qui établit le lemme 18.

D'après ce lemme on a soit $G_p = SL(d,\mathbb{R})$, soit G_p est un réseau. Le dernier cas ne peut être réalisé ici car sinon, G_p étant alors discret, tout élément g du support de p admettrait une puissance égale à l'identité ce qui est contraire à l'hypothèse. Par conséquent, on a $G_p = SL(2,\mathbb{R})$ et puisque G est semi simple

$$\forall g \in G_p \qquad \chi(g) = 1$$

ce qui comme précédemment permet de conclure.

La démonstration de la proposition 7 sous l'hypothèse (P'_2) dans le cas où $d = 2$ est ainsi achevée.

6-2 Remarque : Dans la démonstration du théorème 2, l'hypothèse "il existe deux entiers $n_0 \geq 1$ et $k_0 \geq 1$ tels que les supports de p et de $p^{n_0+k_0}$ se rencontrent sert uniquement à ssurer que $\sigma^2 > 0$. Cette propriété est également vraie dès que la condition (C) est vérifiée, ainsi que le prouve un raisonnement effectué dans [25] .

6-3 Notons $p(u) = \frac{1}{\sqrt{2\pi}} \; e^{-\frac{u^2}{2}} \quad u \in \mathbb{R}$

on a alors le

THEOREME 6

Si les hypothèses (P'_1) *ou* (P'_2) *sont satisfaites pour toute fonction* f *continue à support compact dans* $\mathbb{R} \times P(\mathbb{R}^d)$

$$\lim_{n\to+\infty} \sup_{(u,x)\in\mathbb{R}\times S_{d-1}} \Big| \sqrt{n}\,\sigma\, E[f(u+\mathrm{Log}\|g_n g_{n-1}\cdots g_1 x\| - n\gamma, g_n g_{n-1}\cdots g_1 \bar{x})] - p\left(\frac{u}{\sigma\sqrt{n}}\right) \int_{\mathbb{R}\times P(\mathbb{R}^d)} f(t,\bar{y})\, dt\, \nu(d\bar{y}) \Big| = 0 \quad (58)$$

on en déduit le

Corollaire 2 :

Sous les hypothèses du théorème 6 pour toute fonction f continue à support compact dans $\mathbb{R} \times P(\mathbb{R}^d)$ et pour tout compact C de $\mathbb{R}$ on a

$$\lim_{n\to+\infty} \sup_{(u,x)\in C\times S_{d-1}} \Big| \sqrt{2\pi n}\,\sigma\, E[f(u+\mathrm{Log}\|g_n g_{n-1}\cdots g_1 x\| - n\gamma, g_n g_{n-1}\cdots g_1 \bar{x})] - \int_{\mathbb{R}\times P(\mathbb{R}^d)} f(t,\bar{y})\, dt\, \nu(d\bar{y}) \Big| = 0 .$$

Démonstration du théorème 6

Elle se fait en plusieurs étapes :

a) Soit $\mathcal{H}$ l'espace vectoriel des fonctions boréliennes de $\mathbb{R}$ dans $\mathbb{R}$ telles que $\int_{-\infty}^{+\infty} |h(x)|\, dx < +\infty$ et telles que $h(x) = \frac{1}{2\pi} \int_{-\infty}^{+\infty} e^{-iux}\, \hat{h}(u)du$ où $\hat{h}$ est continue à support compact .

On commence par établir que (58) est vérifié pour les fonctions f de la forme $f = h \otimes \phi \quad h \in \mathcal{H}, \quad \phi \in \mathcal{L}_{\lambda_0}$

On a

$$2\pi\sqrt{n}\,\sigma\, E\{h(u+\mathrm{Log}\|g_n g_{n-1}\cdots g_1 x\| - n\gamma)\,\phi(g_n g_{n-1}\cdots g_1\bar{x})\}$$

$$= \sqrt{n}\,\sigma\int_{-\infty}^{+\infty} e^{-i\lambda u}\,\hat{h}(\lambda)\, E(e^{i\lambda\sigma Z_n(x)} f(g_n g_{n-1}\cdots g_1\bar{x}))\,d\lambda$$

et

$$E(e^{i\lambda\sigma\sqrt{n}Z_n(x)}\,\phi(g_n g_{n-1}\cdots g_1\bar{x})) = e^{-i\lambda n\gamma} P^n(i\lambda)\,\phi(\bar{x}) = e^{-i\lambda n\gamma}[k(i\lambda)]^n N_1(i\lambda)\,\phi(\bar{x})$$

$$+ e^{-i\lambda n\gamma}[Q(i\lambda)]^n\,\phi(\bar{x})$$

D'autre part, on a

$$2\pi\int_{-\infty}^{+\infty} h(t)\,dt\; p\left(\frac{u}{\sigma\sqrt{n}}\right) = \hat{h}(0)\int e^{-i\frac{\lambda u}{\sigma\sqrt{n}}}\, e^{-\frac{\lambda^2}{2}}\,d\lambda$$

Supposons que le support de $\hat{h}$ soit inclus dans $[-\alpha,\alpha]$. Soit $\varepsilon > 0$; il existe un réel $d(\varepsilon) > 0$ tel que $d(\varepsilon) < a, \frac{d(\varepsilon)}{\sigma} < a, d(\varepsilon) < \alpha$ et tel que pour $|\lambda| < d(\varepsilon)\,\sigma\sqrt{n}$, on ait

(59) $$\left|e^{-i\frac{\lambda\sqrt{n}\gamma}{\sigma}}\left(k\left(\frac{i\lambda}{\sigma\sqrt{n}}\right)\right)^n\right| \le e^{-\frac{\lambda^2}{4}}$$

et

(60) $$\left\|N_1\left(\frac{\lambda}{\sigma\sqrt{n}}\right) - \nu\right\|_{\lambda_0} < \varepsilon.$$

Or

(61) $$2\pi\sqrt{n}\,\sigma\, E[h(u+\mathrm{Log}\|g_n g_{n-1}\cdots g_1 x\| - n\gamma)\,\phi(g_n g_{n-1}\cdots g_1\bar{x})] - p\left(\frac{u}{\sqrt{n}}\right)\int_{-\infty}^{+\infty} h(t)\,dt\;\nu(\phi)$$

$$= \left[\sqrt{n}\,\sigma\int_{\{|\lambda|<d(\varepsilon)\}} e^{-i\lambda u}\,\hat{h}(\lambda)\, e^{-in\lambda\gamma}[k(i\lambda)]^n \nu(\phi)\,d\lambda - \hat{h}(0)\int_{\{|\lambda|<d(\varepsilon)\sqrt{n}\sigma\}} e^{-i\frac{\lambda u}{\sigma\sqrt{n}}}\, e^{-\frac{\lambda^2}{2}}\,d\lambda\,\nu(\phi)\right]$$

$$+ \sqrt{n}\,\sigma\int_{\{|\lambda|<d(\varepsilon)\}} e^{-i\lambda u}\,\hat{h}(\lambda)\, e^{-i\lambda n\gamma}[k(i\lambda)]^n [N_1(i\lambda)\,\phi(\bar{x}) - \nu(\phi)]\,d\lambda$$

$$+ \sqrt{n}\,\sigma\int_{\{|\lambda|<d(\varepsilon)\}} e^{-i\lambda u}\,\hat{h}(\lambda)\, e^{-i\lambda n\gamma} Q^n(i\lambda)\phi(\bar{x})\,d\lambda$$

$$+ \sqrt{n}\,\sigma \int_{\{d(\varepsilon)\leq|\lambda|<\alpha\}} e^{-i\lambda u\gamma}\hat{h}(\lambda)\, e^{-i\lambda n\gamma}\, P^n(i\lambda)\,\phi(\bar{x})\, d\lambda$$

$$- \hat{h}(0) \int_{\{|\lambda|\geq d(\varepsilon)\sqrt{n}\sigma\}} e^{-i\frac{\lambda u}{\sigma\sqrt{n}} - \frac{\lambda^2}{2}}\, d\lambda\, \nu(\phi)$$

$$= T_n^{(1)}(u) + T_n^{(2)}(u,x) + T_n^{(3)}(u,x) + T_n^{(4)}(u,x) + T_n^{(5)}(u)$$

Par changement de variable dans la première intégrale de $T_n^{(1)}(u)$, on obtient

$$T_n^{(1)}(u) = \int_{\{|t|<\sigma\sqrt{n}d(\varepsilon)\}} e^{-i\frac{\lambda u}{\sigma\sqrt{n}}} [\, \hat{h}(\tfrac{t}{\sigma\sqrt{n}})\, e^{-i\frac{t\sqrt{n}\gamma}{\sigma}}\, k^n(\tfrac{it}{\sigma\sqrt{n}}) - \hat{h}(0)\, e^{-\frac{t^2}{2}}\,]\, dt \times \nu(\phi)$$

Compte tenu de (59) on en déduit à l'aide du théorème de Lebesgue que

$$(62) \quad \lim_n \sup_{u\in\mathbb{R}} |T_n^{(1)}(u)| = 0$$

De même par changement de variable on a

$$T_n^{(2)}(u,x) = \int_{\{|t|<\sigma\sqrt{n}d(\varepsilon)\}} e^{-i\frac{ut}{\sigma\sqrt{n}}}\, \hat{h}(\tfrac{t}{\sigma\sqrt{n}})\, e^{-i\frac{t\sqrt{n}\gamma}{\sigma\sqrt{n}}}\, k^n(\tfrac{it}{\sigma\sqrt{n}})\, [\, N_1(\tfrac{it}{\sigma\sqrt{n}})\,\phi(\bar{x}) - \nu(\phi)\,]\, dt$$

Compte tenu de (59) et (60) il en résulte que

$$(63) \quad \overline{\lim_n} \sup_{(u,x)\in\mathbb{R}\times S_{d-1}} |T_n^2(u,x)| \leq \varepsilon \sup_{t\in\mathbb{R}} |\hat{h}(t)|\, \|\phi\|_{\lambda_0} \int_{-\infty}^{+\infty} e^{-\frac{u^2}{2}}\, du$$

$$= \sqrt{2\pi}\, \varepsilon \sup_{t\in\mathbb{R}} |\hat{h}(t)|\, \|\phi\|_{\lambda_0}$$

De (9) on en déduit que

$$\sup_{(u,x)\in\mathbb{R}\times S_{d-1}} |T_n^{(3)}(u,x)| \leq c_5\, d(\varepsilon)\, \rho_1^n \sqrt{n}\,\sigma\, \|\phi\|_{\lambda_0} \sup_{t\in\mathbb{R}} |\hat{h}(t)|$$

d'où

$$(64) \quad \lim_n \sup_{(u,x)\in\mathbb{R}\times S_{d-1}} |T_n^{(3)}(u,x)| = 0$$

Pour $d(\varepsilon) < |\lambda| < \alpha$, on a $\forall n \geq 1$ $\|P^n(i\lambda)\|_{\lambda_0} \leq C_8 \beta^n$ avec $0<\beta<1$ et C_8 une constante, du fait de la condition (C), et de la continuité de $\lambda \to P(i\lambda)$.
Par conséquent :

$$\sup_{(u,x)\in\mathbb{R}\times S_{d-1}} |T_n^{(4)}(u,x)| \leq \sqrt{n}\,\sigma \sup_{t\in\mathbb{R}} |\hat{h}(t)|\, C_8\, \beta^n \|\phi\|_\lambda$$

et donc

$$(65) \qquad \lim_n \sup_{(u,x)\in\mathbb{R}\times S_{d-1}} |T_n^{(4)}(u,x)| = 0$$

On a aussi

$$\sup_{u\in\mathbb{R}} |T_n^{(5)}(u)| \leq |\hat{h}(0)|\,|\nu(\phi)| \times \int_{\{|\lambda|>d(\varepsilon)\sqrt{n}\sigma\}} e^{-\frac{\lambda^2}{2}}\, d\lambda$$

et par conséquent

$$(66) \qquad \lim_n \sup_{u\in\mathbb{R}} |T_n^{(5)}(u)| = 0$$

De (61), (62), (63), (64), (65), (66) il résulte que

$$\overline{\lim_n} \sup_{(u,x)\in\mathbb{R}\times S_{d-1}} 2\pi|\sqrt{n}\,\sigma\, E(h(u+\mathrm{Log}\|g_n g_{n-1}\cdots g_1 x\| - n\gamma)\, \phi(g_n g_{n-1}\cdots g_1 \bar{x})$$
$$- p\left(\frac{u}{\sigma\sqrt{n}}\right) \int_{-\infty}^{+\infty} h(t)\,dt \times \nu(\phi)| \leq \sqrt{2\pi}\, \varepsilon \sup_t |\hat{h}(t)|\, \|\phi\|_{\lambda_0}$$

ce qui suffit à établir, ε étant quelconque, que (58) est vérifiée pour $f = h \otimes \phi$ $h \in \mathcal{H}$, $\phi \in \mathcal{L}_{\lambda_0}$

b) Montrons que (58) est également vérifiée pour $f = g \otimes \phi$, où g est une fonction de $\mathbb{R}$ dans $\mathbb{R}$ continue à support compact et $\phi \in \mathcal{L}_{\lambda_0}$. Il suffit d'ailleurs de considérer le cas où $\phi \geq 0$. Soit $\varepsilon > 0$; par un résultat d'approximation connu [2], il existe des fonctions g_ε^- et g_ε^+ de $\mathcal{H}$ telles que

$$g_\varepsilon^- \leq g \leq g_\varepsilon^+$$

et

$$\int_{-\infty}^{+\infty} g_\varepsilon^+(t) - g_\varepsilon^-(t)\, dt < \varepsilon$$

De plus, on a

$$|\sigma\sqrt{n}\, E\,[g(u+\mathrm{Log}\|g_n\cdots g_1 x\| - n\gamma)\, \phi(g_n\cdots g_1\bar{x})] - p\left(\frac{u}{\sigma\sqrt{n}}\right) \int_{-\infty}^{+\infty} g(t)\,dt\, \nu(\phi)|$$
$$\leq \sigma\sqrt{n}\, E\,[g_\varepsilon^+(u+\mathrm{Log}\|g_n\cdots g_1 x\| - n\gamma)\, \phi(g_n\cdots g_1\bar{x})] - \sigma\sqrt{n}\, E(g(u+\mathrm{Log}\|g_n\cdots g_1 x\| - n\gamma)\, \phi(g_n\cdots g_1\bar{x}))$$
$$+ |\sigma\sqrt{n}\, E\,[g^+(u+\mathrm{Log}\|g_n\cdots g_1 x\| - n\gamma)\, \phi(g_n\cdots g_1\bar{x})] - p\left(\frac{u}{\sigma\sqrt{n}}\right) \int_{-\infty}^{+\infty} g_\varepsilon^+(t)\,dt\, \nu(\phi)$$
$$+ |p\left(\frac{u}{\sigma\sqrt{n}}\right) \int_{-\infty}^{+\infty} (g(t) - g_\varepsilon^+(t))\,dt\, \nu(\phi)|$$

Comme

$$\sigma\sqrt{n}\{E[g_\varepsilon^+(u+\mathrm{Log}\|g_n g_{n-1}\cdots g_1\|-n\gamma)\ \phi(g_n g_{n-1}\cdots g_1\bar{x})] - E[g(u+\mathrm{Log}\|g_1\cdots g_1 x\|-n\gamma)\ \phi(g_n\cdots g_1\bar{x})]\}$$

$$\leq \sigma\sqrt{n}\{E[g_\varepsilon^+(u+\mathrm{Log}\|g_n g_{n-1}\cdots g_1 x\|-n\gamma)\ \phi(g_n g_{n-1}\cdots g_1\bar{x})] - E[g^-(u+\mathrm{Log}\|g_n\cdots g_1 x\|-n\gamma)\ \phi(g_n\cdots g_1\bar{x})\}]$$

$$\leq \frac{1}{\sqrt{2\pi}}\int_{-\infty}^{+\infty}[g_\varepsilon^+(t) - g_\varepsilon^-(\varepsilon)]\,dt\ \nu(\phi) + \delta_n(g_\varepsilon^+,\phi,u,x) + \delta_n(g_\varepsilon^-,\phi,u,x)$$

où l'on note

$$\delta_n(\theta,\phi,u,x) = |\sigma\sqrt{n}\ E[\theta(u+\mathrm{Log}\|g_n\cdots g_1 x\| -n\gamma)]\ \phi(g_n g_{n-1}\cdots g_1\bar{x}) - p(\frac{u}{\sigma\sqrt{n}})\int_{-\infty}^{+\infty}\theta(t)\,dt\ \nu(\phi)|.$$

et comme

$$|p(\frac{u}{\sigma\sqrt{n}})\int_{-\infty}^{+\infty}(g(t) - g_\varepsilon^+(t))\,dt\ \nu(\phi)| \leq \frac{1}{\sqrt{2\pi}}\ \varepsilon\ \nu(\phi)$$

on en déduit que

$$\overline{\lim_n}\ \sup_{(u,x)\ \mathbb{R}\times S_{d-1}} |\sigma\sqrt{n}\ E[g(u+\mathrm{Log}\|g_n\cdots g_1 x\|-n\gamma)\ \phi(g_n\cdots g_1\bar{x})] - p(\frac{u}{\sigma\sqrt{n}})\int_{-\infty}^{+\infty}g(t)\,dt\ \nu(\phi)| \leq \frac{2}{\sqrt{2\pi}}\ \varepsilon\ \nu(\phi)$$

ce qui, puisque ε est quelconque, établit le résultat cherché.

c) (58) est également vérifiée pour toute fonction f de la forme $f = g \otimes \psi$ où g est continue à support compact sur $\mathbb{R}$, et $\psi \in \mathscr{C}(P(\mathbb{R}^d))$.

Ceci est une conséquence immédiate de b) ; du fait que $\mathscr{L}_{\lambda_0}$ est dense dans $(\mathscr{C}(P(\mathbb{R}^d)),|\ \ |)$ et de l'inégalité

$$\sup_{(u,x)\in\mathbb{R}\times S_{d-1}} |\sigma\sqrt{n}\ E[g(u+\mathrm{Log}\|g_n g_{n-1}\cdots g_1 x\|-n\gamma)]\ \phi(g_n\cdots g_1\bar{x}) - p(\frac{u}{\sigma\sqrt{n}})\int_{-\infty}^{+\infty}g(t)\,dt\ \nu(\phi)| \leq |\phi|\ C(g)$$

où $C(g) = \frac{1}{\sqrt{2\pi}}\int_{-\infty}^{+\infty}|g(t)|\,dt + \sup_{\substack{(u,x)\in\mathbb{R}\times S_{d-1}\\ n\geq 1}} \sigma\sqrt{n}\ E[|g|\ (u+\mathrm{Log}\|g_n\cdots g_1 x\| -n\gamma)] < +\infty$

d'après b)

d) Soit E l'espace vectoriel engendré par les fonctions considérées au c)

Pour toute fonction continue à support compact, et tout $\varepsilon > 0$ il existe des fonctions f_ε^+ et f_ε^- de E telles que

$$f_\varepsilon^- \leq f \leq f_\varepsilon^+$$

et

$$\int_{\mathbb{R}\times P(\mathbb{R}^d)} [f_\varepsilon^+(u,b) - f_\varepsilon^-(u)t)]\ du\ \nu(dt) < \varepsilon$$

Les résultats de c) et un raisonnement analogue à celui utilisé en b) établissent alors le théorème 6.

Démonstration du corollaire 6 :

Ce corollaire résulte immédiatement du théorème 6.

§ 7) Théorème des grands écarts

7-1 Dans l'énoncé de la proposition 5, on peut supposer $0 < a$ assez petit pour que dès que $\lambda \in \mathbb{R}$, $|\lambda| < a$ on ait $k(\lambda) > 0$. Considérons alors la fonction
$\psi(\lambda) = \mathrm{Log}\, k(\lambda)\cdot\gamma\lambda \quad |\lambda| < a$

Cette fonction est analytique d'après la proposition 5 et on a

$$\psi'(0) = \frac{k'(0)}{k(0)} - \gamma = 0 \quad \text{et} \quad \psi''(0) = \frac{k''(0)\,k(0) - k'(0)^2}{k^2(0)} = \sigma^2 > 0$$

Il existe alors un intervalle $[-A,A]$ $A>0$ sur lequel ψ soit strictement convexe. Nous supposerons de plus A assez petit pour que si $|\lambda| \leq A$ $\lambda \in \mathbb{R}$ et si $e_\lambda \in \mathcal{L}_{\lambda_0}$ $e_\lambda \neq 0$ est une fonction propre vérifiant $P(\lambda)\, e_\lambda = k(\lambda)\, e_\lambda$ alors e_λ est strictement positive sur $P(\mathbb{R}^d)$.

Dans ce paragraphe nous nous proposons d'établir les deux théorèmes suivants:

THEOREME 7 :

Sous les hypothèses du théorème 2, pour tout ε *tel que* $0 < \varepsilon < \frac{\psi(A)}{A}$

$$\lim_n \{ \sup_{x\in S_{d-1}} P(\frac{\mathrm{Log}\|g_n g_{n-1}\cdots g_1 x\|}{n} - \gamma > \varepsilon \}^{1/n} = e^{-c(\varepsilon)}$$

$$0 < c(\varepsilon) = \sup_{0<t\leq A} [t\varepsilon - \psi(t)] = \lambda(\varepsilon)\,\psi'(\lambda(\varepsilon)) - \psi(\lambda(\varepsilon))$$

$\lambda(\varepsilon)$ *désignant l'unique solution de l'équation* $\psi(\lambda) = \varepsilon$.

THEOREME 8 :

Sous l'une ou l'autre des hypothèses (P'_1) *ou* (P'_2) *pour tout* ε *tel que* $0<\varepsilon<\frac{\psi(A)}{A}$, $\forall\, x \in S_{d-1}$ *lorsque* $n \to +\infty$

$$P(\frac{\mathrm{Log}\|g_n g_{n-1}\cdots g_1 x\|}{n} - \gamma>\varepsilon) \sim \frac{e^{-nc(\varepsilon)}}{\sqrt{2\pi n}\; a(\varepsilon)}\, N_1(\lambda(\varepsilon))\, e(\bar{x})$$

$$0<c(\varepsilon) = \sup_{0<t<A} [t\varepsilon - \psi(t)] = \lambda(\varepsilon)\,\psi'(\lambda(\varepsilon)) - \psi(\lambda(\varepsilon))$$

$\lambda(\varepsilon)$ *désignant l'unique solution de l'équation* $\psi'(\lambda) = \varepsilon$ *et* $a(\varepsilon) = \lambda(\varepsilon)\sqrt{\psi''(\lambda(\varepsilon))}$

7-2 Avant de démontrer ces théorèmes précisons quelques notations et énonçons plusieurs lemmes.

On appelle Q le noyau de transition défini sur $\mathbb{R} \times P(\mathbb{R}^d)$ par

$$Qf(u,\bar{x}) = \int f(u+\mathrm{Log}\,\alpha(g,\bar{x}), g\bar{x})\; p(dg) \qquad u \in \mathbb{R} \;,\; \bar{x} \in P(\mathbb{R}^d)$$

Pour $|\lambda| \leq A$ soit $h(u,\bar{x}) = e^{\lambda u} e_\lambda(\bar{x})$ $\quad u \in \mathbb{R}$, $\bar{x} \in P(\mathbb{R}^d)$

On a

$Qh_\lambda = k(\lambda)\, h_\lambda$

Ce qui permet de définir le noyau markovien relativisé λ_Q sur $\mathbb{R} \times P(\mathbb{R}^d)$ par

$$^\lambda Q(f) = \frac{Q(h_\lambda f)}{h_\lambda\, k(\lambda)}$$

On note $({}^\lambda S_n, {}^\lambda X_n)_{n\geq 0}$ la chaine de Markov à valeurs dans $\mathbb{R} \times P(\mathbb{R}^d)$ associée au noyau ${}^\lambda Q$.

On a alors le

LEMME 19 :

Pour tous $\lambda \in \mathbb{R}$, $|\lambda| \leq A$, $x \in S_{d-1}$, $\varepsilon > 0$ *on a :* (67)

$$P\left(\frac{\operatorname{Log}\|g_n g_{n-1}\cdots g_1 x\|}{n} - \gamma > \varepsilon\right) = h_\lambda(0,\bar{x})\, e^{-n[\lambda\varepsilon - \psi(\lambda)]}\, E_{(0,\bar{x})}\left(\frac{e^{-\lambda[{}^\lambda S_n - n(\gamma+\varepsilon)]}}{e_\lambda({}^\lambda X_n)}\, 1_{\{{}^\lambda S_n - n\gamma > n\varepsilon\}}\right)$$

Démonstration du lemme 19 :

On a

$$P\left(\frac{\operatorname{Log}\|g_n g_{n-1}\cdots g_1 x\|}{n} - \gamma > \varepsilon\right) = Q^n\left(1_{\{u/u>n(\gamma+\varepsilon)\}} \otimes 1_{P(\mathbb{R}^d)}\right)(0,\bar{x})$$

$$= h_\lambda(0,\bar{x})\, [k(\lambda)]^n\, ({}^\lambda Q)^n \left(\frac{1}{h_\lambda}\, 1_{\{u/u>n(\gamma+\varepsilon)\}} \otimes 1_{P(\mathbb{R}^d)}\right)(0,\bar{x})$$

$$= h_\lambda(0,\bar{x})\, e^{n\,\psi(\lambda)+\gamma\lambda}\, E_{(0,\bar{x})}\left(\frac{e^{-\lambda\, {}^\lambda S_n}}{e_\lambda({}^\lambda X_n)}\, 1_{\{{}^\lambda S_n > n(\gamma+\varepsilon)\}}\right)(0,\bar{x})$$

L'égalité (67) s'en déduit immédiatement.

Soit $0<\varepsilon<\frac{\psi(A)}{A}$; notons $\lambda(\varepsilon)$ l'unique valeur de λ pour laquelle $0<\lambda(\varepsilon)<A$ et $\varepsilon\,\lambda(\varepsilon) - \psi(\lambda(\varepsilon)) = \sup_{0<t\leq A} (t\varepsilon - \psi(t))$.

Remarquons que (68) $\psi'(\lambda(\varepsilon)) = \dfrac{k'(\lambda(\varepsilon))}{k(\lambda(\varepsilon))} - \gamma = \varepsilon$

Les lemmes qui suivent sont consacrés à l'étude des propriétés asymptotiques de la chaine de Markov $({}^{\lambda(\varepsilon)}S_n, {}^{\lambda(\varepsilon)}X_n)_{n\geq 0}$.

Pour $\lambda \in \mathbb{R}$, $|\lambda| < a$ il est clair que l'opérateur $N_1(\lambda)$ apparaissant dans la proposition 5 peut s'écrire sous la forme

$$N_1(\lambda)\, f = \nu_\lambda\left[\frac{f}{e_\lambda}\right] e_\lambda \qquad f \in \mathcal{L}_{\lambda_0}$$

où ν_λ est une probabilité sur $P(\mathbb{R}^d)$.

On a alors le

LEMME 20 :

Sous les hypothèses du théorème 2, pour tout $\lambda \in \mathbb{R}$ $|\lambda| < a$, *il existe un réel* $r > 0$ *tel que pour* $\alpha \in \mathbb{R}$ $|\alpha| < r$, $f \in \mathcal{L}_{\lambda_0}$, $\bar{x} \in P(\mathbb{R}^d)$ $n \geq 1$

$$E_{(0,\bar{x})}[e^{i\alpha\,{}^{\lambda}S_n} f({}^{\lambda}X_n)] = \frac{P(\lambda+i\alpha)(e_\lambda f)(\bar{x})}{[k(\lambda)]^n \; e_\lambda(\bar{x})} = \{1 + i\alpha\frac{k'(\lambda)}{k(\lambda)}$$

$$- \frac{\alpha^2}{2}\frac{k''(\lambda)}{k(\lambda)} + \alpha^2\,\varepsilon(\alpha)\}^n \{\nu_\lambda(f) + \alpha\frac{N_1^{(1)}(\lambda)[fe_\lambda](\bar{x})}{e_\lambda(\bar{x})}$$

$$- \frac{\alpha^2}{2}\frac{N_1^{(2)}(\lambda)[fe_\lambda]\;(\bar{x})}{e_\lambda(\bar{x})} + \alpha^2\frac{N_1^{(3)}(\lambda+i\alpha)[fe_\lambda]\;(\bar{x})}{e_\lambda(\bar{x})}\}$$

$$+ \frac{Q^n(\lambda+i\alpha)[e_\lambda f](\bar{x})}{k^n(\lambda)\; e_\lambda(\bar{x})}$$

où $\lim_{\alpha\to 0}\varepsilon(\alpha) = 0$, $N_1^{(1)}(\lambda)$, $N_1^{(2)}(\lambda)$, $N_1^{(3)}(\lambda+i\alpha)$ $\mathcal{L}(\mathcal{L}_{\lambda_0},\mathcal{L}_{\lambda_0})$.

Démonstration du lemme 20 :

Ce lemme est une conséquence immédiate de la proposition 5.

On en déduit le

LEMME 21 :

Sous les hypothèses du théorème 2, pour tout réel α, *toute* $f \in \mathcal{L}_{\lambda_0}$ *et tout* $\bar{x} \in P(\mathbb{R}^d)$

1) $$\lim_n E_{(0,\bar{x})}(e^{i\alpha\frac{\lambda(\varepsilon)S_n}{n}}) = e^{i(\gamma+\varepsilon)\alpha}$$

2) $$\lim_n E_{(0,\bar{x})}(e^{i\alpha\frac{\lambda(\varepsilon)S_n-(\gamma+\varepsilon)\alpha}{\sqrt{n}}} f({}^{\lambda(\varepsilon)}X_n)) = e^{-\frac{1}{2}\psi''(\lambda(\varepsilon))\alpha^2}\,\nu_{\lambda(\varepsilon)}(f)$$

Démonstration du lemme 21 :

Ce lemme est une conséquence facile du lemme 20, de la proposition 5 et de l'égalité (68).

On peut préciser les résultats du 2) du lemme précédent en établissant un théorème limite local dans le cas où l'une ou l'autre des hypothèses (P'_1) ou (P'_2) est satisfaite. Pour celà, nous avons tout d'abord besoin du

LEMME 22 :

Sous l'une ou l'autre des hypothèses (P'_1) *ou* (P'_2) *pour tout* $\lambda \in \mathbb{R}, |\lambda| < a$ *et* $\alpha \neq 0$, $\alpha \in \mathbb{R}$ *l'opérateur défini par*

$$f \to \frac{P(\lambda+i\alpha)(fe_\lambda)}{k(\lambda)\; e_\lambda} \qquad f \in \mathcal{L}_{\lambda_0}$$

est de norme spectrale strictement inférieure à 1.

Démonstration du lemme 22 :

Elle se fait de façon analogue à celle de la proposition 7.

Tout d'abord si $f \in \mathcal{L}_{\lambda_0}$ $f \neq 0$ et $\mu \in \mathbb{C}$ sont tels que

$|\mu| = 1$

et

$P(\lambda+i\alpha)(fe_\lambda) = \mu k(\lambda)\, e_\lambda f$

On en déduit que

$$\forall \bar{x} \in P(\mathbb{R}^d) \quad \forall n \geq 1 \quad \frac{P^n(\lambda)(|f|e_\lambda)(\bar{x})}{k^n(\lambda)\, e_\lambda(\bar{x})} \geq |f|(\bar{x})$$

d'où puisque $\lim_n \dfrac{P^n(\lambda)(|f|e_\lambda)(\bar{x})}{(k(\lambda))^n\, e_\lambda(\bar{x})} = \nu_\lambda |f|$

on a

$$\forall \bar{x} \in P(\mathbb{R}^d) \qquad \nu_\lambda |f| \geq |f|(\bar{x})$$

$$(69) \quad \|g\bar{x}\|^{\alpha+i\lambda}\, f(g\bar{x})\, e_\lambda(g\bar{x}) = k(\lambda)\, \mu^n\, f(\bar{x})\, e_\lambda(\bar{x})$$

Remarquons que S_{ν_λ} est stable par T_p et donc aussi qu'en raison de l'irréductibilité de l'action de G_p sur $P(\mathbb{R})^d$, ν_λ n'est portée par aucune sous variété projective de $P(\mathbb{R}^d)$. Ceci permet de conclure par des considérations analogues à celles faites dans la démonstration de la proposition 7 que (69) est impossible sous l'une ou l'autre des hypothèses (P'_1) ou (P'_2), ce qui établit le lemme 21.

En utilisant le lemme 21, et le lemme 22 et en raisonnant comme dans le paragraphe 6.2 nous pouvons conclure que

LEMME 23 :

Sous les hypothèses (P'_1) ou (P'_2) pour toute fonction continue à support compact dans $\mathbb{R}\times P(\mathbb{R}^d)$, pour tout

$$0<\varepsilon<\frac{\psi(A)}{A}$$

$$\lim_{n\to+\infty} \sup_{(u,\bar{x})\in\mathbb{R}\times P(\mathbb{R}^d)} \Big| \sqrt{2\pi n}\, \sqrt{\psi''(\lambda(\varepsilon))}\; E^{\lambda(\varepsilon)}_{(0,\bar{x})}\, f(u+S_n-n(\gamma+\varepsilon), X_n)^{\lambda(\varepsilon)}$$
$$- p\Big(\frac{u}{\sqrt{n}\,\sqrt{\psi''(\lambda(\varepsilon))}}\Big) \int_{-\infty}^{+\infty} f(t,\bar{x})\, dt\; \nu_{\lambda(\varepsilon)}(d\bar{x}) \Big| = 0$$

Démonstration du théorème 7 :

a) Du lemme 23, on déduit que pour $0<\varepsilon<\frac{\psi(A)}{A}$ et $n \geq 1$ on a

$$(70) \Big[\sup_{x\in S_{d-1}} P\Big(\frac{\mathrm{Log}\|g_n g_{n-1}\cdots g_1 x\|}{n} - \gamma>\varepsilon\Big)\Big]^{1/n} \leq |e_{\lambda(\varepsilon)}|^{1/n} \Big|\frac{1}{e_{\lambda(\varepsilon)}}\Big|^{1/n} e^{-[\varepsilon\lambda(\varepsilon)-\psi(\lambda(\varepsilon))]}$$

et par conséquent on a :

$$(71)\quad \overline{\lim_n}\ [\sup_{x\in S_{d-1}} P(\frac{\mathrm{Log}\|g_n g_{n-1}\cdots g_1 x\|}{n} - \gamma > \varepsilon)]^{1/n} \le \exp\{[\sup_{0<t\le A} t\varepsilon - \psi(t)]\}$$

b) Pour $\bar{x} \in P(\mathbb{R}^d)$ et $n \ge 1$ on a

$$[E_{(0,\bar{x})}(\frac{e^{-\lambda(\varepsilon)\,[{}^{\lambda(\varepsilon)}S_n - n(\gamma+\varepsilon)]}}{e_{\lambda(\varepsilon)}({}^{\lambda(\varepsilon)}X_n)}\ 1_{\{{}^{\lambda(\varepsilon)}S_n - n(\gamma+\varepsilon)>0\}})]^{1/n}$$

$$\ge E_{(0,\bar{x})}(\frac{e^{-\lambda(\varepsilon)(\frac{{}^{\lambda(\varepsilon)}S_n}{n} - (\gamma+\varepsilon))}}{[e_{\lambda(\varepsilon)}({}^{\lambda(\varepsilon)}X_n)]^{1/n}}\ 1_{\{{}^{\lambda(\varepsilon)}S_n - n(\gamma+\varepsilon)\}})$$

D'après le lemme 21 la suite ${}^{\lambda(\varepsilon)}S_n/n - (\gamma+\varepsilon)$ $n\ge 1$ converge vers zéro en probabilité $P_{(0,\bar{x})}$. Le théorème de Lebesgue et l'inégalité précédente permettent de conclure que

$$(72)\quad \underline{\lim_n}[E_{(0,\bar{x})}\ (\frac{e^{-\lambda(\varepsilon)({}^{\lambda(\varepsilon)}S_n - n(\gamma+\varepsilon))}}{e_{\lambda(\varepsilon)}({}^{\lambda(\varepsilon)}X_n)}\ 1_{\{{}^{\lambda(\varepsilon)}S_n - n(\gamma+\varepsilon)>0\}})]^{1/n} \ge 1$$

ce qui d'après le lemme 19 établit que

$$(73)\quad \forall x\in S_{d-1}\ \underline{\lim_n}[P(\frac{\mathrm{Log}\|g_n g_{n-1}\cdots g_1 x\|}{n} - \gamma > \varepsilon)]^{1/n} \ge e^{-(\varepsilon\lambda(\varepsilon) - \psi(\lambda(\varepsilon)))},$$

(71) et (73) prouvent le théorème 7.

7-4 Démonstration du théorème 8

La fonction $x \to e^{-\lambda(\varepsilon)x}\ 1_{]0+\infty[}(x)$ est directement Riemann intégrable sur $\mathbb{R}$. Du lemme 23 on déduit alors facilement que

$$\lim_n \sup_{\bar{x}\in P(\mathbb{R}^d)} [\sqrt{2\pi n}\sqrt{\psi''(\lambda(\varepsilon))}\ E_{(0,\bar{x})}(\frac{e^{-\lambda(\varepsilon)[{}^{\lambda(\varepsilon)}S_n - n(\gamma+\varepsilon)]}}{e_{\lambda(\varepsilon)}({}^{\lambda(\varepsilon)}X_n)}\ 1_{\{{}^{\lambda(\varepsilon)}S_n - n\gamma > n\varepsilon\}})]$$

$$= \int_0^{+\infty} e^{-\lambda(\varepsilon)x}\, dx \times \int_{P(\mathbb{R}^d)} \frac{\nu_{\lambda(\varepsilon)}(d\bar{x})}{e_{\lambda(\varepsilon)}(\bar{x})}$$

$$= \frac{1}{\lambda(\varepsilon)}\ \frac{N_1(\lambda(\varepsilon))\ e(\bar{x})}{e_{\lambda(\varepsilon)}(\bar{x})}$$

le théorème 8 résulte alors de l'application du lemme 19 pour $\lambda = \lambda(\varepsilon)$.

REFERENCES

[1] BILLINGSLEY : Convergence of probability measures - John Wiley and Sons, New York, (1968).

[2] BRETAGNOLLE et DACUNHA-CASTELLE : Théorèmes limites à distance finie pour les marches aléatoires - Annales IHP, volume IV n°1 (1968), p. 25-73.

[3] BREIMAN : Probability Addison Wesley (1968).

[4] CREPEL : Loi des grands écarts pour les marches aléatoires sur $\mathbb{R}$ - Séminaire de probabilités de Rennes (1978).

[5] DELAROCHE ET KIRILLOV : Sur les relations entre l'espace dual d'un groupe et la structure de ses sous-groupes fermés - Séminaire Bourbaki (1967-1968) n° 343.

[6] DUNFORD et SCHWARTZ : Linear operators part I - Interscience, New-York (1953).

[7] ESSEEN : Fourier analysis of distribution functions ; a mathematical study of the Laplace Gaussian law. Acta Math 77, p. 1-25.

[8] FURSTENBERG : Non commuting random products - TAMS vol. 108 (1963), p. 377-428.

[9] FURSTENBERG et KESTEN : Products of random matrices - Ann Math Stat vol.31 (1960).

[10] GREENBERG : Discrete groups with dense orbits, flows on homogeneous spaces - Annals of mathematical studies, number 53 Princeton Univ. press (1963).

[11] GUIVARC'H : Etude des produits de matrices aléatoires - Lecture Notes in math 774 (1980), p. 176-250.

[12] GUIVARC'H : Sur les exposants de Lyapounoff des marches aléatoires à pas markovien - C.R.A.S. t. 292, série I, n° 5.

[13] GUIVARC'H et RAUGI : Frontière de Furstenberg propriétés de contraction et théorèmes de convergence (à paraître).

[14] IOSIFESCU et THEODORESCU : Random process and learning - Springer Verlag Band 150, Berlin (1969).

[15] KAIJSER : Some limit theorems for Markov chains with applications to learning models and products of random matrices - Report Institut Mittag Leffler (1972).

[16] KESTEN : Random difference equations and renewal theory for products of random matrices - Acta Math 131, p. 207-248.

[17] NAGAEV : Some limit theorems for stationary Markov chains - Theory of Proba and its applications 2 (1957), p. 378-406.

[18] NEVEU : Bases mathématiques du calcul des probabilités - Masson et Cie (1964).

[19] NORMAN : Markov process and learning models - Academic Press, vol. 84 (1972).

[20] O'CONNOR : A central limit theorem for the disordered harmonic Communications Math Physic 45 (1975), p. 63-77.

[21] RAUGI : Fonctions harmoniques et théorèmes limites pour les marches aléatoires sur les groupes - Bulletin SMF, mémoire 54, (1977).

[22] TUTUBALIN : On limit theorems for the products of random matrices - Theory of Proba.and its applications, vol 10, n°1, (1965) p. 15-27.

[23] TUTUBALIN A variant on the local limit theorem for products of random matrices - Theory of Proba. and its applications vol. XXII (1977), n° 2, p. 203-214.

Pendant la préparation de ce travail, j'ai bénéficié des commentaires et suggestions de Y. GUIVARC'H. Je l'en remercie bien vivement.

Local tightness of convolution semigroups over locally compact groups

M. McCRUDDEN
Department of Mathematics
The University
Manchester M13 9PL
ENGLAND

Let G be a locally compact group and let $M(G)$ denote the topological semigroup of all probability measures on G, where $M(G)$ is given the weak topology and the multiplication is convolution of measures. By a real directed semigroup we shall mean a subsemigroup S of $\mathbb{R}^+$ (the strictly positive reals) under addition, which satisfies the condition that given s, t in S, there exists u in S and positive integers m and n such that $s = nu$, $t = mu$.

Given a homomorphism $\phi : S \to M(G)$, we would like to find conditions on ϕ and G which will ensure that ϕ is locally tight i.e. for each $r \in S$, the set $\{\phi(s) : s \leq r,\ s \in S\}$ is relatively compact in $M(G)$. Our motivation for studying this question is that its solution is likely to be of help in solving the so called embedding problem for infinitely divisible distributions on locally compact groups (see Chapter 3 of the book of Heyer (1) for a comprehensive background account of this problem). If the group G is strongly root compact (see 3.1.10 of (1) for the definition) then any such homomorphism ϕ is indeed locally tight, and this includes for example the case when G is a compactly-generated nilpotent group. We have recently shown (see (5)) that the same result is true when G is a simply-connected solvable Lie group though such a group is certainly not necessarily root compact.

Our objective in this paper is to give a sufficient condition for the local tightness of such a homomorphism $\phi : S \to M(G)$ in terms of the supporting subgroup of ϕ in G, which we define to be the smallest closed subgroup of G which contains $S(\phi(r))$, the support of the measure $\phi(r)$, for every $r \in S$. Denote this subgroup by $G(\phi)$.

Our main result is now expressible as follows.

Theorem 1. Let S be a real directed semigroup and let G be a locally compact group. Suppose $\phi : S \to M(G)$ is a homomorphism such that $G(\phi)$ is a Lie group whose group of components is both finitely generated and has finitely generated centre. Then ϕ is locally tight.

This result has as a consequence the following result, which generalises the main theorem of (5) referred to above.

Theorem 2. Let G be a Lie group whose group of components has all subgroups finitely generated, and let $\phi : S \to M(G)$ be a homomorphism, where S is a real directed semigroup. If for some $r \in S$ the closed subgroup of G generated by $S(\phi(r))$ is solvable, then ϕ is locally tight. In particular if G itself is solvable, then every homomorphism of S into $M(G)$ is locally tight.

§1. The classes $\mathcal{C}$ and $\mathcal{C}^*$

Let A be a subset of a locally compact group G, we write $\mathcal{L}(A, G)$ to denote the set of elements y in G such that, for each neighbourhood U of y there exists a compact set K (depending on U) such that $U \cap aKa^{-1} \neq \emptyset$, for all $a \in A$.

Proposition 3. (i) $\mathcal{L}(A, G)$ is a closed subgroup of G and contains $Z(G)$, the centre of G.

(ii) If $A, B \subseteq G$ with B compact and $A \subseteq BZ(G)$, then $\mathcal{L}(A, G) = G$.

(iii) If $A \subseteq B \subseteq G$ then $\mathcal{L}(B, G) \subseteq \mathcal{L}(A, G)$.

(iv) If $A, B, C \subseteq G$ such that $B \subseteq AC$ and C is compact, then $\mathcal{L}(A, G) \subseteq \mathcal{L}(B, G)$.

(v) If H is a subgroup of G and $A \subseteq H$, then $\mathcal{L}(A, H) \subseteq H \cap \mathcal{L}(A, G)$. If further H is open in G then $\mathcal{L}(A, H) = \mathcal{L}(A, G) \cap H$.

(vi) Let $p : G \to K$ be a continuous homormorphism of locally compact groups. If $A \subseteq G$ then $p(\mathcal{L}(A, G)) \subseteq \mathcal{L}(p(A), K)$.

Proof. Elementary.

We denote by $\mathcal{C}$ the class of all locally compact groups G which have the property that if $A \subseteq G$ satisfies $\mathcal{L}(A, G) = G$, then there is a compact set $B \subseteq G$ such that $A \subseteq BZ(G)$. We denote by $\mathcal{C}^*$ the class of all locally compact groups G which have the property that if $A \subseteq G$ satisfies $\mathcal{L}(A, G) = G$, then A is relatively compact. It is clear that G belongs to $\mathcal{C}^*$ if and only if G belongs to $\mathcal{C}$ and $Z(G)$ is compact.

Our next observation is essentially a rewrite in our notation of a result of Parthasarathy (8).

Proposition 4. Let V be a finite-dimensional real or complex vector space and let $G = SL(V)$. For each unbounded set $A \subseteq G$ there is a proper linear subspace W of V (depending on A) such that for all $x \in \mathcal{L}(A, G)$, $x(W) \subseteq W$.

Proof. Let $\{x_n : n \geq 1\}$ be a sequence in A which has no convergent subsequence. Passing to a subsequence if necessary, we may assume, in view of Proposition 2 of (4), that V has a proper linear subspace W such that every limit of every convergent subsequence of every sequence of the form $\{x_n z_n x_n^{-1} : n \geq 1\}$, where $\{z_n : n \geq 1\}$ is any bounded sequence in G, leaves W invariant.

Now suppose $a \in \mathcal{L}(A, G)$, then for any neighbourhood U of a there is a compact set K such that for all $n \geq 1$, $U \cap x_n K x_n^{-1} \neq \emptyset$. Hence there is a convergent sequence $\{k_r : r \geq 1\}$ in K, and a subsequence $\{x_{n_r} : r \geq 1\}$ such that $\{x_{n_r} k_r x_{n_r}^{-1} : r \geq 1\}$ converges to some $v \in U$. From the paragraph above it follows that $v \in H_W = \{x \in G : x(W) \subseteq W\}$, which means that every neighbourhood of a meets H_W. As H_W is closed, the result follows.

Corollary. Let H be a closed subgroup of $SL(V)$ which acts irreducibly on V. Then $H \in \mathcal{C}^*$.

Proof. If $A \subseteq H$ is not relatively compact in H, it is not relatively compact in $G = SL(V)$, so by Proposition 4, there is a proper linear subspace W of V such that $\mathcal{L}(A, G) \subseteq H_W = \{x \in SL(V) : x(W) \subseteq W\}$. Then by Proposition 3(v), $\mathcal{L}(A, H) \subseteq H \cap \mathcal{L}(A, G) \subseteq H \cap H_W \neq H$, as H acts irreducibly.

Proposition 5. Let $G = \prod_{i=1}^{k} G_i$, where G_i is a locally compact group for all $1 \leq i \leq k$, and let $p_i : G \to G_i$ be the natural projections. Let H be a closed subgroup of G such that $p_i(H) = G_i$ for all $1 \leq i \leq k$, and suppose G_i belongs to $\mathfrak{C}^*$, for all $1 \leq i \leq k$. Then H belongs to $\mathfrak{C}^*$.

Proof. If $A \subseteq H$ satisfies $\mathcal{L}(A, H) = H$, then by Proposition 3(vi), we have $\mathcal{L}(p_i(A), G_i) = G_i$, for all $1 \leq i \leq k$, so since $G_i \in \mathfrak{C}^*$, we see that $p_i(A)$ is relatively compact for all $1 \leq i \leq k$, hence A is relatively compact. We conclude that $H \in \mathfrak{C}^*$.

Corollary. Let $\mathfrak{g}$ be a semisimple Lie algebra, let $G = \mathrm{Int}(\mathfrak{g})$. Then G belongs to $\mathfrak{C}^*$.

Proof. The argument of Proposition 6 of (4) shows that G can be thought of as a closed subgroup of a group of the form $\prod_{i=1}^{k} G_i$, where each G_i is a closed irreducible subgroup of some $SL(V_i)$, for some finite-dimensional vector space V_i. The result now follows from Proposition 5 and Proposition 4, Corollorary.

Proposition 6. Any finitely generated (discrete) group belongs to $\mathfrak{C}$.

Proof. Let G be a finitely-generated discrete group and suppose that $A \subseteq G$ is infinite and that $\mathcal{L}(A, G) = G$. If A is not finite modulo $Z(G)$, there is a sequence $\{x_n : n \geq 1\}$ in A such that for all $n \neq m$, $x_n x_m^{-1} \notin Z(G)$. Since $\mathcal{L}(A, G) = G$ and every point of G is a neighbourhood of itself, we see that for each $a \in G$, $\{x_n^{-1} a x_n : n \geq 1\}$ is finite. If $\{a_1, \ldots, a_k\}$ generates G we can, by passing repeatedly to a subsequence, find a subsequence $\{x_{n_r} : r \geq 1\}$ such that for all $r, s \geq 1$, and all $1 \leq i \leq k$, $x_{n_r}^{-1} a_i x_{n_r} = x_{n_s}^{-1} a_i x_{n_s}$. We conclude that $x_{n_s} x_{n_r}^{-1} \in Z(G)$, which contradicts our above assumption.

Proposition 7. Suppose H is an (open) subgroup of finite index in G. If H belongs to $\mathfrak{C}^*$, so does G.

Proof. Given $A \subseteq G$, let C be a set formed by picking exactly one element from each H-coset that meets A, then C is finite. For each $a \in A$ let c_a be the element of C lying in the same H-coset as a, and let $b_a = a c_a^{-1}$, so that $B = \{b_a : a \in A\}$ is a subset of H. If $\mathcal{L}(A, G) = G$, then by Proposition 3(iv), (v), we have $\mathcal{L}(B, G) = G$ and $\mathcal{L}(B, H) = H \cap \mathcal{L}(B, G)$, so that $\mathcal{L}(B, H) = H$. As H belongs to $\mathfrak{C}^*$ we conclude that B is relatively compact and hence A is relatively compact. This shows that G belongs to $\mathfrak{C}^*$.

Corollary. Let $\mathfrak{g}$ be a semisimple Lie algebra, let G be a subgroup of $\mathrm{Aut}(\mathfrak{g})$ which contains $\mathrm{Int}(\mathfrak{g})$. Then $G \in \mathfrak{C}^*$.

Proof. The index of $\mathrm{Int}(\mathfrak{g})$ in $\mathrm{Aut}(\mathfrak{g})$ is finite, because $\mathrm{Aut}(\mathfrak{g})$ is algebraic and $\mathrm{Int}(\mathfrak{g})$ is its identity component. Hence $\mathrm{Int}(\mathfrak{g})$ has finite index in G, so result follows from Proposition 5, Corollary and Proposition 7.

This brings us to the main result of this section.

Proposition 8. Let G be a compactly-generated semisimple Lie group. Then G belongs to $\mathfrak{C}$.

Proof. Let $\mathfrak{g}$ be the Lie algebra of G, write $p : G \to \mathrm{Aut}(\mathfrak{g})$ for the map Ad_G. Given $A \subseteq G$ such that $\mathcal{L}(A, G) = G$, we see from Proposition 3(vi) that $p(G) = \mathcal{L}(p(A), p(G))$, and since $p(G)$ contains $\mathrm{Int}(\mathfrak{g})$, we conclude by Proposition 7 Corollary that $p(G)$ belongs to $\mathfrak{C}^*$ and hence $p(A)$ is relatively compact. Hence there is a compact set $C \subseteq G$ and a set $B \subseteq \mathrm{Ker}\, p$ such that $A \subseteq BC$ and $B \subseteq AC^{-1}$. Write H for $\mathrm{Ker}\, p$ and note that for all $a \in G^o$ (the connected component of the identity in G) and all $h \in H$ we have $hah^{-1} = a$, which implies that $H \cap G^o = Z(G^o)$. Since $Z(G^o)$ is discrete it follows that H is discrete, and furthermore by Proposition 3(iv), we have $\mathcal{L}(B, G) = \mathcal{L}(A, G) = G$.

Now let U be a compact neighbourhood of 1 in G^o, then for each $y \in G$, yU is a neighbourhood of y and so for all $y \in B$, $yU \cap bKb^{-1} \neq \emptyset$, where K is some compact set in G. But $B \subseteq H$, hence for each $b \in B$, $b^{-1}Ub = U$, whence we obtain $b^{-1}yb \in KU^{-1}$ i.e. the set $\{b^{-1}yb : b \in B\}$ is relatively compact for every $y \in G$. In particular

$$H = \mathcal{L}(B, H) \tag{1}$$

Next observe that $H/_{Z(G^o)} \cong HG^o/_{G^o}$ and the latter group has finite index in $G/_{G^o}$, so since $G/_{G^o}$ is finitely generated it follows that $H/_{Z(G^o)}$ is finitely-generated. But $Z(G^o)$ is finitely-generated, whence H is finitely-generated. It then follows from (1) and Proposition 6 that there is a finite set $D \subseteq H$ and a set $B_1 \subseteq Z(H)$ such that $B \subseteq DB_1$ and $B_1 \subseteq D^{-1}B$. If we write $G_1 = G^oH$, then $B_1 \subseteq Z(G_1)$ and G_1 has finite index in G. Note also that for each $y \in G$, the set $\{b^{-1}yb : b \in B_1\}$ is relatively compact.

We now claim that B_1 must be finite modulo $Z(G)$. For if not then B_1 contains a sequence $\{x_n : n \geq 1\}$ such that for all $m \neq n$ we have $x_n x_m^{-1} \notin Z(G)$. Let $a_1, \ldots, a_k$ be a complete set of coset representatives of G_1 in G and note that for $1 \leq i \leq k$, the set $\{x_n^{-1} a_i x_n a_i^{-1} : n \geq 1\}$ lies in the discrete subgroup H, and is also relatively compact, hence is finite. We may now argue as in the proof of Proposition 6 to produce a subsequence $\{x_{n_r} : r \geq 1\}$ such that for $1 \leq i \leq k$ and for all $r, s \geq 1$, we have

$$x_{n_s} x_{n_r}^{-1} a_i x_{n_r} x_{n_s}^{-1} = a_i.$$

Since $x_n \in Z(G_1)$ for all $n \geq 1$, we conclude that for all $r, s \geq 1$, $x_{n_r} x_{n_s}^{-1} \in Z(G)$. This contradiction completes the proof of our claim.

Finally since B is finite modulo $Z(G)$, so also is B, and hence A is relatively compact modulo $Z(G)$. Hence $G \in \mathfrak{C}$.

Remark. There exists a three-dimensional solvable analytic group that does not belong to $\mathfrak{C}$. Let $\alpha = a+ib \in \mathbb{C}$ with $a > 0$ and let G_α be the semidirect extension of $\mathbb{C}$ by $\mathbb{R}$ defined by

$$(c, t)(c', t') = (c + e^{\alpha t}c', t+t').$$

Taking $A = \{(e^{\alpha n}, n) : n \geq 1\}$ we may check that for every compact set $K \subseteq G_\alpha$, $\bigcup\{a^{-1}Ka : a \in A\}$ is relatively compact, so clearly $G_\alpha = \mathcal{L}(A, G_\alpha)$. But $(e^{\alpha n}, n)$ is not bounded and the centre of G_α is trivial, hence G_α cannot belong to $\mathcal{C}$.

§2. The class $\mathcal{B}$

A probability measure λ on a locally compact group G is said to have large support if and only if the smallest closed subgroup of G which contains the support of λ is G itself. For $\mu \in M(G)$ we denote by $R(\mu, G)$ the root set of μ in G, so that

$$R(\mu, G) = \bigcup_{n\geq 1} \{\nu^k : \nu \in M(G), \nu^n = \mu, 1 \leq k \leq n\}$$

and we say that μ is root compact (on G) if and only if $R(\mu, G)$ is relatively compact in $M(G)$. We denote by $\mathcal{B}$ the class of all locally compact groups G which have the property that every λ in $M(G)$ which has large support on G is also root compact on G.

In (4) we have shown that every connected Lie group belongs to $\mathcal{B}$ and our objective in this section is to extend this result to the class of Lie groups G such that G is compactly generated and $Z(G/G^0)$ is finitely-generated.

The next result explains our earlier interest in the class $\mathcal{C}$.

Proposition 9. Let G belong to $\mathcal{C}$ and suppose that $Z(G)$ is compactly generated. Then G belongs to $\mathcal{B}$.

Proof. Suppose $\lambda \in M(G)$ has large support and let $\{\mu_n : n \geq 1\}$ be a sequence in $R(\lambda, G)$. Then $\mu_n = \nu_{m_n}^{k_n}$, where $1 \leq k_n \leq m_n$ and $\nu_{m_n}^{m_n} = \lambda$. Let $I = \{(n, k) : n \geq 1, 1 \leq k \leq m_n\}$ and for each $i = (n, k) \in I$ write λ_i for $\nu_{m_n}^k$ and $\tilde{\lambda}_i$ for $\nu_{m_n}^{(m_n-k)}$. Then for each $i \in I$, $\lambda_i\tilde{\lambda}_i = \tilde{\lambda}_i\lambda_i = \lambda$, so by Theorem 2.1 of Chapter III of (9) there exists $\{x_i : i \in I\}$ in G such that $\{x_i\tilde{\lambda}_i : i \in I\}$ and $\{\lambda_i x_i^{-1} : i \in I\}$ are relatively compact in $M(G)$, and so $\{x_i\lambda x_i^{-1} : i \in I\}$ is also relatively compact.

Let y lie in the support of λ, then given a neighbourhood U of y, we have $\lambda(U) = \delta > 0$, and by compactness we also have a compact set K in G such that $\lambda(x_i^{-1}Kx_i) > 1-\delta$, for all $i \in I$. We conclude that $V \cap x_i^{-1}Kx_i \neq \emptyset$, for every $i \in I$, so writing $A = \{x_i^{-1} : i \in I\}$, we see that $y \in \mathcal{L}(A, G)$. Since $\mathcal{L}(A, G)$ is closed and λ has large support, it follows that $G = \mathcal{L}(A, G)$.

Since G belongs to $\mathcal{C}$, we conclude that there is a set $\{y_i : i \in I\}$ in $Z(G)$ such that $\{\lambda_i y_i^{-1} : i \in I\}$ is relatively compact in $M(G)$. Hence there is a compact set $K' \subseteq G$ such that for all $i \in I$, $\lambda_i(K'y_i) > 2/3$ and note that if $i = (n, m_n)$ we may take y_i to be 1, because in that case $\lambda_i = \lambda$, so we simply take K' big enough to ensure that $\lambda(K') > 2/3$.

Then for all $n \geq 1$ and for $1 \leq k+\ell \leq m_n$, we have

$$\nu_{m_n}^{k+\ell}(K'y_{n,k}\,K'y_{n,\ell}) > {}^4/_9 \quad\text{and}\quad \nu_{m_n}^{k+\ell}(K'y_{n,k+\ell}) > {}^2/_3$$

which implies, because $y_i \in Z(G)$, that

$$y_{n,k}\,y_{n,\ell}\,y_{n,k+\ell}^{-1} \in (K')^{-2}K', \quad\text{with } y_{n,m_n} = 1.$$

But $Z(G)$, being compactly generated, is strongly root compact by 3.1.12 of (1), so there is a compact set $D \subseteq Z(G)$ such that $y_i \in D$, for all $i \in I$. We conclude that $\{\lambda_i : i \in I\}$ is relatively compact, hence $\{\mu_n : n \geq 1\}$ is relatively compact. It follows that $R(\lambda, G)$ is relatively compact and the proof is complete.

<u>Corollary</u>. The following groups belong to $\mathcal{B}$.

(i) Any finitely-generated group whose centre is finitely-generated.

(ii) Any compactly generated semisimple Lie group whose centre is finitely-generated (the centre of such a group is necessarily discrete).

<u>Proof</u>. Immediate from Propositions 6, 8, 9.

<u>Proposition 10</u>. Let G be a locally compact group and H a compact normal subgroup. If ${}^G/_H$ belongs to $\mathcal{B}$, then so does G.

<u>Proof</u>. Routine.

<u>Proposition 11</u>. Let G be a locally compact group, let N be a closed normal subgroup of G such that N is a vector group and let $p : G \to {}^G/_N = K$ be the natural projection. If $A \subseteq G$ such that $p(A)$ is relatively compact but A is not relatively compact, then there exists some $1 \neq v \in N$ such that $\mathcal{L}(A, G) \subseteq Z(v, G)$, where $Z(v, G)$ denotes the centraliser of v in G. Hence every element of $\mathcal{L}(A, G)$ commutes with all elements $\lambda v \in N$, for all $\lambda \in \mathbb{R}$.

<u>Proof</u>. The proof of Proposition 9 of (4) shows that we can find a sequence $\{y_m : m \geq 1\}$ in A and an element $v \in N$ such that whenever $\{z_r : r \geq 1\}$ is a convergent sequence in G with $\{y_{m_r} z_r y_{m_r}^{-1} : r \geq 1\}$ convergent to some $b \in G$, then $b \in Z(v, G)$. (In the notation of the proof of Proposition 9 of (4), this v is aua^{-1}.)

The argument of the second paragraph of the proof of Proposition 4 above now shows that if $x \in \mathcal{L}(A, G)$ then any neighbourhood of x meets $Z(v, G)$, so since this group is closed, we conclude that $x \in Z(v, G)$.

We now have the main result of this section.

<u>Proposition 12</u>. Let G be a compactly generated Lie group and suppose that $Z({}^G/_{G^o})$ is finitely-generated. Then G belongs to $\mathcal{B}$.

<u>Proof</u>. Suppose the result is not true and let G be a counterexample of minimum dimension. Then in view of Proposition 9, Corollary, we know that $\dim G \geq 1$.

We note that if H is a connected subgroup of G of positive dimension, then $K = {}^G/_H$ belongs to $\mathcal{B}$, because K is a compactly generated Lie group, $\dim K < \dim G$, and ${}^K/_{K^o} \cong {}^G/_{G^o}$, whence $Z({}^G/_{G^o}) \cong Z({}^K/_{K^o})$ and so the latter group is finitely-generated as the former is finitely-generated.

In view of Proposition 9, Corollary, we may also assume that G is not semisimple, so by lemma 3.6 of page 185 of (2), G contains a normal subgroup N that is either a toroid or a vector group. But by Proposition 10 and the previous paragraph, N cannot be a toroid, hence N is a vector group. We can also say that G has no connected centre, otherwise $G/Z^0(G)$ lies in $\mathcal{B}$ and so G lies in $\mathcal{B}$ by Proposition 8 of (4).

Now suppose $\lambda \in M(G)$ has large support on G, let $\{\lambda_n : n \geq 1\}$ be a sequence in $R(\lambda, G)$ and let $p : G \to G/N$ be the natural projection. As usual we may find $\{x_n : n \geq 1\}$ in G such that $\{\lambda_n x_n : n \geq 1\}$ and $\{x_n^{-1}\lambda x_n : n \geq 1\}$ are relatively compact. As $p(\lambda)$ has large support on G/N and G/N belongs to $\mathcal{B}$, we conclude that $\{p(x_n) : n \geq 1\}$ is relatively compact in G/N. Furthermore the argument of the second paragraph of the proof of Proposition 9 shows that, because λ has large support on G, then $\mathcal{L}(A, G) = G$, where $A = \{x_n : n \geq 1\}$. But we know that G has no connected centre, so by Proposition 11 it follows that A must be relatively compact. Hence $\{\lambda_n : n \geq 1\}$ is relatively compact, which is enough to show that $R(\lambda, G)$ is relatively compact.

This contradicts the assumption that G does not lie in $\mathcal{B}$.

Problem. It is an open question whether there is a necessary and sufficient topological group condition for a group G to belong to the class $\mathcal{B}$.

§3. Local tightness of convolution semigroups

Let H be a locally compact group, then for $\lambda \in M(H)$ we write $S(\lambda)$ for the support of λ and $H(\lambda)$ for the smallest closed subgroup of H that contains the support of λ.

Proposition 13. Let H be a locally compact group and let $\mu, \nu \in M(H)$ such that $\nu^n = \mu$. Then $H(\mu) \subseteq H(\nu) \subseteq N(H(\mu))$, where $N(H(\mu))$ is the normaliser of $H(\mu)$ in H. Further for any $a \in S(\nu)$, $a^n \in S(\mu)$ and $S(\nu) \subseteq aH(\mu)$.

Proof. Recall from (1.2.1) of (1) that

$$S(\mu) = \overline{\underbrace{S(\nu)\ S(\nu)\ \dots\ S(\nu)}_{n}}$$

from which it is immediate that $H(\mu) \subseteq H(\nu)$ and that for any $a \in S(\nu)$, $a^n \in S(\mu)$.

Now let $y \in S(\mu)$, $a \in S(\nu)$ and observe that

$$aya^{-1} = aya^{n-1}a^{-n} \in \overline{\underbrace{S(\nu)\ S(\nu)\ \dots\ S(\nu)}_{2n}}\ H(\mu) \subseteq H(\mu).$$

Similarly $a^{-1}ya \in H(\mu)$, hence $S(\nu) \subseteq N(H(\mu))$ and so $H(\nu) \subseteq N(H(\mu))$. If $f : N(H(\mu)) \to N(H(\mu))/H(\mu)$ is the natural projection, then $f(\nu)^n = 1$, so

$f(\nu)$ is a point mass. Hence $S(\nu) \subseteq aH(\mu)$ for any $a \in S(\nu)$.

Let S be a real directed semigroup. A <u>constrained convolution S-semigroup</u> is a pair (H, ϕ), where H is a locally compact group and ϕ is a homomorphism of S into $M(H)$, such that the smallest closed subgroup of H that contains $\bigcup\{S(\phi(r)) : r \in S\}$ is H itself. For each $r \in S$ we write $H(r)$ for $H(\phi(r))$.

<u>Proposition 14</u>. Let S be a real directed semigroup and let (H, ϕ) be a constrained convolution S-semigroup.

(i) Given $r, s \in S$, there exists $t \in S$ such that $H(r) \cup H(s) \subseteq H(t)$.

(ii) $\bigcup\{H(r) : r \in S\}$ is a dense subgroup of H.

(ii) For each $r_0 \in S$, $\bigcup\{H(r) : r \in S\} = \bigcup\{H(r) : r \in S,\ r = nr_0,$ some $n \in \mathbb{N}\}$.

(iv) For each $r \in S$, $H(r)$ is normal in H.

(v) For each $r \in S$, $H/_{H(r)}$ is an Abelian group in which the subgroup of all elements of finite order is dense. If H is compactly generated then $H/_{H(r)}$ is compact.

(vi) If N is a normal subgroup of H and $p : H \to H/_N = K$ is the natural projection, and we define $\tilde{\phi} : S \to M(K)$ by $\tilde{\phi}(r) = p(\phi(r))$, then $(K, \tilde{\phi})$ is a constrained convolution S-semigroup.

<u>Proof</u>. (i) Given $r, s \in S$, there exists $t \in S$ such that $r = nt$ and $s = mt$, for $m, n \in \mathbb{N}$. Then $\phi(t)^n = \phi(r)$ so by Proposition 13, $H(r) \subseteq H(t)$, and similarly for $H(s)$.

(ii) The fact that $\bigcup\{H(r) : r \in S\}$ is a subgroup follows from (i). Its closure contains $S(\phi(r))$ for all $r \in S$, so must be all of H.

(iii) Let $s \in S$, then there exists $t \in S$ such that $r_0 = nt$ and $s = mt$, with $m, n \in \mathbb{N}$. By Proposition 13,

$$H(s) \subseteq H(t) \subseteq \bigcup\{H(r) : r \in S,\ r_0 = kr \text{ some } k \in \mathbb{N}\}.$$

(iv) By Proposition 13, for each $r \in S$, $H(r)$ is normal in $\bigcup\{H(s) : s \in S,\ r = ns,$ some $n \in \mathbb{N}\}$, which gives the result by (ii) and (iii).

(v) By (ii) and (iii) $\bigcup\{{}^{H_s}/_{H_r} : s \in S$ such that $ns = r$, some $n \in \mathbb{N}\}$ is a dense subgroup of $H/_{H_r}$, and the groups in this union are all finite cyclic by Proposition 13. These cyclic subgroups are directed by (i), hence $H/_{H_r}$ is Abelian. The rest of (v) follows from the structure theorem for compactly generated locally compact abelian groups.

(vi) Proof is routine.

<u>Proposition 15</u>. Let S be a real directed semigroup and let (H, ϕ) be a constrained convolution S-semigroup. If H is finitely-generated and has finitely-generated centre, then (H, ϕ) is locally tight.

<u>Proof</u>. By Proposition 14 (ii) we have $H = \bigcup\{H(r) : r \in S\}$ and since H is finitely-generated, it follows from Proposition 14 (i) that there exists $r_0 \in S$ such that $H = H(r_0)$.

If $s \in S$ with $s \leq r_0$, there exists $m, n \in \mathbb{N}$ and $t \in S$ such that $r_0 = nt$, $s = mt$ and since $s \leq r_0$, then $1 \leq m \leq n$. Hence $\phi(s) = \phi(t)^m$ and $\phi(r_0) = \phi(t)^n$ which implies that $\phi(s) \in R(\phi(r_0), H)$. Since $\phi(r_0)$ has large support on H and H lies in $\mathcal{B}$ by Proposition 9, Corollary, we conclude that $\{\phi(s) : s \in S, s \leq r_0\}$ is relatively compact in $M(H)$.

Since S is a real directed semigroup it follows that for each $r \in S$ there exists some $k \in \mathbb{N}$ such that

$$\{\phi(s) : s \in S, s \in r\} \subseteq \{\phi(s) : s \in S, s \leq r_0\}^k.$$

Hence ϕ is locally tight as claimed.

Proposition 16. Let S be a real directed semigroup and let (H, ϕ) be a constrained convolution S-semigroup. If H is a compactly generated semisimple Lie group whose centre is finitely-generated, then (H, ϕ) is locally tight.

Proof. Given $r \in S$, the group $H/H(r)$ is abelian by Proposition 14 (v), while H^o is semisimple, so since a connected semisimple Lie group cannot have a non-trivial abelian image, we conclude that $H^o \subseteq H(r)$ for every $r \in S$. Hence each $H(r)$ is open and closed in H, so by Proposition 14 (ii), $H = \bigcup\{H(r) : r \in S\}$. As H is compactly generated it follows from Proposition 14 (i) that there is some $r_0 \in S$ such that $H = H(r_0)$. Then $\phi(r_0)$ has large support on H, so since H lies in $\mathcal{B}$ by Proposition 9, Corollary, an argument as in the proof of Proposition 15 shows that for each $r \in S$, $\{\phi(s) : s \in S, s \leq r\}$ is relatively compact, and (H, ϕ) is locally tight.

Proposition 17. Let S be a real directed semigroup and let (H, ϕ) be a constrained convolution S-semigroup. If N is a compactly generated central closed subgroup of H such that the induced semigroup $(H/N, \tilde{\phi})$ (as defined in Proposition 14 (vi)) is locally tight, then (H, ϕ) is locally tight.

Proof. Let $p : H \to H/N$ be the natural projection and let $r_0 \in S$. In the usual way we may find $\{x_r : r \in S, r \leq r_0\}$ in H such that $\{\phi(r)x_r : r \in S, r \leq r_0\}$ is relatively compact in $M(H)$. Projecting via p into H/N and using the local tightness of $(G/N, \tilde{\phi})$ we see that we may assume that x_r lies in N, for each $r \leq r_0$, with $r \in S$.

Arguing as in the proof of lemma 5 of (3) we see that $\{x_r : r \in S, r \leq r_0\}$ satisfies the condition that the set $\{x_{r+s}x_r^{-1}x_s^{-1} : r, s \in S, r, s, r+s \leq r_0\}$ is relatively compact in N. But N being compactly generated abelian, it follows from the structure theorem and the arguments of lemma 4 and its corollary of (3) that $\{x_r : r \in S, r \leq r_0\}$ is relatively compact. Hence $\{\phi(r) : r \in S, r \leq r_0\}$ is relatively compact and (H, ϕ) is locally tight as claimed.

Proposition 18. Let S be a real directed semigroup and let (H, ϕ) be a constrained convolution S-semigroup. If H is a compactly generated Lie group such that H/H^o has finitely-generated centre, then (H, ϕ) is locally tight.

Proof. Suppose the result is not true and let (H, ϕ) be a counterexample for which $\dim H$ is minimum. By Proposition 15, $\dim H \geq 1$ and by Proposition 16 we may assume that H is not semisimple (note that $Z(H)$ is finitely-generated because $H/_{H^o}$ has finitely generated centre). Then by Lemma 3.6 of Chapter XV of (2), H contains a normal subgroup N such that N is either a toroid or a vector group.

If N is a toroid and $p : H \to H/_N$ is the natural projection, then by Proposition 14 (vi) and the minimality of H, we see that $(H/_N, \tilde{\phi})$ is locally tight, which by an obvious argument and the compactness of N implies that (H, ϕ) is locally tight. We deduce that N is not a toroid, hence is a vector group. We may also assume, in view of Proposition 17, that H has no connected centre.

For each $r \in S$ write $Z(r) = \{x \in N : xy = yx, \forall y \in H(r)\}$ and choose $r_0 \in S$ for which $\dim Z(r)$ is minimum. If $r \in S$ such that $nr = r_0$, some $n \in \mathbb{N}$, then by Proposition 13, $H(r_0) \subseteq H(r)$, so $Z(r) \subseteq Z(r_0)$, hence $\dim Z(r) \leqslant \dim Z(r_0)$. By minimality of $Z(r_0)$, we conclude that $Z^o(r_0) \subseteq Z(r)$, hence by Proposition 14 (ii), (iii), $Z^o(r_0) \subseteq Z(H)$, and since H has no connected centre we deduce that $Z^o(r_0) = \{1\}$.

Let $p : H \to H/_N = K$ be the natural projection, then by minimality of H the induced semigroup $(K, \tilde{\phi})$ is locally tight. So in the usual way we may find a set $A = \{x_r : r \in S, r \leqslant r_0\}$ in H such that $\{\phi(r)x_r : r \in S, r \leqslant r_0\}$ and $\{x_r^{-1}\phi(r_0-r) : r \in S, r \leqslant r_0\}$ are relatively compact in $M(H)$. Since $(K, \tilde{\phi})$ is locally tight we see that $\{p(x_r) : r \in S, r \leqslant r_0\}$ is relatively compact in K, and we also have $\{x_r^{-1}\phi(r_0)x_r : r \in S, r \leqslant r_0\}$ relatively compact in $M(H)$. By an argument as in the proof of the second paragraph of Proposition 9, we conclude that $H(r_0) \subseteq \mathcal{L}(A, H)$, where $A = \{x_r : r \in S, r \leqslant r_0\}$.

But if A is not relatively compact then by Proposition 11 there exists $1 \neq v \in N$ such that $\mathcal{L}(A, H) \subseteq Z(v, H)$, so $v \in Z(r_0)$ and indeed $\lambda v \in Z(r_0)$, for each $\lambda \in \mathbb{R}$. This contradicts the fact that $Z^o(r_0) = 1$, and shows that A is indeed relatively compact. Hence $\{\phi(r) : r \in S, r \leqslant r_0\}$ is relatively compact which is sufficient to show that (H, ϕ) is locally tight.

<u>Proof of Theorem 1.</u> Just apply Proposition 18 to the constrained convolution S-semigroup $(G(\phi), \phi)$.

<u>Proof of Theorem 2.</u> If S is a real directed semigroup and G is a Lie group, and $\phi : S \to M(G)$ is a homomorphism, then $H = G(\phi)$ is a Lie group and if $H(r)$ is solvable for some $r \in S$, then by Proposition 14 (v), H is solvable. Hence $H \cap G^o$ is a solvable closed subgroup of the connected Lie group G^o, so by a theorem of Mostow (7), $(H \cap G^o)/_{(H \cap G^o)^o}$ is a polycyclic group, and since $(H \cap G^o)^o = H^o$, we see that $H/_{H^o}$ is an extension of a polycyclic group by a subgroup of $G/_{G^o}$. So if we are also given that $G/_{G^o}$ has all its subgroups finitely-generated, it follows that every subgroup of $H/_{H^o}$ is finitely-generated, and theorem 1 may be applied to give the result.

<u>Concluding Remarks.</u> (i) We have no example of a homormorphism of say $\mathbb{Q}^+$ (positive rationals) into $M(G)$, where G is a connected Lie group, which fails to

be locally tight. Certainly if we are given a homomorphism $\phi : S \to M(G)$ (S real directed) such that each $\phi(r)$ is a point mass on G, then ϕ is necessarily locally tight, because the closed subgroup H of G supporting $\phi(S)$ has to be of the form $V \times T \times D$, where V is a vector group, T is a torus, and D is a (discrete) finitely-generated abelian group.

(ii) Any homomorphism $\phi : S \to M(G)$, where $G = SL(2, \mathbb{C})$ or $SL(2, \mathbb{R})$, is necessarily locally tight. This is because for each $r \in S$, either $H(r)$ acts irreducibly on $\mathbb{C}^2$ (or $\mathbb{R}^2$), or else $H(r)$ is solvable (Lemma 5 of (6)). In the latter case the conclusion follows from Theorem 2 above, in the former case the result is immediate from Lemma 4 of (6). Incidentally, every infinitely divisible probability measure on G is indeed continuously embedded, as is shown in (6).

(iii) It is clear that any real submonogeneous semigroup (see p.180 of (1)) is a real directed semigroup in the sense defined above. The question as to whether the converse is true has been raised in conversation by Professor L. Schmetterer, and the answer is affirmative. This is seen by observing that a real directed semigroup S is necessarily countable, and that every finite set of elements lies in a common cyclic subsemigroup. An inductive construction now shows that S is the union of an increasing sequence of cyclic subsemigroups, which is essentially the definition of submonogeneous.

References

(1) Heyer, H. "Probability measures on locally compact groups", Ergebnisse der Mathematik und ihrer Grenzgebiete 94, Berlin-Heidelberg-New York, Springer 1977.

(2) Hochschild, G. "The structure of Lie groups", San Francisco, Holden-Day 1965.

(3) McCrudden, M., Burrell, Q.L. "Infinitely divisible distributions on connected nilpotent Lie groups", J. Lond. Math. Soc. II, Ser. 7, 193-196 (1974).

(4) McCrudden, M. "Factors and roots of large measures on connected Lie groups", Math. Zeit. 177, 315-322 (1981).

(5) McCrudden, M. "On embedding infinitely divisible distributions on simply-connected solvable Lie groups". To appear in J. Lond. Math. Soc.

(6) McCrudden, M. "Infinitely divisible probabilities on SL(2, $\mathbb{C}$) are continuously embedded", (Preprint, June 1981).

(7) Mostow, G.D. "On the fundamental group of a homogeneous space", Ann. of Math. 66, 249-255 (1957).

(8) Parthasarathy, K.R. "Infinitely divisible distributions in SL(k, $\mathbb{C}$) or SK(k, $\mathbb{R}$) may be embedded in diadic convolution semigroups" in "Probability measures on groups" (Ed. H. Heyer) Springer Lecture Notes in Mathematics No.706.

(9) Parthasarathy, K.R. "Probability measures on metric spaces", Academic Press, London 1967.

Convergence of Nonhomogeneous Stochastic Chains with Countable States: An Application to Measures on Semigroups

By

Arunava Mukherjea and Anastase Nakassis

0. In this paper we continue the work started in [5,6,7]. Let (P_n) be a sequence of countably infinite dimensional stochastic matrices such that for each positive integer k, the sequence $P_{k,n}$ defined by

$$P_{k,n}=P_{k+1}P_{k+2}\cdots P_n \qquad (k < n)$$

converges to a stochastic matrix Q_k pointwise. The sequence (P_n) will be called a convergent stochastic chain. See [3-7] for the finite case. In these papers efforts were made to understand how the convergence behavior was connected to ergodicity of certain related subchains. Maksimov started this study in [3]. In [3,5,6], measure-theoretic (elementary) methods were used, but found to be of limited scope. In [7], our earlier methods were abandoned in favor of an even simpler and more effective method. We will further exploit this method in this paper.

Professor H. Cohn, after studying our papers, has mentioned to us that some of our results follow from those in his papers, where he considers a completely different method (using properties of the tail sigma-field of the associated Markov chain and martingale convergence theorems) in a general framework. We strongly encourage any serious reader of our works also to look into Cohn's papers. The authors of [5,6] were not aware of Cohn's 1976-77 papers at the time [5,6] went into print. In our context, however, our results are best possible and our methods, though very simple, can be considered in the general framework of nonnegative matrices. Though this will be clarified in details elsewhere, let us mention why similar results can be expected for products of nonnegative matrices. The key inequality that works in our papers is an inequality for backward or forward products of nonnegative matrices. Though it will be described later via Markov chains, let us briefly digress at this point to demonstrate a quick derivation of this inequality for general nonnegative matrices.

Let P_n's be nonnegative matrices and $P_{k,n}$'s, $k < n$, as before. Let i,j,k,n_1,n_2,n be fixed positive integers such that $k< n_1< n_2< n$. Then we claim that

$$(P_{k,n})_{ij} \geq \sum_{m=n_1}^{n_2} (P_{k,m})_{ii}(P_{m+1})_{ij}(P_{m+1,n})_{jj}$$

$$- \sum_{n_1\leq m< m'\leq n_2} (P_{k,m})_{ii}(P_{m+1})_{ij}(P_{m+1,m'})_{ji}(P_{m'+1})_{ij}(P_{m'+1,n})_{jj}.$$

(We assume here, of course, that $(P_{k,n})_{ij}$ is finite.)
To prove this claim, we proceed as follows. For each n-k-1 tuple of positive integers $(s_1,s_2,\ldots,s_{n-k-1})$, consider an element $x(s_1,s_2,\ldots,s_{n-k-1})$ and let all such elements form the set A. Let β be a discrete measure on A such that

$$\beta(\{x(s_1,s_2,\ldots,s_{n-k-1})\})=(P_{k+1})_{is_1}(P_{k+2})_{s_1s_2}\cdots(P_n)_{s_{n-k-1}j}.$$

For $k < m < n$, define the sets A_m by

$$A_m= \{x(s_1,s_2,\ldots,s_{n-k-1}) \varepsilon A: s_{m-k}=i \text{ and } s_{m-k+1}=j\}.$$

Then the above inequality follows immediately from the observation that

$$\beta(A) \geq \beta(\bigcup_{m=n_1}^{n_2} A_m) \geq \sum_{m=n_1}^{n_2} \beta(A_m) - \sum_{n_1\leq m< m'\leq n_2} \beta(A_m\cap A_{m'}).\square$$

The above inequality plays the central role here as well as in [7]. Results for backward chains are also obtainable via this inequality. Backward chains are studied in details in [7]. The results given here are all best possible as verified by various examples.

Let us now get into the differences between the finite and the infinite case. This will then clarify the need for a separate paper to deal with the infinite case. First, the result (see Theorem 8,[5]) that the restriction of a convergent chain with no T class in its basis to any of its C classes is ergodic is false in the infinite case. The important result, due to Hajnal, that two finite state chains, when equivalent, always converge or diverge together (and have the same basis in case of convergence) turns out to be false in the infinite case. All these and other differences are shown via suitable examples in section 2. Some of our results in the finite case depend upon the concept of equivalence. Thus, there is a need to answer the question of validity of such or similar results in the infinite case. Our present paper is motivated in this respect.

The organization of this paper is as follows. All matrices, in what follows, are (unless otherwise mentioned) infinite dimensional stochastic matrices. (Though similar results can be obtained by our methods for nonnegative matrices, we will not consider them here any more since much more space is needed for proper statements of such results and examples determining domains of validity of such results.) In section 1, we establish the existence of a basis of a convergent chain. Unlike in the finite case, this is less evident here. In section 2, most of our examples are given. In section 3, we present our main results. Finally, in section 4, an application to measures on semigroups is given.

1. Basis of a convergent stochastic chain:

Establishing the existence of a basis in the infinite case is not immediate. The reason is, of course, that the pointwise limit of a sequence of stochastic matrices need not be a stochastic matrix any more and also, in this case, the joint continuity of matrix multiplication is no longer true. Let us first establish a few simple (but necessary) lemmas.

Lemma 1.1. Suppose (B_n) and (A_n) are two sequences of stochastic matrices converging pointwise to respectively, the matrices B and A. Suppose also that B is stochastic. Then, A is substochastic, and the sequence $(B_n A_n)$ converges pointwise to the matrix BA. Unless A is stochastic, the sequence $(A_n B_n)$ need not converge to the limit AB.)□

Proof. First, we observe that

$$\left|(B_n A_n)_{ij} - (BA)_{ij}\right| \leq \sum_t (A_n)_{tj} \cdot \left|(B_n)_{it} - B_{it}\right| + \sum_t B_{it} \left|(A_n)_{tj} - A_{tj}\right| . \qquad \ldots \ldots (1)$$

Given $\varepsilon > 0$, let N be so that $\sum_{t=1}^{N} B_{it} > 1-\varepsilon$ and n_o such that $n \geq n_o$ implies

$$\sum_{t=1}^{N} (B_n)_{it} > 1-\varepsilon \ ;$$

then for $n \geq n_o$,

$$\sum_{t=N+1}^{\infty} (B_n)_{it} < \varepsilon.$$

Now it follows from (1) that

$$\left|(B_n A_n)_{ij} - (BA)_{ij}\right| \leq \sum_t \left|(B_n)_{it} - B_{it}\right| + \sum_{t=1}^{N} \left|(A_n)_{tj} - A_{tj}\right| + \sum_{t=N+1}^{\infty} B_{it}.$$

This inequality implies that $\lim_n B_n A_n = BA$.

The last assertion is now illustrated by the following example:

Let $B_n = B$ for each n, where each row in B is the same as

$$(1/2,\ 1/2^2,\ 1/2^3,\ \ldots\ );$$

let A_n be defined by

$$\begin{aligned}(A_n)_{ij} &= \tfrac{1}{n} \text{ for } 1 \leq i,j \leq n \ ; \\ &= 0 \text{ for } 1 \leq i \leq n < j \text{ and } 1 \leq j \leq n < i; \\ &= 1 \text{ for } i=j > n.\end{aligned}$$

Then $A_n \to 0$ (the zero matrix), but $A_n B = B$. □

Lemma 1.2. Suppose that for each positive integer k

$$\lim_n P_{k,n} = Q_k \quad \text{(pointwise)}$$

where each P_n as well as each Q_k is a stochastic matrix. Let Q' be a limit point of the Q_k's with respect to pointwise convergence. Then for each k, we have:

$$Q_k Q' = Q_k. \quad \square \qquad \ldots \qquad \ldots. \ (2)$$

Proof. For $k < n < m$, $P_{k,m} = P_{k,n}P_{n,m}$. By Lemma 1.1,

$$Q_k = P_{k,n}Q_n.$$

Using Lemma 1.1 again, (2) follows. □

Lemma 1.3. In Lemma 1.2 let k and t be such that $(Q_k)_{it} > 0$ for some i. Then we have:

$$\sum_{j=1}^{\infty} (Q')_{tj} = 1. \quad \square$$

Proof. The lemma follows immediately from the following equality that follows from (2):

$$1 = \sum_{j=1}^{\infty}(Q_k)_{ij} = \sum_{j=1}^{\infty} \sum_{s=1}^{\infty} (Q_k)_{is}Q'_{sj} = \sum_{s=1}^{\infty}(Q_k)_{is} \cdot \left[\sum_{j=1}^{\infty} Q'_{sj} \right]. \quad \square$$

Lemma 1.4. Let Q' be as in Lemma 1.2. Then, the entries in the j th row of Q' add up to 1, provided that the j th column of Q' is not a zero column. □

Proof. Immediate from Lemma 1.3. □

Lemma 1.5. Let Q' be as in Lemma 1.2. Let

$$T = \{j : Q'_{ij} = 0 \text{ for each } i\},$$

that is, T is the set of zero columns of Q'. Then, the matrix $(Q')^+$, defined as the restriction of Q' to the complement of T, is a stochastic idempotent matrix with no zero columns. (By Lemma 1.3, the complement of T is non-empty.) □

Proof. Note that because of equation (2), the i th column of each Q_k, for i in T, is a zero column. Also, note that for $j \notin T$ and $k \notin T$, we have:

$$(Q_p)_{jk} = (Q_pQ')_{jk} = \sum_{t=1}^{\infty} (Q_p)_{jt}Q'_{tk} = \sum_{t\notin T} (Q_p)_{jt}Q'_{tk}$$

so that for each positive integer p,

$$(Q_p)^+ \cdot (Q')^+ = (Q_p)^+ \; . \qquad \ldots\ldots \quad \ldots\ldots (3)$$

Now notice that for any given $t \notin T$, there is a Q_p (for some p) such that the t th column of Q_p is not a zero column. It follows from (3), like in Lemma 1.3, that for each $t \notin T$,

$$\sum_{s\notin T} (Q')_{ts} = 1.$$

Then it follows from equation (3) and Lemma 1.1 that $(Q')^+$ is a stochastic idempotent matrix. Finally, we show that this matrix has no zero columns. To this end, let T' be the set of zero columns of $(Q')^+$. Of course, $T' \subset T^c$ (the complement of T). For $j \in T'$ and for every i and k,

$$(Q_k)_{ij} = (Q_kQ')_{ij} = \sum_{t=1}^{\infty} (Q_k)_{it}Q'_{tj} = \sum_{t\notin T}(Q_k)_{it}Q'_{tj} = 0,$$

since $j \in T'$. This means that $Q'_{ij} = 0$ for each i so that $j \in T$. This is a contradiction. □

__Lemma 1.6.__ Let the Q_k's be as in Lemma 1.2. Let Q' and Q'' be any two (pointwise) limit points of the Q_k's. Then, the sets

$$\{j:\ Q'_{ij}=0 \text{ for each } i\} \text{ and } \{j:\ Q''_{ij}=0 \text{ for each } i\}$$

are the same. If we denote these sets by T, then Q' and Q'', when restricted to the complement of T, coincide. Each of these restrictions is a stochastic idempotent matrix with no zero columns. □

__Proof.__ Notice that by equation (2),

$$Q'_{ij}=0 \text{ for each } i \quad \text{iff } (Q_k)_{ij} = 0 \text{ for each } i \text{ and } k$$

for any limit point Q' of the Q_k's. This means that the ''T'' set is the same for both Q' and Q''. It follows from equation (3), Lemma 1.1 and Lemma 1.5 that when restricted to the complement of T,

$$Q''Q'=Q'' \text{ and } Q'Q''=Q'.$$

By using the structure theorem for stochastic idempotent matrices, it can be easily verified that $Q' = Q''$, when restricted to T^c. □

Because of Lemma 1.6, we can now define the basis of a convergent non-homogeneous stochastic chain.

Definition 1.7. Suppose that (P_n) is a convergent stochastic chain and that each Q_k, where $Q_k = \lim_n P_{k,n}$ (pointwise limit), is stochastic. Let Q' be a (pointwise) limit point of the Q_k's and let

$$T=\{j: Q'_{ij} = 0 \text{ for each } i\}.$$

Then Q', restricted to T^c, is a stochastic idempotent matrix with no zero columns. By the structure theorem for stochastic idempotent matrices (see [5]), there exists a partition $\{C_1, C_2, \ldots\}$ of T^c such that

$$\begin{aligned} Q'_{ij} &= 0 \text{ for } i, j \text{ in different C-classes;} \\ &= Q'_{kj}\ (>0) \text{ for } i, j \text{ and } k \text{ in the same C-class.} \end{aligned}$$

The partition $\{T, C_1, C_2, \ldots\}$ remains the same for all limit points Q' of the Q_k's. This will be called the basis of the convergent chain (P_n). □

Remark 1.8. The pointwise convergence considered above is certainly weaker than the usual norm convergence, where

$$\|P-Q\| = \sup_i \sum_{j=1}^{\infty} |P_{ij}-Q_{ij}| ,$$

even when the pointwise limit is a stochastic matrix. For example, consider (P_n) defined by

$$\begin{aligned} (P_n)_{ij} &= 1-(1/n^{1/i}) \text{ for } j=i \leq n ; \\ &= 1/n^{1/i} \text{ for } j=i+1 \leq n+1 ; \\ &= 1 \text{ for } j=i > n ; \\ &= 0, \text{ otherwise.} \end{aligned}$$

Then P_n converges to the identity matrix I pointwise, but $P_n \not\to I$ in norm as n tends to infinity.

Remark 1.9. There are a number of interesting differences in the convergence behavior of forward and backward chains. Here we point out only one. Consider the chain (P_n) defined by

$$(P_n)_{ij} = 0 \text{ for } j=1,2,\dots,n \text{ and all } i \geq 1 ;$$
$$= 1/2^k \text{ for } j=n+k \text{ and all } i \geq 1.$$

For $n > k$, define $P_{n,k} = P_n P_{n-1} \cdots P_{k+1}$. Then, $P_{n,k} = P_k$ for all $n > k$. In this case, unlike in the forward chain case,

$$\lim_k \lim_n P_{n,k} = 0.$$

2. Examples.

(a) In the finite case one important result for convergent chains where the ''T'' set in its basis is empty is that each subchain obtained by normalizing the restriction of the original chain to any particular C-class is strongly ergodic. See Theorem 8 in [5]. This example shows that this result is false in the infinite situation. Define P_n as

$$(P_n)_{11} = (P_n)_{22} = 1-\tfrac{1}{n};$$
$$(P_n)_{1,n+2} = (P_n)_{2,n+2} = 1/n ;$$
$$(P_n)_{ii} = 1,\ 2 < i \leq n;\ (P_n)_{i1} = (P_n)_{i2} = \tfrac{1}{2} \text{ for } i \geq n+1.$$

It then follows that $\lim_n P_{k,n} = Q_k$ and $Q_k \to Q$. (We will give its proof in section 3 when we again discuss this example in the context of condition (U).) The matrices Q_k and Q are given by

$$Q = \left|\begin{array}{cc|c} \frac{1}{2} & \frac{1}{2} & \\ \frac{1}{2} & \frac{1}{2} & 0 \\ \hline \multicolumn{2}{c|}{0} & I \end{array}\right| , \quad \begin{array}{l} (Q_k)_{11} = (Q_k)_{12} = (Q_k)_{21} = (Q_k)_{22} = \frac{1}{2} ; \\ (Q_k)_{i1} = (Q_k)_{i2} = \frac{1}{2} \text{ for } i > k+1 ; \\ (Q_k)_{ii} = 1 \text{ for } 2 < i \leq k+1 . \end{array}$$

so that the basis of the chain is

$$\{C_1 = \{1,2\},\ C_2 = \{3\},\ C_3 = \{4\},\ \dots\dots \}.$$

Note that there is no T class here. But the chain restricted to the C_1-class is not ergodic since the restriction of P_n to C_1, after being normalized, is $\begin{vmatrix} 1 & 0 \\ 0 & 1 \end{vmatrix}$. □

(b) This example shows that Theorem 2 in [6] is false in the infinite dimensional case. Consider the chain (P_n) given by

$$(P_n)_{11}=(P_n)_{22}= 1-\frac{1}{n} ;$$
$$(P_n)_{1,n+2}=(P_n)_{2,n+2}= 1/n ;$$
$$(P_n)_{i1}=(P_n)_{i2}= 1/2 \text{ for } i > 2.$$

It can be verified that for each positive integer k,

$$\lim_n P_{k,n} = Q,$$

where Q is a matrix with identical rows and the rows are

$$(1/2, 1/2, 0, 0, \ldots).$$

The basis of the chain is

$$\{T=\{3,4,5,\ldots\}, C=\{1,2\}\}.$$

Note that for $i \in C$ and $j \in T$, $\sum_n (P_n)_{ij} < \infty$. But the chain obtained from normalized restriction of P_n to the C-class is not ergodic. □

(c) Here we give an example of a strongly ergodic chain where the convergence is convergence in norm. It also serves a purpose similar to that of example (b). Consider the chain (P_n) where the first n rows of P_n are $(0,1/2, 1/2^2,\ldots)$ and all other rows are $(1,0,0,\ldots)$. It can be verified that for each positive integer k,

$$\lim_n P_{k,n} = Q,$$

where Q has identical rows and each row is $(0,1/2,1/2^2,\ldots)$. Notice that in the product $P_{k,n}$, the first k rows are identical and all the other remaining rows are also identical. It follows that $P_{k,n} \rightarrow Q$ in the norm. The basis of the chain is, of course,

$$\{T=\{1\}, C=\{2,3,4,\ldots\}\}.$$

For $i \in C$ and $j \in T$, $\sum_n (P_n)_{ij} < \infty$. But we cannot get a stochastic matrix by normalizing the restriction of P_n to the C-class. □

(d) This example will show that two stochastic chains, though equivalent, may behave differently; one of them may be convergent while the other may not converge. Recall that the chains (P_n) and (P'_n) are called equivalent if for every i and j, the series

$$\sum_{n=1}^{\infty} \left|(P_n)_{ij} - (P'_n)_{ij}\right| < \infty.$$

Consider a sequence a_k of positive integers such that $a_k = 2^{4(k-1)}$ for $k \geq 1$. For $a_k \leq n < a_{k+1}$, let us define:

$$\begin{aligned}(P_{3n-2})_{ij} &= 1/2^j \text{ if } i=0 \text{ and } j=1,2,3,\ldots ;\\ &= 1-(1/2^k) \text{ if } i=j=1,2,\ldots,k ;\\ &= 1/2^k \text{ if } j=n \text{ and } i=1,2,\ldots,k ;\\ &= 1 \text{ if } i=j=k+1,\ k+2,\ldots.\end{aligned}$$

$$\begin{aligned}(P_{3n-1})_{ij} &= 1 \text{ if } i=j=0,1,\ldots,k ;\\ &= 1 \text{ if } j=n \text{ and } i > k\end{aligned}$$

and finally,

$$\begin{aligned}(P_{3n})_{ij} &= 1 \text{ if } i=n \text{ and } j=0 ;\\ &= 1 \text{ if } i=j \neq n.\end{aligned}$$

Write: $P'_n = P_{3n-2}P_{3n-1}P_{3n}$. We claim that for any probability vector $u=(u_0, u_1, u_2,\ldots\ldots)$,

$$\lim_n uP'_{m,n} = x \quad \text{(for each positive integer m)}$$

where

$$x=(0,1/2,1/2^2,\ldots\ldots).$$

Note that the claim, when proven, will imply that the chain (P'_n) is convergent (in fact, strongly ergodic). But suppose we redefine (P_{3n}) as simply identity matrix ; then the resulting chain will not converge. One way to see this is the following: Write $P^*_n=P_{3n-2}P_{3n-1}I$ and suppose that for each k, $\lim_n P^*_{k,n}$ is a stochastic matrix. Note that then (because of what we know about the basis of a chain)

$$\limsup_m \left[\lim_n (P^*_{m,n})_{jj} \right]$$

cannot be zero for each j. But this is exactly what happens in this case since for $m \geq a_k$ and n large, $(P^*_{m,n})_{jj}$ will be less than

$$\left(1-\frac{1}{2^k}\right)^{a_{k+1}-a_k} \leq \exp\left(-\frac{15}{16}\, 2^{3k}\right) \qquad \ldots\ldots \quad \ldots\ldots (4)$$

which goes to zero as k goes to infinity.

All this means is the following: Suppose that we define the chains (S_n) and (S_n^*) in the following manner:

$$S_n = P_{3m-2} \text{ for } n=3m-2;$$
$$= P_{3m-1} \text{ for } n=3m-1;$$
$$= P_{3m} \text{ for } n=3m,$$

and

$$S_n^* = P_{3m-2} \text{ for } n=3m-2;$$
$$= P_{3m-1} \text{ for } n=3m-1;$$
$$= I \text{ for } n=3m.$$

Then the chain (S_n) will converge (notice that $\lim P_{3n-1} = I$ and $x.(\lim_n P_{3n-2}) = x$, x as given before), whereas the chain (S_n^*) is not convergent. But these two chains are clearly equivalent.

To complete this example, it remains for us to establish our claim. Let $n > m$ and write: $u_n = uP'_{m,n}$ so that $u_n = u_{n-1}P'_n$. Let $a_k \leq n < a_{k+1}$. It is verified easily that

$$(u_n)_j = 0 \text{ for } j > k;$$
$$= (u_{n-1})_o.(1/2^j) + (u_{n-1})_j.(1- \frac{1}{2^k}) \text{ for } j=1,2,\ldots,k.$$

Summing both sides of $(u_n)_j$ over $j=1,2,\ldots,k$, we see that $(u_n)_o = \frac{1}{2^k}$. Then it follows that for $a_k < n < a_{k+1}$ and $1 \leq j \leq k$,

$$(u_n)_j - \frac{1}{2^j} = (1- \frac{1}{2^k})[(u_{n-1})_j - \frac{1}{2^j}] \quad \ldots\ldots \quad \ldots (5)$$

Also, it is easy to see that for $n=a_{k+1}$ and $1 \leq j \leq k$,

$$(u_n)_j - \frac{1}{2^j} = (1- \frac{1}{2^{k+1}})[(u_{n-1})_j - \frac{1}{2^j}] + (1/2^{j+k+1}) \ldots\ldots (6)$$

From (4), (5) and (6), it follows easily that $\lim_n u_n = x$. □

(e) This example serves the same purpose as that of the previous example. But this is nicer and serves an additional purpose. We will only slightly modify the previous example. The a_k's are as before, leave P_{3n-1} as before and change $1/2^k$ by $1/4^k$ in the definition of P_{3n-2}. We make a new definition for P_{3n} as follows:

$$(P_{3n})_{n,0} = \frac{1}{2^k},\ (P_{3n})_{nn} = 1 - \frac{1}{2^k} \qquad (a_k \leq n < a_{k+1})$$
$$\text{and } (P_{3n})_{ii} = 1 \text{ for } i \neq n.$$

As in the previous example, we will again show that for any probability vector u and any positive integer m,

$$\lim_n uP'_{m,n} = x \qquad \ldots\ldots \quad \ldots\ldots \quad (7)$$

where $P'_n = P_{3n-2}P_{3n-1}P_{3n}$ and $x=(0,1/2,1/2^2,\ldots)$ Defining the chains (S_n) and (S^*_n) as before (in example (d)), it will follow again that (S_n) is convergent with basis

$$\{T=\{0\},\ C=\{1,2,3,\ldots\}\},$$

whereas (S^*_n), though equivalent to (S_n), is not convergent. Notice that in this case $\sum_{n=1}^{\infty}(S_n)_{i0} \to 0$ as i tends to infinity and yet the chain obtained by restricting S_n to the C-class will not converge. This shows the extreme nature of the difference between the finite and infinite dimensional situations as exhibited in example (d).

Now to establish (7), write $u_n=u_{n-1}P'_n$. For $a_k \leq n < a_{k+1}$, we have:

(i) if $j \neq n$ and $j>k$, then $(u_n)_j=0$;

$$(u_n)_n=(u_{n-1})_o\left(\frac{1}{2^k}-\frac{1}{4^k}\right)+\sum_{i=1}^{k}(u_{n-1})_i\left(\frac{1}{4^k}-\frac{1}{8^k}\right)+(u_{n-1})_{n-1}\left(1-\frac{1}{2^k}\right);$$

$$(u_n)_o=(u_{n-1})_o\cdot\frac{1}{4^k}+\sum_{i=1}^{k}(u_{n-1})_i\cdot\frac{1}{8^k}+(u_{n-1})_{n-1}\cdot\frac{1}{2^k};$$

(ii) $(u_n)_j = (u_{n-1})_o\cdot\frac{1}{2^j} + (u_{n-1})_j\cdot\left(1-\frac{1}{4^k}\right)$, for $j=1,2,\ldots,k$.

From the last two equalities in (i), for $a_k \leq n-1<n<a_{k+1}$, we obtain:

$$(u_n)_n \leq (1/4^{k-1})+(u_{n-1})_{n-1}\cdot[1-(1/2^k)].$$

It follows easily using (4) that $\lim_n (u_n)_n = 0$. Now write: $X_n=\sum_{j=1}^{k}(u_n)_j$. Then again adding the last two equalities in (i),

$$1-X_n=(u_{n-1})_o(1/2^k)+X_{n-1}(1/4^k)+(u_{n-1})_{n-1}. \qquad \ldots\ldots \quad \ldots\ldots \quad (8)$$

It follows that $\lim_n X_n = 1$. Now we have from (ii) that

$$X_n=(u_{n-1})_o\cdot[1-(1/2^k)]+X_{n-1}\cdot[1-(1/4^k)]. \qquad \ldots\ldots \quad \ldots\ldots \quad (9)$$

Now let i be any positive integer and $k > i$. It follows from (ii) that

$$\sum_{j=i}^{k}[2(u_n)_{j+1}-(u_n)_j] = (1-\frac{1}{4^k}).\sum_{j=i}^{k}[2(u_{n-1})_{j+1}-(u_{n-1})_j].$$

Simplifying, we have:

$$\sum_{j=i+1}^{k}(u_n)_j - (u_n)_i = (1-\frac{1}{4^k}).[\sum_{j=i+1}^{k}(u_{n-1})_j - (u_{n-1})_i].$$

Notice that

$$(1-\frac{1}{4^k})^{a_{k+1}-a_k} \to 0 \text{ as } k \text{ goes to } \infty$$

so that

$$\lim_n \sum_{j=i+1}^{k}(u_n)_j - (u_n)_i = 0. \quad .... \quad .... (10)$$

Taking i=1 in (10), we then have as $n \to \infty$

$$X_n - 2(u_n)_1 \to 0 \text{ or } \lim_n (u_n)_1 = 1/2.$$

Similarly, taking i=2,3,...etc., we easily obtain $\lim (u_n)_j = 1/2^j$.

This establishes (7). □

3. Main results.

Our first theorem is a straightforward extension of a similar result in [7].

Theorem 3.1. Let (P_n) be a convergent stochastic chain with basis $\{T,C_1,C_2,...\}$. Then for i and j in two different C-classes,

$$\sum_{n=1}^{\infty}(P_n)_{ij} < \infty. \quad □$$

The proof is omitted since the proof given in [7] also works here.

Though the above result is nice, a stronger version of this result is desirable specially when there are an infinite number of C-classes. We would like to have for $i \in C_k$,

$$\sum_{n=1}^{\infty}\sum_{s \notin C_k \cup T}(P_n)_{is} < \infty. \quad .... \quad .... (11)$$

If we assume some kind of uniformity in convergence, then (11) is possible. That some condition is surely needed to obtain (11) follows from example (a) in section 2. The inequality (11) does not hold there. Let us revisit this example and prove its convergence.

In example (a), for $n > k$, each $i > 2$ and $i \neq n+2$, and $s \geq 2$

$$(P_{k,n})_{1i} = 0 = (P_{k,n})_{k+s,i} \quad ; \qquad \ldots\ldots \quad (12)$$

Also, we have

$$(P_{k,n+1})_{11} = (P_{k,n})_{11} \cdot (1-\tfrac{1}{n+1}) + \tfrac{1}{2} \cdot (P_{k,n})_{1,n+2} \,, \qquad \ldots\ldots \quad (12A)$$

$$(P_{k,n+1})_{12} = (P_{k,n})_{12} \cdot (1-\tfrac{1}{n+1}) + \tfrac{1}{2} \cdot (P_{k,n})_{1,n+2} .$$

It follows from (12) and (12A) that $\lim\ (P_{k,n})_{1,n+2} = 0$, and also

$$(P_{k,n+1})_{11} - (P_{k,n+1})_{12} = (1-\tfrac{1}{n+1})[(P_{k,n})_{11} - (P_{k,n})_{12}]. \qquad \ldots\ldots \quad (13)$$

It follows easily from (12), (12A) and (13) that

$$\lim\ (P_{k,n})_{11} = \lim\ (P_{k,n})_{12} = 1/2.$$

Similarly, for each $s \geq 2$, we have:

$$\lim\ (P_{k,n})_{k+s,1} = \lim\ (P_{k,n})_{k+s,2} = 1/2.$$

This proves the convergence of the chain in example (a). We will now show that (11) is guaranteed by the following uniformity condition, which holds trivially for finite dimensional matrices.

<u>Condition (U)</u>: A convergent chain (P_n) is said to satisfy condition (U), if for each j in each C-class

$$\lim_{k}\ \sup_{i \notin T}\ |(Q_k)_{ij} - Q'_{ij}| = 0,$$

where, as usual, Q' is a limit point of the Q_k's with $Q_k = \lim P_{k,n}$. □

Note that the chain in example (a) does not satisfy condition (U). Now we present our next theorem. It is worthwhile to point out that the proof of Theorem 3.1, though not given here, will be clear from the proof of this theorem.

<u>Theorem 3.2.</u> For every convergent stochastic chain, condition (U) implies (11). □

We will prove a lemma before we present its proof.

Lemma 3.3. Let (P_n) be a convergent stochastic chain with basis $\{T,C_1,C_2,\ldots\}$. Then the following hold:

(a) Let $i \in C_s$. Let $D_s = \cup\{C_p : p \neq s\}$. Then, $\lim_n \sum_{t \in D_s \cup T} (P_n)_{it} = 0$.

(b) Suppose condition (U) holds. Then given $\varepsilon > 0$, there exist integers $k(\varepsilon)$ and $n(\varepsilon)$ such that for $k \geq k(\varepsilon)$ and $n > \max\{k,n(\varepsilon)\}$,

$$\sum_{t \in D_s \cup T} (P_{k,n})_{dt} > 1-\varepsilon$$

for each $d \in D_s$. □

Proof. (a) Suppose there exists $c > 0$ such that there is a subsequence (n_p) and for each p, $\sum_{t \in D_s \cup T} (P_{n_p})_{it} > c$. Since for each j,

$$(P_{k,n_p-1})_{ii}(P_{n_p})_{ij} \leq (P_{k,n_p})_{ij},$$

we have:

$$(P_{k,n_p-1})_{ii} \cdot \sum_{j \in D_s \cup T} (P_{n_p})_{ij} \leq \sum_{j \in D_s \cup T} (P_{k,n_p})_{ij}. \quad \ldots\ldots (14)$$

Notice that given $\varepsilon > 0$, there exists $k(\varepsilon)$ such that for $k \geq k(\varepsilon)$,

$$(Q_k)_{ii} > \varepsilon \text{ and } \sum_{j \in D_s \cup T} (Q_k)_{ij} < 2\varepsilon c. \quad \ldots\ldots (15)$$

It follows from (14) and (15) that for large p, $\sum_{t \in D_s \cup T} (P_{n_p})_{it} < c$. This is a contradiction and (a) is established.

(b) For $k < m$, $Q_k = P_{k,m}Q_m$. Then for $i \in C_s$ and $d \in D_s$, we have:

$$(Q_k)_{di} \geq \sum_{j \in C_s} (P_{k,m})_{dj}(Q_m)_{ji}.$$

Choose $0 < \varepsilon < Q'_{ii}$. Notice that $Q'_{di} = 0$. Therefore, by condition (U), there exist integers $k(\varepsilon)$ and $n(\varepsilon)$ such that $n > n(\varepsilon)$, $n > k > k(\varepsilon)$ imply that

$$\tfrac{1}{2}\varepsilon \cdot Q'_{ii} \geq \tfrac{1}{2} \cdot Q'_{ii} \cdot \sum_{j \in C_s} (P_{k,n})_{dj}.$$

Part (b) of the lemma now follows. □

Proof of Theorem 3.2. Let (P_n) be a convergent chain with basis $\{T,C_1,C_2,\ldots\}$. Let $i \in C_j$ and D_j be $(T \cup C_j)^c$. Let (X_n) be the Markov chain induced by (P_n). Then for $k < n_1 < n_2 < n$, we have:

$$(P_{k,n})_{is} = \Pr(X_n = s \mid X_k = i) \geq \Pr\Big(\bigcup_{m=n_1}^{n_2} \{X_n = s, X_{m+1} \varepsilon\ D_j, X_m = i\} \Big| X_k = i\Big)$$

$$\geq \sum_{m=n_1}^{n_2} \Pr(X_n = s, X_{m+1} \varepsilon D_j, X_m = i \mid X_k = i)$$

$$- \sum_{n_1 \leq m < m' \leq n_2} \Pr(X_n = s, X_{m'+1} \varepsilon D_j, X_{m'} = i, X_{m+1} \varepsilon D_j, X_m = i \mid X_k = i)$$

$$= \sum_{\substack{m=n_1 \\ t \varepsilon D_j}}^{n_2} (P_{k,m})_{ii} (P_{m+1})_{it} (P_{m+1,n})_{ts}$$

$$- \sum_{\substack{n_1 \leq m < m' \leq n_2 \\ t, t' \varepsilon D_j}} (P_{k,m})_{ii} (P_{m+1})_{it} (P_{m+1,m'})_{ti} (P_{m'+1})_{it'} (P_{m'+1,n})_{t's}.$$

It is now clear that given $\varepsilon > 0$, there is a $c > 0$ such that for k, n_1, n_2 and n sufficiently large, we have:

$$\varepsilon > \sum_{s \varepsilon D_j \cup T} (P_{k,n})_{is}$$

$$\geq c. \sum_{\substack{m=n_1 \\ t \varepsilon D_j}}^{n_2} (P_{m+1})_{it} \Big(\sum_{s \varepsilon D_j \cup T} (P_{m+1,n})_{ts} \Big)$$

$$- \sum_{\substack{n_1 \leq m < m' \leq n_2 \\ t, t' \varepsilon D_j}} (P_{m+1})_{it} (P_{m'+1})_{it'}.$$

Now we use condition (U). By Lemma 3.3(b), we now have: Given $\varepsilon > 0$, there exists a positive integer $N(\varepsilon)$ such that for any $n_2 > n_1 > N(\varepsilon)$,

$$\varepsilon \geq \frac{c}{2}. \sum_{\substack{m=n_1 \\ t \varepsilon D_j}}^{n_2} (P_{m+1})_{it} - \frac{1}{2}. \Big(\sum_{\substack{m=n_1 \\ t \varepsilon D_j}}^{n_2} (P_{m+1})_{it} \Big)^2.$$

From the above inequality, it is not difficult to prove using Lemma 3.3(a) that (11) holds. □

Now we present another result of the same type as (11) using a condition different from condition (U).

<u>Theorem 3.4.</u> Let (P_n) be a convergent stochastic chain with basis $\{T, C_1, C_2, \ldots\}$. Suppose that there is a $i \varepsilon T$ and some C_j such that $\liminf_m \sum_{t \notin C_j} (Q_m)_{it} > 0$. Then for each t in C_j, we have:

$$\sum_{n=1}^{\infty} (P_n)_{ti} < \infty.$$

If the previous lim inf condition is strenghthened by

$$\liminf_{m} \left[\inf_{k\varepsilon T} \sum_{t\notin C_j} (Q_m)_{kt} \right] > 0,$$

then we can conclude that for each t in C_j,

$$\sum_{n=1}^{\infty} \sum_{k\varepsilon T} (P_n)_{tk} < \infty. \quad \square$$

We omit the proof since it follows easily from the fact that

$$\lim_{n} \sum_{s\varepsilon T} (P_n)_{ts} = 0$$

for each t in any C-class and an argument similar to the one used in the proof of Theorem 3.2.

In the finite dimensional case for a convergent chain (P_n), when we replace $(P_n)_{ij}$, for each n, by zero for all i,j belonging to two different C-classes, the resulting chain remains convergent with the same basis. This is a consequence of equivalence. Though the idea of equivalence does not carry over to the infinite case as our examples (d) and (e) showed, a similar result on convergence exists under condition (U). Our next theorem illustrates this.

<u>Theorem 3.5.</u> Let (P_n) be a convergent chain with condition (U). Define for each n, the nonnegative matrix P_n^* (not necessarily stochastic) by:

$$(P_n^*)_{ij} = 0 \text{ if } i,j \text{ belong to two different C-classes;}$$
$$= (P_n)_{ij}, \text{ otherwise.}$$

Then, for every positive integer k, $P_{k,n}^* = P_{k+1}^* P_{k+2}^* \cdots P_n^*$ converges to some Q_k^* as $n \to \infty$ and for every i and j,

$$\lim_{k\to\infty} [(Q_k)_{ij} - (Q_k^*)_{ij}] = 0. \quad \square$$

<u>Proof.</u> Let (X_n) be the Markov chain induced by (P_n). Now for any i,j, $0 \leq (P_{k,n}^*)_{ij} \leq (P_{k,n})_{ij}$ so that for $j\varepsilon T$, $\lim_{n} (P_{k,n}^*)_{ij} = 0$. Assume now that j belongs to some C-class C and let $D=(C\cup T)^c$. Notice that for any i, we can write:

$$(P_{k,n})_{ij} = \Pr(X_n=j \mid X_k=i)$$
$$= \Pr(X_n=j, X_s \in D \text{ for some } s,\ k<s<n \mid X_k=i)$$
$$+ \Pr(X_n=j, X_s \notin D \text{ for each } s \text{ with } k<s<n \mid X_k=i)$$
$$= (P_{k,n})^D_{ij} + \Sigma\, (P_{k+1})_{is_1}(P_{k+2})_{s_1s_2}\cdots(P_n)_{s_{n-k-1},j}\ ,$$

where the Ist term represents the probability of transition from i to j through a state in D and the summation in the second term is over all $s_1, s_2, \ldots, s_{n-k-1}$ in CUT. It follows that

(i) for $i \in CUT$,

$$(P_{k,n})_{ij} - (P^*_{k,n})_{ij} \leq (P_{k,n})^D_{ij} \qquad \ldots\ldots \quad \ldots\ldots (16)$$

(ii) for $i \in D$, $Q^!_{ij} = 0$ so that

$$\lim_n (P^*_{k,n})_{ij} \leq \lim_n (P_{k,n})_{ij} \to 0 \text{ as } k \to \infty.$$

Now it can be verified that

$$(P_{k,n})^D_{ij} \leq \sum_{m=k+1}^{n-1} (P_{k,m})^f_{iD} \cdot \sup_{d\in D} (P_{m,n})_{dj} \qquad \ldots\ldots \quad \ldots (17)$$

where

$$(P_{k,m})^f_{iD} = \Pr(X_m \in D,\ X_t \notin D \text{ for } k<t<m \mid X_k=i).$$

Since $\sum_{m=k+1}^{\infty} (P_{k,m})^f_{iD} \leq 1$, given $\varepsilon > 0$, there exists N such that

$$(P_{k,n})^D_{ij} \leq \sum_{m=k+1}^{N} (P_{k,m})^f_{iD} \cdot \sup_{d\in D} (P_{m,n})_{dj} + \varepsilon \quad \text{(for all n)}$$

It follows by condition (U) that if k is sufficiently large, then there exists a positive integer n(k) such that for $n > n(k)$,

$$(P_{k,n})^D_{ij} < 2\varepsilon. \qquad \ldots\ldots \quad \ldots\ldots \quad (18)$$

From (16) and (18), and from Theorem 3.2, we now have:

''Given $\varepsilon > 0$, there exists k_o such that for $k > k_o$ and $n > n(k)$, $(P_{k,n})_{ij} - (P^*_{k,n})_{ij} < \varepsilon$. The integers k_o and n(k)'s depend on i and j. ''

It is now clear that the theorem will be proven once we show that for each k, $\lim_n P^*_{k,n} = Q^*_k$ exists. To this end, notice that

$$(P^*_{k,n} - P^*_{k,n'})_{ij} = \sum_s (P^*_{k,m})_{is}[P^*_{m,n} - P^*_{m,n'}]_{sj} \qquad \ldots\ldots \quad \ldots\ldots (19)$$

Given $\varepsilon > 0$, there exists m_o and N such that $m \geq m_o$ =>

$$\sum_{s=1}^{N} (P_{k,m})_{is} > 1-(\varepsilon/2),$$

which means

$$\sum_{s=N+1}^{\infty} (P^*{}_{k,m})_{is} < \varepsilon/2. \qquad \text{.... .... (20)}$$

By the assertion following (18), there also exists $m_1 \geq m_o$ such that for $m > m_1$ and $n > n(m)$, we have:

$$(P_{m,n})_{sj}-(P^*_{m,n})_{sj} < \varepsilon/2^N \qquad \text{.... .... (21)}$$

for s=1,2,...,N.

From (19), (20), (21) and the fact that $\lim_n P_{m,n}$ exists, the convergence of $P^*_{k,n}$ as $n \to \infty$ follows. □

4. An application to measures on semigroups.

Though the discussion below can be given in the context of countable infinite semigroups, for simplicity we restrict ourselves to a finite semigroup.

Let(β_n) be a sequence of probability measures on a finite semigroup S. Suppose that for each $k \geq 1$, the sequence $\beta_{k,n}=\beta_{k+1}\cdots\beta_n$ (convolution being the measure multiplication here) converges weakly to π_k as $n \to \infty$. Any weak limit point of the sequence (π_n) will be called a tail limit. For any two tail limits π' and π'', $\pi'\pi''=\pi'$. Also, the union A of the supports of all the tail limits is a completely simple subsemigroup of S. (See Theorem 4.4, [4].) When S is a group, A is a group and $\lim_n \pi_n$ exists and $\sum_{n=1}^{\infty} \beta_n(S-A) < \infty$. This is no longer true when S is a semigroup and the situation is less trivial. This case was studied in [5] when S has an identity and it was assumed there that $\lim_n \pi_n = \pi$ exists and S_π, the support of π, contains the identity. Here we relax these assumptions and study the general situation.

Definition 4.1. Let π' be an arbitrary (but fixed in our discussion) tail limit of the π_n's. Let $H=S_{\pi'}$. For a,b in S, we will say aRb iff $a^{-1}b \cap H$ is non-empty. Here, $a^{-1}b=\{x \in S: ax=b\}$. □

Let $S_T=\{x\varepsilon S:(x,x)\notin R\}$. Then $S-S_T$ is a non-empty left ideal of S; non-empty since H, being completely simple, contains idempotents e so that $e^{-1}e\cap H$ is non-empty. The relation R is transitive in all of S, and in $S-S_T$, this relation is also symmetric. The symmetry property needs some explanation as follows:

Let $S=\{a_1,a_2,\ldots,a_s\}$. With no loss of generality, we can assume that S has an identity; for, if not, we can always add an identity. For each $a_k \varepsilon S$, we correspond the 0-1 stochastic matrix (s by s) A_k such that $(A_k)_{ij}=1$ if $a_ia_k=a_j$, $=0$ otherwise. Then the correspondence $a_k \rightarrow A_k$ is an isomorphism from S into the semigroup of s by s stochastic matrices. We define the stochastic matrix Q by

$$Q=\sum_{i=1}^{s}\pi'(a_i).A_i.$$

Then Q is idempotent with basis, say, $\{T,C_1,C_2,\ldots,C_p\}$. Notice that $Q_{kt}=\pi'(a_k^{-1}a_t)$ and therefore, $\pi'(a_k^{-1}a_t)=0$ whenever k and t belong to two different C-classes. Now we define the sets $S_T, S_1,\ldots,S_p$ by

$$S_T=\{a_k:k\varepsilon T\} \text{ and } S_i=\{a_k:k\varepsilon C_i\},\ 1\leq i\leq p.$$

Note that this S_T happens to be the same S_T defined earlier. Since for $i\notin T$ and $j\notin T$, $Q_{ij} > 0$ iff $Q_{ji} > 0$, it is clear that the relation R is symmetric in $S-S_T$. Clearly, the partition of $S-S_T$ by the equivalence relation R is $\{S_1,S_2,\ldots,S_p\}$. (Let us remark that if we started with a tail limit different from π', we would still get the same partition since the idempotent matrix corresponding to the new tail limit will also have the same basis.)

Let us now consider the partition $\{S_T,S_1,S_2,\ldots,S_p\}$. Let S^* be defined by

$$S^*=\{x\varepsilon S:\ S_ix \subset S_i\cup S_T \text{ for each } i,\ 1\leq i\leq p\}.$$

We claim that

$$A \subset S^*. \qquad \ldots\ldots \quad \ldots\ldots (22)$$

To verify (22), let $x \varepsilon A$. Then x is in the support of some tail limit π''. If $x \notin S^*$, there exist i,j $(i\neq j)$, $1\leq i,j\leq p$, such that $S_ix\cap S_j$ is non-empty so that $x \varepsilon a_k^{-1}a_t$ for some $k \varepsilon C_i$ and some $t \varepsilon C_j$.

This means that

$$\pi''(x) \leq \pi''(a_k^{-1}a_t) = 0,$$

since the same partition, as we have remarked above, works for all tail limits. This gives us a contradiction and (22) follows.

We now make our final claim that

$$\sum_{n=1}^{\infty} \beta_n(S-S^*) < \infty. \qquad \ldots\ldots \qquad \ldots\ldots \qquad (23)$$

To verify (23), we follow the same method as before and for each β_n, we define a stochastic matrix P_n such that

$$(P_n)_{kt} = \beta_n(a_k^{-1}a_t).$$

Then it is easy to verify that the matrix products $P_{k,n}$ correspond to $\beta_{k,n}$, and the chain (P_n) becomes a convergent stochastic chain with the same partition $\{T, C_1, C_2, \ldots, C_p\}$ as the basis. Now (23) follows immediately from Theorem 3.1.

We make one final remark. It is now clear that it is possible to obtain many similar results more general than presented above by following the same method and utilizing our other results in section 3.

REFERENCES

0. Isaacson,D. and Madsen,R.(1976). Markov Chains: Theory and Applications, Wiley, New York.

1. Blackwell,D.(1945). Finite nonhomogeneous Markov chains. Ann. Math. 46, 594-599.

2. Cohn,H.(1970). On the tail sigma-algebra of finite inhomogeneous Markov chains. Ann. Math. Statist. 41, 2175-2176.

----.(1974). A ratio limit theorem for finite nonhomogeneous Markov chains. Israel J. Math. 19, 329-334.

----.(1976). Finite nonhomogeneous Markov chains: Asymptotic Behavior. Adv. in App. Probab. 8, 502-516.

----. (1977). Countable nonhomogeneous Markov chains: Asymptotic Behavior. Adv. in App. Probab. 9, 542-552.

3. Maksimov, V.M.(1970). Convergence of nonhomogeneous bistochastic Markov chains. Theor. Probability Appl. 15, 604-618.

4. Mukherjea, A.(1979). Limit theorems: Stochastic matrices, ergodic Markov chains, and measures on semigroups, Prob. Analysis and Related Topics, Vol. 2, Academic Press, 143-203.

5. ----.(1980). A new result on the convergence of nonhomogeneous stochastic chains. Trans. Amer. Math. Soc. 262, No.2, 505-520.

6. ---- and R. Chaudhuri. (1981). Convergence of nonhomogeneous stochastic chains II. Math. Proc. Cambridge Phil. Soc.

7. ---- and A. Nakassis. (1981). Convergence of nonhomogeneous stochastic chains III. Submitted.

8. Senchenko, D.V.(1972). The final sigma-algebra of an inhomogeneous Markov chain with a finite number of states. Math. Notes 12, 610-3.

9. Seneta, E.(1973). Nonnegative matrices. Wiley, New York.

A. Mukherjea
University of South Florida
Tampa, Fla 33620, USA

A. Nakassis
8107 Fallow Drive
Gaithersburg, MD 20877, USA

INFINITE CONVOLUTION AND SHIFT-CONVERGENCE OF MEASURES ON TOPOLOGICAL GROUPS

Imre Z. Ruzsa
Mathematical Institute of the
Hungarian Academy of Sciences,
Budapest, Hungary.

ABSTRACT. Kloss' "general principle of convergence", until now available for first-countable groups, is established for a wider class including e.g. all locally compact commutative groups. The proof combines Csiszár's method with a Fourier analytic technique.

0. Conventions

We shall deal with tight probability measures on Hausdorff topological groups; they will be called distributions. The topology on them will always be the weak topology.

We shall use nets, which are indexed by elements of a directed set. Directed sets will generally be denoted by D or I, and their elements by k,m,n. Directed sets will be assumed to have a minimal element, which will be denoted by 0.

The convolution of distributions will be denoted simply by juxtaposition. The point mass at g is denoted by $\delta(g)$, the normed Haar measure on a compact group H by $\omega(H)$.

The unit in any group will be denoted by 1.

The right concentration function Q_r is defined by

$$Q_r(\mu,X) = \sup_{g\in G} \mu(gX) = \sup_{g\in G}(\delta(g)\mu)(X),$$

where G is the group at issuse and X is any Borel subset of G. The left concentration function Q_ℓ is defined analogously.

A topological space is M_1, (resp. M_2) if every point (resp. the whole space) has a countable basis of neighbourhoods (first and second axiom of countability).

1. Introduction

The main object of this paper is to investigate the convergence of an infinite convolution of distributions on a Hausdorff topological group G. Let

$$\nu_n = \mu_1\mu_2\cdots\mu_n.$$

The infinite convolution (i.e. the sequence ν_n) may be a) convergent, b) divergent due to wrong centering, when

$$\nu_n\delta(g_n) \to \nu \tag{1.1}$$

for some distribution ν with suitable $g_n \in G$, and c) essentially divergent, when no centering makes it converge. Certainly this is the case if for every compact $K \subset G$ we have

$$\lim Q_\ell(\nu_n, K) = 0. \tag{1.2}$$

That is the only possibility for such an essential divergence was proved for locally compact M_2 groups, generalizing a result of Kloss (1961), by Csiszár 1966 . Tortrat (1970) proved (1.1) assuming the uniform tightness of the measures ν_n (which is not much stronger than the negation of (1.2)) for arbitrary M_1 groups.

I arrived at this problem independently and found a Fourier-analytic proof of this result for compact commutative groups. This idea is exposed in section 9. Later I noticed that by combining my method with Csiszár's it is possible to achieve a satisfactory generality among commutative groups, namely to prove the theorem for a class of groups that contains all M_1 and all locally compact groups and is closed under (finite) direct product. I am quite sure that it is valid for every group. For noncommutative groups I can scarcely go beyond M_1 groups; the problem is left open even for compact non-commutative groups.

The core of the problem is the following. Let $\mathcal{D}$ be the convolution semigroup of distributions and (in the commutative case) $\mathcal{D}^*$ its factor-semigroup with respect to the group E of point masses, $\phi:\mathcal{D} \to \mathcal{D}^*$ the natural homomoprhism. By the classical methods we can prove the convergence of $\psi(\nu_n)$ in $\mathcal{D}^*$. Now we arrive to the central

PROBLEM OF SHIFT-CONVERGENCE. If ν_n is a sequence of distributions and $\psi(\nu_n) \to \psi(\nu)$, does it follow that $\nu_n\delta(g_n) \to \nu$ for a suitable choice of $g_n \in G$?

This innocent-looking problem is probably deep. It is easy for M_1 group and can be proved for compact commutative groups analytically. Even the case $\nu = \delta(1)$ is not trivial. It can be reformulated in the following way. If $Q_\ell(\nu_n, U) \to 1$ for every open U, does it follow that $\nu_n\delta(g_n) \to \delta(1)$ for suitable g_n ? Another equivalent form of the condition is $\nu_n\bar{\nu}_n \to \delta(1)$, where $\bar{\mu}$ is the conjugate measure of μ defined by

$$\bar{\mu}(X) = \mu(X^{-1}). \tag{1.3}$$

One has the feeling that one must simply to choose the center g_n at random with distribution $\bar{\nu}_n$; this, however, does not work (at least not trivially) unless the group is M_1.

The structure of the paper is the folloiwng. After stating our main result exactly (Th.1. of Sec.2) and summing up some preliminaries (Sec.2-4), we build up the space $\mathcal{D}^*$ in the noncommutative case (Sec. 5-6) and show how a partial answer to the problem of shift-convergence can be applied to infinite convolutions (Th. 2 of Sec.8). This partial answer (Th. 3 of Sec. 10) is given in Sections 8-10.

2. The main result

We shall consider a generalized version of infinite products. By a D-product, where D is a directed set, we mean a system $(a_m^n)_{m<n,m,n\in D}$ of elements (from a set with an operation called multiplication) satisfying

$$a_k^m a_m^n = a_k^n \quad (k<m<n).$$

In the case also a topology is defined on this set, we call the above D-product convergent, if the net $(a_m^n)_{n\in D}$ is convergent for every fixed $m\in D$. Note that for ordinary products this is a bit stronger than usual convergence (this is called "composition convergence" by Heyer (1977), def. 2.3.3.).

Two D-products of measures, (μ_m^n) and (ν_m^n), will be called associates, if for suitable $g_n \in G$ the relation

$$\nu_m^n = \delta(g_m^{-1})\mu_m^n\delta(g_n)$$

holds for all m,n.

THEOREM 1. Let G be a topological group and (ν_m^n) a D-product of distributions on G. Suppose that G has a normal commutative compact subgroup H such that G/H is M_1. Then either some associate product of (ν_m^n) is convergent, or

(2.1) $$\lim_n Q_\ell(\nu_m^n, K) = 0$$

for every fixed $m\in D$ and compact $K\subset G$.

If (2.1) holds, we shall call the product dispersing.

ADDENDUM. Locally compact commutative groups fall within the scope of Theorem 1, i.e. every locally compact commutative group G has a compact subgroup H such that G/H is M_1.

This Addendum is well-known and easily follows from the fundamental theorem of compactly generated groups, see e.g. Hewitt-Ross (1963), Th.9.6.

3. Preliminaries

A net $(y_k)_{k\in I}$ is a subnet of the net $(x_n)_{n\in D}$ if there is a function $f:I\to D$ such that $f(k)\to\infty$ as $k\to\infty$ and $y_k = x_{f(k)}$. If in a topological space x is a cluster point of the net (x_n), then there is a subnet converging to x; moreover, the function f can be chosen to be increasing. See Kelley (1955).

A set M of distributions in tight, if for every $\varepsilon>0$ there exists a compact $K_\varepsilon\subset G$ such that $\mu(K_\varepsilon)>1-\varepsilon$ for all $\mu\in M$.

(3.1) A tight set is relatively compact. (See Topsøe (1970).)

(3.2) If $\lambda_i=\mu_i\nu_i$ for $i\in I$ (I an arbitrary indexing set) and two of the sets $\{\mu_i\}$, $\{\nu_i\}$, $\{\lambda_i\}$ is tight, then so is the third.

A net (μ_n) of distributions is tight (quasitight) if for every $\varepsilon>0$ there is a compact $K_\varepsilon\subset G$ such that

$$\liminf \mu_n(K_\varepsilon) > 1-\varepsilon,$$

resp.if

$$\liminf \mu_n(U) > 1-\varepsilon$$

for every open $U\supset K_\varepsilon$.

(3.3) A quasitight net has cluster points.

(3.4) A net is convergent if and only if it is quasitight and has at most one cluster point.

(3.5) If $\lambda_n=\mu_n\nu_n$ and two of the nets $(\mu_n),(\nu_n),(\lambda_n)$ is (quasi) tight, then so is the third.

For (3.3-5), see Siebert (1976).

(3.6) If $\mu\nu=\nu$ (μ,ν distributions), then there is a compact subgroup H such that $\omega(H)\nu=\nu$ and $\mu\omega(H)=\omega(H)$. See Tortrat (1965).

(3.7) COROLLARY. The only idempotent distributions are the $\omega(H)$'s.

4. Divisibility in $\mathcal{D}$

We say that μ is a right divisor of ν and write $\mu|_r\nu$ if $\rho\mu = \nu$ for some distribution ρ. We say that μ and ν are right associates and write $\mu\sim_r\nu$ if both $\mu|_r\nu$ and $\nu|_r\mu$.

REMARK. All our "right" concepts and results have, of course, their "left" analogues, which will be sometimes used without any speical mentioning. In fact, the right and left structures of $\mathcal{D}$ are isomorphic; an isomorphy is established by the involution $\mu \to \bar{\mu}$, defined by (1.3).

There is no natural principle to decide which of the corresponding concepts should be the "right" and which the "left". In chosing the names I tried to be consistent (e.g. the right concentration coincides for right associates); to this end I sometimes departed from the common usage.

(4.1) LEMMA. *$\mu\sim_r\nu$ if and only if $\mu = \delta(g)\nu$ for some $g\in G$.*

PROOF. Immediately follows from (3.6).

Now we establish the continuity of divisibility.

(4.2) PROPOSITION. *If for two nets (μ_n) and (ν_n) we have $\mu_n|_r\nu_n$, $\mu_n \to \mu$ and $\nu_n \to \nu$, then $\mu|_r\nu$.*

PROOF. We use (3.3-5). (μ_n) and (ν_n), being convergent, are quasitight, thus if $\rho_n\mu_n = \nu_n$, so is (ρ_n) and hence it has a cluster point ρ. Now we have obviously $\rho\mu = \nu$.

(4.3) REFORMULATION. *The set $\{(\mu,\nu):\mu|_r\nu\}$ is closed in $\mathcal{D}\times\mathcal{D}$.*

(4.4) COROLLARY. *The set $\{(\mu,\nu):\mu\sim_r\nu\}$ is closed in $\mathcal{D}\times\mathcal{D}$. For a fixed ν the sets $\{\mu:\mu|_r\nu\}$ and $\{\mu:\nu|_r\mu\}$ are closed in $\mathcal{D}$.*

5. The factor-space $\mathcal{D}_r^*$

By regarding the equivalence classes of $\sim_r$ we obtain the (right) factor-space $\mathcal{D}_r^*$. We consider the factor-topology on it. Let $\phi_r:\mathcal{D} \to \mathcal{D}_r^*$ be the natural homomorphism.

Convolution is, generally, incompatible with $\sim_r$ (unless G is commutative), thus $\mathcal{D}_r^*$ will not be a semigroup. Right divisibility is, however, compatible, thus we can define a divisibility in $\mathcal{D}_r^*$: $\alpha|\beta$ if $\mu|_r\nu$ for $\phi_r(\mu) = \alpha$, $\phi_r(\nu) = \beta$. Divisibility is antisymmetric on $\mathcal{D}_r^*$ by definition: if $\alpha|\beta$ and $\beta|\alpha$, then $\alpha = \beta$.

(5.1) LEMMA. *ϕ_r is open.*

PROOF. Let $V \subset \mathcal{D}$ be open. We have

$$\phi_r^{-1}(\phi_r(V)) = \bigcup_{g\in G} \delta(g)V ,$$

also an open set.

REMARK. It is not closed even when G is the additive group of reals.

(5.2) LEMMA. <u>The sets</u>

$$\{(\alpha,\beta):\alpha,\beta\in\mathcal{D}_r^*,\ \alpha|\beta\},\quad \{(\alpha,\alpha):\alpha\in\mathcal{D}_r^*\}$$

<u>and for a fixed</u> $\beta\in\mathcal{D}_r^*$ <u>the sets</u>

$$\{\alpha:\alpha|\beta\},\quad \{\alpha:\beta|\alpha\}$$

<u>are closed</u>.

PROOF. This follows from (4.3-4).

(5.3) LEMMA. $\mathcal{D}_r^*$ <u>is Hausdorff.</u>

PROOF. This is a reformulation of the second assertion of the previous lemma.

A concentration function can also be defined on $\mathcal{D}_r^*$: let $Q(\alpha,X)$ be the common value of $Q_r(\mu,X)$ for $\psi_r(\mu)=\alpha$. Call a set $A\subset\mathcal{D}_r^*$ <u>tight,</u> if for every $\varepsilon>0$ there is a compact $K_\varepsilon\subset G$ such that $Q(\alpha,K_\varepsilon)>1-\varepsilon$ for all $\alpha\in A$.

(5.4) LEMMA. $A\subset\mathcal{D}_r^*$ <u>is tight if and only if there is a tight</u> $M\subset\mathcal{D}$ <u>such that</u> $A=\psi_r(M)$.

PROOF. Sufficiently is obvious. To prove the necessity, write $K_{1/2}=K$ and for each $\alpha\in A$ choose a $\mu\in\psi_r^{-1}(\alpha)$ such that $\mu(K)>$ $>1/2$; let M be the set of these μ 's. For $\mu\in M$ and $\varepsilon<1/2$ there is a g such that $\mu(gK_\varepsilon)>1-\varepsilon$. K and gK_ε cannot be disjoint, hence $g\in KK_\varepsilon^{-1}$ and thus $gK_\varepsilon\subset KK_\varepsilon^{-1}K_\varepsilon=K'_\varepsilon$. K'_ε is compact and $\mu(K'_\varepsilon)>1-\varepsilon$, which just shows the tightness of M.

(5.5) COROLLARY. <u>A tight subset of</u> $\mathcal{D}_r^*$ <u>is relatively compact</u>.

PROOF. A consequence of the previous and (3.1).

(5.6) PROPOSITION. <u>The set of divisors of a fixed</u> $\alpha\in\mathcal{D}_r^*$ <u>is compact</u>.

PROOF. Since $\beta|\alpha$ implies $Q(\beta,X)\geq Q(\alpha,X)$, it is tight, thus relatively compact, and it is closed by (5.2).

Now we show that convergence in $\mathcal{D}_r^*$ is closely connected with a phenomenon that frequently occurs in papers on infinite convolution.

(5.7) PROPOSITION. <u>Let</u> $(\mu_n)_{n\in D}$ <u>be a quasitight net of distributions</u>. <u>The following conditions are equivalent:</u>

(a) $\psi_r(\mu_n)\to\psi_r(\mu)$,

(b) <u>every cluster point of</u> (μ_n) <u>is of the form</u> $\delta(g)\mu$.

PROOF. (a) ⇒ (b) evidently. Now assume (b). If $\psi_r(\mu_n)$ does not tend to $\psi_r(\mu)$, then there is a subnet $(\nu_i)_{i\in I}$ of (μ_n) such that $\psi_r(\mu)$ is not a cluster point of $\psi_r(\nu_i)$. (ν_i), being quasitight, has a convergent subnet $(\lambda_j)_{j\in J}$. By (b), $\lim\lambda_j=\delta(g)\mu$ for some

g and hence $\psi_r(\lambda_j) \to \psi_r(\mu)$, contradicting the choice of (ν_i).

(5.8) PROBLEM. If (α_n) is a convergent net in $\mathcal{D}_r^*$, can one select $\mu_n \in \psi_r^{-1}(\alpha_n)$ to form a quasitight net? This is weaker than the problem of shift-convergence.

One can define quasitight nets in $\mathcal{D}_r^*$ by the condition that for every $\varepsilon > 0$ there is a compact $K_\varepsilon \subset G$ such that

$$\lim\inf Q(\alpha_n, U) > 1-\varepsilon$$

for every open $U \supset K_\varepsilon$. Is every quasitight net in $\mathcal{D}_r^*$ the image of a quasitight net of $\mathcal{D}$?

6. Monotonic nets in $\mathcal{D}_r^*$

Call a net (α_n) in $\mathcal{D}_r^*$ dispersing, if

$$\lim Q(\alpha_n, K) = 0$$

for every compact $K \subset G$. A net (α_n) is decreasing, resp. increasing, if $\alpha_n | \alpha_m$, resp. $\alpha_m | \alpha_n$ whenever $m<n$. The aim of this section is to prove the following

(6.1) PROPOSITION. An increasing net in $\mathcal{D}_r^*$ is either convergent or dispersing; a decreasing net is always convergent.

REMARK. If (ν_m^n) is a D-product of distributions, for a fixed m the net $(\psi_\ell(\nu_m^n))$ is increasing in $\mathcal{D}_\ell^*$. Hence the above proposition and a positive answer to the problem of shift-convergence yield the convergence of $\nu_m^n \delta(g_n)$ for a fixed m with suitable g_n. A deeper connection of this type will be established in the following section.

(6.2) LEMMA. Let $K \subset G$ be compact and $Q_r(\mu, K) \leq c$. Then there exists a compact $K' \subset G$ such that for every distribution ν and $g \in G$ we have

$$\nu\mu(gK) \leq c-c(1-\nu(gK'))/2.$$

This is a reformulation of Lemma 1.2 of Csiszár (1966).

(6.3) LEMMA. An increasing net in $\mathcal{D}_r^*$ is either dispersing or forms a tight set.

PROOF. Let $\mu_n \in \psi_r^{-1}(\alpha_n)$ and

$$\lim Q_r(\mu_n, K) = c > 0$$

for some compact $K \subset G$. Choose an $\varepsilon > 0$ and an m such that

$$Q_r(\mu_m, K) < c+\varepsilon .$$

For $n>m$ write $\mu_n = \rho_n \mu_m$. Applying Lemma (6.2) we obtain

$$c \leq Q_r(\mu_n, K) \leq (c+\varepsilon)(1-(1-Q_r(\rho_n, K'))/2).$$

whence

$$Q_r(\rho_n, K') \leq (c-\varepsilon)/(c+\varepsilon) .$$

Let K'' be a compact set with $\mu_m(K'') > 1-\varepsilon$ and $K^* = K'K''$. From the previous formula

$$Q_r(\mu_n, K^*) \geq (c-\varepsilon)/(c+\varepsilon)-\varepsilon = 1-\delta \qquad (n>m) \tag{6.4}$$

where $\delta \to 0$ as $\varepsilon \to 0$. μ_n being increasing, $Q_r(\mu_n, X)$ is decreasing for arbitrary X thus (6.4) holds for all n and the proof is completed.

PROOF of the Proposition. Let (α_n) be an increasing and not dispersing net. Its terms form a tight set by the preceding lemma, thus it is sufficient to prove that it has at most one cluster point. Let β and γ be two cluster points. For a fixed m, β is a cluster point of the net $(\alpha_n)_{n>m}$, thus it is divisible by α_m. This holds for every m, hence by the continuity of divisibility $\gamma | \beta$. Similarly we obtain $\beta | \gamma$ and hence $\beta = \gamma$.

If the net is decreasing, it forms a tight net obviously and the proof can be completed similarly.

7. Shift-convergence and convergent products

We say that a pair (G,D), where G is a topological group and D a directed set, has the property of shift-convergence if the following assumption holds:

(SC) If $(\mu_n)_{n \in D}$ is a net of distributions on G and $\psi_r(\mu_n) \to \psi_r(\mu)$, then $\delta(g_n)\mu_n \to \mu$ for suitable $g_n \in G$.

We shall consider also two weaker versions of (SC).

(SCH):(SC) when μ is a Haar measure on a compact subgroup of G.

(SCDH):(SC) when μ is Haar and moreover (μ_n) is right-decreasing.

Obviously (SC) $\Rightarrow$ (SCH) $\Rightarrow$ (SCDH). Our aim is to prove

THEOREM 2. *If (SCDH) holds for a pair* (G,D), *then every D-product*

<u>of distributions on G is either dispersing or associate to a convergent product.</u>

REMARK. I can prove the converse when D is linearly ordered. Namely in that case for a given right-decreasing net (μ_n) one can find a D-product (ν_m^n) satisfying

$$\nu_m^n \mu_n = \mu_m \quad (m<n).$$

The problem is not the size but the structure of D; if such ν's can be found for the finite parts of D, then a compactness argument provides them for the whole D.

PROOF. Regard a D-product (ν_m^n). The net (ν_o^n) is left--increasing, thus from Lemmas (6.3) and (5.4) it follows that if it is not dispersing, then for suitable g_n the set $\{\nu_o^n \delta(g_n)\}$ is tight. Now replace (ν_m^n) by its associate $(\delta(g_m^{-1})\nu_m^n \delta(g_n))$; for simplicity we keep the same symbol (ν_m^n) for this new product. We know that $\{\nu_o^n\}$ is a tight set and hence using (3.5) easily follows that so is the set $\{\nu_m^n : m, n \in D, m<n\}$.

Now, like the proof of Theorem 5.2.1. of Csiszár (1966), we can conclude that there exists a directed set I and a (monotonic, cofinal) function $f: I \to D$ such that the limits

$$\lim_{i \in I} \nu_m^{f(i)} = \tilde{\nu}_m, \qquad \lim_{i \in I} \tilde{\nu}_{f(i)} = \tilde{\nu}$$

exist and

$$\nu_k^m \tilde{\nu}_m = \tilde{\nu}_k, \quad \tilde{\nu}_k \tilde{\nu} = \tilde{\nu}_k, \quad \tilde{\nu}^2 = \tilde{\nu}.$$

Thus $(\tilde{\nu}_n)$ is a right-decreasing net, $\tilde{\nu}$ is a Haar-measure and it is a cluster point of this net; hence by Proposition (6.1) we obtain

$$\phi_r(\tilde{\nu}_n) \to \phi_r(\tilde{\nu})$$

and from the assumed (SCDH)

$$\delta(g_n)\tilde{\nu}_n \to \tilde{\nu}$$

for suitable $g_n \in G$.

Now define

$$\lambda_m^n = \delta(g_m)\nu_m^n \delta(g_n^{-1}), \qquad \tilde{\lambda}_n = \delta(g_n)\tilde{\nu}_n, \qquad \tilde{\lambda} = \tilde{\nu}.$$

We know that $\tilde{\lambda}$ is a Haar measure,

(7.1) $$\lambda_m^n \tilde{\lambda}_n = \tilde{\lambda}_m \quad (m<n),$$

(7.2) $$\lambda_k \tilde{\lambda} = \tilde{\lambda}_k, \quad \tilde{\lambda}_k \to \tilde{\lambda} .$$

Our aim is to deduce the convergence

(7.3) $$\lim_n \lambda_m^n = \tilde{\lambda}_m$$

for every fixed $m \in D$.

$(\tilde{\lambda}_n)$, being convergent, is quasitight, thus by (7.1) and (3.5), so is (λ_m^n) for a fixed m. Hence to prove (7.3) it is sufficient to show that it cannot have other cluster points than $\tilde{\lambda}_m$.

For a fixed m (λ_m^n) is left-increasing, hence $\psi_\ell(\lambda_m^n)$ converges. We have

$$\psi_\ell(\lambda_m^n) = \psi_\ell(\delta(g_m)\nu_m^n\delta(g_n^{-1})) = \psi_\ell(\delta(g_m)\nu_m^n).$$

$\tilde{\nu}_m$ is a cluster point of the net $(\nu_m^n)_{n\in D}$, hence $\tilde{\lambda}_m$ is a cluster point of $(\delta(g_m)\nu_m^n)$, therefore we must have

$$\psi_\ell(\lambda_m^n) \to \psi_\ell(\tilde{\lambda}_m) .$$

Consequently any cluster point of (λ_m^n) must be of the form $\rho = \tilde{\lambda}_m\delta(g)$. (7.1) yields $\rho\tilde{\lambda} = \tilde{\lambda}_m$, i.e.

$$\tilde{\lambda}_m\delta(g)\tilde{\lambda} = \tilde{\lambda}_m .$$

By (3.6) this means the existence of a compact subgroup H such that

$$\tilde{\lambda}_m\omega(H) = \tilde{\lambda}_m , \quad \omega(H)\delta(g)\tilde{\lambda} = \omega(H) .$$

The latter equality shows $g \in H$, hence by the first

$$\rho = \tilde{\lambda}_m\delta(g) = \lambda_m\omega(H)\delta(g) = \tilde{\lambda}_m\omega(H) = \tilde{\lambda}_m .$$

This completes the proof of Theorem 2.

8. Shift-convergence in first-countable groups

Now we start the final part of the paper, whose aim is to establish (SC) and (SCH) for certain groups.

(8.1) PROPOSITION. (SC) holds in first-countable groups.

(8.2) LEMMA. If G is M_1, so is $\mathcal{D}(G)$.

PROOF. An M_1 group is metrizable, and fixing a metric we can regard the induced Prohorov metric on the set of measures. By a theorem of Varadarajan (1961) (see also Billingsley (1968)), on the space of separable measures (i.e. measures with a separable support; this is a weaker condition than tightness) this induces the weak topology, thus $\mathcal{D}(G)$ is even metrizable.

REMARK. It is not clear whether the weak topology on the set of distributions on an arbitrary M_1 space is M_1; I doubt it. The analogous statement for M_2 property is valid, see e.g. Topsøe (1970).

PROOF of the Proposition. Assume $\psi_r(\mu_n) \to \psi_r(\mu)$. Let $\{U_1, U_2, \ldots\}$ be a base of neighbourhoods of μ. By (5.1) $\{\psi_r(U_j)\}$ is a base of neighbourhoods of $\psi_r(\mu)$. Put $U_\infty = \{\mu\}$. For every $n \in D$ there exists a maximal $j = j(n)$ (a natural number or ∞) with the property that

$$\psi_r(\mu_n) \in \psi_r(U_j) .$$

Thus there exists a $g_n \in G$ such that

$$\delta(g_n)\mu_n \in U_{j(n)}$$

and with this g_n we have obviously

$$\delta(g_n)\mu_n \to \mu .$$

9. An asymptotical equation for characters

Let G be a commutative group and $(f_n)_{n\in D}$ a net of complex-valued functions on G satisfying

$$\text{(9.1)} \qquad \liminf_n |f_n(g)| > 0$$

for every $g \in G$. We want to find characters γ_n of G such that the limit

$$\text{(9.2)} \qquad \lim_n f_n(g)\gamma_n(g) = \chi(g)$$

exists for all $g \in G$.

REMARK. The group for which the solution to the above problem will be applied will be the group of characters of a commutative subgroup of the original G. This is really a method of finding elements in a compact commutative group with prescribed asymptotical properties. I do not know whether it can be modified to produce continuous characters, which would make it possible to locate elements in locally compact commutative groups.

(9.3) LEMMA. <u>A necessary and sufficient condition for the existence of a net (γ_n) of characters such that the limits (9.2) exist is the convergence of</u>

$$f_n(g_1)f_n(g_2)\overline{f_n(g_1g_2)} \tag{9.4}$$

<u>for all</u> $g_1, g_2 \in G$. <u>The</u> γ_n <u>'s can be chosen so that</u> $\chi(g)=1$ <u>for all</u> g <u>if and only if the limit of (9.4) is identically one</u>.

PROOF. Necessity is obvious.

To prove sufficiency, first setting $g_1=g_2=1$ in (9.4) we obtain that $f_n(1)$ converges. The limit is different from zero by (9.1), thus we may replace $f_n(g)$ by $f_n(g)/f_n(1)$, hence we may assume $f_n(1)=1$. Next substituting $g_2=1$, g_1 arbitrary, we obtain the convergence of $|f_n(g)|$, hence we may change $f_n(g)$ into $f_n(g)/|f_n(g)|$ and consequently assume $|f_n(g)|=1$. (If the denominator happens to be 0, put $f_n(g)=1$.) Setting $g_2=g_1^{-1}$ we see that $f_n(g)f_n(g^{-1})$ converges. Now it is easy to check that if $\chi(g)$ exists for g_1 and g_2, then so does it for $g_1g_2^{-1}$, i.e. the g 's for which $\chi(g)$ exists with a given (γ_n) form a group.

Now consider the pairs $W=(S, (\eta_n)_{n\in D})$, where S is a subgroup of G and the η_n 's are characters of S such that

$$\lim_n f_n(g)\eta_n(g)$$

exists for all $g \in S$. (Such is e.g. $S=\{1\}$, $\eta_n=1$.) If $W'=(S',(\eta_n'))$ is another such system, write $W<W'$ if $S\subset S'$ and each η_n' is an extension of η_n. By Zorn's lemma there is a maximal W ; we have to show $S=G$. If not, let $g \in G\setminus S$ and let S' be the group generated by $S\cup\{g\}$. We are going to extend η_n to S' ; it is sufficient to define it for g. If $g^r \notin S$ for all natural numbers r, $\eta_n'(g)$ may

be prescribed arbitrarily; putting

$$\eta'_n(g) = \overline{f_n(g)}$$

$\chi(g)$ exists for g and therefore by the above remark over the whole S'.

If this is not the case, let r be the smallest exponent for which $g^r \in S$; $\eta'_n(g)$ must always be an r'th root of $\eta_n(g^r)$. We know that $\eta_n(g^r)f_n(g^r)$ converges and by an iterated application of (9.4) we can see that so does $f_n(g)^r f_n(g^r)$. Multiply these and write

$$A_n = \eta_n(g^r)f_n(g)^r , \quad A = \lim A_n .$$

Now fix an r'th root B of A and let B_n be the r'th root of A_n that is nearest to B. Then $B_n \to B$ and we may put

$$\eta'_n(g) = B_n\overline{f_n(g)} .$$

To prove the final assertion of the lemma observe that the limit of (9.4) is equal to

$$\chi(g_1)\chi(g_2)\overline{\chi(g_1g_2)} ,$$

thus it is identically one if and only if χ is a character. In this case replacing γ_n by $\gamma_n\bar{\chi}$, the obtained new χ will be identically one as well.

10. Shift-convergence in some not M_1 groups

To dot out i now we prove

THEOREM 3. *If the group G has a compact commutative normal subgroup H such that G/H is M_1, then (SCH) holds in G.*

PROOF. Suppose

$$\psi_r(\mu_n) \to \psi_r(\mu), \quad \mu = \omega(G_1) , \tag{10.1}$$

where G_1 is a compact subgroup of G. Regarding the induced measures μ_n/H on the factor-group G/H we may apply Proposition (8.1) and make it converge with a suitable centering. Therefore without restricting the generality we may assume

$$\mu_n/H \to \mu/H . \tag{10.2}$$

(10.3) LEMMA. <u>The net</u> (μ_n) <u>is quasitight and its every cluster point is of the form</u> $\delta(h)\mu$, $h \in H$.

PROOF. (μ_n/H), being convergent, is quasitight, thus so is (μ_n) itself. (10.1) implies that every cluster point is of the form $\delta(g)\mu$. By (10.2)

$$(\delta(g)\mu)/H = \mu/H ,$$

and hence, comparing the supports, we obtain $g \in HG_1$. Let $g=hg_1$, $h \in H$, $g_1 \in G_1$; then

$$\delta(g)\mu = \delta(h)\delta(g_1)\mu = \delta(h)\mu .$$

Now we define a pseudo-Fourier-transform. Let Γ denote the group of those characters of H which assume 1 on $G_1 \cap H$. We shall extend the elements of Γ to the whole G. First if $h \in H$, $g \in G_1$, put

$$\gamma'(hg) = \gamma(h) .$$

An element may have several representations of the form hg, but the fact that γ is constant on a fixed coset of $G_1 \cap H$ assures that the value of $\gamma(h)$ is uniquely determined.

Next, γ' being a uniformly continuous function, it has uniformly continuous extensions to G; let γ'' be any of these extensions satisfying the condition

$$|\gamma''(g)| \leq 1 \qquad (g \in G) .$$

Finally, for $\nu \in \mathcal{D}$ and $\gamma \in \Gamma$ write

$$\hat{\nu}(\gamma) = \int \gamma'' d\nu .$$

Since these mock-characters γ'' in general will not be homomorphisms (nor will even the γ'), this transformation will not be multiplicative, thus we must be a bit careful.

(10.4) LEMMA. <u>For</u> $h \in H$ <u>we have</u>

$$(\delta(h)\mu)^{\wedge}(\gamma) = \gamma(h) .$$

PROOF.

$$(\delta(h)\mu)^{\wedge}(\gamma) = \int \gamma'' d\delta(h)\mu$$

$$= \int \gamma''(hy)d\mu(y) = \gamma(h),$$

since for $y \in \operatorname{supp} \mu = G_1$ we have

$$\gamma''(hy) = \gamma'(hy) = \gamma(h) .$$

Now put

$$f_n(\gamma) = \hat{\mu}_n(\gamma) .$$

In order to apply Lemma (9.3) we shall prove that

$$(10.5) \qquad A_n = \overline{f_n(\gamma_1\gamma_2)}f_n(\gamma_1)f_n(\gamma_2) \to 1$$

for arbitrary fixed $\gamma_1, \gamma_2 \in \Gamma$. Suppose the contrary; then for some $\varepsilon>0$ the set

$$E = \{n \in D : |A_n - 1| > \varepsilon\}$$

is cofinal in D. Consider the net $(\mu_n)_{n \in E}$; it is quasitight, thus it has a convergent subnet $(\mu_{p(i)})_{i \in I}$, whose limit, according to (10.3), is of the form $\rho = \delta(h)\mu$, $h \in H$. Hence

$$A_{p(i)} \to \overline{\hat{\lambda}(\gamma_1\gamma_2)}\hat{\lambda}(\gamma_1)\hat{\lambda}(\gamma_2)$$

$$= \overline{\gamma_1(h)\gamma_2(h)}\gamma_1(h)\gamma_2(h) = 1$$

by (10.4), a contradiction.

Now apply Lemma (9.3). It provides us some characters of Γ, which are induced by elements $h_n \in H$, such that

$$(10.6) \qquad f_n(\gamma)\gamma(h_n) \to 1$$

for all $\gamma \in \Gamma$. We are going to deduce

$$(10.7) \qquad \delta(h_n)\mu_n \to \mu .$$

Suppose that (10.7) does not hold. Then there is a subnet $(\mu_{p(i)})$

such that μ is not a cluster point of $\delta(h_{p(i)})\mu_i$. We may assume (this can be achieved by selecting further subnets) that the nets $(\mu_{p(i)})$ and $(h_{p(i)})$ converge (the first by the quasitightness of (μ_n), the second by the compactness of H). Thus

$$\mu_{p(i)} \to \delta(h)\mu, \quad h_{p(i)} \to h', \quad h,h' \in H .$$

Hence for every γ, in view of (10.4),

$$f_{p(i)}(\gamma) \to \gamma(h), \quad \gamma(h_{p(i)}) \to \gamma(h'),$$

thus by (10.6) we have

$$\gamma(hh') = 1 .$$

Since this holds for all $\gamma \in \Gamma$, we have $hh' \in G_1 \cap H$, therefore

$$\delta(h_{p(i)})\mu_{p(i)} \to \delta(hh')\mu = \mu ,$$

a contradiction.

This concludes the proof of Theorem 3, which, together with Theorem 2, implies also Theorem 1.

11. Further problems

(i) To remove the condition imposed on the groups G on which the measures are defined.

(ii) To weaken the condition of tightness to τ-regularity. In that case the condition (2.1) should be changed into

$$\inf_U \lim_n Q_\ell(\nu_m^n, U) = 0 ,$$

where U runs over the open sets. On the group R^{ω_1} I have of an (ordinary, countable) product of tight measures that is dispersing in the sense (2.1), thus "essentially divergent" inside the set of tight measures, and still converges to a τ-regular measure. The main difficulty is that we do not know such a comfortable condition for a set of τ-regular measures to be compact as uniform tightness for tight measures.

The following connected problems deserve some attention. Is the set of a) divisors, b) multiples, c) translates of a fixed τ-regular measures closed? For a fixed μ, is there a compact set M

of measures such that every divisor of μ is the translate of an element of M? In general, the answer is negative; I think these properties hold if the group is complete, which in this context seems to be a natural restriction.

(iii) Let G be commutative. Call a (countable) product absolutely convergent, if each of its rearrangements converges to the same limit. Is every convergent product associated with an absolutely convergent one? I can prove this for compact groups with Fourier--analytic methods. For the case G = additive group of reals this is an immediate consequence of Kolmogorov's three-series-theorem.

ACKNOWLEDGEMENT

I am grateful to I. Csiszár for a number of interesting talks on the subject.

REFERENCES

Billingsley,P. (1968), Convergence of probability measures, New York, Wiley.

Ciszár,I. (1966), On infinite products of random elements and infinite convolutions of probability distributions on locally compact groups, Z. Wahrscheinlichkeitstheorie verw. Geb. 5, 279-295.

Hewitt, E. and Ross, K.A. (1963), Abstract Harmonic Anylsis I., Springer,

Heyer,H. (1977), Probability measures on locally compact groups, Springer.

Kelley,J.L. (1955), General topology, New York, D.Van Nostrand.

Kloss,B.M. (1961), Limiting distributions on compact Abelian groups (in Russian), Teor. Veroyatn. Primen. 6, 392-421.

Siebert,E. (1976), Convergence and convolutions of probability measures on a topological group. Ann. Probab. 4, 433-443.

Topsøe,F. (1970b), Topology and measure, Springer.

Topsøe,F. (1970), Compactness in spaces of measures, Studia Math. 36, 192-212.

Tortrat,A. (1965), Lois de probabilité sur un espace topologique complètement régulier et produits infinis à termes indépendants dans un groupe topologique, Ann.Inst.H. Poincaré 1, 217-237.

Tortrat,A. (1970), Convolutions dénombrables équitendues dans un groupe topologique X, Proc.of the conference "Les probabilités sur les structures algébriques", Clermont-Ferrand 1969, Paris, CNRS.

Varadarajan, V.S. (1961), Measures on topological spaces (in Russian), Matem. Sbornik 55, 35-100.

IRREDUCIBLE AND PRIME DISTRIBUTIONS

Imre Z. Ruzsa and Gábor J. Székely

Mathematical Institute of the Hungarian Academy of Sciences

Department of Probability Theory Eötvös Loránd University Budapest

Abstract: The complete list of prime distributions will be given on locally compact Abelian groups.

1. Introduction

By distribution we mean a tight probability measure on a Hausdorff topological group G. These distributions form a semigroup $\mathcal{D}(G)$ with the operation of convolution. In this paper we consider only locally compact commutative groups G. The convolution of distribution will be denoted by $*$.

DEFINITION. Let F denote a nondegenerate distribution. F is irreducible if $F=F_1*F_2$ implies that either F_1 or F_2 is degenerate (degenerate distributions play the role of units). F is weakly irreducible if $F=F_1*F_2$ implies that either F_1 or F_2 is an associate of F (two distributions are associate if they divide each other). F is prime if $F|F_1*F_2$ implies $F|F_1$ or $F|F_2$ ($F|F_1$ denotes that F is a convolution factor of F_1 i.e. $F=F_1*F_2$ for some distribution F_0). F is anti-irreducible if it has no irreducible factors.

Evidently irreducible and prime distributions are always weakly irreducible, irreducible distributions are never anti-irreducible ; we shall see that these are the only connections between these concepts.

For the additive group of real numbers $G=(R,+)$ Hinčin proved that every distribution on G is the convolution of an anti-irreducible and a finite or countable infinite convolution of irreducible distributions. He also noted that this decomposition is not unique; his example also showed that not every distribution is prime. (The decomposition in a semigroup may not be unique even when every irreducible is prime.) Hinčin's theorem was extended to locally compact second countable groups by Parthasarathy-Rao-Varadhan (1963).

We proved (Ruzsa-Székely (198?a)) that for $G=(R,+)$ there are no primes. Now we investigate the situation for locally compact groups.

THEOREM 1. _In_ $\mathcal{D}(G)$ _there are primes only if_ G _is a cyclic group of order_ 2,3,4,5 _or_ 8. _For_ $G=C_4$ _there are two primes (apart from associates); in the other cases there is only one._

As an hors d'Oeuvre we prove

LEMMA 1.1. _If_ $G=C_2$ _or_ C_3, _then the only prime in_ $\mathcal{D}(G)$ _is the Haar measure on_ G.

PROOF. Let γ be the only nonprincipal character of C_2, resp. one of the two conjugate characters of C_3. Let $\hat{F}$ denote the Fourier transform of an $F \in \mathcal{D}$. The homomorphism $F \to \hat{F}(\gamma)$ maps $\mathcal{D}(C_2)$ onto the multiplicative semigroup $[-1,1]$ and $\mathcal{D}(C_3)$ onto the equilateral triangle whose vertices are the third roots of unity. Now the statement follows easily.

REMARK. The Haar measure $\omega(C_2)$ is antiirreducible; $\omega(C_3)$ is neither irreducible nor antiirreducible; though they are weakly irreducible, still they are infinite products of distributions, none of which is their associate.

2. Big ordinary primes

Let G be (as always throughout this paper) a locally compact commutative group and suppose that $\pi \in \mathcal{D}(G)$ is a prime. For $F \in \mathcal{D}$ define

$$\varphi(F) = \sup\{n : \pi^n | F\} .$$

φ is a mapping from $\mathcal{D}$ into $N \cup \{\infty\}$, where N is the set of non-negative integers. We may divide the primes into subclasses:

π is _big_, if always $\varphi(F) < \infty$; _small_ otherwise;

π is _ordinary_, if φ is a homomorphism into the additive semigroup $N \cup \{\infty\}$; _extravagant_ otherwise.

THEOREM 2. _There is no nontrivial homomorphism of_ $\mathcal{D}$ _into the additive group of integers._

COROLLARY 2.1. _There is no big ordinary prime in_ $\mathcal{D}$.

If S is an arbitrary semigroup, $ss_1=ss_2$ for some $s,s_1,s_2 \in S$ and φ is a homomorphism of S into a group or a cancellative semigroup, then, of course, $\varphi(s_1) = \varphi(s_2)$.

DEFINITION. _Call two elements_ $s_1,s_2 \in S$ _sisters and write_ $s_1 \sim s_2$ _if there is an_ $s \in S$ _such that_ $ss_1=ss_2$.

The relation $\sim$ is evidently a congruence-relation and it is natural to consider the factor-semigroup $S_o=S/\sim$. S_o is cancellative, thus in the Abelian case it can be embedded into its quotient group H. This H will be called the quotient group of S.

THEOREM 3. The quotient group of $\mathcal{D}$ is divisible.

Theorem 3 obviously implies Thoerem 2: from a divisible group no nontrivial homomorphism goes into the additive group of integers.

LEMMA 2.2. Two distributions F_1 and F_2 are sisters if and only if their characteristic functions coincide in a neighbourhood of the principal character 1.

PROOF. If $F_1*F=F_2*F$, then, since $\hat{F}$ is continuous and assumes 1 at 1, $\hat{F}_1$ and $\hat{F}_2$ must coincide near 1. On the other hand, if $\hat{F}_1$ and $\hat{F}_2$ coincide in a neighbourhood of 1, then the existence of an F such that $F_1*F=F_2*F$ follows by V.1.2. of Reiter (1968).

According to this lemma, to prove Theorem 3 we have to find, for any characteristic function χ and natural number n, two characteristic functions χ_1 and χ_2 such that

$$\chi(\gamma) = (\chi_1(\gamma)/\chi_2(\gamma))^n$$

in a neighbourhood of 1.

This will be a consequence of the following two lemmas.

LEMMA 2.3. Let χ be the Fourier transform of a function from $L_1(G)$ and f a function, analytic in a neighbourhood of $\chi(1)$. Then there is another $h \in L_1$ whose Fourier transform coincides with $f(\chi(\gamma))$ in a neighbourhood of 1.

This is a Wiener-Lèvy type theorem. For a proof see Reiter (1968). We shall need this result in the case $f(t)=t^{1/n}$.

LEMMA 2.4. Suppose $h \in L_1$ and $\int_G h d\omega > 0$, where ω denotes the Haar measure on G. Then one can find a nonnegative $k \in L_1$, $k \neq 0$ such that $h*k$ is also nonnegative.

The proof of this lemma will be published separately, see Ruzsa-Székely (198?b).

PROOF of Theorem 3. Let $K_1 \subset \operatorname{int} K_2$ be compact neighbourhoods of 1. The uniform distribution on K_i $(i=1,2)$ is the Fourier transform of a function from $L_2(G)$, thus their convolution η is the Fourier transform of a function from L_1, and it is equal to 1 near 1. Hence $\chi\eta$ is the Fourier transform of a function from L_1 as well. Applying Lemma 2.3 for $f(t)=t^{1/n}$ we obtain an $h \in L_1$ whose Fourier transform ψ satisfies

$$\psi(\gamma) = (\chi(\gamma)\eta(\gamma))^{1/n} = (\chi(\gamma))^{1/n}$$

in a small neighbourhood of 1. This h satisfies the requirement of Lemma 2.4, since

$$\int_G h\,d\omega = \psi(1) = \chi(1)^{1/n} = 1.$$

Lemma 2.4 provides us a nonnegative k; we may assume it to be a density function, since this can be achieved by a constant factor, and then

$$\int h*k = (\int h)(\int k) = 1,$$

thus so is $k_1 = h*k$.

3. Small ordinary primes

LEMMA 3.1. π <u>is small if and only if</u> $\operatorname{supp}\pi$ <u>is contained in a coset of a compact subgroup.</u> $\varphi(F) = \infty$ <u>if and only if</u> $\omega_H | F$, <u>where</u> H <u>is the smallest such subgroup</u>.

This can easily be shown by Fourier transform, or even for noncommutative groups by Csiszár's method (see Csiszár (1966)), and is probably known; we suppress the proof.

LEMMA 3.2. <u>Let</u> γ_o <u>be a nonprincipal character of</u> G <u>and</u> U <u>a neighbourhood of unity in the group of characters. There exists a distribution</u> F <u>such that</u> $\hat{F}(\gamma_o) \neq 0$ <u>and</u> $\hat{F}(\gamma) = 0$ if $\gamma \notin U \cup U\gamma_o \cup U\overline{\gamma_o}$.

PROOF. Let F_o be a distribution whose Fourier transform is nonnegative and vanishes outside U (cf. Reiter (1968), p.110). Let F be the measure whose Radon-Nikodym derivative with respect to F_o is $1+(\gamma_o+\overline{\gamma_o})/2$. Then

$$\hat{F}(\gamma) = \hat{F}_o(\gamma) + (\hat{F}_o(\gamma\gamma_o) + \hat{F}_o(\gamma\overline{\gamma_o}))/2 .$$

$\hat{F}(1) = \hat{F}_o(1) = 1$ if $\gamma_o \notin U$, which may be assumed, thus F is also a distribution, $\hat{F}$ obviously vanishes outside the region required and finally $\hat{F}(\gamma_o) > 0$ since $\hat{F}_o(\gamma_o\overline{\gamma_o}) = \hat{F}_o(1) = 1$ and the other terms are nonnegative.

LEMMA 3.3. <u>A small ordinary prime is the associate of a Haar measure</u> ω_H <u>on a two or three element subgroup</u> H <u>of</u> G.

PROOF. We may assume $e \in \operatorname{supp}\pi$ (e is the unit of G), since this can be achieved by a translation. Let H be the closed subgroup generated by $\operatorname{supp}\pi$; H is compact by Lemma 4.1. π is a prime also in $\mathcal{D}(H)$. If $|H| \leq 3$, we refer to Lemma 1.1. If H has at least 4 elements, then in its character group Γ we can find a $\gamma_1 \neq 1$ and a $\gamma_2 \neq 1, \gamma_1, \overline{\gamma_1}$. By the previous lemma there are distributions F_1 and

F_2 such that $\hat{F}_j(\gamma_j) \neq 0$ and $\hat{F}_j$ vanishes outside $\{1, \gamma_j, \overline{\gamma_j}\}$ $(j=1,2)$ (Γ is discrete). Then $\varphi(F_j) < \infty$ but $F_1 * F_2 = \omega_H$, thus $\varphi(F_1 * F_2) = \infty$, a contradiction.

4. Aggressive distributions

Write $F \prec G$ if every root of $\hat{F}$ is also a root of $\hat{G}$. Evidently, $F \prec G$ if $F|G$. Now call F aggressive, if $F \prec G$ implies $F|G$. Haar measures are obviously aggressive.

LEMMA 4.1. Every prime is aggressive.

PROOF. For ordinary primes this follows from Corollary 2.1 and Lemma 3.3. Now let π be an extraordinary prime and suppose that there exists an F such that $\pi \prec F$ but $\pi \nmid F$. We are going to show that $\varphi(F_1 * F_2) = \varphi(F_1) + \varphi(F_2)$. Let

$$\varphi(F_1) = m, \quad \varphi(F_2) = n, \quad F_1 = \pi^{*m} * F_1', \quad F_2 = \pi^{*n} * F_2' .$$

Then

$$F_1 * F_2 = \pi^{*(m+n)} * F_1' * F_2' = \pi^{*k} * F_3, \quad k = m+n, \quad F_3 = F_1' * F_2' .$$

$\pi \nmid F_3$ by primality; we have to show $\pi^{*(k+1)} \nmid \pi^{*k} * F_3$. If $\pi^{*k} * F_3 = \pi^{*(k+1)} * F_3'$, then $\hat{\pi}^k(\hat{F}_3 - \hat{\pi}\hat{F}_3') = 0$, hence by $\pi \prec F$ we obtain $\hat{F}(\hat{F}_3 - \hat{\pi}\hat{F}_3') = 0$, thus $F_3 * F = F_3' * \pi * F$, $\pi \mid F_3 * F$ though $\pi \nmid F_3$ and $\pi \nmid F$, a contradiction.

LEMMA 4.2. An aggressive distribution π is prime if and only if $\hat{\pi}$ has exactly one root or two conjugate roots.

PROOF. If $\hat{\pi}$ has no root, then $\pi \prec \delta$, where δ is a degenerate distribution, thus π is a unit. Now suppose that $\hat{\pi}$ has two different, not conjugate roots γ_1 and γ_2. Let V be a small neighbourhood of unity in the character group such that $\hat{\pi}$ does not vanish on V and

$$(V\gamma_1 \cup V\overline{\gamma_1}) \cap (V\gamma_2 \cup V\overline{\gamma_2}) = \emptyset .$$

Construct the measures F_1, F_2 by Lemma 3.2, with γ_1, resp. γ_2 in the place of γ_0. Then $\pi \nmid F_j$ $(j=1,2)$, but $\hat{F}_1\hat{F}_2$ vanishes outside U, thus $\pi \prec F_1 * F_2$, a contradiction.

Finally if the roots of $\hat{\pi}$ are γ and $\overline{\gamma}$ (which may concide) and $\pi | F_1 * F_2$, then either $\hat{F}_1$ or $\hat{F}_2$ vanishes at γ, thus $\pi \prec F_1$ or $\pi \prec F_2$, hence $\pi | F_1$ or $\pi | F_2$ by the assumed aggressivity.

It would be interesting to determine all the aggressive distributions.

Now we are able to describe the ordinary primes.

THEOREM 4. <u>The only ordinary primes are the following:</u> $\pi_2=\omega(C_2)$ <u>on</u> C_2, $\overline{\pi_2}=\omega(C_2)$ <u>on</u> C_4 <u>and</u> $\pi_3=\omega(C_3)$ <u>on</u> C_3.

PROOF. By Corollary 2.1 and Lemma 3.3 we may assume that this prime is $\pi=\omega(H)$, where H is a two or three element subgroup of G. The number of roots of $\hat{\pi}$ is infinite if $|G|=\infty$ and it is $|G|-|G|/|H|$ if $|G|<\infty$. Thus by the above lemmas we have $|G|-|G|/|H|\leq 2$, hence either $G=H$ or $|G|=4$ and $|H|=2$. In this case there are two possibilities: $G=C_4$ or $G=C_2\times C_2$. However, in $C_2\times C_2$ the two roots of $\hat{\pi}$ are not conjugates.

5. <u>Extravagant primes</u>

We shall several times use the following idea: if $F=F_1*F_2$, then $\text{supp } F = \text{supp } F_1 \cdot \text{supp } F_2$, thus if $F_1|F$, then $\text{supp } F$ contains a translate of $\text{supp } F_1$. If moreover $|\text{supp } F_1| = |\text{supp } F| < \infty$, then $\text{supp } F_1$ and $\text{supp } F$ must be translates of each other.

Now let π be an extravagant prime, γ and $\overline{\gamma}$ the only roots of $\hat{\pi}$.

LEMMA 5.1. <u>If</u> γ <u>assumes</u> -1, <u>then</u> $|\text{supp } \pi| = 2$, γ <u>is of order</u> 2 <u>or</u> 4 <u>and</u> $|\gamma^{-1}(1)| \leq 2$.

PROOF. Suppose $\gamma(g)=-1$. If $F(\{e\}) = F(\{g\}) = 1/2$, then $F(\gamma)=0$; hence $\text{supp } \pi$ is a translate of $\text{supp } F = \{e,g\}$.

If γ is or order >4, then it assumes a value ϑ in the quarter circle between i and -1, say $\gamma(h)=\vartheta$. Now let x,y be the solutions of $x+2y=1$, $x+2y \text{ Re } \vartheta = 0$ (they are positive) and let $F_1(\{e\})=x$, $F_1(\{h\})=F_1(\{h^{-1}\})=y$. Then $\hat{F}_1(\gamma)=0$, still $\text{supp } F$ does not contain any translatesof $\{e,g\}$.

Finally write $G_o=\gamma^{-1}(1)$. If g_o is any element of G_o, then a distribution F_2 can be constructed with $\text{supp } F_2 = \{g_o,g\}$ and $\hat{F}_2(\gamma)=0$, thus $\{g_o,g\}$ must be a translate of $\{e,g\}$. This means

$$g_o g^{-1} \in \{g,g^{-1}\}, \quad g_o=e \text{ or } g^2, \quad |G_o| \leq 2.$$

LEMMA 5.2. <u>If</u> γ <u>does not assume</u> -1, <u>then</u> $|\text{supp } \pi| = 3$, γ <u>is of order</u> 3 <u>or</u> 5 <u>and</u> $|\gamma^{-1}(1)| = 1$.

PROOF. Obviously $|\text{supp } \pi| \geq 3$. Let $g,h \in G$ such that $\text{Re } \gamma(g) < 0$, $\gamma(h) = \overline{\gamma(g)}$. One can easily construct a measure F with $\text{supp } F = \{e,g,h\}$ such that $\hat{F}(\gamma)=0$; hence $|\text{supp } \pi| = |\text{supp } F| = 3$.

If the order of γ is at least 7, then it assumes at least two different values in the quarter circle $(i,-1)$, and by the above method one can construct two distributions F_1 and F_2 such that $\hat{F}_1(\gamma) = \hat{F}_2(\gamma) = 0$ but $\text{supp } F_1$ and $\text{supp } F_2$ are not translates of each other.

Finally we have to show $|G_o| = 1$, where again $G_o = \gamma^{-1}(1)$. If g_o is an arbitrary element of G_o, then a distribution F_o can be constructed with $\operatorname{supp} F_o = \{g_o, g, h\}$ and $\hat{F}_o(\gamma)=0$, thus $\{g_o, g, h\}$ must be a translate of $\{e,g,h\}$. This means

$$g_o g^{-1} \in \{g, h, g^{-1}, h^{-1}, gh^{-1}, g^{-1}h\}, \quad |G_o| \le 6.$$

The number of triplets in G whose γ-image is the same as that of $\{e,g,h\}$ is $|G_o|^3$. All these triplets must be translates of $\{e,g,h\}$: some of e,g or h must go into G_o, thus the number of admissible translations is $\le 3|G_o|$. Hence $|G_o|^3 \le 3|G_o|$, therefore (G_o is finite) $|G_o|=1$.

THEOREM 5. <u>The extravagant primes are given by the following formulas:</u>

(i) $\pi_4(\{e\})=\pi_4(\{g\})=1/2$ <u>on</u> C_4,

(ii) $\bar{\pi}_4(\{e\})=\bar{\pi}_4(\{g^2\})=1/2$ <u>on</u> C_8;

(iii) $\pi_5(\{e\})=a$, $\pi_5(\{g^2\}) = \pi_5(\{g^3\})=b$ <u>on</u> C_5, <u>where</u> $a+2b=1$, $a=2b\cos 36^\circ$,

<u>and</u> g <u>denotes always a generator of the corresponding group</u>.

PROOF. Lemmas 5.1 and 5.2 show that a distribution π can be prime on G only if $|\operatorname{supp} \pi| = 2$ and $|G| = 2$, 4 or 8, or $|\operatorname{supp} \pi|=3$ and $|G|=3$ or 5. If $|G|=2$ or 3, then the only prime on G (the Haar measure) is ordinary. If $|G|=5$, then the conditions $|\operatorname{supp} \pi| = 3$ and $\hat{\pi}(\gamma)=0$ with a fixed character γ determine a distribution uniquely; it is easy to check that it is aggressive, thus prime. If $|G|=4$ or 8, then π must be uniform on a two-element support. Now it needs only a routine calculation to show that such a π can be prime only if G is cyclic, and on C_4 the only primes are the ordinary π_2 and the extravagant π_4, on C_8 the extravagant $\bar{\pi}_4$.

REMARK. One can easily see that

(i) all primes are weakly irreducible,

(ii) π_2 and $\bar{\pi}_2$ are antiirreducible, π_4, $\bar{\pi}_4$ and π_5 are irreducible and π_3 is neither irreducible nor antiirreducible.

6. Further problems

(i) Find the prime distributions on some not locally compact groups (especially Hilbert and Banach spaces) and on semigroups.

(ii) Find the primes within some interesting subsemigroups of $\mathcal{D}(G)$ (e.g. among distributions with a compact support). Probably no new prime will appear, but the present proof does not work.

(iii) Call F_1 and $F_2 \in \mathcal{D}(G)$ <u>coprime</u>, if $F_1|F$ and $F_2|F$ implies $F_1 * F_2|F$. E.g. in $\mathcal{D}(C_4)$ $\overline{\pi}_2$ and π_4^k are coprimes. Is there any other example?

REFERENCES

Csiszár,I. (1966), On infinite products of random elements and infinite convolutions of probability distributions on locally compact groups, Z.Wahrscheinlichkeitstheorie verw. Geb. 5, 279-295.

Parthasarathy,K.R., Ranga Rao, R. and Varadhan, S.R.S. (1963), Probability distributions on locally compact groups, Illinois J. Math. 7, 337-369.

Reiter,H. (1968), Classical harmonic analysis and locally compact groups, Clarendon, Oxford.

Ruzsa,I.Z. and Székely,G.J. (198?a), No distribution is prime, submitted to the Z.Wahrscheinlichkeitstheorie verw. Geb.

Ruzsa,I.Z. and Székely, G.J. (198?b), Convolution quotients of non-negative functions, submitted to the Monathshefte f. Math.

CONTINUOUS HEMIGROUPS OF PROBABILITY MEASURES ON A LIE GROUP

Eberhard Siebert

Introduction

Let $(X_t)_{0\leq t\leq 1}$ be a stochastically continuous process with independent increments taking its values in a locally compact group G. Then the distributions $\mu(s,t)$ of the (left) increments $X_s^{-1}X_t$, $0 \leq s \leq t \leq 1$, of the process form a continuous (convolution) hemigroup in the sense of $\mu(r,s) * \mu(s,t) = \mu(r,t)$ if $0 \leq r \leq s \leq t \leq 1$.

Continuous hemigroups have been studied by various authors with the aid of different techniques. For a detailed discussion of the development of this field and for the relevant references we recommend Sections R 4.6 and R 5.6 of the monograph [4] and the survey article [5].

In the present paper we investigate in some detail the generation, characterization and representation of continuous hemigroups on a Lie group G. The restriction to Lie groups would not be necessary everywhere. But it allows a unified treatment of the problem and avoids some merely technical difficulties. The starting point for our work was the unpublished paper [9] of D.F.Wehn which is an improved version of his thesis [7]. (The results of [7] were published in [8].) A characteristic aspect of [9] is its rather elementary technique: namely the systematic application of convolution operators and differentiable functions. But on a non-compact and non-commutative Lie group it turned out that working with convolution operators alone was only possible under rather restrictive and artificial technical assumptions for the hemigroups.

It was tempting to free Wehn's results from these restrictions. For this it became important to treat the classes of left and right differentiable functions on G separately. This distinction will penetrate the whole paper.

A decisive role in the investigation of hemigroups is played by the generating functionals of continuous convolution semigroups. Their general theory is by now fairly complete (cf. [4], IV). This enabled us to attack the problem of hemigroups with some success by purely analytical tools.

Let us give a summary of the present paper. The first three sections (on differentiable functions on a Lie group, on generating functionals, and on infinitesimal systems of probability measures) are preparatory and fairly technical. The main result of Section 4 is Theorem 4.3 that characterizes Lipschitz continuous hemigroups: They have a generating family $(A(t))_{o\leq t\leq 1}$ (of generating functionals A(t) of continuous convolution semigroups) such that a (strong) forward or a (weak) backward evolution equation is satisfied. The existence and uniqueness of a Lipschitz continuous hemigroup for a given generating family $(A(t))_{o\leq t\leq 1}$ is proved in Section 5 (Theorems 5.1 and 5.7). In section 6 we show that a large subclass of the Lipschitz continuous hemigroups admits a product integral representation (Theorem 6.2).

An extension of some of our results from Lipschitz continuous hemigroups to hemigroups of bounded variation or to absolutely continuous hemigroups is indicated in Section 7. Hemigroups of bounded variation seem to constitute the most general class of hemigroups that can be investigated by our methods. In the final Section 8 we are concerned with the problem (important for the solution of the evolution equations) under what conditions the backward evolution equation holds in the strong version. We show that in general this can be only done on Lie groups possessing a relatively compact adjoint group. We characterize the class $\mathfrak{A}$ of these groups (Theorems 8.5 and 8.7) and prove that every maximally almost periodic Lie group belongs to $\mathfrak{A}$ (Theorem 8.6).

Part of our work (in particular of Sections 4 and 5) is parallel to Ph.Feinsilver's paper [2]. However Feinsilver deals with the additive process associated with a hemigroup. Fundamental are sophisticated methods from the theory of processes and martingales. In [2] the forward evolution equation appears as an assertion about martingales whereas the backward evolution equation apparently plays no role.

The author would like to thank Professor H.Heyer who made accessible to him the unpublished paper [9] and who encouraged him to work through it.

Preliminaries

$\mathbb{N}, \mathbb{Z}, \mathbb{Q}, \mathbb{R}, \mathbb{C}$ denote the sets of positive integers, integers, rational numbers, real numbers, and complex numbers respectively. We define $\mathbb{R}_+ = \{r \in \mathbb{R} : r \geq 0\}$ and $\mathbb{R}^* = \{r \in \mathbb{R} : r \neq 0\}$. Let I denote the unit interval $[0,1]$ in $\mathbb{R}$ and S the unit simplex $\{(s,t): 0 \leq s \leq t \leq 1\}$ in $\mathbb{R}^2$. We furnish I with the σ-field $\mathfrak{M}$ of Lebesgue measurable subsets. Let m denote the Lebesgue measure on $(I, \mathfrak{M})$. Measurability and integrability of functions defined on I without further specification refer to the measure space $(I, \mathfrak{M}, m)$. A subset Z of I is said to be a zero set if $Z \in \mathfrak{M}$ and $m(Z) = 0$. A decomposition $\Delta = \{c_o, c_1, \ldots, c_n\}$ of a real interval $[a,b]$ is a subset of $\mathbb{R}$ such that $a = c_o < c_1 < \ldots < c_n = b$. We put $|\Delta| = \max\{c_j - c_{j-1}: j=1,\ldots,n\}$.

Let X be a locally compact space and A a subset of X. Then A^- denotes the closure of A and 1_A the indicator function of A. By $\mathcal{C}^b(X)$ we denote the space of real valued bounded continuous functions on X furnished with the norm $\|f\| = \sup\{|f(x)|: x \in X\}$. $\mathcal{K}(X)$ and $\mathcal{C}^o(X)$ are the subspaces of functions with compact support and of functions vanishing at infinity respectively. The support of a function f in $\mathcal{C}^b(X)$ is denoted by supp(f). If E is a linear subspace of $\mathcal{C}^b(X)$ we define $\overline{E} = E \oplus \mathbb{R}.1_X$ and $E_+ = \{f \in E: f \geq 0\}$. Let $(f_n)_{n\geq o}$ be a sequence in $\mathcal{C}^b(X)$. We will write $f_n \uparrow f_o$ if $f_n \leq f_{n+1}$ for all $n \in \mathbb{N}$ and if $\sup\{f_n(x): n \in \mathbb{N}\} = f_o(x)$ for all $x \in X$.

$\mathcal{M}^b(X)$ denotes the space of real valued bounded Radon measures on X and $\mathcal{M}^1(X)$ the subset of probability measures on X. The Dirac measure in $x \in X$ is denoted by ε_x. The weak topology $\mathfrak{T}_w$ on $\mathcal{M}^b(X)$ is by definition the topology of pointwise convergence on $\mathcal{C}^b(X)$.

By G we always denote a Lie group. $\mathfrak{W}(G)$ is the system of all neighbourhoods of the identity e in G. If f is a function defined on G and if $y \in G$ then by ${}_yf$, f_y, f^* we denote the functions on G defined by ${}_yf(x) = f(yx)$, $f_y(x) = f(xy)$, $f^*(x) = f(x^{-1})$ (all $x \in G$).

By λ we denote a left Haar measure on G. Let $(L^1(G), \|.\|_1)$ denote the Banach space of λ-integrable real valued functions on G. $\mathcal{M}^b(G)$ and $L^1(G)$ are algebras with respect to convolution $*$ as multiplication operation. If $\mu \in \mathcal{M}^b(G)$ the associated convolution operator T_μ on $\mathcal{C}^o(G)$ is defined by $T_\mu f(x) = \int f(xy)\,\mu(dy)$ (all $x \in G$).

A triangular system $\mathfrak{J} = (\mu_{nk})_{k=1,\ldots,n;n\geq 1}$ in $\mathcal{M}^1(G)$ is said to be infinitesimal if $\lim_{n\geq 1} \max\{\mu_{nk}(\complement U): 1 \leq k \leq n\} = 0$ for all open $U \in \mathfrak{V}(G)$. Our restriction to systems $\mathfrak{J}$ with length n of the n-th row is convenient but by no means necessary.

A continuous (convolution) <u>hemigroup</u> in $\mathcal{M}^1(G)$ is a family $(\mu(s,t))_{(s,t)\in S}$ in $\mathcal{M}^1(G)$ satisfying the following conditions:

$\mu(r,s) * \mu(s,t) = \mu(r,t)$ for all $(r,s),(s,t) \in S$;

$\mu(s,s) = \varepsilon_e$ for all $s \in I$;

the mapping $(s,t) \longrightarrow \mu(s,t)$ of S into $\mathcal{M}^1(G)$ is $\mathfrak{T}_w$-continuous.

The restriction of parameters s and t to the unit interval I is merely for convenience. The results also hold mutatis mutandis for hemigroups with parameters in $\mathbb{R}_+$.

Obviously every continuous hemigroup in $\mathcal{M}^1(G)$ is supported by a σ-compact subgroup of G. Hence we may always assume without loss of generality that G is σ-compact and consequently has a countable basis of its topology.

1. Differentiable functions on a Lie group

1.1. Let G be a σ-compact Lie group of dimension $d \geq 1$ with Lie algebra $\mathfrak{g}$ and exponential mapping exp of $\mathfrak{g}$ into G. We choose a basis $\{X_1,\ldots,X_d\}$ of $\mathfrak{g}$. By $\mathcal{D}(G)$ we denote the space of functions in $\mathcal{K}(G)$ which are differentiable infinitely often. Furthermore let $\mathfrak{t}_2(G) = \{f \in \mathcal{C}^o(G): f$ is twice continuously differentiable in some $U \in \mathfrak{W}(G)\}$. If $f \in \mathcal{C}^b(G)$ is continuously differentiable in some neighbourhood of an $x \in G$ then for every $X \in \mathfrak{g}$ there exist the left and right derivatives of f in x with respect to X defined by:

$$Xf(x) = \lim_{t\to o} \frac{1}{t} \{f((\exp tX)x) - f(x)\},$$

$$\tilde{X}f(x) = \lim_{t\to o} \frac{1}{t} \{f(x(\exp tX)) - f(x)\}.$$

1.2. For every $k \in \mathbb{N}$ the Banach spaces $(\mathcal{C}_k(G), |\cdot|_k)$ and $(\mathcal{C}^{\sim}_k(G), |\cdot|^{\sim}_k)$ of differentiable functions on G are defined as in [4], 4.1.6 but with the notable difference to contain only functions vanishing at infinity. Thus the following chain of inclusions does hold:

$$\mathcal{D}(G) \subset \mathcal{C}_2(G) \cap \mathcal{C}^{\sim}_2(G) \subset \mathcal{C}_2(G) \cup \mathcal{C}^{\sim}_2(G) \subset \mathfrak{t}_2(G) \subset \mathcal{C}^o(G).$$

For $\mu \in \mathcal{M}^1(G)$ and $f \in \mathcal{C}_2(G)$ we have $T_\mu f \in \mathcal{C}_2(G)$. Moreover if $X \in \mathfrak{g}$ then $X(T_\mu f) = T_\mu(Xf)$. Consequently $|T_\mu f|_2 \leq |f|_2$.

1.3. With the basis $\{X_1,\ldots,X_d\}$ of $\mathfrak{g}$ there are associated a compact canonical neighbourhood U_o of $e \in G$ and a canonical coordinate system $\{x_1,\ldots,x_d\}$ in $\mathcal{D}(G)$ (valid in U_o). Moreover let $\varphi \in \bar{\mathcal{D}}(G)$ be a Hunt function for G i.e. $\varphi(x) = x_1(x)^2 + \ldots + x_d(x)^2$ for all $x \in U_o$, $0 < \varphi(x) \leq 1$ for all $x \in G-\{e\}$, and $\varphi(\complement V) = 1$ for some compact $V \in \mathfrak{W}(G)$ (cf. [4], 260). The datas $\{X_1,\ldots,X_d\}$, U_o, $\{x_1,\ldots,x_d\}$ and φ of G will be held fixed throughout this paper.

Now for every $\mu \in \mathcal{M}^1(G)$ we define the important characteristic

$$q(\mu) = \left|\int x_1 \, d\mu\right| + \ldots + \left|\int x_d \, d\mu\right| + \int \varphi \, d\mu.$$

1.4. If G is a Lie group of dimension zero i.e. if G is discrete we put $U_o = \{e\}$, $\varphi = 1_{G\sim\{e\}}$ and $q(\mu) = \int \varphi \, d\mu = \mu(G\sim\{e\})$. Furthermore let $\mathcal{D}(G) = \mathcal{K}(G)$ and $\mathcal{C}_k(G) = \mathcal{C}_k^{\sim}(G) = \ell_2(G) = \mathcal{C}^o(G)$ for all $k \in \mathbb{N}$. With these conventions the case of a discrete group will be contained in our considerations.

1.5. We need one more space of differentiable functions on G:

$$\mathcal{C}_{2,2}(G) = \{f \in \mathcal{C}_2(G) \cap \mathcal{C}_2^{\sim}(G): Xf,\ XYf \in \mathcal{C}_2^{\sim}(G) \text{ for all } X,Y \in \mathfrak{g}\}.$$

On $\mathcal{C}_{2,2}(G)$ we define a norm by

$$|f|_{2,2} = |f|_2^{\sim} + \sum_{1\le i\le d} |X_i f|_2^{\sim} + \sum_{1\le i,j\le d} |X_i X_j f|_2^{\sim} .$$

Obviously $\mathcal{D}(G)$ is a subspace of $\mathcal{C}_{2,2}(G)$.

1.6. LEMMA. There exists a constant $b > 0$ such that the following inequalities hold for all $\mu \in \mathcal{M}^1(G)$:

(A) $\quad |\int f \, d\mu - f(e)| \le b\, |f|_2\, q(\mu) \qquad (f \in \mathcal{C}_2(G))$;

(B) $\quad \| T_\mu f - f \| \le b\, |f|_2^{\sim}\, q(\mu) \qquad (f \in \mathcal{C}_2^{\sim}(G))$;

(C) $\quad | T_\mu f - f |_2 \le b\, |f|_{2,2}\, q(\mu) \qquad (f \in \mathcal{C}_{2,2}(G))$.

PROOF. 1. There exists a constant $c_1 > 0$ such that $1_{G\sim U_o} \le c_1 \varphi$. Hence we have for all $f \in \mathcal{C}^b(G)$:

$$|\int 1_{G\sim U_o}(f - f(e))d\mu| \le 2\| f \| \mu(G\sim U_o)$$
$$\le 2\| f \| c_1 \int \varphi \, d\mu \le 2c_1 \| f \| q(\mu).$$

2. Let $f \in \mathcal{C}_2(G)$. Then for $y \in U_o$ the following (left-hand) Taylor expansion is valid (cf. [4], 4.1.8):

$$f(y) - f(e) = \sum_{i=1}^{d} x_i(y) X_i f(e) + \frac{1}{2} \sum_{i,j=1}^{d} x_i(y) x_j(y) X_i X_j f(\xi(y))$$

with $\xi(y) \in U_o$. This yields (with some appropriate constant $c_2 > 0$):

$$|\int 1_{U_o}(f - f(e))d\mu| \le$$
$$\sum_{i=1}^{d} |X_i f(e)|\, |\int 1_{U_o} x_i \, d\mu| + \sum_{i,j=1}^{d} \| X_i X_j f \| \int \varphi \, d\mu \le |f|_2\, c_2\, q(\mu).$$

Taking into account 1. we obtain (A).

3. Let $f \in \mathcal{C}_2^{\sim}(G)$. For $x \in G$ and $y \in U_o$ we have this time the follow-

ing (right-hand) Taylor expansion:

$$f(xy) - f(x) = \sum_{i=1}^{d} x_i(y)\tilde{X}_i f(x) + \frac{1}{2}\sum_{i,j=1}^{d} x_i(y)x_j(y)\tilde{X}_i\tilde{X}_j f(x\,\eta(y))$$

with $\eta(y) \in U_o$. This yields (with some appropriate constant $c_3 > 0$):

$$\left|\int 1_{U_o}(y)(f(xy) - f(x))\,\mu(dy)\right| \leq$$

$$\sum_{i=1}^{d}|X_i f(x)|\left|\int 1_{U_o} x_i\, d\mu\right| + \sum_{i,j=1}^{d}\| X_i X_j f\| \int \varphi\, d\mu \leq |f|_2^{\sim} c_3\, q(\mu).$$

Taking into account 1. we obtain (B).

4. Taking into account 1.2. inequality (C) follows immediately from inequality (B). ┘

1.7. LEMMA. (i) If $f \in \mathcal{C}_1(G)$ and if g is a real valued function on G such that $g^* \in L^1(G)$ then we have $f * g \in \mathcal{C}_1(G)$, $X(f * g) = (Xf) * g$ $\| X(f * g)\| \leq \| Xf\| \| g^*\|_1$.

(ii) If $f \in L^1(G)$ and if $g \in \mathcal{C}_1^{\sim}(G)$ then we have $f * g \in \mathcal{C}_1^{\sim}(G)$, $\tilde{X}(f * g) = f * (\tilde{X}g)$ and $\| \tilde{X}(f * g)\| \leq \| f\|_1 \| \tilde{X}g\|$.

(iii) There exists a sequence $(f_n)_{n\geq 1}$ in $\mathcal{D}_+(G)$ with the following properties: $\sup\{|f_n|_{2,2}: n \in \mathbb{N}\} < \infty$, $f_n \uparrow 1_G$, and for every compact $K \in \mathfrak{W}(G)$ there is an $n \in \mathbb{N}$ such that $f_n(K) = 1$.

PROOF. (i) and (ii) follow by direct calculations.

(iii) Let $(C_n)_{n\geq 1}$ be a sequence of compact subsets of G such that $C_1^{-1} = C_1$, $e \in C_1$, and $1_{C_n} \uparrow 1_G$. Moreover let $(g_n)_{n\geq 1}$ be a sequence in $\mathcal{K}_+(G)$ such that $g_n \uparrow 1_G$ and $g_n(C_1 C_n C_1) = 1$ for all $n \in \mathbb{N}$. Finally let $f \in \mathcal{D}_+(G)$ such that $f = f^*$, $\mathrm{supp}(f) \subset C_1$, and $\| f\|_1 = 1$. We put $f_n = f * g_n * f$ for all $n \in \mathbb{N}$. By (i) and (ii) we have $f_n \in \mathcal{C}_{2,2}(G)$ and

$$|f_n|_{2,2} \leq |f|_2^{\sim}\{\| f\|_1 + \sum_{i=1}^{d}\| X_i f\|_1 + \sum_{i,j=1}^{d}\| X_i X_j f\|_1\}.$$

Hence the sequence $(f_n)_{n\geq 1}$ has the first of the desired properties. Obviously $0 \leq f_n \leq f_{n+1}$ for all $n \in \mathbb{N}$. Moreover for all $x \in G$:

$$f_n(x) = f * g_n * f(x) = \iint f(z)g_n(z^{-1}y)f(y^{-1}x)\,\lambda(dy)\,\lambda(dz) \leq$$

$$\iint f(z)f(y^{-1}x)\,\lambda(dy)\,\lambda(dz) = \{\int f(z)\,\lambda(dz)\}\{\int f(y)\,\lambda(dy)\} = 1.$$

Finally if $x \in C_n$, $y^{-1}x \in C_1$ and $z \in C_1$ we have $z^{-1}y \in C_1 C_n C_1$ and hence $f_n(x) = \iint f(z)f(y^{-1}x)\,\lambda(dy)\,\lambda(dz) = 1$. Thus $f_n(C_n) = 1$. Consequently the sequence $(f_n)_{n\geq 1}$ has all the desired properties. ┘

1.8. LEMMA. There exists a countably generated linear subspace $\mathfrak{V}$ of $\mathcal{D}(G)$ containing $x_1,\ldots,x_d$ and $1_G - \varphi$ which has the following density property: For every $f \in \mathcal{L}_2(G)$ and for every $\varepsilon > 0$ there exists a $g \in \mathfrak{V}$ such that $|f - g| \leq \varepsilon\varphi$.

PROOF. Since the Banach space $\mathcal{C}^o(G)$ is separable there exists a dense sequence $(f_n)_{n\geq 1}$ such that $f_n \in \mathcal{D}(G)$ for all $n \in \mathbb{N}$ (cf. (i) and (ii) of Lemma 1.7.). Furthermore let $(U_n)_{n\geq 1}$ be a basis of open neighbourhoods for $e \in G$ such that $U^-_{n+1} \subset U_n \subset U_o$ for all $n \in \mathbb{N}$. For every $n \in \mathbb{N}$ we choose a $g_n \in \mathcal{D}(G)$ such that $1_{U_{n+1}} \leq g_n \leq 1_{U_n}$. Let now $\mathfrak{V}$ be the algebra generated by the functions $f_1, f_2,\ldots;\ g_1, g_2,\ldots;\ x_1,\ldots,x_d;\ 1_G - \varphi$. Obviously $\mathfrak{V}$ is then also a countably generated linear subspace of $\mathcal{D}(G)$.

Let $f \in \mathcal{L}_2(G)$ and $\varepsilon > 0$ be given. Then there exists at first an $n(o) \in \mathbb{N}$ such that f is twice continuously differentiable in $U_{n(o)}$ and such that $|X_i X_j f(x) - X_i X_j f(e)| \leq 2\varepsilon/d^2$ for all $x \in U_{n(o)}$ and for $i,j = 1,\ldots,d$. We put

$$h = f(e)g_{n(o)-1} + \sum_{i=1}^{d} X_i f(e)x_i + \frac{1}{2}\sum_{i,j=1}^{d} X_i X_j f(e)x_i x_j.$$

Then $h \in \mathfrak{V}$; and the (left-hand) Taylor expansion of f in e (cf. part 2 of the proof of Lemma 1.6.) yields for $x \in U_{n(o)}$:

$$|h(x) - f(x)| \leq (\varepsilon/d^2)\sum_{i,j=1}^{d} x_i(x)x_j(x) \leq \varepsilon\varphi(x).$$

Moreover there exists a $c > 0$ such that $\varphi(x) \geq c$ for $x \in G \sim U_{n(o)+1}$. Let $n(1) \in \mathbb{N}$ such that $\| f - f_{n(1)} \| \leq \varepsilon c$. We put $g = g_{n(o)}h + (1_G - g_{n(o)})f_{n(1)}$. Then $g \in \mathfrak{V}$ and we have:

$$|f(x) - g(x)| = |f(x) - h(x)| \leq \varepsilon\varphi(x) \qquad (x \in U_{n(o)+1}),$$

$$|f(x) - g(x)| = |f(x) - f_{n(1)}(x)| \leq \varepsilon c \leq \varepsilon\varphi(x) \qquad (x \in G \sim U_{n(o)}).$$

Finally for $x \in U_{n(o)} \setminus U_{n(o)+1}$ we have:

$$|f(x) - g(x)| \leq$$

$$g_{n(o)}(x)|f(x) - h(x)| + (1 - g_{n(o)}(x))|f(x) - f_{n(1)}(x)| \leq$$

$$g_{n(o)}(x)\,\varepsilon\varphi(x) + (1 - g_{n(o)}(x))\,\varepsilon\varphi(x) = \varepsilon\varphi(x).$$

This proves the lemma. ┘

2. Generating functionals

2.1. Let G be a Lie group as in 1.1. and let E be a linear subspace of $\overline{\mathcal{L}^o}(G)$. A linear functional A on E is said to be almost positive if $A(f) \geq 0$ for all $f \in E_+$ such that $f(e) = 0$. Obviously A has then the following (equivalent) property:

(P) $|A(f)| \leq A(g)$ for all $f,g \in E$ such that $|f| \leq g$ and $g(e) = 0$.

2.2. Let us call a sequence $(f_n)_{n\geq 1}$ in E_+ a norming sequence if $f_n \uparrow 1_G$ and if for every compact $K \in \mathfrak{W}(G)$ there is an $n \in \mathbb{N}$ such that $f_n(K) = 1$. An almost positive linear functional A on a subspace E of $\overline{\mathcal{L}^o}(G)$ containing a norming sequence $(f_n)_{n\geq 1}$ is said to be normed if $\sup\{A(f_n): n \in \mathbb{N}\} = 0$. Obviously this definition is independent of the special choice of the norming sequence. By $\mathbb{A}(G)$ we denote the space of almost positive and normed linear functionals on $\mathfrak{D}(G)$. Since $\mathbb{A}(G)$ is a convex cone in the algebraic dual of $\mathfrak{D}(G)$ it defines a pre-order in this space.

Every $A \in \mathbb{A}(G)$ admits an extension to an almost positive linear $\overline{A}$ on $\mathcal{E}_2(G)$ with the property $\overline{A}(1_G) = 0$. In view of Lemma 1.8. this extension is unique. Hence we will also denote this extension simply by A.

2.3. The elements of $\mathbb{A}(G)$ are exactly the generating functionals (restricted to $\mathfrak{D}(G)$) of the continuous convolution semigroups in $\mathcal{M}^1(G)$ ([4], 4.4.16 and 4.4.18). The convolution semigroup corresponding to $A \in \mathbb{A}(G)$ will be denoted by $(\mathrm{Exp}\, tA)_{t\geq 0}$. Furthermore by $(N_A f)(x) =$

$A({}_x f)$ there is defined a linear mapping of $\mathcal{C}_2^{\sim}(G)$ into $\mathcal{C}^o(G)$. N_A is the restriction to $\mathcal{C}_2^{\sim}(G)$ of the infinitesimal generator of the strongly continuous semigroup $(T_{\text{Exp } tA})_{t\geq 0}$ of convolution operators on $\mathcal{C}^o(G)$ associated with $(\text{Exp } tA)_{t\geq 0}$ ([4], 4.2.4, 4.2.5). Hence we will also use the notation $\text{Exp } tN_A$ instead of $T_{\text{Exp } tA}$.

2.4. Every $A \in \mathbb{A}(G)$ admits on $\overline{\mathcal{C}}_2(G)$ the following representation:

$$\text{(LLC)} \qquad A(f) = \sum_{i=1}^{d} a_i (X_i f)(e) + \sum_{i,j=1}^{d} a_{ij} (X_i X_j f)(e)$$

$$+ \int \{f - f(e) - \sum_{i=1}^{d} (X_i f)(e) x_i\} \, d\eta$$

where $a_1, \ldots, a_d \in \mathbb{R}$, $(a_{ij})_{1\leq i,j\leq d}$ a real, symmetric, positive-semidefinite matrix, and η a Levy measure on $G \smallsetminus \{e\}$ ([4], 4.2.4). Thus on the Banach spaces $\mathcal{C}_2(G)$ and $\mathcal{C}_2^{\sim}(G)$ the linear functional A is continuous. Let $|A|_2$ and $|A|_2^{\sim}$ denote the corresponding norms. Moreover one easily sees that the linear mapping N_A of $\mathcal{C}_2^{\sim}(G)$ into $\mathcal{C}^o(G)$ (cf. 2.3.) is continuous and that its norm is just $|A|_2^{\sim}$.

2.5. From (LLC) and from the Taylor expansions of functions in $\mathcal{C}_2(G)$ respectively in $\mathcal{C}_2^{\sim}(G)$ (cf. the proof of Lemma 1.6.) one easily derives the following result:

LEMMA. For every $A \in \mathbb{A}(G)$ let $\| A \| = |A(x_1)| + \ldots + |A(x_d)| + A(\varphi)$. Then there exist positive constants c_1 and c_2 such that

$$c_1 |A|_2 \leq \| A \| \leq c_2 |A|_2 \quad \text{and} \quad c_1 |A|_2^{\sim} \leq \| A \| \leq c_2 |A|_2^{\sim}$$

for all $A \in \mathbb{A}(G)$.

2.6. Let $t \longrightarrow A(t)$ be a mapping of I into $\mathbb{A}(G)$ usually denoted by $A(.)$. The mapping $A(.)$ is said to be measurable (integrable, bounded) if the functions $A(.)(f)$ are Lebesgue measurable (Lebesgue integrable, bounded) for all $f \in \mathcal{D}(G)$. In view of Lemma 1.8. the functions $A(.)(f)$ are then also measurable (integrable, bounded) for all $f \in \overline{\mathcal{C}}_2(G)$. In view of Lemma 2.5. the boundedness of $A(.)$ is equivalent to the boundedness of $|A(.)|_2$ or of $|A(.)|_2^{\sim}$. For abbreviation let us call the mapping A(.) admissible if it is measurable and bounded.

For every $t \in I$ let $a_i(t)$, $a_{ij}(t)$ $(1 \leq i,j \leq d)$ and η_t be the datas associated with A(t) via (LLC) (cf. 2.4.). It is easy to check that A(.) is measurable (integrable, bounded) if and only if the following functions are measurable (integrable, bounded): $a_i(.)$, $a_{ij}(.)$ $(1 \leq i,j \leq d)$; $t \longrightarrow \int f\, d\eta_t$ for all $f \in \mathfrak{D}_+(G)$ such that $f(e) = 0$.

2.7. LEMMA. Let A(.) be an admissible mapping of I into $\mathbb{A}(G)$. For every $t \in I$ and for every $f \in \mathfrak{D}(G)$ we put

$$\text{(I)} \qquad B(t)(f) = \int_0^t A(\tau)(f)d\tau .$$

Then the functionals B(t) are in $\mathbb{A}(G)$; and equation (I) holds for all $f \in \overline{\mathfrak{L}}_2(G)$. The mapping B(.) of I into $\mathbb{A}(G)$ is increasing with $B(0) = 0$. Finally $|A(t)|_2 \leq \beta$ for all $t \in I$ implies $|B(t) - B(s)|_2 \leq \beta(t - s)$ for all $(s,t) \in S$.

The functional B(t) will be denoted by $\int_0^t A(\tau)d\tau$.

PROOF. Obviously every B(t) is an almost positive linear functional on $\mathfrak{D}(G)$. Let $(f_n)_{n\geq 1}$ be a norming sequence in $\mathfrak{D}(G)$ (cf.2.2.). Then for every $\tau \in [0, t]$ the sequence $(A(\tau)(f_n))_{n\geq 1}$ ascends to 0 and $A(\tau)(f_1) \geq -\beta|f_1|_2$. Hence in view of the theorem of monotonic convergence B(t) is normed. The validity of (I) for all $f \in \overline{\mathfrak{L}}_2(G)$ follows now easily taking into account Lemma 1.8. and property (P) of 2.1. Arguing as above $B(t) - B(s) = \int_s^t A(\tau)d\tau$ is in $\mathbb{A}(G)$ for all $(s,t) \in S$ hence B(.) is increasing. $B(0) = 0$ is obvious. The last assertion now follows immediately. ┘

2.8. LEMMA. Let B(.) be an increasing mapping of I into $\mathbb{A}(G)$ with $B(0) = 0$. Let there exist a constant $\beta > 0$ such that $|B(t) - B(s)|_2 \leq \beta(t - s)$ for all $(s,t) \in S$.

Then there exists an admissible mapping A(.) of I into $\mathbb{A}(G)$ such that $|A(t)|_2 \leq \beta$ and $B(t) = \int_0^t A(\tau)d\tau$ for all $t \in I$. Moreover there exists a zero set Z in I such that $\frac{d}{dt} B(t)(f) = A(t)(f)$ for all $t \in I - Z$ and for $f \in \bar{\mathcal{L}}_2(G)$.

<u>PROOF</u>. 1. Let $\mathcal{V}$ be a linear subspace of $\mathcal{D}(G)$ as in Lemma 1.8. For every $f \in \mathcal{D}(G)$ and for all $(s,t) \in S$ we have $|B(t)(f) - B(s)(f)| \leq \beta |f|_2 (t - s)$. Since $\mathcal{V}$ is countably generated there is a zero set Z_1 in I such that $A_1(t)(f) = dB(t)(f)/dt$ exists for all $t \in I - Z_1$ and for all $f \in \mathcal{V}$. For $t \in Z_1$ and $f \in \mathcal{V}$ let $A_1(t)(f) = 0$. Then every $A_1(t)$ is an almost positive linear functional on $\mathcal{V}$ such that $|A_1(t)(f)| \leq \beta |f|_2$ for all $f \in \mathcal{V}$. Moreover $A_1(.)(f)$ is measurable; and we have $B(t)(f) = \int_0^t A_1(\tau)(f)d\tau$ for all $t \in I$ (and for all $f \in \mathcal{V}$).

2. Without loss of generality we may assume that $\mathcal{V}$ contains a norming sequence $(f_n)_{n\geq 1}$ (cf. 2.2.). Since $A_1(t)$ is almost positive the sequence $(A_1(t)(f_n))_{n\geq 1}$ is increasing; moreover $A_1(t)(f_1) \geq -\beta |f_1|_2$ (all $t \in I$). Hence the theorem of monotonic convergence and the normedness of $B(1)$ yield: $0 = \sup\{B(1)(f_n): n \in \mathbb{N}\} = \int_0^1 (\sup\{A_1(\tau)(f_n): n \in \mathbb{N}\})d\tau$. But $A_1(t)(f_n) \leq 0$ for all $t \in I$ and for all $n \in \mathbb{N}$.

[Let $t \notin Z_1$. Since $B(.)$ is increasing we have $B(t)(f_n) - B(s)(f_n) = \{B(t) - B(s)\}(f_n) \leq 0$ for $0 \leq s < t$ and thus

$$A_1(t)(f_n) = \lim_{s\uparrow t} \frac{1}{t - s} \{B(t)(f_n) - B(s)(f_n)\} \leq 0.\ ]$$

Hence there exists a zero set Z_2 in I such that $A_1(t)$ is normed for every $t \in I - Z_2$. If $t \in Z_2$ we substitute $A_1(t)$ by the zero functional on $\mathcal{V}$. Let $Z = Z_1 \cup Z_2$. Every $A_1(t)$ can be extended to an almost positive linear functional $\bar{A}_1(t)$ on $\tilde{\mathcal{V}}$ such that $\bar{A}_1(t)(1_G) = 0$.

3. Every functional $\bar{A}_1(t)$ has the property (P) of 2.1. Thus taking into account Lemma 1.8. every $\bar{A}_1(t)$ can be uniquely extended to an almost positive linear functional $A(t)$ on $\bar{\mathcal{L}}_2(G)$. Moreover it can be easily checked that $A(.)$ inherits all the properties of $\bar{A}_1(t)$. Hence the proof of the lemma is complete. ⌟

<u>2.9.</u> Let us identify admissible mappings which are equal almost everywhere. Then taking into account Lemma 2.7 and Lemma 2.8 the equation (I) constitutes a bijection between admissible mappings A(.) and in-

creasing mappings $B(.)$ of I into $\mathbb{A}(G)$ which are Lipschitz continuous with $B(0) = 0$.

2.10. LEMMA. Let X be a locally compact space having a countable basis of its topology. Let $\chi(.)$ be a mapping of I into the Banach space $\mathcal{C}^{0}(X)$ such that the functions $\chi(.)(x)$ are measurable for all $x \in X$. Then $\chi(.)$ is strongly measurable (cf. [10], 130).

PROOF. Since the Banach space $\mathcal{C}^{0}(X)$ is separable it suffices in view of Pettis' theorem ([10], 131) to prove that $\chi(.)$ is weakly measurable. But $\mathcal{M}^{b}(X)$ is the topological dual of $\mathcal{C}^{0}(X)$. Thus given $\mu \in \mathcal{M}^{b}(X)$ we have to show that $\int \chi(.)d\mu$ is measurable. Since X has a countable basis of its topology there exists a sequence $(\mu_n)_{n\geq 1}$ in $\mathcal{M}^{b}(X)$ which converges to μ with respect to $\mathfrak{T}_w$ and such that every μ_n has finite support. By assumption every mapping $\int \chi(.)d\mu_n = \sum_{x\in X} \mu_n(\{x\})\chi(.)(x)$ is measurable. Thus $\lim_{n\geq 1} \int \chi(t)d\mu_n = \int \chi(t)d\mu$ for all $t \in I$ yields the assertion. ⌟

2.11. COROLLARY. Let $A(.)$ be a measurable mapping of I into $\mathbb{A}(G)$. Then for every $f \in \mathcal{C}_2^{\sim}(G)$ the mapping $N_{A(.)}f$ of I into $\mathcal{C}^{0}(G)$ (cf. 2.3.) is strongly measurable.

PROOF. This follows immediately from Lemma 2.10. applied to $X = G$ and $\chi(.) = N_{A(.)}f$ since $\chi(.)(x) = A(.)({}_xf)$ is measurable for every $x \in G$ by assumption. ⌟

3. Infinitesimal systems of probability measures

Let $\mathfrak{J} = (\mu_{nk})_{k=1,\dots,n;n\geq 1}$ be always an infinitesimal (triangular) system of probability measures on the Lie group G.

3.1. LEMMA. The following two conditions are equivalent:

(S) There exist a constant $\beta > 0$, a sequence $(\varepsilon_n)_{n\geq 1}$ in $\mathbb{R}_+$ converging to 0, and a sequence of increasing mappings $v(n;.)$ of I into $\{0,1,\dots,n\}$ satisfying $v(n;0) = 0$, $v(n;1) = n$ such that the following estimations hold:

$$\sum_{v(n;s)<j\leq v(n;t)} q(\mu_{nj}) \leq \beta(t-s) + \varepsilon_n \qquad ((s,t)\in S;\ n\in\mathbb{N}).$$

(S') There exists a constant $\gamma > 0$ such that $\sum_{1\leq j\leq n} q(\mu_{nj}) \leq \gamma$ for all $n \in \mathbb{N}$.

PROOF. '(S) $\Longrightarrow$ (S')' This is obvious in view of $v(n;0) = 0$ and $v(n;1) = n$ defining $\gamma = \beta + \sup\{\varepsilon_n : n \in \mathbb{N}\}$.

'(S') $\Longrightarrow$ (S)' Let $\varepsilon_n = \max\{q(\mu_{nj}): 1\leq j\leq n\}$ for all $n \in \mathbb{N}$. The infinitesimality of $\mathfrak{I}$ implies $\lim \varepsilon_n = 0$. Without loss of generality we may assume $q_n = \sum_{1\leq j\leq n} q(\mu_{nj}) > 0$ for all $n \in \mathbb{N}$. For $k = 0,1,\ldots,n$ let $r_n(k) = (\sum_{1\leq j\leq k} q(\mu_{nj}))/q_n$. Since $0 = r_n(0) \leq r_n(1) \leq \ldots \leq r_n(n) = 1$ there exists for every $t \in]0,1[$ a unique $v(n;t) \in \{1,\ldots,n\}$ such that $r_n(v(n;t)-1) < t \leq r_n(v(n;t))$. Furthermore let $v(n;0) = 0$ and $v(n;1) = n$. Then we have for all $(s,t) \in S$:

$$r_n(v(n;t)) - r_n(v(n;s)) \leq (t-s) + \varepsilon_n/q_n.$$

[Let $t \in]0,1[$ and $s \in [0,t]$. The definitions of $v(n;.)$ and r_n yield $-t + r_n(v(n;t)) \leq r_n(v(n;t)) - r_n(v(n;t)-1) = q(\mu_{n,v(n;t)})/q_n \leq \varepsilon_n/q_n$ and $-r_n(v(n;s)) \leq -s$. This shows the assertion.]

Taking into account $q_n \leq \gamma$ this implies:

$$\sum_{v(n;s)<j\leq v(n;t)} q(\mu_{nj}) = q_n\{r_n(v(n;t)) - r_n(v(n;s))\} \leq$$

$$q_n\{(t-s) + \varepsilon_n/q_n\} = q_n(t-s) + \varepsilon_n \leq \gamma(t-s) + \varepsilon_n.$$

This proves (S). ┘

3.2. For the rest of this section let us assume that $\mathfrak{I}$ has property (S). For all $(s,t) \in S$ and $n \in \mathbb{N}$ we define:

$$\mu_n(s,t) = \mu_{n,v(n;s)+1} * \ldots * \mu_{n,v(n;t)};$$

$$\varkappa_n(s,t) = \sum_{v(n;s)<j\leq v(n;t)} (\mu_{nj} - \varepsilon_e);$$

$$\eta_n(s,t) = \sum_{v(n;s)<j\leq v(n;t)} \mu_{nj}.$$

Futhermore let T_{nk} and $T_n(s,t)$ denote the convolution operators corresponding to μ_{nk} and $\mu_n(s,t)$ respectively.

Obviously we have for all (r,s), $(s,t) \in S$:

$$\mu_n(r,s) * \mu_n(s,t) = \mu_n(r,t) ;$$

$$\varkappa_n(r,s) + \varkappa_n(s,t) = \varkappa_n(r,t) ;$$

$$\eta_n(r,s) + \eta_n(s,t) = \eta_n(r,t).$$

3.3. LEMMA. For all $(s,t) \in S$, $n \in \mathbb{N}$ and $f \in \mathcal{C}_{2,2}(G)$ we have:

$$|\int f\, d\mu_n(s,t) - f(e) - \int f\, d\varkappa_n(s,t)| \leq b^2\ |f|_{2,2}\ (\beta(t-s) + \varepsilon_n)^2$$

and $q(\mu_n(s,t)) \leq c((t-s) + \varepsilon_n)$ where $c > 0$ is an appropriate constant.

PROOF. 1. For $j = 0,1,\ldots,v(n;t)-1$ we have (by 1.2. and 1.6.(C)):

$$|T_{n,j+1}\cdots T_{n,v(n;t)}f - f|_2 =$$

$$|\sum_{j<k\leq v(n;t)} T_{n,j+1}\cdots T_{n,k-1}(T_{nk} - I)f|_2 \leq$$

$$\sum_{j<k\leq v(n;t)} |T_{nk}f - f|_2 \leq b|f|_{2,2} \sum_{j<k\leq v(n;t)} q(\mu_{nk}) .$$

2. By successive application of 1.6.(A), 1. and (S) we obtain:

$$|\int f\, d\mu_n(s,t) - f(e) - \int f\, d\varkappa_n(s,t)| =$$

$$|T_{n,v(n;s)+1}\cdots T_{n,v(n;t)}f(e) - f(e) - \sum_{v(n;s)<j\leq v(n;t)} (T_{nj}f(e) - f(e))| =$$

$$|\sum_{v(n;s)<j\leq v(n;t)} (T_{nj} - I)T_{n,j+1}\cdots T_{n,v(n;t)}f(e) - \sum_{v(n;s)<j\leq v(n;t)} (T_{nj}f(e) - f(e))| =$$

$$|\sum_{v(n;s)<j\leq v(n;t)} (T_{nj} - I)(T_{n,j+1}\cdots T_{n,v(n;t)}f - f)(e)| \leq$$

$$\sum_{v(n;s)<j\leq v(n;t)} b\, q(\mu_{nj})|T_{n,j+1}\cdots T_{n,v(n;t)}f - f|_2 \leq$$

$$\sum_{v(n;s)<j\leq v(n;t)} b\, q(\mu_{nj})(b\ |f|_{2,2} \sum_{j<k\leq v(n;t)} q(\mu_{nk})) \leq$$

$$b^2\ |f|_{2,2}\ (\sum_{v(n;s)<j\leq v(n;t)} q(\mu_{nj}))^2 \leq$$

$$b^2\ |f|_{2,2}\ (\beta(t-s) + \varepsilon_n)^2.$$

3. The inequality just proved yields together with 1.6.(B) and (S):

$$\left|\int f\, d\mu_n(s,t) - f(e)\right| \le$$

$$b^2 \, |f|_{2,2} \{\beta(t-s) + \varepsilon_n\}^2 + \sum_{v(n;s)<j\le v(n;t)} \left|\int f\, d\mu_{nj} - f(e)\right| \le$$

$$b^2 \, |f|_{2,2} \{\beta(t-s) + \varepsilon_n\}^2 + \sum_{v(n;s)<j\le v(n;t)} b\, |f|_{2,2}\, q(\mu_{nj}) \le$$

$$b\, |f|_{2,2} \{\beta(t-s) + \varepsilon_n\}\{b(\beta(t-s) + \varepsilon_n) + 1\}.$$

Choosing successively $f = x_1,\dots,x_d,\ 1_G - \varphi$ yields now the second assertion of the lemma. ⌟

3.4. LEMMA. Let the infinitesimal system $\mathfrak{J}$ satisfy in addition to (S) the following tightness condition:

(T) For every $\varepsilon > 0$ there exists a compact subset C_ε of G such that

$$\mu_{n1}(\complement C_\varepsilon) + \dots + \mu_{nn}(\complement C_\varepsilon) < \varepsilon \text{ for all } n \in \mathbb{N}.$$

Then the family $\{\mu_n(s,t) : (s,t) \in S,\ n \in \mathbb{N}\}$ of probability measures (cf. 3.2.) is uniformly tight.

PROOF. Without loss of generality we may assume $e \in C_\varepsilon$ for all $\varepsilon > 0$. For every $f \in \mathcal{C}_{2,2}(G)$ Lemma 3.3. supplies us with the following inequality (*):

$$\int f\, d\mu_n(s,t) - f(e) \le \int f\, d\eta_n(s,t) + b^2\, |f|_{2,2} \{\beta(t-s) + \varepsilon_n\}^2 .$$

Let $\varepsilon > 0$. Then there exist a compact subset K of G and an $f \in \mathcal{C}_{2,2}(G)$ such that $1_{\complement K} \le f - f(e) \le 1_{\complement C_\varepsilon}$. Moreover we choose a $p \in \mathbb{N}$ such that $(b\beta)^2\, |f|_{2,2}/p \le \varepsilon$. Finally $\lim \varepsilon_n = 0$ yields the existence of an $n_o \in \mathbb{N}$ such that $b^2\, |f|_{2,2}\, (2\beta + 1)\, \varepsilon_n \le \varepsilon$ and $\varepsilon_n \le 1/p$ for $n > n_o$. Let $(s,t) \in S$ and put $c_j = s + j(t-s)/p$ for $j = 0,1,\dots,p$. In view of (*) and of

$$\mu_n(s,t) = \mu_n(c_o,c_1) * \dots * \mu_n(c_{p-1},c_p) ,$$

$$\eta_n(s,t) = \eta_n(c_o,c_1) + \dots + \eta_n(c_{p-1},c_p)$$

(cf.3.2.) we obtain for all $n > n_o$:

$$\mu_n(s,t)(\complement K^p) \leq \sum_{j=1}^{p} \mu_n(c_{j-1},c_j)(\complement K) \leq \sum_{j=1}^{p} \int f \, d\mu_n(c_{j-1},c_j) \leq$$

$$\sum_{j=1}^{p} \{\int f \, d\eta_n(c_{j-1},c_j) + b^2 \, |f|_{2,2} \, [\beta(c_j - c_{j-1}) + \varepsilon_n]^2\} =$$

$$\int f \, d\eta_n(s,t) + pb^2 \, |f|_{2,2} \, (\beta(t-s)/p + \varepsilon_n)^2 \leq$$

$$\eta_n(s,t)(\complement C_\varepsilon) + pb^2 \, |f|_{2,2} \, \{\beta^2/p^2 + 2\beta\varepsilon_n/p + \varepsilon_n^2\} \leq$$

$$\sum_{j=1}^{n} \mu_{nj}(\complement C_\varepsilon) + b^2 \, |f|_{2,2} \, \{\beta^2/p + 2\beta\varepsilon_n + (p\varepsilon_n)\varepsilon_n\} \leq 3\varepsilon .$$

Since the family $\{\mu_n(s,t): (s,t) \in S, 1 \leq n \leq n_o\}$ is finite the assertion is proved. ┘

3.5. __REMARK__. If $(\omega(s,t))_{(s,t)\in S}$ is a continuous hemigroup in $\mathcal{M}^1(G)$ and if $(\Delta_n)_{n\geq 1}$ is a sequence of decompositions $\Delta_n = \{c_{no}, c_{n1}, \ldots, c_{nn}\}$ of I then the system $(\omega(c_{n,k-1}, c_{nk}))_{k=1,\ldots,n;n\geq 1}$ satisfies condition (T) (cf. [2], Lemma 6.1.).

3.6. __THEOREM__. We consider an infinitesimal system $\mathfrak{I}$ as above satisfying condition (S) and keep the notations of 3.2. Let there exist a universal subnet $(n(\alpha))_{\alpha\in J}$ of $\mathbb{N}$ and measures $\mu(s,t) \in \mathcal{M}^1(G)$ such that $\mu(s,t) = \mathfrak{T}_w\text{-}\lim_{\alpha\in J} \mu_{n(\alpha)}(s,t)$ for all $(s,t) \in S$.

Then there exists a constant $c > 0$ such that the following assertions are valid:

(i) $(\mu(s,t))_{(s,t)\in S}$ is a continuous hemigroup satisfying

$q(\mu(s,t)) \leq c\,(t-s)$ for all $(s,t) \in S$.

(ii) For every $t \in I$ there exists a $B(t) \in \mathbb{A}(G)$ such that

$B(t)(f) = \lim_{\alpha\in J} \int f \, d\varkappa_{n(\alpha)}(0,t)$ for all $f \in \mathcal{L}_2(G)$.

The mapping $B(.)$ of I into $\mathbb{A}(G)$ is increasing with $B(0) = 0$.

(iii) (L) $|B(t) - B(s)|_2 \leq c\,(t-s)$ for all $(s,t) \in S$.

(iv) For all $f \in \mathcal{L}_{2,2}(G)$ and for all $(s,t) \in S$

(E) $|\int f \, d\mu(s,t) - f(e) - \{B(t)(f) - B(s)(f)\}| \leq c \, |f|_{2,2} \, (t-s)^2$.

PROOF. 1. In view of Lemma 3.1. and of Lemma 1.6.(A) we have

$$(*)\quad \left|\int f\, d\varkappa_n(s,t)\right| \le b\, |f|_2\, \{\beta(t-s) + \varepsilon_n\}$$

for all $f \in \mathcal{C}_2(G)$, $(s,t) \in S$ and $n \in \mathbb{N}$. Taking into account Lemma 3.3. this yields (i).

Moreover in view of $(*)$ there exist almost positive linear functionals $B(s,t)$ on $\mathcal{D}(G)$ such that $\lim\limits_{\alpha\in J} \int f\, d\varkappa_{n(\alpha)}(s,t) = B(s,t)(f)$ for all $f \in \mathcal{D}(G)$. Taking into account $(*)$ we have $|B(s,t)(f)| \le b\, |f|_2\, \beta(t-s)$ for all $f \in \mathcal{D}(G)$. In view of 3.2. we have $B(r,s) + B(s,t) = B(r,t)$ for all (r,s), $(s,t) \in S$.

Finally in view of Lemma 3.3. we have for all $f \in \mathcal{D}(G)$ (with $\gamma = (\beta b)^2$):

$$(**)\quad \left|\int f\, d\mu(s,t) - f(e) - B(s,t)(f)\right| \le \gamma\, |f|_{2,2}\, (t-s)^2.$$

2. For every pair $(s_o,t_o) \in S$ the linear functional $B(s_o,t_o)$ is normed. [Let $(f_n)_{n\ge 1}$ be a norming sequence in $\mathcal{D}(G)$ such that $|f_n|_{2,2} \le c < \infty$ for all $n \in \mathbb{N}$ with some appropriate constant $c > 0$ (Lemma 1.7.). Then $(**)$ yields for all $(s,t) \in S$:

$$(***)\quad 0 \ge B(s,t)(f_n) \ge -\gamma c\,(t-s)^2 + \left(\int f_n\, d\mu(s,t) - 1\right).$$

For some $p \in \mathbb{N}$ let $c_j = s_o + (j/p)(t_o - s_o)$ $(j = 0,1,\dots,p)$. Then by $(***)$ we have for all $n \in \mathbb{N}$:

$$0 \ge B(s_o,t_o)(f_n) = \sum_{j=1}^{p} B(c_{j-1},c_j)(f_n) \ge$$

$$-\gamma c \sum_{j=1}^{p} (c_j - c_{j-1})^2 + \sum_{j=1}^{p} \left(\int f_n\, d\mu(c_{j-1},c_j) - 1\right).$$

If n tends to infinity this yields:

$$0 \ge \sup\{B(s_o,t_o)(f_n) : n \in \mathbb{N}\} \ge$$

$$-\gamma c p[(t_o - s_o)/p]^2 = -(\gamma c/p)(t_o - s_o)^2.$$

Since $p \in \mathbb{N}$ was arbitrary we conclude $\sup\{B(s_o,t_o)(f_n) : n \in \mathbb{N}\} = 0$.]

3. In view of 1. and 2. we have for all $(s,t) \in S$:

$$\lim_{\alpha\in J} \int \varphi\, d\varkappa_{n(\alpha)}(s,t) = \lim_{\alpha\in J} \int (\varphi - 1_G)\, d\varkappa_{n(\alpha)}(s,t) =$$

$B(s,t)(\varphi - 1_G) = B(s,t)(\varphi)$.

4. Let $B(t) = B(0,t)$ for all $t \in I$. Then we obviously have $B(s,t) = B(t) - B(s)$ for all $(s,t) \in S$.

Taking into account 1., 2., 3. together with Lemma 1.8. we finally obtain the remaining assertions (ii), (iii) and (iv). ┘

3.7. COROLLARY. Let the infinitesimal system $\mathfrak{J}$ satisfy conditions (S) and (T). Then the assumptions of Theorem 3.6. are fulfilled for every universal subnet of $\mathbb{N}$.

In particular the sequence $(\mu_{n1} * \cdots * \mu_{nn})_{n\geq 1}$ of row products of $\mathfrak{J}$ is uniformly tight and each of its limit points can be embedded into a continuous hemigroup possessing the properties described in Theorem 3.6.

PROOF. This is a consequence of Lemma 3.4. and Theorem 3.6. ┘

3.8. REMARK. A related result to Corollary 3.7. has been obtained by D.Wehn [9]. But the conditions imposed on the system $\mathfrak{J}$ by Wehn are considerably stronger than ours and are therefore not repeated here.

4. Characterization of Lipschitz continuous hemigroups

4.1. DEFINITION. A hemigroup $(\omega(s,t))_{(s,t)\in S}$ in $\mathcal{M}^1(G)$ is said to be *Lipschitz continuous* (with constant $c > 0$) if $q(\omega(s,t)) \leq c\,(t-s)$ for all $(s,t) \in S$.

For example the limit hemigroups of Theorem 3.6. are Lipschitz continuous.

4.2. LEMMA. Let $(\omega(s,t))_{(s,t)\in S}$ be a Lipschitz continuous hemigroup. Then there exists a constant $\gamma > 0$ such that for all $(s,t) \in S$ and for all $h,k \in \mathbb{R}$ with $(s,t+k),(s+h,t) \in S$ the following inequalities hold:

(i) $\left|\int f\,d\omega(s+h,t) - \int f\,d\omega(s,t)\right| \leq \gamma |f|_2\,|h|$ $\quad (f \in \mathcal{C}_2(G))$;

(ii) $\| T_{\omega(s,t+k)}f - T_{\omega(s,t)}f \| \leq \gamma |f|_2^{\sim}\,|k|$ $\quad (f \in \mathcal{C}_2^{\sim}(G))$.

PROOF. This follows readily by 1.6.(A) (and 1.2.) and by 1.6.(B). ┘

4.3. THEOREM. Let $H = (\omega(s,t))_{(s,t)\in S}$ be a continuous hemigroup in $\mathcal{M}^1(G)$. Then the following assertions are equivalent:

(i) H is Lipschitz continuous (with constant c).

(ii) There exists an increasing mapping B(.) of I into $\mathbb{A}(G)$ with $B(0) = 0$ such that B(.) and H satisfy conditions (L) and (E) of Theorem 3.6.

(iii) There exists an admissible mapping A(.) of I into $\mathbb{A}(G)$ such that the following equations hold:

(EE 1) $$T_{\omega(s,t)} f - f = \int_s^t T_{\omega(s,\tau)} N_{A(\tau)} f \, d\tau$$

for all $f \in \mathcal{C}_2^{\sim}(G)$ and for all $(s,t) \in S$.

(EE 2) $$\int f \, d\omega(s,t) - f(e) = \int_s^t A(\sigma)(T_{\omega(\sigma,t)} f) \, d\sigma$$

for all $f \in \mathcal{C}_2(G)$ and for all $(s,t) \in S$.

(iv) There exists an admissible mapping $A_1(.)$ of I into $\mathbb{A}(G)$ such that H and $A_1(.)$ satisfy (EE 1).

(v) There exists an admissible mapping $A_2(.)$ of I into $\mathbb{A}(G)$ such that H and $A_2(.)$ satisfy (EE 2).

PROOF. '(i)$\Longrightarrow$(ii)' Let $\mu_{nk} = \omega(\frac{k-1}{n}, \frac{k}{n})$ for all $k \in \{1,\dots,n\}$ and $n \in \mathbb{N}$. By the continuity of H the system $\mathfrak{I} = (\mu_{nk})_{k=1,\dots,n; n\geq 1}$ is infinitesimal. For $t \in I$ and $n \in \mathbb{N}$ let $[nt]$ denote the largest integer $\leq nt$. Since H is Lipschitz continuous with constant $c > 0$ we have:

$$\sum_{[ns]<k\leq[nt]} q(\mu_{nk}) \leq c\,(t-s) + \frac{c}{n} \qquad ((s,t) \in S,\ n \in \mathbb{N}).$$

Hence $\mathfrak{I}$ satisfies condition (S) of Lemma 3.1. With the notations of 3.2. we have $\mu_n(s,t) = \omega([ns]/n, [nt]/n)$. Now $\lim_{n\geq 1} [nr]/n = r$ (all $r \in I$) yields $\lim_{n\geq 1} \mu_n(s,t) = \omega(s,t)$ (all $(s,t) \in S$). Therefore the assumptions of Theorem 3.6. are satisfied. Hence the assertion.

'(ii)$\Longrightarrow$(iii)' Since (L) holds Lemma 2.8. can be applied to B(.). Hence there exist a zero set Z in I and an admissible mapping A(.) such that $dB(t)(f)/dt = A(t)(f)$ for all $f \in \mathcal{C}_2(G)$ and $t \in I - Z$. Combined with (E) this yields for $s \notin Z$ and $f \in \mathcal{L}_{2,2}(G)$ (and $h,k \in \mathbb{R}_+$):

$$(*) \qquad \lim_{h+k\downarrow 0} \frac{1}{h+k} \{\int f \, d\omega(s-h,s+k) - f(e)\} = A(s)(f).$$

Lemma 1.8. shows that (*) even holds for all $f \in \mathcal{L}_2(G)$. Moreover (L) and (E) imply that H is a Lipschitz continuous hemigroup. Finally we put $T(s,t) = T_{\omega(s,t)}$ for all $(s,t) \in S$.

Proof of (EE 1): Let $f \in \tilde{\mathcal{L}}_2(G)$ and $t \notin Z$. Then for $k > 0$ we have:

$$\frac{1}{k} \{\int f \, d\omega(s,t+k) - \int f \, d\omega(s,t)\} = \int \frac{1}{k} \{T(t,t+k)f - f\} \, d\omega(s,t).$$

Lemma 4.2.(ii) yields $|\{T(t,t+k)f(x) - f(x)\}/k| \leq \gamma |f|_2^{\sim}$ for all $x \in G$; and by (*) we have:

$$\lim_{k\downarrow 0} \frac{1}{k} \{T(t,t+k)f(x) - f(x)\} = \lim_{k\downarrow 0} \frac{1}{k} \{\int {}_x f \, d\omega(t,t+k) - {}_x f(e)\} =$$

$$A(t)({}_x f) = N_{A(t)} f(x) \qquad \text{(all } x \in G).$$

Hence by Lebesgue's theorem we obtain:

$$\lim_{k\downarrow 0} \frac{1}{k} \{\int f \, d\omega(s,t+k) - \int f \, d\omega(s,t)\} =$$

$$\int N_{A(t)} f \, d\omega(s,t) = T(s,t) N_{A(t)} f(e).$$

In particular $t \longrightarrow T(s,t)N_{A(t)}f(e)$ is measurable.

Again by Lemma 4.2.(ii) the mapping $t \longrightarrow \int f \, d\omega(s,t)$ of $[s,1]$ into $\mathbb{R}$ is Lipschitz continuous and hence almost everywhere differentiable. But as shown above its derivative coincides almost everywhere with $T(s,t)N_{A(t)}f(e)$. This yields for $t \in [s,1]$:

$$(**) \quad T(s,t)f(x) - f(x) = \int_s^t T(s,\tau) N_{A(\tau)} f(x) \, d\tau$$

first of all for $x = e$ and then for all $x \in G$ (replacing f by ${}_x f$). The measurability of $t \longrightarrow T(s,t)N_{A(t)}f(x)$ for all $x \in G$ yields the strong measurability of $t \longrightarrow T(s,t)N_{A(t)}f$ (Lemma 2.10.). Finally in view of 2.4. and 2.6. this mapping is norm bounded; hence Bochner integrable ([10], 133, Th.1). Thus (**) implies (EE 1).

Proof of (EE 2): Let $f \in \mathcal{L}_2(G)$ and $s \notin Z$. Then for $h > 0$ we have: $\int f \, d\omega(s-h,t) = T(s-h,s)(T(s,t)f)(e)$. Since $T(s,t)f \in \mathcal{L}_2(G)$ (cf.1.2) relation (*) yields:

$$\lim_{h\downarrow 0} \frac{1}{h} \{\int f\, d\omega(s-h,t) - \int f\, d\omega(s,t)\} = A(s)(T(s,t)f).$$

In particular $s \longrightarrow A(s)(T(s,t)f)$ is measurable. By Lemma 4.2.(i) the mapping $s \longrightarrow \int f\, d\omega(s,t)$ of $[0,t]$ into $\mathbb{R}$ is Lipschitz continuous and hence almost everywhere differentiable. As just has been shown its derivative coincides with $A(s)(T(s,t)f)$ almost everywhere. Thus (EE 2) holds.

'(iii)$\Longrightarrow$(iv)' and '(iii)$\Longrightarrow$(v)' are trivial.

'(iv)$\Longrightarrow$(i)' and '(v)$\Longrightarrow$(i)' Let $|A_1(t)|_2^{\sim} \leq \gamma'$ (respectively $|A_2(t)|_2 \leq \gamma$) for all $t \in I$. Then (EE 1) (respectively (EE 2)) implies $|\int f\, d\omega(s,t)| \leq \gamma |f|_2^{\sim} (t-s)$ for all $f \in \mathcal{C}_2^{\sim}(G) \oplus \mathbb{R}.1_G$ (respectively $\leq \gamma |f|_2 (t-s)$ for all $f \in \mathcal{C}_2(G) \oplus \mathbb{R}.1_G$) such that $f(e) = 0$ (all $(s,t) \in S$). Choosing successively $f = x_1,\dots,x_d, \varphi$ now yields (i). ⌋

4.4. <u>REMARKS</u>. 1. The equations (EE 1) and (EE 2) in Theorem 4.3.(iii) are called the (forward respectively backward) <u>evolution equations</u> (in integral form). In general (EE 2) can not be replaced by its strong version (but see Section 8).

2. If $(\mu_t)_{t\geq 0}$ is a continuous convolution semigroup in $\mathcal{M}^1(G)$ then by $\omega(s,t) = \mu_{t-s}$, $(s,t) \in S$, there is defined a continuous hemigroup $H = (\omega(s,t))_{(s,t)\in S}$ in $\mathcal{M}^1(G)$. If A is the generating functional of $(\mu_t)_{t\geq 0}$ (cf.2.3.) then (EE 1) holds with the admissible mapping $A(.) \equiv A$ (cf. [10], 239). Hence in view of Theorem 4.3. H is an example for a Lipschitz continuous hemigroup.

4.5. <u>COROLLARY</u>. (i) Let (EE 1) hold. Then for every $f \in \mathcal{C}_2^{\sim}(G)$ there exists a zero set $Z(f)$ in I such that $\partial T_{\omega(s,t)}f/\partial t = T_{\omega(s,t)}N_{A(t)}f$ for all $t \in [s,1] \sim Z(f)$ and for $s \in I$.

(ii) Let (EE 2) hold. Then there exists a zero set Z_o in I such that $\partial\{\int f\, d\omega(s,t)\}/\partial s = -A(s)(T_{\omega(s,t)}f)$ for all $f \in \mathcal{C}_2(G)$,

$s \in [0,t] \setminus Z_o$ and $t \in I$.

PROOF. Let $T(s,t) = T_{\omega(s,t)}$ for all $(s,t) \in S$.

(i) Let $(s,t) \in S$ be fixed and let $k \in \mathbb{R}^*$ such that $(s,t+k) \in S$. Since $T(s,t)$ is a bounded operator with norm ≤ 1 (EE 1) yields:

$$\left\| \frac{1}{k} \{T(s,t+k)f - T(s,t)f\} - T(s,t)N_{A(t)}f \right\| =$$

$$\frac{1}{|k|} \left\| \int_t^{t+k} \{T(s,\tau)N_{A(\tau)}f - T(s,t)N_{A(t)}f\} \, d\tau \right\| \leq$$

$$\frac{1}{k} \int_t^{t+k} \| N_{A(\tau)}f - N_{A(t)}f \| \, d\tau \; +$$

$$\sup\{ \|(T(s,\tau) - T(s,t))N_{A(t)}f\| : |t-\tau| \leq |k|\}.$$

Since the hemigroup H is continuous the second summand of the last term tends to zero if k tends to zero. But $N_{A(.)}f$ is Bochner integrable (by Corollary 2.11. and [10], 133, Th.1). Hence there exists a zero set $Z(f)$ in I such that the first summand of the last term tends to zero if k tends to zero and if $t \notin Z(f)$ ([6], 88, Cor.2). This proves (i).

(ii) Let $(s,t) \in S$ be fixed and let $h \in \mathbb{R}^*$ such that $(s+h,t) \in S$. (EE 2) yields:

$$\left| \frac{1}{h} \{\int f \, d\omega(s+h,t) - \int f \, d\omega(s,t)\} + A(s)(T(s,t)f) \right| =$$

$$\left| -\frac{1}{h} \int_s^{s+h} \{A(\sigma)(T(\sigma,t)f) - A(s)(T(s,t)f)\} \, d\sigma \right| \leq$$

$$(\sup\{|A(\tau)|_2 : \tau \in I\})(\sup\{|T(\sigma,t)f - T(s,t)f|_2 : |s-\sigma| \leq |h|\}) \; +$$

$$\frac{1}{h} \int_s^{s+h} |A(\sigma)(T(s,t)f) - A(s)(T(s,t)f)| \, d\sigma .$$

If h tends to zero the first summand of the last term tends to zero too by the continuity of H. In view of [6], 88, Corollary 2 and of Lemma 1.8. there exists a zero set Z_o in I such that

$$\lim_{h\to 0} \frac{1}{h} \int_s^{s+h} |A(\sigma)(g) - A(s)(g)| \, d\sigma = 0$$

for all $g \in \mathcal{L}_2(G)$ and $s \notin Z_o$. This proves (ii). ⌟

4.6. COROLLARY. The mappings $A(.)$, $A_1(.)$, $A_2(.)$ and $B(.)$ of I into $\mathbb{A}(G)$ figuring in Theorem 4.3. are uniquely determined by the hemigroup H, and $A(.) = A_1(.) = A_2(.)$ almost everywhere, $B(t) = \int_0^t A(\tau)d\tau$ for all $t \in I$.

More precisely we have:

(i) $$B(t)(f) = \lim_{n\geq 1} \sum_{1\leq k\leq [nt]} \{\int f \, d\omega(\tfrac{k-1}{n},\tfrac{k}{n}) - f(e)\}$$

for all $f \in \mathscr{C}_2(G)$ and $t \in I$.

(ii) $$\frac{\partial}{\partial t}\Big|_{t=s} \{\int f \, d\omega(s,t)\} = A_1(s)(f)$$

for all $f \in \tilde{\mathscr{C}}_2(G)$ and $s \in I \sim Z(f)$.

(iii) $$\frac{\partial}{\partial s}\Big|_{s=t} \{\int f \, d\omega(s,t)\} = - A_2(t)(f)$$

for all $f \in \mathscr{C}_2(G)$ and $t \in I \sim Z_0$.

PROOF. Let $B(.)$ and $B'(.)$ be two mappings as in Theorem 4.3.(ii). Then (E) implies:

(*) $$|\{B(t)(f) - B(s)(f)\} - \{B'(t)(f) - B'(s)(f)\}| \leq c\, |f|_{2,2}\, (t - s)^2$$

for all $f \in \mathscr{C}_{2,2}(G)$ and $(s,t) \in S$. Fix $t \in I$ and choose $p \in \mathbb{N}$. Then (*) implies:

$$|B(t)(f) - B'(t)(f)| =$$

$$|\sum_{j=1}^{p} \{[B(\tfrac{j}{p}t)(f) - B(\tfrac{j-1}{p}t)(f)] - [B'(\tfrac{j}{p}t)(f) - B'(\tfrac{j-1}{p}t)(f)]\}| \leq$$

$$pc(\tfrac{t}{p})^2\, |f|_{2,2} = \tfrac{c}{p}\, t^2\, |f|_{2,2}.$$

Since p was arbitrary this proves $B(.) = B'(.)$.

Now the proof of '(i)$\Longrightarrow$(ii)' in Theorem 4.3. together with Theorem 3.6.(ii) yield (i).

If (iv) (respectively (v)) of Theorem 4.3. is satisfied then 4.5.(i) (respectively 4.5.(ii)) yields (ii) (respectively (iii)). Hence $A(.) = A_1(.) = A_2(.)$ almost everywhere (in view of Theorem 4.3.). Finally $B(t) = \int_0^t A(\tau)d\tau$ by the proof of '(ii)$\Longrightarrow$(iii)' in Theorem 4.3.⌟

5. Existence and uniqueness of Lipschitz continuous hemigroups

5.1. THEOREM. For every admissible mapping $A(.)$ of I into $\mathbb{A}(G)$ there exists a continuous hemigroup in $\mathcal{M}^1(G)$ satisfying (EE 1) and (EE 2).

5.2. For the proof of the theorem we need some preparations: Let $c = \sup\{|A(t)|_2^\sim : t \in I\}$. For every pair $(s,t) \in S$ let $B(s,t) = \int_s^t A(\sigma)\, d\sigma$ in the sense of Lemma 2.7. By $(\omega_r(s,t))_{r\geq 0}$ we denote the continuous convolution semigroup corresponding to $B(s,t)$ i.e. $\omega_r(s,t) = \operatorname{Exp}(r\, B(s,t))$ for all $r \in \mathbb{R}_+$ (cf. 2.3.).

Now let $(\Delta_n)_{n\geq 1}$ be a sequence of decompositions $\Delta_n = \{c_{no}, c_{n1}, \ldots, c_{nn}\}$ of I such that $\lim |\Delta_n| = 0$. For every $r \in I$ let $v(n;r) \in \{0,1,\ldots,n\}$ be defined by $c_{n,v(n;r)-1} < r \leq c_{n,v(n;r)}$. For abbreviation we put $c_n(r) = c_{n,v(n;r)}$. Since $|c_n(r) - r| \leq |\Delta_n|$ we have $\lim_{n\geq 1} c_n(r) = r$ uniformly in $r \in I$. Finally let

$$\mu_{nk} = \omega_1(c_{n,k-1}, c_{nk}) = \operatorname{Exp} B(c_{n,k-1}, c_{nk}) .$$

We study the triangular system $\mathfrak{J} = (\mu_{nk})_{k=1,\ldots,n;n\geq 1}$.

5.3. LEMMA. (i) There exists a constant $\gamma > 0$ such that $q(\omega_r(s,t)) \leq \gamma(t-s)r$ for all $(s,t) \in S$ and $r \in \mathbb{R}_+$.

(ii) $\sum_{v(n;s)<j\leq v(n;t)} q(\mu_{nj}) \leq \gamma\{(t-s) + 2|\Delta_n|\}$

for all $(s,t) \in S$ and $n \in \mathbb{N}$.

(iii) The system $\mathfrak{J}$ is infinitesimal and satisfies condition (S) of 3.1. (with the mappings $v(n;.)$ defined in 5.2. and with $\varepsilon_n = 2\gamma|\Delta_n|$).

PROOF. (i) For all $f \in \mathcal{C}_2^\sim(G)$ we have (cf. [10], 239):

$$(*) \qquad \int f\, d\omega_r(s,t) - f(e) = \int_0^r \{\int N_{B(s,t)} f\, d\omega_\tau(s,t)\}\, d\tau .$$

Since $\| N_{B(s,t)} f \| \leq c\, |f|_2^\sim (t-s)$ (cf. 2.4. and Lemma 2.7.) equation (*) yields $|\int f\, d\omega_r(s,t) - f(e)| \leq c\, |f|_2^\sim (t-s)r$. Now (i) follows immediately by choosing successively $f = x_1, \ldots, x_d, 1_G - \varphi$.

(ii) In view of (i) we have:

$$\sum_{v(n;s)<j\leq v(n;t)} q(\mu_{nj}) =$$

$$\sum_{v(n;s)<j\leq v(n;t)} q(\omega_1(c_{n,j-1},c_{nj})) \leq$$

$$\sum_{v(n;s)<j\leq v(n;t)} \gamma(c_{nj} - c_{n,j-1}) = \gamma(c_n(t) - c_n(s)).$$

But $c_n(t) - c_n(s) \leq (t - s) + 2|\Delta_n|$ proves (ii).

(iii) is an immediate consequence of (i) and (ii). ⌟

5.4. LEMMA. Let $\varkappa_n(s,t) = \sum_{v(n;s)<j\leq v(n;t)} (\mu_{nj} - \varepsilon_e)$ for all $(s,t) \in S$ and $n \in \mathbb{N}$ (cf.3.2.). Then for all $f \in \mathcal{C}_{2,2}(G)$ we have

$$\lim_{n\geq 1} \int f\, d\varkappa_n(s,t) = B(s,t)(f) \quad \text{uniformly in } (s,t) \in S.$$

PROOF. Let $f \in \mathcal{C}_{2,2}(G)$. Since $XN_{B(s,t)}f = N_{B(s,t)}Xf$ for all $X \in \mathfrak{g}$ (cf. 1.2.) we have $N_{B(s,t)}f \in \mathcal{C}_2(G)$ and $|N_{B(s,t)}f|_2 \leq c\,|f|_{2,2}\,(t - s)$ (all $(s,t) \in S$). Together with Lemma 1.6.(A) and Lemma 5.3.(i) (and formula (*) in its proof) this yields:

$$|\int f\, d\omega_1(s,t) - f(e) - B(s,t)(f)| =$$

$$|\int_0^1 \{\int (N_{B(s,t)}f - N_{B(s,t)}f(e))\, d\omega_r(s,t)\}\, dr| \leq$$

$$|\int_0^1 b|N_{B(s,t)}f|_2\, q(\omega_r(s,t))\, dr \leq$$

$$bc\,|f|_{2,2}\,(t - s)\int_0^1 q(\omega_r(s,t))\, dr \leq bc\gamma|f|_{2,2}\,(t - s)^2 .$$

It follows:

$$|\int f\, d\varkappa_n(s,t) - B(c_n(s),c_n(t))(f)| =$$

$$|\sum_{v(n;s)<j\leq v(n;t)} (\int f\, d\omega_1(c_{n,j-1},c_{nj}) - f(e) - B(c_{n,j-1},c_{nj})(f))| \leq$$

$$bc\gamma|f|_{2,2}\sum_{v(n;s)<j\leq v(n;t)} (c_{nj} - c_{n,j-1})^2 \leq bc\gamma|f|_{2,2}\,|\Delta_n|.$$

Together with $|B(c_n(s),c_n(t))(f) - B(s,t)(f)| \leq 2c\,|f|_{2,2}\,|\Delta_n|$ (cf. 5.2.) this proves our assertion. ⌟

5.5. LEMMA. The system $\mathfrak{J}$ satisfies property (T) (cf. 3.4.).

PROOF. Let $(f_m)_{m\geq 1}$ be a norming sequence in $\mathfrak{D}(G)$ (cf. 2.2.), $g_m = 1_G - f_m$, and $K_m = \text{supp}(f_m)$ ($m \in \mathbb{N}$). Taking into account Lemma 5.4. we have for all $m \in \mathbb{N}$:

$$\overline{\lim_{n\geq 1}} \sum_{k=1}^{n} \mu_{nk}(\complement K_m) = \overline{\lim_{n\geq 1}}\ \varkappa_n(0,1)(\complement K_m) \leq \lim_{n\geq 1} \int g_m \, d\varkappa_n(0,1) =$$

$$- \lim_{n\geq 1} \int f_m \, d\varkappa_n(0,1) = -B(0,1)(f_m).$$

But the sequence $(B(0,1)(f_m))_{m\geq 1}$ ascends to 0 since $B(0,1)$ is normed.┘

5.6. PROOF OF THEOREM 5.1. Lemma 5.3. and Lemma 5.5. justify the application of Corollary 3.7. (and hence of Theorem 3.6.) to our system $\mathfrak{J}$. Taking into account Lemma 5.4., Theorem 4.3. and Corollary 4.6. this yields the assertion. ┘

5.7. THEOREM. For every admissible mapping $A(.)$ of I into $\mathbb{A}(G)$ there exists exactly one continuous hemigroup in $\mathcal{M}^1(G)$ satisfying (EE 2).

PROOF. The existence has been proved in Theorem 5.1. The proof of the uniqueness will be carried out with the aid of the following lemma which is a modification of Lemma 1.5.14 in [4]:

5.8. LEMMA. Let $J = [0,a]$ be a compact interval, E a compact space and $F: J \times E \longrightarrow \mathbb{R}$ a continuous function. Let there exist a second function $f: J \times E \longrightarrow \mathbb{R}$ such that $f(.,x)$ is Lebesgue integrable for every $x \in E$ and such that $F(t,x) = \int_0^t f(s,x)ds$ for all $(t,x) \in J \times E$. Furthermore for every pair $(s,y) \in J \times E$ such that $F(s,y) = \min\{F(s,x): x \in E\}$ let there exist a function $\chi: J \longrightarrow \mathbb{R}_+$ such that $\lim_{t\to s} \chi(t) = 0$ and $f(t,y) \geq -\chi(t)$ for all $t \in J$.

Then we have $F(t,x) \geq 0$ for all $(t,x) \in J \times E$.

[For all $(t,x) \in J \times E$ we put $F_1(t,x) = F(t,x) + 1$ and $Q(t,x) = e^{ct} F_1(t,x)$ where $c > 0$ is arbitrary. It suffices to prove $Q \geq 1$ for $c \downarrow 0$ yields $F_1 \geq 1$ and hence $F \geq 0$.

Let us assume the contrary. Then by the compactness of $J \times E$

there exist $q \in]0,1[$ and $(s,y) \in J \times E$ such that $Q(s,y) = q$ and $Q(t,x) > q$ for all $(t,x) \in [0,s[\times E$. Since $Q(0,x) = 1$ for all $x \in E$ we have $s > 0$. Moreover $Q(s,y) = \min\{Q(s,x): x \in E\}$ and hence $F(s,y) = \min\{F(s,x): x \in E\}$. Let χ be the function associated with (s,y) by our assumption. Then there exists a $\delta \in]0,s]$ such that $\chi(t) \leq \frac{1}{2}cqe^{-ca}$ for all $t \in [s-\delta,s]$. Futhermore by assumption there is a zero set Z in J such that $\partial F_1(t,y)/\partial t = f(t,y)$ if $t \notin Z$. Then we have for all $t \in [s-\delta,s] \setminus Z$:

$$\frac{\partial}{\partial t} Q(t,y) = e^{ct} \frac{\partial}{\partial t} F_1(t,y) + cQ(t,y) = e^{ct} f(t,y) + cQ(t,y) \geq$$

$$-e^{ct} \chi(t) + cQ(t,y) \geq -e^{ca} \tfrac{1}{2}cqe^{-ca} + cq = \tfrac{1}{2}cq.$$

Together with $F(.,y)$ also $Q(.,y)$ is absolutely continuous. This yields

$$Q(s,y) - Q(s-\delta,y) = \int_{s-\delta}^{s} \frac{\partial}{\partial t} Q(t,y)\, dt \geq \tfrac{1}{2}cq\delta > 0.$$

Hence $Q(s,y) > Q(s-\delta,y)$. But this is a contradiction to the choice of (s,y). Thus we have $Q \geq 1$.]

5.9. <u>CONTINUATION OF THE PROOF OF THEOREM 5.7</u>. Let $E = G \cup \{\omega\}$ be the Alexandrov compactification of G. We define $\omega x = x\omega = \omega$ for all $x \in E$. We extend every $g \in \mathcal{C}^0(G)$ to a continuous function on E by $g(\omega) = 0$. Finally we choose an $a \in [0,1]$ and put $J = [0,a]$.

Now let $(\mu(s,t))_{(s,t)\in S}$ and $(\nu(s,t))_{(s,t)\in S}$ be two continuous hemigroups in $\mathcal{M}^1(G)$ satisfying (EE 2) with the given admissible mapping $A(.)$. We fix $f \in \mathcal{L}_2(G)$ and define for all $(t,x) \in J \times E$ (cf.1.2):

$$g\{t\} = T_{\mu(a-t,a)}f - T_{\nu(a-t,a)}f;$$

$$f(t,x) = A(a-t)({}_x g\{t\}); \quad F(t,x) = g\{t\}(x).$$

Then F is continuous on $J \times E$ ([4], Th.1.5.5), and we have for all $(t,x) \in J \times E$:

$$F(t,x) = \int {}_x f\, d\mu(a-t,a) - \int {}_x f\, d\nu(a-t,a) =$$

$$\int_{a-t}^{a} A(\sigma)[T_{\mu(\sigma,a)}({}_x f) - T_{\nu(\sigma,a)}({}_x f)]\, d\sigma =$$

$$\int_{a-t}^{a} A(\mathfrak{s})\, (_x(T_{\mu(\mathfrak{s},a)}f - T_{\nu(\mathfrak{s},a)}f))\, d\mathfrak{s} =$$

$$\int_{o}^{t} A(a-s)\, (_xg\{s\})\, ds = \int_{o}^{t} f(s,x)\, ds.$$

[Here the fourth equality follows by the transformation $\mathfrak{s} = a - s$.]
Now let $(s,y) \in J \times E$ such that $F(s,y) = \min\{F(s,x): x \in E\}$. Then we have for all $x \in E$:

$$(_yg\{s\})(e) = g\{s\}(y) = F(s,y) \leq F(s,yx) = (_yg\{s\})(x).$$

Since every functional $A(t)$ is almost positive and normed we obtain for all $t \in J$:

$$f(t,y) = A(a-t)(_yg\{t\}) = A(a-t)(_yg\{t\} - _yg\{s\}) + A(a-t)(_yg\{s\}) \geq$$

$$A(a-t)(_yg\{t\} - _yg\{s\}) \geq -|A(a-t)|_2 \; |_yg\{t\} - _yg\{s\}|_2 \geq$$

$$-c \; |(T_{\mu(a-t,a)} - T_{\mu(a-s,a)} - T_{\nu(a-t,a)} + T_{\nu(a-s,a)})(_yf)|_2.$$

[Here we have put $c = \sup\{|A(t)|_2: t \in J\}$. Moreover we have used that $f \in \mathcal{L}_2(G)$ implies $_yf \in \mathcal{L}_2(G)$ for all $y \in G$.] Let

$$\chi(t) = c \, |(T_{\mu(a-t,a)} - T_{\mu(a-s,a)} - T_{\nu(a-t,a)} + T_{\nu(a-s,a)})(_yf)|_2$$

for all $t \in J$. Obviously we have $\chi \geq 0$ and $\lim_{t \to s} \chi(t) = 0$. Hence all the assumptions of Lemma 5.8. are fulfilled. It follows:

$$0 \leq F(t,e) = \int f\, d\mu(a-t,a) - \int f\, d\nu(a-t,a),$$

hence $\int f\, d\nu(a-t,a) \leq \int f\, d\mu(a-t,a)$ for all $t \in J$. By exchanging the hemigroups we obtain the inverse inequality.

Thus $\int f\, d\mu(a-t,a) = \int f\, d\nu(a-t,a)$ for all $f \in \mathcal{L}_2(G)$, $a \in \;]0,1]$, and $t \in [0,a]$. This proves the uniqueness of the hemigroup. ⌟

5.10. REMARK. A combination of Theorem 4.3., Corollary 4.6. and Theorem 5.7. yields the following results:

1. For every admissible mapping $A(.)$ of I into $\mathbb{A}(G)$ there exists exactly one continuous hemigroup in $\mathcal{M}^1(G)$ satisfying (EE 1).

2. For every increasing mapping $B(.)$ of I into $\mathbb{A}(G)$ such that $B(0) = 0$ and 3.6.(L) are fulfilled there exists exactly one continuous hemigroup in $\mathcal{M}^1(G)$ satisfying 3.6.(E).

6. Product integral representation of Lipschitz continuous hemigroups

6.1. Let $A(.)$ be an admissible mapping of I into $\mathbb{A}(G)$ and let $H = (\mu(s,t))_{(s,t)\in S}$ be the associated hemigroup constructed in Section 5. For a fixed pair $(s,t) \in S$ let $\mathfrak{Z}(s,t)$ denote the directed set of decompositions $Z = \{z(0), z(1), \ldots, z(m_Z)\}$ of $[s,t]$. Then we have

$$T_{\mu(s,t)} = \lim_{Z\in\mathfrak{Z}(s,t)} \prod_{j=1}^{m_Z} \operatorname{Exp}\left(\int_{z(j-1)}^{z(j)} N_{A(\tau)}\, d\tau\right)$$

with respect to the strong operator topology (cf. 2.3.).

[Let $(Z_n)_{n\geq 1}$ be a sequence in $\mathfrak{Z}(s,t)$ such that $\lim_{n\geq 1} |Z_n| = 0$. Every Z_n can be extended to a decomposition $\Delta_n = \{c_{no}, c_{n1}, \ldots, c_{nk(n)}\}$ of I such that $\lim_{n\geq 1} |\Delta_n| = 0$. With the notations of 5.2. we have $c_n(s) = s$ and $c_n(t) = t$ for all $n \in \mathbb{N}$. (In fact our general assumption $k(n) = n$ on the length of the n-th decomposition is nowhere necessary.) But then the construction of the hemigroup H (carried out in Sections 3 and 5) together with the uniqueness of H (Theorem 5.7.) yields:

$$T_{\mu(s,t)} = \lim_{n\geq 1} \prod_{j=1}^{m_{Z(n)}} \operatorname{Exp} N_{B(z(j-1),z(j))}, \quad B(\sigma,\tau) = \int_{\sigma}^{\tau} A(r)\, dr$$

for all $(\sigma,\tau) \in S$ (taking into account [4], 1.5.5). This proves our assertion.]

For a particular class of admissible mappings the associated (Lipschitz continuous) hemigroups admit a product integral representation. In fact product integrals have been often applied to represent hemigroups (cf. [4], 4.6 and R 4.6).

6.2. THEOREM. Let $A(.)$ be a mapping of I into $\mathbb{A}(G)$ such that $A(.)(f)$ is Riemann integrable for every $f \in \mathfrak{D}(G)$ and let $H = (\mu(s,t))_{(s,t)\in S}$ be the associated hemigroup in $\mathcal{M}^1(G)$. Then we have for all $(s,t) \in S$:

$$T_{\mu(s,t)} = \lim_{Z\in\mathfrak{Z}(s,t)} \prod_{j=1}^{m_Z} \operatorname{Exp}(N_{A(z(j))}(z(j) - z(j-1)))$$

with respect to the strong operator topology.

PROOF. Of course $A(.)$ is an admissible mapping. Hence H is well defined in view of Theorem 5.7. We have $M = \sup\{|A(t)|_2^\sim : t \in I\} < \infty$. Being Riemann integrable $A(.)(f)$ is continuous almost everywhere for all $f \in \mathfrak{D}(G)$. Hence in view of Lemma 1.8. there exists a zero set Z in I such that $A(.)(f)$ is continuous in all points of $I \setminus Z$ for every $f \in \mathcal{E}_2(G)$.

Let $\alpha(t;r) = \operatorname{Exp} rA(t)$ for all $r \in \mathbb{R}_+$ and $t \in I$ (cf. 2.3.). Let $(\Delta_n)_{n\geq 1}$, $v(n;t)$ and $c_n(t)$ as in 5.2. We put for all $n \in \mathbb{N}$:

$$\nu_{nk} = \alpha(c_{nk};c_{nk} - c_{n,k-1}) \qquad (k = 1,\dots,n);$$

$$d_n(t) = c_{n,v(n;t)} - c_{n,v(n;t)-1} \qquad (t \in I);$$

$$F_n(t)(f) = \{\int f\, d\nu_{n,v(n;t)} - f(e)\}/d_n(t) \qquad (f \in \mathcal{E}_2(G),\ t \in I).$$

1. We have $|F_n(t)|_2^\sim \leq M$ for all $t \in I$ and $n \in \mathbb{N}$; and $\lim_{n\geq 1} F_n(t)(f) = A(t)(f)$ for all $f \in \mathcal{E}_2(G)$ and $t \in I \setminus Z$.

[Since $(\alpha(t;r))_{r\geq 0}$ is the continuous convolution semigroup with generating functional $A(t)$ we have for all $r \in \mathbb{R}_+$:

$$(*) \qquad T_{\alpha(t;r)}f - f = \int_0^r T_{\alpha(t;s)} N_{A(t)} f\, ds \qquad (f \in \mathcal{E}_2^\sim(G));$$

$$(**) \qquad \int f\, d\alpha(t;r) - f(e) = \int_0^r A(t)(T_{\alpha(t;s)}f)\, ds \qquad (f \in \mathcal{E}_2(G)).$$

(Of course this is a particular case of (EE 1) and (EE 2) in 4.3.; namely for a constant admissible mapping; cf. 4.4.2.)

First of all (*) yields our first assertion. For the proof of the second assertion let $f \in \mathcal{E}_{2,2}(G)$. Then by (**) we have:

$$|F_n(t)(f) - A(t)(f)| \leq R_n(t) + S_n(t) =$$
$$|\int_0^{d_n(t)} \{A(c_n(t))(T_{\alpha(c_n(t);s)}f - f)\}\, ds| \,/\, d_n(t) \;+$$
$$|A(c_n(t))(f) - A(t)(f)|.$$

Since $\lim_{n\geq 1} c_n(t) = t$ (cf. 5.2.) we have $\lim_{n\geq 1} S_n(t) = 0$ for all $t \in I \setminus Z$ (by the choice of Z). $R_n(t)$ can be majorized by

$(\sup\{|A(t)|_2 : t \in I\})(\sup\{|T_{\alpha(c_n(t);s)}f - f|_2 : 0 \leq s \leq d_n(t)\})$.

But (*) implies:

(***) $\quad |T_{\alpha(t;r)}f - f|_2 \leq rM\,|f|_{2,2} \qquad (r \in \mathbb{R}_+)$.

Hence we have $\lim_{n\geq 1} R_n(t) = 0$ for all $t \in I$ (since $\lim_{n\geq 1} d_n(t) = 0$). This proves the second assertion if $f \in \mathcal{L}_{2,2}(G)$. The general case now follows by an application of Lemma 1.8.]

2. The system $\mathfrak{I} = (\nu_{nk})_{k=1,\dots,n;n\geq 1}$ is infinitesimal and satisfies condition (S) of 3.1. (with the mappings $v(n;.)$ defined in 5.2.).

[From (***) we derive $q(\alpha(t;r)) \leq r\gamma$ with an appropriate constant $\gamma > 0$. This yields our assertions (cf. the proof of Lemma 5.3.).]

3. Let $\lambda_n(s,t) = \sum_{v(n;s)<j\leq v(n;t)} \{\nu_{nj} - \varepsilon_e\}$ and $B(s,t) = \int_s^t A(\sigma)\,d\sigma$ for all $(s,t) \in S$ (cf. 2.7.). Then we have for $f \in \mathcal{L}_2^\sim(G)$:

$$\lim_{n\geq 1} \int f\,d\lambda_n(s,t) = B(s,t)(f) \quad \text{uniformly in } (s,t) \in S.$$

[In view of 1. we have $\lim_{n\geq 1} \int_s^t F_n(\sigma)(f)\,d\sigma = B(s,t)(f)$ uniformly in $(s,t) \in S$. On the other hand we have (with $c(n;r) = c_n(r)$):

$$\int_s^t F_n(\sigma)(f)\,d\sigma =$$

$$\int_s^{c(n;s)} F_n(\sigma)(f)d\sigma + \int_{c(n;s)}^{c(n;t)} F_n(\sigma)(f)d\sigma - \int_t^{c(n;t)} F_n(\sigma)(f)d\sigma.$$

But $\left|\int_r^{c(n;r)} F_n(\sigma)(f)d\sigma\right| \leq |\Delta_n|M|f|_2^\sim$ and $\int_{c(n;s)}^{c(n;t)} F_n(\sigma)(f)d\sigma = \int f\,d\lambda_n(s,t)$ yield now the assertion.]

4. As in Lemma 5.5. one proves that $\mathfrak{I}$ satisfies property (T).

5. Let $\nu_n(s,t) = \nu_{n,v(n;s)+1} * \dots * \nu_{n,v(n;t)}$ for all $(s,t) \in S$. Taking into account 2.,3. and 4. we can apply Corollary 3.7. (and hence Theorem 3.6.), Theorem 4.3. and Corollary 4.6. to our system $\mathfrak{I}$. Hence for every universal subnet $(n(\alpha))_{\alpha\in J}$ of $\mathbb{N}$ there exists a continuous hemigroup $(\nu(s,t))_{(s,t)\in S}$ in $\mathcal{M}^1(G)$ which satisfies (EE 2) with re-

spect to the given A(.) such that $\tilde{J}_w\text{-}\lim_{\alpha\in J} \nu_{n(\alpha)}(s,t) = \nu(s,t)$ for all $(s,t) \in S$. But Theorem 5.7. yields $\nu(s,t) = \mu(s,t)$ and consequently $\tilde{J}_w\text{-}\lim_{n\geq 1} \nu_n(s,t) = \mu(s,t)$ for all $(s,t) \in S$.

6. Finally let $(s,t) \in S$ be fixed. Considering only decompositions $\Delta_n = \{c_{no}, c_{n1}, \ldots, c_{nn}\}$ of I which contain s and t (i.e. $c_n(s) = s$ and $c_n(t) = t$) we obtain:

$$T_{\nu_n}(s,t) = \prod_{s\leq c_{n,j-1}<c_{nj}\leq t} \operatorname{Exp}(N_{A(c_{nj})}(c_{nj} - c_{n,j-1})).$$

Hence the assertion of the theorem. ┘

7. Hemigroups of bounded variation

7.1. DEFINITION. Let $H = (\mu(s,t))_{(s,t)\in S}$ be a continuous hemigroup in $\mathcal{M}^1(G)$. H is said to be *of bounded variation* if there exists a constant $\gamma > 0$ such that $\sum_{1\leq j\leq n} q(\mu(c_j,c_{j-1})) \leq \gamma$ for every decomposition $\{c_o, c_1, \ldots, c_n\}$ of I.

H is said to be *absolutely continuous* if for every $\varepsilon > 0$ there exists a $\delta > 0$ such that $\sum_{1\leq k\leq n} q(\mu(s_k,t_k)) < \varepsilon$ for every sequence $0 \leq s_1 < t_1 \leq s_2 < \ldots \leq s_n < t_n \leq 1$ satisfying $\sum_{1\leq k\leq n} (t_k - s_k) < \delta$.

7.2. Let H be a hemigroup of bounded variation. For every $t \in I$ let u(t) denote the supremum over all numbers $\sum_{1\leq j\leq n} q(\mu(c_{j-1},c_j))$ where $\{c_o, c_1, \ldots, c_n\}$ is an arbitrary decomposition of $[0,t]$. As can be easily seen the function u has the following properties:

(1) u is an increasing and continuous mapping of I onto $[0,u(1)]$.

(2) $u(s) = u(t)$ iff $\mu(\sigma,\tau) = \varepsilon_e$ for all $(\sigma,\tau) \in S$ with $s \leq \sigma \leq \tau \leq t$.

(3) $q(\mu(s,t)) \leq u(t) - u(s)$ for all $(s,t) \in S$.

(4) u is absolutely continuous iff H is absolutely continuous.

7.3. Let $U(t) = u(t)/u(1)$ for all $t \in I$ and let dU denote the Lebesgue-Stieltjes measure defined by U on the Borel subsets of I. We have

$$(*) \qquad \int_0^t f(U(\tau))\, dU(\tau) = \int_0^{U(t)} f(\sigma)\, d\sigma$$

for every bounded Borel measurable function f on I and for every $t \in I$.

Let $U^{-1}(t) = \sup\{s \in I: U(s) \leq t\}$ for every $t \in I$. Then U^{-1} is a strictly increasing and left continuous mapping of I into itself such that $U(U^{-1}(t)) = t$ and $U^{-1}(U(t)) \leq t$ for all $t \in I$. Moreover we have $\mu(U^{-1}(U(s)),U^{-1}(U(t))) = \mu(s,t)$ for all $(s,t) \in S$.

7.4. Taking into account these observations it is straightforward to extend the validity of Theorem 4.3.: If $H = (\mu(s,t))_{(s,t)\in S}$ is a hemigroup of bounded variation then $H' = (\mu(U^{-1}(s),U^{-1}(t)))_{(s,t)\in S}$ is a Lipschitz continuous hemigroup to which Theorem 4.3. applies. Performing now the inverse transformation U to H' this yields the desired extension. In the evolution equations (EE 1) and (EE 2) this means the substitution of the Lebesgue measure m by the Lebesgue-Stieltjes measure dU (in view of (*) in 7.3.).

If the hemigroup H is absolutely continuous one can keep the Lebesgue measure in the evolution equations if in turn one admits integrable mappings A(.) (cf.2.6.) instead of admissible ones.

7.5. Theorems 5.1. and 5.7. can be extended to integrable mappings A(.) of I into $\mathbb{A}(G)$ by means of the following procedure: Let $w(t) = \int_0^t |A(\tau)|_2\, d\tau$, $z(t) = \sup\{s \in I: w(s) \leq t\, w(1)\}$ and $B'(t) = \int_0^{z(t)} A(\tau)\, d\tau$ for all $t \in I$. Then we clearly have $|B'(t) - B'(s)|_2 \leq w(1)(t - s)$ for all $(s,t) \in S$. In view of Lemma 2.8., Theorem 5.7., Theorem 4.3. and Corollary 4.6. there exists exactly one (Lipschitz) continuous hemigroup $(\omega'(s,t))_{(s,t)\in S}$ in $\mathcal{M}^1(G)$ such that

$$\left|\int f\, d\omega'(s,t) - f(e) - [B'(t)(f) - B'(s)(f)]\right| \leq c\, |f|_{2,2}\, (t - s)^2$$

for all $f \in \mathcal{C}_{2,2}(G)$ and $(s,t) \in S$. It is now easy to check that $\omega(s,t) = \omega'(w(s)/w(1),w(t)/w(1))$, $(s,t) \in S$, is an absolutely continuous hemigroup which satisfies the evolution equations (EE 1) and (EE 2) with respect to A(.).

7.6. Let $H = (\mu(s,t))_{(s,t)\in S}$ be a continuous hemigroup in $\mathcal{M}^1(G)$ and let χ be a continuous mapping of I into G. Then obviously $H_\chi = (\varepsilon_{\chi(s)} * \mu(s,t) * \varepsilon_{\chi(t)^{-1}})_{(s,t)\in S}$ is also a continuous hemigroup. It may happen that for a hemigroup H of bounded variation and for a mapping χ the hemigroup H_χ is not of bounded variation.

Conversely however it has been pointed out in [2] that for every continuous hemigroup H in $\mathcal{M}^1(G)$ there exists a continuous mapping χ of I into G such that H_χ is of bounded variation. But for the proof of this fact it seems to be necessary to work with the additive process associated with the hemigroup H.

8. Lie groups with a relatively compact adjoint group

8.1. If one is interested in the solution of evolution equations on a Lie group G one needs to know under what conditions the backward evolution equation (EE 2) in Theorem 4.3. can be strengthened to

(SEE 2) $$T_{\omega(s,t)}f - f = \int_s^t N_{A(\sigma)} T_{\omega(\sigma,t)} f \, d\sigma$$

for all $f \in \mathcal{L}_2(G)$ and for all $(s,t) \in S$.

On a general Lie group G this will only hold true for particular integrable mappings A(.) of I into $\mathbb{A}(G)$; for example if all the A(t) generate a Poisson semigroup.

But if $\mathcal{L}_2(G) \subset \mathcal{L}_2^\sim(G)$ holds then (SEE 2) is obviously true for every integrable mapping A(.) (cf. 1.2. and 2.11.). Hence the problem arises to characterize the Lie groups G satisfying $\mathcal{L}_2(G) \subset \mathcal{L}_2^\sim(G)$ or equivalently $\mathcal{L}_2(G) = \mathcal{L}_2^\sim(G)$ (since $f \longrightarrow f^*$ is an isometry of $\mathcal{L}_2(G)$ onto $\mathcal{L}_2^\sim(G)$). In the course of these investigations G need not be σ-compact.

8.2. Let Ad_G or simply Ad denote the adjoint representation of G into the (topological) automorphism group aut($\mathfrak{g}$) of its Lie algebra $\mathfrak{g}$.

If G is discrete then we have $\mathfrak{g} = \{0\}$ and aut($\mathfrak{g}$) consists only of the identity mapping on $\mathfrak{g}$. As can be easily seen all of our results hold for a discrete group. Hence we can assume without loss of generality that G has dimension $d \geq 1$.

By $(Ad_{ij}(x))_{1\leq i,j\leq d}$ we denote the matrix representing Ad(x) with respect to the basis $\{X_1,\ldots,X_d\}$ of $\mathfrak{g}$. Moreover we have

$$\exp(Ad(x)X) = x(\exp X)x^{-1} \quad \text{and} \quad \tilde{X}f(x) = (Ad(x)X)f(x)$$

for all $x \in G$, $X \in \mathfrak{g}$, and $f \in \mathcal{C}^o(G)$ continuously differentiable in a neighbourhood of x.

8.3. By $\mathfrak{A}$ let us denote the class of all Lie groups G such that the adjoint group Ad(G) is relatively compact (in aut($\mathfrak{g}$)). Clearly every Abelian or compact Lie group is in $\mathfrak{A}$. As is well known the relative compactness of Ad(G) is equivalent with the boundedness of the set $\{(Ad_{ij}(x))_{1\leq i,j\leq d} : x \in G\}$ in $\mathbb{R}^{d\times d}$ (cf. [1], Chap.7, § 3.1, Prop.1).

8.4. LEMMA. Let $k \in \mathbb{N}$. If $\mathcal{C}_k(G) = \tilde{\mathcal{C}}_k(G)$ then the norms $|.|_k$ and $|.|_k^{\sim}$ (cf. 1.2.) are equivalent.

PROOF. The identity mapping T of $(\mathcal{C}_k(G),|.|_k)$ onto $(\tilde{\mathcal{C}}_k(G),|.|_k^{\sim})$ is closed. [$\lim |f_n - f|_k = 0$ and $\lim |f_n - g|_k^{\sim} = 0$ imply $\lim \|f_n - f\| = 0$ and $\lim \|f_n - g\| = 0$. Hence $f = g$.] The closed graph theorem yields the continuity of T. Hence there exists a constant $c > 0$ such that $|f|_k^{\sim} \leq c\ |f|_k$ for all $f \in \mathcal{C}_k(G)$. The converse inequality follows analogously. ⌋

8.5. THEOREM. The following assertions are equivalent:

(i) $\mathcal{C}_k(G) = \tilde{\mathcal{C}}_k(G)$ for some $k \in \mathbb{N}$.

(ii) $\mathcal{C}_k(G) = \tilde{\mathcal{C}}_k(G)$ for all $k \in \mathbb{N}$.

(iii) $G \in \mathfrak{A}$.

PROOF. '(i)$\Longrightarrow$(iii)' By Lemma 8.4. we have $|.|_k^{\sim} \leq c\ |.|_k$ with some constant $c > 0$. Let $f \in \mathcal{C}_k(G)$, $x \in G$ and $i \in \{1,\ldots,d\}$. Then we have for all $y \in G$:

$$|(\mathrm{Ad}(x)X_i)f(xy)| = |(\mathrm{Ad}(x)X_i)f_y(x)| = |(\tilde{X}_i f_y)(x)| \leq$$

$$|f_y|_k^{\sim} \leq c\ |f_y|_k = c\ |f|_k.$$

Hence $\|(\mathrm{Ad}(x)X_i)f\| \leq c\ |f|_k$. With $f = x_j$ this yields $|\mathrm{Ad}_{ji}(x)| \leq c\ |x_j|_k$ for all $x \in G$ $(i,j = 1,\dots,d)$. This proves $G \in \mathfrak{A}$.

'(iii) $\Longrightarrow$ (ii)' Let $|\mathrm{Ad}_{ij}(x)| \leq c$ for all $x \in G$ and $i,j = 1,\dots,d$. Let $k \in \mathbb{N}$ and $f \in \mathcal{C}_k(G)$. Then we have for all $x \in G$:

$$\tilde{X}_i f(x) = (\mathrm{Ad}(x)X_i)f(x) = \sum_{j=1}^{d} \mathrm{Ad}_{ji}(x)X_j f(x).$$

Since $\mathrm{Ad}_{ji}(\cdot) \in \mathcal{C}^b(G)$ we conclude $\tilde{X}_i f \in \mathcal{C}^o(G)$. In particular $\tilde{X}_i f$ is uniformly continuous. By Taylor's formula we have for all $x \in G$:

$$\left|\frac{1}{t}\,(f(x(\exp tX_i)) - f(x)) - \tilde{X}_i f(x)\right| \leq$$

$$|\tilde{X}_i f(x(\exp s(x)X_i)) - \tilde{X}_i f(x)|$$

for all $t \in \mathbb{R}^*$ and some $s(x) \in [-|t|,|t|]$. This proves $f \in \mathcal{C}_1^{\sim}(G)$. Iterating this procedure (with f successively replaced by $\tilde{X}_{i(1)}f$, $\tilde{X}_{i(1)}\tilde{X}_{i(2)}f,\dots,\tilde{X}_{i(1)}\dots\tilde{X}_{i(k-1)}f$ for every choice of $i(1),\dots,i(k-1)$ in $\{1,\dots,d\}$ yields $f \in \mathcal{C}_k^{\sim}(G)$. Hence $\mathcal{C}_k(G) \subset \mathcal{C}_k^{\sim}(G)$ and consequently $\mathcal{C}_k(G) = \mathcal{C}_k^{\sim}(G)$.

'(ii) $\Longrightarrow$ (i)' is trivial. ⌋

8.6. THEOREM. Every maximally almost periodic Lie group G is in $\mathfrak{A}$.

PROOF. G admits a locally faithful representation D into a unitary group H ([4], 4.3.5). Hence the differential dD of D is a faithful representation of $\mathfrak{g}$ into the Lie algebra of H. An easy calculation shows $dD(\mathrm{Ad}_G(x)X) = \mathrm{Ad}_H(D(x))dD(X)$ for all $x \in G$ and $X \in \mathfrak{g}$. Since H is compact $\{\mathrm{Ad}_H(D(x))dD(X) : x \in G\}$ is bounded (all $X \in \mathfrak{g}$). The injectivity of dD then yields the boundedness of $\{\mathrm{Ad}_G(x)X : x \in G\}$ (all $X \in \mathfrak{g}$). This proves our assertion. ⌋

A locally compact group G is said to have small invariant neighbourhoods if there exists a basis of $\mathfrak{W}(G)$ consisting of sets invariant under all inner automorphisms of G.

8.7. THEOREM. For every Lie group G the following assertions are equivalent:

(i) $G \in \mathfrak{A}$.

(ii) G has small invariant neighbourhoods.

PROOF. '(i) $\Longrightarrow$ (ii)' Let $U \in \mathfrak{W}(G)$ be of the form $U = \exp V$ for some $V \in \mathfrak{W}(\mathfrak{g})$. There exists a $W \in \mathfrak{W}(\mathfrak{g})$ such that $\mathrm{Ad}(G)W \subset V$. One easily shows that $\exp(\mathrm{Ad}(G)W)$ is an invariant neighbourhood of $e \in G$contained in U.

'(ii) $\Longrightarrow$ (i)' There exists an open and bounded $V \in \mathfrak{W}(\mathfrak{g})$ such that $U = \exp V$ is invariant (under the inner automorphisms of G) and such that exp is an analytic isomorphism of V onto U. By log we denote the inverse of exp on U.

Let $X \in \mathfrak{g}$ such that $tX \in V$ for $|t| < 1$. We fix $x \in G$. The invariance of U yields $(\exp t\mathrm{Ad}(x)X) \in U$ for $|t| < 1$ (cf. 8.2.). Hence by $X(t) = \log \exp t\mathrm{Ad}(x)X$ there is defined an analytic mapping X(.) of $]-1,1[$ into V. Obviously there exists a $\delta > 0$ such that $t\mathrm{Ad}(x)X \in V$ and consequently $t\mathrm{Ad}(x)X = X(t)$ if $|t| < \delta$. Since X(.) is analytic this implies $t\mathrm{Ad}(x)X = X(t)$ for $|t| < 1$. Hence $\mathrm{Ad}(x)X$ is in the closure of V. Since x was arbitrary $\mathrm{Ad}(G)X$ is bounded. This proves our assertion. ⌋

8.8. REMARKS. 1. Let $G \in \mathfrak{A}$. Then the connected component G_o of the identity e of G is a maximally almost periodic group.

[G_o has small invariant neighbourhoods by Theorem 8.7. Hence [3], Theorem 2.9 yields the assertion.]

2. A connected Lie group is in $\mathfrak{A}$ if and only if the group is maximally almost periodic.

[This follows from Remark 1 and Theorem 8.6.]

3. Obviously every discrete group is in $\mathfrak{A}$ (cf. 8.2.). Hence there exist groups in $\mathfrak{A}$ which are not maximally almost periodic.

4. Let G be a Lie group and let C denote the centralizer of the connected component G_o (of the identity e of G) in G. Then G is in $\mathfrak{A}$ if and only if the factor group G/C is maximally almost periodic.
[Obviously C is the kernel of the adjoint representation Ad of G. It can be easily seen that the adjoint group of G/C is relatively compact if and only if Ad(G) is relatively compact. Hence the assertion taking into account Theorem 8.6.]

5. A maximally almost periodic (locally compact) group G is Lie projective if and only if G has small invariant neighbourhoods.
[The necessity follows by Theorems 8.6. and 8.7; the sufficiency is a result of S.Grosser and M.Moskowitz ([3], Th.2.11).]

List of notations and symbols (not explained in the Preliminaries)

References

1. Bourbaki,N.: Éléments de Mathématique XXIX: Livre VI: Intégration. Chapitres 7-8. Actual.Scient.Ind.1306. Paris: Hermann 1963

2. Feinsilver,Ph.: Processes with independent increments on a Lie group. Trans.Amer.Math.Soc.242, 73-121 (1978)

3. Grosser,S., Moskowitz,M.: Compactness conditions in topological groups. J.Reine Angew.Math.246, 1-40 (1971)

4. Heyer,H.: Probability measures on locally compact groups. Berlin -Heidelberg-New York: Springer 1977

5. Heyer,H.: Stetige Hemigruppen von Wahrscheinlichkeitsmassen und additive Prozesse auf einer lokalkompakten Gruppe. Nieuw Arch. Wiskunde 27, 287-340 (1979)

6. Hille,E., Phillips,R.S.: Functional analysis and semigroups. Amer.Math.Soc.Colloquium Publications, Vol.31. Revised edition. Providence, R.I., Amer.Math.Soc. 1957

7. Wehn,D.F.: Limit distributions on Lie groups. Thesis, Yale 1959

8. Wehn,D.F.: Probabilities on Lie groups. Proc.Nat.Acad.Sci.USA 48, 791-795 (1962)

9. Wehn,D.F.: Limit distributions on Lie groups. Manuscript (130 pages), 1967

10. Yosida,K.: Functional analysis. 3rd ed. Berlin-Heidelberg-New York: Springer 1971

Eberhard Siebert
Mathematisches Institut
der Universität
Auf der Morgenstelle 10
D-7400 Tübingen 1
Bundesrepublik Deutschland

POTENTIAL THEORY FOR RECURRENT SYMMETRIC INFINITELY DIVISIBLE PROCESSES

by

CHARLES J. STONE

Department of Mathematics, UCLA

Los Angeles, CA 90024 (USA)

SUMMARY

Let X be a locally compact, second countable Abelian group. Let $\xi(t)$, $t \geq 0$, be an irreducible, recurrent, symmetric infinitely divisible Hunt process on X such that for $t > 0$, $\xi(t) - \xi(0)$ has a bounded continuous density $p(t,\cdot)$ with respect to Haar measure on X. Potential theory is developed for the kernel $k = \int_0^1 p(t,\cdot)dt \cdot \int_1^\infty (p(t,\cdot) - p(t,0))dt$. In particular, balayage and equilibrium problems corresponding to an arbitrary relatively compact Borel set are formulated and solved and the solutions are characterized in terms of energy. Logarithmic potential theory is included as the special case corresponding to planar Brownian motion.

POTENTIAL THEORY FOR RECURRENT SYMMETRIC INFINITELY DIVISIBLE PROCESSES

by CHARLES J. STONE*

1. Introduction

Let X be a locally compact, second countable Abelian group (which may be either compact or noncompact). Let $\xi(t)$, $t \geq 0$, be an irreducible, recurrent, symmetric, infinitely divisible Hunt process on X such that for $t > 9$, $\xi(t) - \xi(0)$ has a bounded continuous density $p(t,\cdot)$ with respect to Haar measure dx on X. These assumptions correspond to the following set of assumptions on $p(t,x)$, $t > 0$ and $x \in X$:

(i) (bounded continuous probability density) for $t > 0$, $p(t,\cdot)$ is a nonnegative bounded continuous function on X such that

$$\int p(t,x)dx = 1;$$

(ii) (semigroup property) for $s,t > 0$

$$p(s + t,x) = \int p(s,y)p(t,x - y)dy, \quad x \in X;$$

(iii) (symmetry) for $t > 0$

---

*Research partially supported by National Science Foundation grant GP-MCS 72-04591.

$$p(t,-x) = p(t,x), \qquad x \in X;$$

(iv) (recurrence) for some compact set C

$$\int_0^\infty \left(\int_C p(t,x)dx \right)dt = \infty;$$

(v) (irreducibility) if X_0 is a proper closed subgroup of X and $t > 0$, then

$$\int_{X_0} p(t,x)dx < 1.$$

After a number of preliminary results are stated in Section 2, the potential kernel k is defined in Section 3 by

$$k(x) = \int_0^1 p(t,x)dt + \int_1^\infty (p(t,x) - p(t,0))dt, \qquad x \in X.$$

The second integral is shown to be continuous in x, so the first integral contains the singularities, if any, of the kernel. The balayage and equilibrium problems associated with this kernel are formulated and solved in Section 3. In Section 4 the concept of thinness is studied. A number of potential principles are derived in Section 5. Finally in Section 6 energy is defined and various potential principles involving energy are obtained.

Logarithmic potential theory is included as the special case corresponding to planar Brownian motion. This special case is treated in Sections 3.4 and 6.7 of Port and Stone [5], where references to the literature are given. There is considerable overlap between proofs of the results in this paper and those in [5]. As a compromise between brevity and clarity, the results in Section 2 are stated without proof, while complete proofs are given for the results in the remainder of the paper.

2. Preliminaries

In this section a number of results which will be used freely later on are stated without proof. The purely analytic results are standard or contained in Hewitt and Ross [2]. The probalistic results are contained in Chapter I of Blumenthal and Getoor [1] or follow by essentially the same proofs as the corresponding results in Chapters 1 and 2 of Port and Stone [5].

Throughout this paper subsets of X are understood to be Borel sets (i.e., in the σ-algebra generated by the open sets or equivalently, the compact sets in X). Similarly a function on X or a subset of X is understood to be a Borel function on its domain. A real-valued function is allowed to assume the values ∞ and $-\infty$ except as otherwise noted; but "continuous" will be taken to mean finite-valued and continuous.

A function f on X is said to be lower semicontinuous if $f > -\infty$ on X and $\{x \in X : f(x) > c\}$ is open for each $c \in \mathbb{R}$. If f is lower semicontinuous and C is a compact subset of X, there is an $x \in C$ such that $f(x) = \min_{y \in C} f(y)$. In particular a lower semicontinuous function is bounded below on compacts, and a positive lower semicontinuous function is bounded away from zero on compacts. An increasing limit of lower semicontinuous functions is lower semicontinuous. A function f on X is said to be upper semicontinuous if $f < \infty$ on X and $\{x \in X : f(x) < c\}$ is open for each $c \in \mathbb{R}$ or, equivalently, if $-f$ is lower semicontinuous. Observe that f is continuous if and only if it is both lower semicontinuous and upper semicontinuous.

Let μ be a <u>Radon</u> measure on X; i.e., such that $\mu(C) < \infty$ for every compact set C. Then

$$\mu(B) = \sup[\mu(C) : C \text{ is a compact subset of } B]$$
$$= \inf[\mu(U) : U \text{ is an open set containing } B]$$

A property is said to hold <u>almost everywhere with respect to</u> μ (a.e. (μ)) on B if there is a set A such that $\mu(A) = 0$ and the property holds on $B \setminus A$ ($B \setminus A = B \cap A^c$, where A^c denotes the complement of A). If μ is Haar measure, "a.e. (μ)" will be abbreviated to "a.e." and if $B = X$ the phrase "on B" will be omitted. The measure μ is said to be <u>concentrated</u> on B if $\mu(B^c) = 0$. Given a set B, the measure $\mu|_B$ is defined by $\mu|_B(A) = \mu(A \cap B)$ for $A \subset X$. Clearly $\mu|_B$ is concentrated on B. Let μ_n, $n \geq 1$, and μ be finite measures on X. Then μ_n is said to <u>converge completely</u> to μ if $\lim_n \int \varphi d\mu_n = \int \varphi d\mu$ for every bounded continuous function φ on X.

Let $\mathcal{M}^+$ denote the collection of finite measures μ on X which are concentrated on some compact subset of X (which set is allowed to depend on μ). Let $\mathcal{M}^+(B)$ denote the collection of measures $\mu \in \mathcal{M}^+$ which are concentrated on B. Suppose C is compact, $\mu_n \in \mathcal{M}^+(C)$ for $n \geq 1$ and $\mu_n(C)$ is bounded in n. Then there is a measure $\mu \in \mathcal{M}^+(C)$ and a strictly increasing sequence $\{n_j\}$ of positive integers such that μ_{n_j} converges completely to μ.

Let $\mathcal{M}$ denote the vector space of signed measures $\mu = \mu_1 - \mu_2$ where $\mu_1, \mu_2 \in \mathcal{M}^+$. For $\mu \in \mathcal{M}$ let $|\mu|$ denote the total variation measure corresponding to μ. Then μ can be decomposed uniquely as $\mu = \mu^+ - \mu^-$ where $\mu^+, \mu^- \in \mathcal{M}^+$ and $|\mu| = \mu^+ + \mu^-$. Let $\mathcal{M}(B)$

denote the collection of signed measures $\mu \in \mathcal{m}$ such that $|\mu|$ is concentrated on B.

Let $L(x,\cdot)$ be a measure or signed measure on X for each $x \in X$. Under appropriate conditions on L and a function f on X, Lf is defined by $Lf(x) = \int L(x,dy)f(y)$, $x \in X$; and under appropriate conditions on a measure or signed measure μ on X, μL is defined by $\mu L(A) = \int \mu(dy)L(y,A)$ for $A \subset X$. If M is of the same form, then ML is defined by

$$ML(x,A) = \int M(x,dz)L(z,A), \qquad A \subset X,$$

or $MLf = M(Lf)$.

Suppose L is of the form $L(x,dy) = L(x,y)dy$. Then $Lf(x) = \int L(x,y)f(y)dy$. Also $ML(x,dy) = ML(x,y)dy$, where $ML(x,y) = \int M(x,dz)L(z,y)$. Given a measure or signed measure μ on X, the L-potential $L\mu$ of μ is defined by $L\mu(x) = \int L(x,y)\mu(dy)$. If $L(x,y)$ is symmetric in x and y and Fubini's theorem is applicable, then $\int L\mu d\nu = \int L\nu d\mu$.

Observe that

$$\begin{aligned} p(t,0) &= \int p(t/2,z)p(t/2,-z)dz \\ &= \int (p(t/2,z))^2 dz, \qquad t > 0 \text{ and } x \in X, \end{aligned}$$

and

$$p(t,x) = \int p(t/2,z)p(t/2,x-z)dz, \quad t > 0 \text{ and } x \in X;$$

so by Schwarz's inequality

$$p(t,x) \leq p(t,0), \quad t > 0 \text{ and } x \in X.$$

Consequently by the semigroup property

$$p(s,x) \le p(t,0), \qquad s \ge t > 0 \text{ and } x \in X.$$

If $t > 0$, $x \in X$ and $p(t,x) > 0$, then $p(s,x) > 0$ for $s \ge t$. Set

$$p^t(x,y) = p(t,x,y) = p(t,y-x), \; t > 0 \text{ and } x,y \in X.$$

A <u>character</u> θ on X is a complex-valued function, written as $\theta(x) = \langle\theta,x\rangle$, $x \in X$, such that $|\langle\theta,x\rangle| = 1$ for $x \in X$ and $\langle\theta, x_1 + x_2\rangle = \langle\theta,x_1\rangle \langle\theta,x_2\rangle$ for $x_1, x_2 \in X$. The "zero" character is given by $\langle 0, x\rangle = 1$ for $x \in X$. Given characters θ_1 and θ_2, their sum is defined by $\langle\theta_1 + \theta_2, x\rangle = \langle\theta_1, x\rangle \langle\theta_2, x\rangle$ for $x \in X$. The collection of continuous characters forms a locally compact Abelian group Θ, called the <u>dual group</u> to X. Given an integrable function f on X, its <u>Fourier transform</u> $\hat{f}$ on Θ is defined by $\hat{f}(\theta) = \int \langle\theta,x\rangle f(x)dx$, $\theta \in \Theta$; the function $\hat{f}$ is continuous and vanishes at infinity. Haar measure $d\theta$ on Θ is chosen so that if f is continuous and integrable on X and $\hat{f}$ is integrable on Θ, then $f(x) = \int \overline{\langle\theta,x\rangle}\hat{f}(\theta)d\theta$ for $x \in X$. If f is both integrable and square-integrable on X, then $\hat{f}$ is square-integrable on Θ and $\int|\hat{f}(\theta)|^2 d\theta = \int |f(x)|^2 dx$.

A function f on X is said to be <u>symmetric</u> if $f(-x) = f(x)$, $x \in X$, and a similar definition applies for functions on Θ. If f is real-valued and symmetric, so is $\hat{f}$.

For $t > 0$ let $\hat{p}(t,\cdot)$ denote the Fourier transform of $p(t,\cdot)$. For proofs of the following properties see Section 3 of Port and Stone [3] and Section 16 of Port and Stone [4]: There is a nonnegative continuous symmetric function Ψ on Θ such that $\hat{p}(t,\cdot) = \exp(-t\Psi)$. Moreover $\Psi > 0$ on $\Theta \setminus \{0\}$. If A is a compact subset of X and B is a compact subset of Θ, then $\Psi^{-1}(\theta)\,(\Re\langle x,\theta\rangle - 1)$ is bounded for $x \in A$ and $\theta \in B \setminus \{0\}$.

For $t > 0$, $p(t/2,\cdot)$ is bounded and hence square-integrable on X, so $\exp(-t\Psi/2)$ is square-integrable on Θ. Consequently $\exp(-t\Psi)$ is integrable on Θ and therefore $p(t,x) = \int e^{-t\Psi(\theta)} \langle\theta,x\rangle d\theta$, $t > 0$ and $x \in X$. Thus $p(t,x)$ is jointly continuous for $(t,x) \in (0,\infty) \times X$.

Observe that for $\lambda > 0$

$$\int_1^\infty e^{-\lambda t}(p(t,x) - p(t,0))dt = e^{-\lambda} \int \frac{e^{-\Psi(\theta)}}{\lambda + \Psi(\theta)} (\Re \langle \theta,x\rangle - 1)d\theta, \quad x \in X.$$

It follows by letting $\lambda \to 0$ that

$$\int_1^\infty(p(t,x) - p(t,0))dt = \int_{\Theta\setminus\{0\}} \frac{e^{-\Psi(\theta)}}{\Psi(\theta)} (\Re\langle\theta,x\rangle - 1)d\theta, \quad x \in X.$$

(Note that if Θ is noncompact, then Ψ^{-1} vanishes at infinity since $\hat{p}(t,\cdot)$ vanishes at infinity.) Since the process $\xi(t)$, $t \geq 0$, is recurrent, it now follows that $\int_1^\infty p(t,\cdot)dt = \infty$ on X. For every compact subset C of X, there is a $t > 0$ such that $p(s,\cdot) > 0$ on C for $s \geq t$.

The process $\xi(t)$, $t \geq 0$, is assumed to be a Hunt process as described in Blumenthal and Getoor [1]; in particular it is a strong Markov Process. Let τ_B denote the hitting time of the set B, defined by $\tau_B = \inf[t > 0\colon\ \xi(t) \in B]$ if $\xi(t) \in B$ for some $t > 0$ and $\tau_B = \infty$ otherwise. Then τ_B is a stopping time. Clearly if $A \subset B$, then $\tau_A \geq \tau_B$. For $t > 0$ the function $P_\cdot(\tau_B \leq t)$ is lower semicontinuous on X. If $x \in X$, then $P_x(\tau_B = 0)$ equals zero or one and $P_x(\tau_B = t) = 0$ for $t > 0$. If $\{B_n\}$ is a decreasing sequence of sets, then for each $x \in X$

$$\xi(\tau_{B_n}) \to \xi(\lim_n \tau_{B_n}) \quad \text{a.e.} \quad (P_x) \quad \text{on} \quad \{\lim_n \tau_{B_n} < \infty\}$$

(this is a special case of "quasi-left-continuity"). There is an increasing sequence $\{B_n\}$ of compact subsets of B such that $P_{\cdot}(\tau_{B_n} \downarrow \tau_B) = 1$ on X.

A set B is said to be polar if $P_{\cdot}(\tau_B < \infty) = 0$ on X. A countable union of polar sets is polar. Suppose in the remainder of this paragraph that B is nonpolar. There is a nonpolar compact subset of B. If C is compact, then

$$\min_{x \in C} P_x(\tau_B \leq t) > 0 \quad \text{for } t \text{ sufficiently large.}$$

Since the process $\xi(t)$, $t \geq 0$, is recurrent, $P_{\cdot}(\tau_B < \infty) = 1$ on X; also $\lim_{t \to \infty} P_{\cdot}(\tau_B \geq t) = 0$ uniformly on compacts.

If μ is a measure on X such that $\mu(A) = 0$ whenever A is polar, it is said that μ does not change polar sets. A property is said to hold quasi-everywhere (q.e.) on B if there is a polar set A such that the property holds on $B \setminus A$. Now every polar set has zero Haar measure, so if a property holds q.e. on B, it holds a.e. on B. If a property holds q.e. on B and μ is a measure on X which does not change polar sets, the property holds a.e. (μ) on B.

A point $x \in X$ is regular for a set B if $P_x(\tau_B = 0) = 1$ and irregular for B otherwise (i.e. if $P_x(\tau_B = 0) = 0$). Let B^r denote the collection of points $x \in X$ which are regular for B. Then $(B^r)^c$ is an F_σ set (i.e. a countable union of closed sets) and $\mathring{B} \subset B^r \subset \bar{B}$, where $\mathring{B}$ and $\bar{B}$ denote respectively the interior and closure of B. The set $B \setminus B^r$ is polar, and B is polar if and only if B^r is empty.

Let μ be a Radon measure on X such that $\mu(B \setminus B^r) = 0$. Then there is a decreasing sequence $\{U_n\}$ of open sets containing B such that $P_{\cdot}(\tau_{U_n} \uparrow \tau_B) = 1$ a.e. (μ).

Set $h_B(x,A) = P_x(\tau_B < \infty,\ \xi(\tau_B) \in A)$ for $x \in X$ and $A \subset X$. If B is nonpolar, then $h_B(x,\cdot)$ is a probability measure on X for each $x \in X$. If $x \in B^r$, then $h_B(x,A)$ is the probability measure δ_x concentrated at x. If $x \notin B^r$, then $h_B(x,\cdot)$ is concentrated on B^r and does not charge polar sets. If $A \subset B$, then $h_A = h_A h_B = h_B h_A$. Note that

$$h_B f = E_{\cdot}(f(\xi(\tau_B));\tau_B < \infty) \quad \text{for} \quad f \geq 0,$$

where $E_{\cdot}(Y;A) = \int_A Y(\omega)P_{\cdot}(d\omega)$.

Set

$$\begin{aligned} q_B^t(x,y) &= q_B(t,x,y) \\ &= p(t,x,y) - E_x(p(t - \tau_B, \xi(\tau_B),y);\tau_B < t), \quad t > 0 \text{ and } x,y \in X. \end{aligned}$$

Then

$$P_x(\tau_B > t, \xi(t) \in A) = \int_A q_B(t,x,y)dy, \quad t > 0,\ x \in X \text{ and } A \subset X.$$

For $t > 0$, $q_B(t,x,y)$ is upper semicontinuous in x and y separately. Also

$$\begin{aligned} &q_B(t,x,y) \geq 0, \quad t > 0 \text{ and } x,y \in X, \\ &q_B(t,x,y) = q_B(t,y,x), \quad t > 0 \text{ and } x,y \in X, \\ &q_B(t,x,y) = 0, \quad t > 0 \text{ and either } x \in B^r \text{ or } y \in B^r, \end{aligned}$$

and

$$\int q_B(s,x,z)q_B(t,z,y)dz = q_B(s + t,x,y), \quad s,t > 0 \text{ and } x,y \in X.$$

If $A \subset B$, then

$$q_A(t,x,y) \geq q_B(t,x,y), \quad t > 0 \text{ and } x,y \in X.$$

If $\{B_n\}$ is an increasing sequence of subsets of B such that

$P_{\cdot}(\tau_{B_n} \downarrow \tau_B) = 1$ on X, then

$$q_{B_n}(t,x,y) \downarrow q_B(t,x,y), \quad t > 0 \text{ and } x,y \in X.$$

Let B be nonpolar. Set $g_B(x,y) = \int_0^\infty q_B(t,x,y)dt$ for $x,y \in X$. Then $g_B(x,y)$ is symmetric in x and y and vanishes if either $x \in B^r$ or $y \in B^r$. If C is compact, then $\sup_x g_B(x,C) < \infty$. Note that $g_B f = E_{\cdot} \int_0^{\tau_B} f(\xi(t))dt$ for $f \geq 0$.

3. Potentials, equilibrium problem and balayage problem

In this section the potential kernel k is defined and studied and the corresponding equilibrium and balayage problems are formulated and solved.

Recall from Section 2 that

$$(1) \qquad p(t,x) = \int e^{-t\psi(\theta)}\langle\theta,x\rangle d\theta, \quad t > 0 \text{ and } x \in X.$$

Let $\lambda > 0$. The λ-potential kernel g^{λ} is defined by $g^{\lambda}(x) = \int_0^{\infty} e^{-\lambda t}p(t,x)dt$, $x \in X$. This function is positive, lower semicontinuous and integrable. Let g_0^{λ} and g_1^{λ} be defined on X by

$$g_0^{\lambda}(x) = \int_0^1 e^{-\lambda t}p(t,x)dt, \qquad x \in X,$$

and

$$g_1^{\lambda}(x) = \int_1^{\infty} e^{-\lambda t}p(t,x)dt, \qquad x \in X.$$

Then $g^{\lambda} = g_0^{\lambda} + g_1^{\lambda}$. The function g_0^{λ} is nonnegative, integrable and lower semicontinuous. The function g_1^{λ} is positive, integrable and continuous. By (1)

$$(2) \qquad g_1^{\lambda}(x) = e^{-\lambda} \int \frac{e^{-\psi(\theta)}}{\lambda + \psi(\theta)} \langle\theta,x\rangle d\theta, \quad x \in X.$$

Set $k^{\lambda} = g^{\lambda} - g_1^{\lambda}(0)$, $k_0^{\lambda} = g_0^{\lambda}$ and $k_1^{\lambda} = g_1^{\lambda} - g_1^{\lambda}(0)$. Then $k^{\lambda} = k_0^{\lambda} + k_1^{\lambda}$.

It follows from (2) that

$$(3) \qquad k_1^{\lambda}(x) = e^{-\lambda} \int \frac{e^{-\psi(\theta)}}{\lambda + \psi(\theta)} (\Re\langle\theta,x\rangle - 1)d\theta, \quad x \in X.$$

Let $\lambda \downarrow 0$. Then k_0^λ increases to the function k_0 given by

$$k_0(x) = \int_0^1 p(t,x)dt, \qquad x \in X.$$

This function is nonnegative, integrable and lower semicontinuous. According to (3), k_1^λ converges uniformly on compacts to the continuous function k_1 given by

(4) $$k_1(x) = \int_{\Theta\setminus\{0\}} \frac{e^{-\psi(\theta)}}{\psi(\theta)} (\Re\langle\theta,x\rangle - 1)\, d\theta, \qquad x \in X.$$

Since $k_1^\lambda(x) = \int_0^\infty e^{-\lambda t}(p(t,x) - p(t,0))dt$ and $p(t,x) \leq p(t,0)$ for $t > 0$,

(5) $$k_1(x) = \int_1^\infty (p(t,x) - p(t,0)dt, \qquad x \in X.$$

In particular $k_1 \leq 0$ on X. Set $k = k_0 + k_1$, so that

(6) $$k(x) = \int_0^1 p(t,x) + \int_1^\infty (p(t,x) - p(t,0))dt, \qquad x \in X.$$

The function k, called the potential kernel, is lower semicontinuous and integrable on compacts. If $k(0) < \infty$, then $\lim_{t\to 0} \int_0^t p(s,\cdot)ds = 0$ uniformly on X and hence k is continuous on X. Since $\int_1^\infty p(t,0)dt = \infty$, $p(t,x) \leq p(t,0)$ for $t > 0$ and $x \in X$, and $\int_1^\infty (p(t,x) - p(t,0))dt$ is bounded on compacts

(7) $$\lim_{\lambda\to 0} \frac{g_1^\lambda}{g_1^\lambda(0)} = 1 \text{ uniformly on compacts.}$$

Set $k(x,y) = k(y - x)$ and let a similar notation be used for k_0, k_1, k^λ, g^λ etc.. Given $\mu \in \mathcal{M}^+$, the function $k\mu$ is called the potential of μ. This function is lower semicontinuous and integrable on compacts.

THEOREM 3.1 *Let* $\mu,\nu \in \mathcal{M}^+$ *and* $\alpha \in \mathbb{R}$. *If* $k\nu \leq (=) k\mu + \alpha$ *a.e. on an open set* U, *then* $k\nu \leq (=) k\mu + \alpha$ *on* U. *If* $k\nu = k\mu + \alpha$ *a.e. on* X, *then*

$$(8) \qquad \nu = \mu + |X|^{-1}(\nu(X) - \mu(X)) \cdot \text{Haar measure},$$

where $|X|$ *is the Haar measure of* X *(which is finite and positive if* X *is compact and infinite otherwise).*

Proof. Observe first that $p(s,0) = \int e^{-s\psi(\theta)}d\theta \to \beta$ as $s \to \infty$, where β denotes the Haar measure of the origin of Θ (which is zero unless X is compact). Consequently

$$(9) \qquad p^t k_1\mu = k_1\mu - \int_1^{1+t} p^s\mu ds + \beta t\mu(X), \qquad t > 0,$$

which implies that

$$(10) \qquad p^t k\mu = k_1\mu + \int_t^1 p^s\mu ds + \beta t\mu(X), \qquad t > 0 .$$

It follows from (9) and (10) respectively that

$$(11) \qquad \lim_{t\to 0} p^t k_1\mu = k_1\mu$$

and

$$(12) \qquad \lim_{t\to 0} p^t k\mu = k\mu.$$

Suppose U is open and $k\nu \leq k\mu + \alpha$ a.e. on U. Choose $x \in U$ and $-\infty < c < k\nu(x) - \alpha$. Since $k\nu$ is lower semicontinuous, there is a relatively compact open subset V of U containing x such that $k\nu - \alpha \geq c$ on V. Now $k\mu \geq c$ a.e. on V, so

$$(13) \qquad \underline{\lim}_{t\to 0}\, p^t(I_V k\mu) \geq c \quad \text{on } V.$$

(Here I_A denotes the indicator of the set A, defined by $I_A(x) = 1$ for $x \in A$ and $I_A(x) = 0$ for $x \notin A$.) Clearly

(14) $$\underline{\lim}_{t\to 0}\, P^t(I_{V^c}\, k_0\mu) \geq 0 \quad \text{on } X.$$

Since $k_1\mu$ is continuous

(15) $$\lim_{t\to 0} P^t(I_V k_1\mu) = k_1\mu \quad \text{on } V.$$

By (11) and (15)

(16) $$\lim_{t\to 0} P^t(I_{V^c}\, k_1\mu) = 0 \quad \text{on } V.$$

It follows from (13), (14) and (16) that

(17) $$\underline{\lim}_{t\to 0} P^t k\mu \geq c \quad \text{on } V.$$

By (12) and (17) $k\mu \geq c$ on V. In particular $k\mu(x) \geq c$. Consequently $k\mu(x) \geq k\nu(x) - \alpha$. Since x is an arbitrary point of U, $k\nu \leq k\mu + \alpha$ on U. Suppose $k\nu = k\mu + \alpha$ a.e. on U. Then $k\nu \leq k\mu + \alpha$ on U and $k\mu \leq k\nu - \alpha$ on U and hence $k\nu = k\mu + \alpha$ on U.

It follows from (10) that for $t > 0$, $P^t k\mu$ is finite-valued and

(18) $$\int_0^t P^s \mu ds = k\mu - P^t k\mu + \beta t\mu(X).$$

Suppose $k\nu = k\mu + \alpha$ a.e. on X. Then $k\nu = k\mu + \alpha$ on X by the first result, so by (18)

$$\int_0^t P^s \nu ds = \int_0^t P^s \mu ds + \beta t(\nu(X) - \mu(X)), \qquad t > 0$$

Let φ be a nonnegative continuous function having compact support. Then

$$\int \left(\frac{1}{t} \int_0^t p^s \varphi ds \right) d\nu = \int \left(\frac{1}{t} \int_0^t p^s \varphi ds \right) d\mu + \beta(\nu(X) - \mu(X)) \int \varphi(x) dx$$

for $t > 0$. It follows by letting $t \to 0$ that

$$\int \varphi d\nu = \int \varphi d\mu + \beta(\nu(X) - \mu(X)) \int \varphi(x) dx.$$

Consequently

$$\nu = \mu + \beta(\nu(X) - \mu(X)) \cdot \text{Haar measure.} \tag{19}$$

Since $\mu, \nu \in \mathfrak{M}^+$, it follows easily from (19) that $\nu = \mu$ unless X is compact, in which case (8) holds. This completes the proof of the theorem.

By the definition of q_B

$$p(t,x,y) = q_B(t,x,y) + E_x(p(t - \tau_B, \xi(\tau_B), y); \tau_B < t), \ t > 0 \text{ and } x,y \in X. \tag{20}$$

Let $\lambda > 0$. Set $g_B^\lambda = \int_0^\infty e^{-\lambda t} q_B^t dt$ and let h_B^λ be defined by

$$h_B^\lambda f = E_\cdot(e^{-\lambda\tau_B} f(\xi(\tau_B)); \tau_B < \infty) \quad \text{for } f \geq 0.$$

It follows from (20) that

$$g^\lambda = g_B^\lambda + h_B^\lambda g^\lambda. \tag{21}$$

By (20) and the symmetry of $p(t,x,y)$ and $q_B(t,x,y)$ in x and y

$$\int h_B^\lambda(x,dz) g^\lambda(z,y) = \int h_B^\lambda(y,dz) g^\lambda(z,x), \quad x,y \in X. \tag{22}$$

Now $q_A \geq q_B$ if $A \subset B$, so by (20)

$$h_A^\lambda g^\lambda \leq h_B^\lambda g^\lambda \quad \text{if } A \subset B. \tag{23}$$

Set $W_B^\lambda = g_1^\lambda(0)(1 - E_.(e^{-\lambda\tau_B}))$. Then $W_B^\lambda \geq 0$ and $W_B^\lambda = 0$ on B^r. If $A \subset B$, then $W_A^\lambda \geq W_B^\lambda$. By (21)

(24)
$$k^\lambda = g_B^\lambda + h_B^\lambda k^\lambda - W_B^\lambda.$$

THEOREM 3.2. Let B be relatively compact and nonpolar. Then $W_B = \lim_{\lambda\to 0} W_B^\lambda$ exists and the convergence is uniform on compacts. Also

(25)
$$k = g_B + h_B k - W_B.$$

Proof. Let C be a compact set having positive Haar measure $|C|$. By (24)

(26)
$$k^\lambda(\cdot,C) = g_B^\lambda(\cdot,C) + h_B^\lambda k^\lambda(\cdot,C) - |C| W_B^\lambda.$$

Now $\sup_y g_B(y,C) < \infty$, so by the Markov property for $s > 0$ and $x \in X$

$$0 \leq g_B(\cdot,C) - g_B^\lambda(\cdot,C) = E_. \int_0^{\tau_B} (1 - e^{-\lambda t}) I_C(\xi(t))dt$$
$$\leq \sup_y g_B(y,C)(1 - e^{-\lambda s} + P_.(\tau_B \geq s)).$$

Since $\lim_{s\to\infty} P_.(\tau_B \geq s) = 0$ uniformly on compacts,

(27) $\lim_{\lambda\to 0} g_B^\lambda(\cdot,C) = g_B(\cdot,C)$ uniformly on compacts.

Observe also that

(28) $\lim_{\lambda\to 0} k^\lambda(\cdot,C) = k(\cdot,C)$ uniformly on compacts.

By the definition of h_B^λ

(29) $\lim_{\lambda\to 0} h_B^\lambda k^\lambda(\cdot,C) = h_B k(\cdot,C)$ uniformly on compacts.

By (26)-(29) there is a function W_B on X such that

(30) $$\lim_{\lambda\to 0} W_B^\lambda = W_B \text{ uniformly on compacts.}$$

Clearly W_B is nonnegative and bounded on compacts. Now

(31) $$\lim_{\lambda\to 0} k^\lambda = k$$

and

(32) $$\lim_{\lambda\to 0} g_B^\lambda = g_B.$$

Also $\lim_{\lambda\to 0} h_B^\lambda k_0^\lambda = h_B k_0$ and $\lim_{\lambda\to 0} h_B^\lambda k_1^\lambda = k_1$, so

(33) $$\lim_{\lambda\to 0} h_B^\lambda k^\lambda = h_B k;$$

moreover $h_B k$ is bounded below on compacts. It now follows from (24) and (30)-(33) that (25) holds. This completes the proof of the theorem.

Let B be relatively compact and nonpolar. Equation (25) is called the <u>fundamental identity</u>. It follows from Theorem 3.2 that $W_B \geq 0$, $W_B = 0$ on B^r and W_B is bounded on compacts; if $A \subset B$ and A is nonpolar, then $W_A \geq W_B$.

THEOREM 3.3. *Let B be relatively compact and nonpolar and let $\mu, \nu \in \mathcal{M}^+(B^r)$ be such that $\nu(X) = \mu(X)$ and $k\nu = k\mu + \alpha$ on B^r some $\alpha \in \mathbb{R}$. Then $\nu = \mu$.*

Proof. Now $h_B(x,\cdot)$ is a probability measure concentrated on B^r for $x \in X$, so by the fundamental identity

$$k\nu = h_B k\nu - \nu(X)W_B = h_B k\mu + \alpha - \mu(X)W_B = k\mu + \alpha.$$

Consequently $\nu = \mu$ by Theorem 3.1.

Let B be relatively compact and nonpolar and let $\mu \in \mathcal{M}^+$. The balayage problem is to find a measure $\nu \in \mathcal{M}^+(B^r)$ such that $\nu(X) = \mu(X)$ and $k\nu = k\mu + \alpha$ on B^r for some $\alpha \in \mathbb{R}$. It follows from Theorem 3.3 that there is at most one solution to the balayage problem. To see that there is a solution, observe that by (22) for $\lambda > 0$

$$(34)\quad \int h_B^\lambda(x,dz)k^\lambda(z,y) - W_B^\lambda(x) = \int h_B^\lambda(y,dz)k^\lambda(z,x) - W_B^\lambda(y), \qquad x,y \in X.$$

It follows by letting $\lambda \to 0$ that

$$(35)\quad \int h_B(x,dz)k(z,y) - W_B(x) = \int h_B(y,dz)k(z,x) - W_B(y), \quad x,y \in X.$$

Now $\mu h_B \in \mathcal{M}^+(B^r)$ and $\mu h_B(X) = \mu(X)$. By (35)

$$(36)\qquad k(\mu h_B) = h_B k\mu + \int W_B d\mu - \mu(X)W_B.$$

In particular

$$(37)\qquad k(\mu h_B) = k\mu + \int W_B d\mu \quad \text{on} \quad B^r.$$

Thus μh_B is the unique solution to the balayage problem. (In particular $h_B(x,\cdot)$ is the unique solution to the balayage problem when $\mu = \delta_x$.) It follows from the fundamental identity that

$$(38)\qquad k\mu = g_B\mu + k(\mu h_B) - \int W_B d\mu$$

and hence that

$$(39)\qquad k(\mu h_B) \le k\mu + \int W_B d\mu.$$

The equilibrium problem for the relatively compact nonpolar set B is to find a probability measure $\mu \in \mathcal{M}^+(B^r)$ such that

$k\mu$ is constant on B^r. (The constant value is necessarily finite. For if $k\mu = \infty$ on B^r, then $k\mu = \infty$ on X by the fundamental identity. But this is impossible since $k\mu$ is integrable on compacts.) By Theorem 3.3 there is at most one solution to the equilibrium problem. If it exists it is denoted by μ_B and called the <u>equilibrium measure</u> of B, the constant value $R(B)$ of $k\mu_B$ on B^r is called the <u>Robin constant</u> of B and $k\mu_B$ is called the equilibrium potential of B; by the fundamental identity.

$$k\mu_B = R(B) - W_B. \tag{40}$$

The equilibrium problem will be solved by first solving the corresponding problem for λ-potentials.

Let $\lambda > 0$. Given a Radon measure μ on X, the function $g^\lambda\mu$ is called the <u>λ-potential</u> of μ. It is lower semicontinuous. Observe that

$$\int g^\lambda(\cdot,y)\lambda dy = 1 \text{ on } X. \tag{41}$$

Set $\mu_B^\lambda = \int \lambda dy h_B^\lambda(y,\cdot)$. It follows from (22) and (41) that

$$g^\lambda\mu_B^\lambda = h_B^\lambda 1 = E_\cdot(e^{-\lambda\tau_B}), \tag{42}$$

where $e^{-\infty} = 0$. The measure μ_B^λ is concentrated on B^r and its λ-potential equals 1 on B^r. It is called the <u>λ-equilibrium measure</u> of B, its total mass $C^\lambda(B) = \mu_B^\lambda(X)$ is called <u>λ-capacity</u> of B and its λ-potential $g^\lambda\mu_B^\lambda = E_\cdot(e^{-\lambda\tau_B})$ is called the <u>λ-equilibrium potential</u> of B. Observe that B is polar if and only if $C^\lambda(B) = 0$. For if $C^\lambda(B) = 0$, then $P_\cdot(\tau_B = \infty) = 1$ by (42) and hence B is polar. Conversely if B is polar, then $\mu_B^\lambda = 0$, so $C^\lambda(B) = \mu_B^\lambda(X) = 0$.

THEOREM 3.4. *Let B be relatively compact and nonpolar. Then B has an equilibrium measure.*

Proof. Let A be a relatively compact set having positive Haar measure $|A|$. Now $P_\cdot(\tau_B < \infty) = 1$, so $\lim_{\lambda\to 0} E_\cdot(e^{-\lambda\tau_B}) = 1$. Thus by (42)

$$(43) \qquad \lim_{\lambda\to 0} \int g^\lambda(\cdot,A)d\mu_B^\lambda = |A|.$$

Since $\lim_{\lambda\to 0} g_1^\lambda(0) = \int_1^\infty p(t,0)dt = \infty$, it follows from (7) and (43) that

$$(44) \qquad \lim_{\lambda\to 0} C^\lambda(B)g_1^\lambda(0) = 1.$$

By (42)

$$k^\lambda\mu_B^\lambda = E_\cdot(e^{-\lambda\tau_B}) - g_1^\lambda(0)C^\lambda(B).$$

Observe that $\nu_B^\lambda = (C^\lambda(B))^{-1}\mu_B^\lambda$ is a probability measure in $\mathcal{M}^+(B^r)$ and that

$$(45) \qquad k^\lambda\nu_B^\lambda = (C^\lambda(B))^{-1} - g_1^\lambda(0) - (C^\lambda(B)g_1^\lambda(0))^{-1}W_B^\lambda.$$

Let $\{\lambda_n\}$ be a sequence of positive numbers such that $\lambda_n \downarrow 0$ and $\mu_n = \nu_B^{\lambda_n}$ converges completely to a probability measure $\mu \in \mathcal{M}^+(\bar{B})$. Observe that

$$(46) \quad \int_A k^{\lambda_n}\mu_n(x)dx = \int k^{\lambda_n}(\cdot,A)d\mu_n \to \int k(\cdot,A)d\mu = \int_A k\mu(x)dx.$$

Since $\lim_{\lambda\to 0} \int_A W_B^\lambda(x)dx = \int_A W_B(x)dx$, it follows from (44)-(46) that $C^{\lambda_n}(B)^{-1} - g_1^{\lambda_n}(0)$ has a finite limit L as $n \to \infty$ and

$$\int_A k\mu(x)dx = L|A| - \int_A W_B(x)dx.$$

Consequently $k\mu = L - W_B$ a.e.. Since $W_B = 0$ on $B^r \supset \mathring{B}$, $k\mu = L$ a.e. on $\mathring{B}$. Therefore $k\mu = L$ on $\mathring{B}$ by Theorem 3.1.

Let C be a compact set containing $\bar{B}$ in its interior. By the result of the previous paragraph there is a probability measure $\mu \in \mathcal{M}^+(C)$ such that $k\mu = L$ on $\bar{B}$ for some $L \in \mathbb{R}$. Then $\mu h_B \in \mathcal{M}^+(B^r)$ and by (37), $k(\mu h_B) = L + \int W_B d\mu$ on B^r. Thus B has equilibrium measure μh_B, as desired.

THEOREM 3.5. *Suppose A and B are relatively compact and nonpolar and that $A \subset B$. Then $\mu_A = \mu_B h_A$ and $R(A) = R(B) + \int W_A d\mu_B$.*

Proof. By (36)

$$k(\mu_B h_A) = R(B) + \int W_A d\mu_B - W_A,$$

so $\mu_B h_A$ is the equilibrium measure of A and $R(B) + \int W_A d\mu_B$ is the Robin constant of A.

Let B be relatively compact. It follows from Theorem 3.5 that if B is nonpolar, then μ_B does not charge polar sets. If B is polar, set $R(B) = \infty$ and $W_B = \infty$ on X.

THEOREM 3.6. *Let A and B be relatively compact. Then*

$$W_{A\cup B} + W_{A\cap B} \geq W_A + W_B$$

<u>and</u>

$$R(A \cup B) + R(A \cap B) \geq R(A) + R(B).$$

<u>Proof</u>. The results are trivially true if $A \cap B$ is polar, so assume that this set is nonpolar. Note that $\tau_{A\cup B} = \tau_A \wedge \tau_B$ and $\tau_{A\cap B} \geq \tau_A \vee \tau_B$. Let $\lambda > 0$. Then

$$e^{-\lambda\tau_{A\cup B}} + e^{-\lambda\tau_{A\cap B}} \leq e^{-\lambda\tau_A} + e^{-\lambda\tau_B},$$

so

$$E_.(e^{-\lambda\tau_{A\cup B}}) + E_.(e^{-\lambda\tau_{A\cap B}}) \leq E_.(e^{-\lambda\tau_A}) + E_.(e^{-\lambda\tau_B})$$

and hence

$$W^\lambda_{A\cup B} + W^\lambda_{A\cap B} \geq W^\lambda_A + W^\lambda_B .$$

The first conclusion of the theorem now follows by letting $\lambda \to 0$. Let C be a compact set containing $A \cup B$. Then

$$R(A) = R(C) + \int W_A d\mu_C$$

by Theorem 3.5 and the same formula holds with A replaced by B, $A \cup B$ or $A \cap B$. Thus the second conclusion follows from the first conclusion.

<u>THEOREM 3.7</u>. <u>Suppose</u> B <u>is relatively compact and</u> $\{B_n\}$ <u>is an increasing sequence of subsets of</u> B <u>such that</u> $P_.(\tau_{B_n} \downarrow \tau_B) = 1$ <u>on</u> X. <u>Then</u> $W_{B_n} \downarrow W_B$ <u>on</u> X <u>and</u> $R(B_n) \downarrow R(B)$.

<u>Proof</u>. Let f be a continuous nonnegative function having compact support and such that $\int f(x)dx = 1$. Since

$$kf(x) = \int k(x,y)f(y)dy = \int k(y)f(x+y)dy,$$

kf is continuous.

In proving the theorem it can be assumed that B and B_n, $n \geq 1$, are nonpolar. By the fundamental identity

$$kf = g_{B_n} f + h_{B_n} kf - W_{B_n} \tag{47}$$

and

$$kf = g_B f + h_B kf - W_B. \tag{48}$$

Since $g_{B_n} f = E_{\cdot} \int_0^{\tau_{B_n}} f(\xi(t))dt$ and $g_B f = E_{\cdot} \int_0^{\tau_B} f(\xi(t))dt < \infty$, it follows from the dominated convergence theorem that

$$\lim_n g_{B_n} f = g_B f. \tag{49}$$

Now B is necessarily nonpolar, so by the continuity of kf and right-continuity of paths

$$\lim_n h_{B_n} kf = h_B kf. \tag{50}$$

It follows from (47)-(50) that $W_{B_n} \downarrow W_B$, which proves the first conclusion of the theorem. By Theorem 3.5

$$R(B_n) = R(B) + \int W_{B_n} d\mu_B \downarrow R(B) + \int W_B d\mu_B = R(B),$$

which completes the proof of the theorem.

<u>THEOREM 3.8</u>. <u>Let</u> $\mu \in \mathfrak{M}^+$ <u>and let</u> $\{B_n\}$ <u>be an increasing</u>

<u>sequence of subsets of the relatively compact set</u> B <u>such that</u> $P_{\cdot}(\tau_{B_n} \downarrow \tau_B) = 1$. <u>Then</u> $\lim_n h_{B_n} k\mu = h_B k\mu$.

<u>Proof</u>. It can be assumed that B_n is nonpolar for each n and hence that B is nonpolar. Since $k\mu$ is lower semicontinuous it is bounded below on $\overline{B}$, so $\underline{\lim}_n h_{B_n} k\mu \geq h_B k\mu$ by right-continuity of paths and Fatou's lemma. Thus it suffices to show that

$$\overline{\lim}_n h_{B_n} k\mu \leq h_B k\mu. \tag{51}$$

To this end observe that by (23)

$$h_{B_n}^{\lambda} k^{\lambda} - W_{B_n}^{\lambda} \leq h_B^{\lambda} k^{\lambda} - W_B^{\lambda}.$$

It follows by letting $\lambda \to 0$ that

$$h_{B_n} k - W_{B_n} \leq h_B k - W_B$$

and hence that

$$h_{B_n} k\mu - \mu(X) W_{B_n} \leq h_B k - \mu(X) W_B.$$

Now $W_{B_n} \downarrow W_B$ by Theorem 3.7, so (51) holds as desired.

<u>THEOREM 3.9</u>. <u>Let</u> $x \in X$ <u>and let</u> $\{B_n\}$ <u>be a decreasing sequence of relatively compact sets each containing</u> B <u>such that</u> $P_x(\tau_{B_n} \uparrow \tau_B) = 1$. <u>Then</u> $W_{B_n}(x) \uparrow W_B(x)$.

<u>Proof</u>. Let f be as in the proof of Theorem 3.7. In proving the desired result it can be assumed that each set B_n is nonpolar.

If B is nonpolar the proof is the same as that of the first result of Theorem 3.7. Suppose B is polar. By the fundamental identity

$$kf(x) = g_{B_n} f(x) + h_{B_n} kf(x) - W_{B_n}(x), \tag{52}$$

and by the monotone convergence theorem

(53) $$g_{B_n} f(x) \uparrow \infty.$$

Now $h_{B_n} kf(x)$ is bounded below in n, so by (52) and (53), $W_{B_n}(x) \uparrow \infty$, which completes the proof of the theorem.

Given a relatively compact (Borel) set B, define $\underline{R}(B)$ and $\overline{R}(B)$ by

$$\underline{R}(B) = \inf[R(A) : A \text{ is a compact subset of } B]$$

and

$$\overline{R}(B) = \sup[R(U) : U \text{ is a relatively compact open set containing } B].$$

Clearly $\underline{R}(B) \geq R(B) \geq \overline{R}(B)$. The set B is said to be <u>capacitable</u> if $\underline{R}(B) = R(B) = \overline{R}(B)$.

<u>THEOREM 3.10</u>. *Let B be relatively compact. Then B is capacitable.*

<u>Proof</u>. Without loss of generality it can be assumed that B is nonempty. There is an increasing sequence $\{B_n\}$ of compact subsets of B such that $P_{\cdot}(\tau_{B_n} \downarrow \tau_B) = 1$. Now $R(B_n) \downarrow R(B)$ by Theorem 3.7, so $\underline{R}(B) = R(B)$.

Let C be a compact nonpolar set containing B in its interior. Since μ_C does not charge polar sets, $\mu_C(B \setminus B^r) = 0$. Let $\{U_n\}$ be a decreasing sequence of open sets such that $B \subset U_n \subset C$ for each n and $P_{\cdot}(\tau_{U_n} \uparrow \tau_B) = 1$ a.e. (μ_C). Then $W_{U_n} \uparrow W_B$ a.e. (μ_C) by Theorem 3.9. Consequently by Theorem 3.5.

$$R(U_n) = R(C) + \int W_{U_n} d\mu_C \uparrow R(C) + \int W_B d\mu_C = R(B),$$

so $\overline{R}(B) = R(B)$. This completes the proof of the theorem.

4. Thinness

In this section analytic criteria for irregularity and polarity are obtained.

THEOREM 4.1. Let $\mu \in \mathcal{M}^+$. Then $k\mu < \infty$ q.e. on X.

Proof. Suppose otherwise. Then there is a compact nonpolar subset B of $\{x \in X: k\mu(x) = \infty\}$. By the fundamental identity

$$k\mu = g_B\mu + h_B k\mu - \mu(X)W_B = \infty \quad \text{on} \quad X,$$

which is impossible since $k\mu$ is integrable on compacts.

THEOREM 4.2. Let B be a relatively compact polar set, let $\nu \in \mathcal{M}^+$ be such that $\nu(B) = 0$, and let U be an open set containing $\bar{B}$. Then there is a measure $\mu \in \mathcal{M}^+(U)$ such that $k\mu = \infty$ on B and $\int k\mu d\nu < \infty$.

Proof. Without loss of generality it can be assumed that μ is not the zero measure. Choose $\lambda > 0$. Let $\{U_n\}$ be a decreasing sequence of relatively compact open sets such that $B \subset U_n \subset \bar{U}_n \subset U$ for each n and $P_\cdot(\tau_{U_n} \uparrow \tau_B = \infty) = 1$ a.e. (ν). Then

$$g^\lambda \mu^\lambda_{U_n} = E_\cdot(e^{-\lambda\tau_{U_n}}) = 1 \text{ on } B, \quad g^\lambda \mu^\lambda_{U_n} \downarrow 0 \text{ a.e. } (\nu) \text{ and hence}$$

$\int g^\lambda \mu^\lambda_{U_n} d\nu \downarrow 0$. It can be assumed (take a subsequence if necessary) that $\Sigma_n \int g^\lambda \mu^\lambda_{U_n} d\nu < \infty$. Since g^λ is lower semicontinuous and strictly positive, it is bounded away from zero on compacts and hence $\Sigma_n \mu^\lambda_{U_n}(X) < \infty$. Set $\mu = \Sigma_n \mu^\lambda_{U_n}$. Then $\mu \in \mathcal{M}^+(U)$, $g^\lambda\mu = \Sigma_n 1 = \infty$ on B and $\int g^\lambda \mu d\nu < \infty$. It now follows from (3.6) that $k\mu = \infty$ on B and $\int k\mu d\nu < \infty$. Thus μ is the desired measure.

The following analytic criterion for a relatively compact set to be polar is an immediate consequence of Theorems 4.1 and 4.2.

<u>THEOREM 4.3</u>. *A relatively compact set* B *is polar if and only if there is a measure* $\mu \in \mathfrak{M}^+$ *such that* $k\mu = \infty$ *on* B.

If $k(0) < \infty$, then k is continuous and hence every nonempty set is nonpolar. If $k(0) = \infty$, every one-point set is polar. Linear Brownian motion and linear symmetric stable processes of order $\alpha \in (1,2)$ are examples of processes for which $k(0) < \infty$. Planar Brownian motion and the linear symmetric stable process of order $\alpha = 1$ (Cauchy process) are examples of processes for which $k(0) = \infty$.

A set B is said to be *thin* at x if any of the following is true: (i) $x \notin \overline{B}$; (ii) x is an isolated point of $\overline{B}$ and $\{x\}$ is polar; (iii) x is a nonisolated point of $\overline{B}$, $\{x\}$ is polar and there is a measure $\mu \in \mathfrak{M}^+$ such that

$$\varliminf_{\substack{y \to x \\ y \in B}} k\mu(y) > k\mu(x).$$

If $k(0) < \infty$, then B is thin at x if and only if $x \notin \overline{B}$. If $k(0) = \infty$, then B is thin at x if and only if one of the following is true: $x \notin \overline{B}$; x is an isolated point of $\overline{B}$; (iii) holds.

Analytic criteria for irregularity and polarity will now be obtained. It follows in particular from the next result that if $k(0) < \infty$, then $B^r = \overline{B}$.

<u>THEOREM 4.4</u>. *A point* x *is irregular for a set* B *if and only if* B *is thin at* x. *The set* B *is polar if and only if it is thin everywhere*.

<u>Proof</u>. Suppose first that $x \in (B\backslash\{x\})^r$. Let U be a relatively compact open set containing x and set $A = (B\backslash\{x\}) \cap U$. Then A is a relatively compact nonpolar set and $x \in A^r$. Let $\{B_n\}$ be an increasing sequence of compact nonpolar subsets of A such that $P_\cdot(\tau_{B_n} \downarrow \tau_A) = 1$. By Theorem 3.7, $W_{B_n}(x) \downarrow W_A(x) = 0$. Choose $\mu \in \mathfrak{M}^+$. According to the fundamental identity

$$k\mu(x) \geq h_{B_n} k\mu(x) - \mu(X)W_{B_n}(x) \geq \inf[k\mu(y) : y \in A] - \mu(X)W_{B_n}(x).$$

Consequently $k\mu(x) \geq \inf[k\mu(y) : y \in (B \setminus \{x\}) \cap J]$ and therefore

$$k\mu(x) \geq \underline{\lim}_{\substack{y \to x \\ y \in B}} k\mu(y).$$

Thus B is not thin at x. Suppose next that $x \in B^r$, but $x \notin (B \setminus \{x\})^r$. Then x is regular for $\{x\}$, so $\{x\}$ is nonpolar and hence B is not thin at x.

Suppose $x \notin \overline{B}$. Then $x \notin B^r$ and B is thin at x.

Suppose $x \in \overline{B} \setminus B^r$. Then x is irregular for $\{x\}$ and hence $\{x\}$ is polar. Therefore if x is an isolated point of B, then B is thin at x. Assume instead that x is a nonisolated point of $\overline{B}$. Set $A = B \setminus \{x\}$. Then $\overline{A} = \overline{B}$, $A^r = B^r$ and x is a nonisolated point of A. Choose $\lambda > 0$. Now $g^\lambda \mu_A^\lambda = E_\cdot(e^{-\lambda \tau_A}) = 1$ on A^r and $g^\lambda \mu_A^\lambda(x) = E_x(e^{-\lambda \tau_A}) < 1$. By the proof of Theorem 4.2 there is a measure $\nu \in \mathcal{M}^+$ such that $g^\lambda \nu = \infty$ on the polar set $A \setminus A^r$ and $g^\lambda \nu(x) < \infty$. Set $\mu = \mu_A^\lambda + \nu$. Then

$$\underline{\lim}_{\substack{y \to x \\ y \in B}} g^\lambda \mu(y) = \underline{\lim}_{\substack{y \to x \\ y \in A}} g^\lambda \mu(y) > g^\lambda \mu(x)$$

It now follows from (3.6) that

$$\underline{\lim}_{\substack{y \to x \\ y \in B}} k\mu(y) > k\mu(x)$$

and hence that B is thin at x. This completes the proof of the first conclusion of the theorem.

The set B is polar if and only if B^r is empty; by the first conclusion this is true if and only if B is thin everywhere. Thus the second conclusion of the theorem is valid.

5. Potential Principles

The following result, called the maximum principle for λ-potentials, is valid: if $\lambda > 0$ and $\mu \in \mathcal{M}^+(B)$, then $\sup_x g^\lambda\mu(x) = \sup_{x\in B} g^\lambda\mu(x)$.

In proving this result it can be assumed that $M = \sup_{x\in B} g^\lambda\mu(x) < \infty$. Choose $\varepsilon > 0$ and set $A = \{x \in X: g^\lambda\mu(x) \leq M + \varepsilon\}$. Then A is closed since $g^\lambda\mu$ is lower semicontinuous. Observe that for $t > 0$

$$g^\lambda\mu \geq \int_t^\infty e^{-\lambda s}P^s\mu ds = e^{-\lambda t}\int_0^\infty e^{-\lambda s}P^{s+t}\mu ds = e^{-\lambda t}P^t g^\lambda\mu$$

and hence $g^\lambda\mu \geq (M+\varepsilon)e^{-\lambda t}P_\cdot(\tau_A > t)$. Consequently $g^\lambda\mu \geq (M+\varepsilon)P_\cdot(\tau_A > 0)$, so $g^\lambda\mu \geq M + \varepsilon$ on $(A^r)^c$ and therefore $B \subset A^r$. It now follows from (3.21) that $g^\lambda\mu = h_A^\lambda g^\lambda\mu \leq M + \varepsilon$. Since ε can be made arbitrarily small $g^\lambda\mu \leq M$, which yields the desired conclusion.

THEOREM 5.1. Suppose $\mu \in \mathcal{M}^+(B)$ and $k\mu < \infty$ on B. Then μ does not charge polar sets.

Proof. Choose $\lambda > 0$. Then $g^\lambda\mu < \infty$ on B. It suffices to show that if A is a compact polar subset of B and $g^\lambda\mu \leq M$ on A, then $\mu(A) = 0$. Now $g^\lambda\mu|_A \leq M$ on A, so $g^\lambda\mu|_A \leq M$ on X by the maximum principle for λ-potentials. By the proof Theorem 4.2 there is a measure $\nu \in \mathcal{M}^+$ such that $g^\lambda\nu = \infty$ on A. Since

$$\int_A g^\lambda\nu d\mu = \int g^\lambda\mu|_A d\nu \leq M\nu(X) < \infty,$$

$\mu(A) = 0$ as desired.

The next result is called the maximum principle.

THEOREM 5.2. Let $\mu \in \mathcal{M}^+(B)$. Then $\sup_x k\mu(x) = \sup_{x\in B} k\mu(x)$.

Proof. It can be assumed that $M = \sup_{x\in B} k\mu(x) < \infty$. Then by Theorem 5.1, μ does not charge polar sets. Choose $\varepsilon > 0$ and set $A = \{x \in X: k\mu(x) \le M + \varepsilon\}$. Then A is closed since $k\mu$ is lower semicontinuous, and μ is concentrated on A^r. Also A is nonpolar. There is a compact nonpolar subset C of A such that μ is concentrated on C^r. By the fundamental identity $k\mu \le h_C k\mu \le M + \varepsilon$. Since ε can be made arbitrarily small $k\mu \le M$, which yields the desired result.

The next result is called the continuity principle.

THEOREM 5.3. *Let B be compact and let $\mu \in \mathcal{M}^+(B)$ be such that the restriction of $k\mu$ to B is continuous on B. Then $k\mu$ is continuous on X.*

Proof. Choose $\lambda > 0$. It follows from (3.6) that the restriction of $g^\lambda\mu$ to B is continuous on B. In particular $g^\lambda\mu$ is bounded on B, so $g^\lambda\mu$ is bounded on X by the maximum principle for λ-potentials. Choose α with $\lambda < \alpha < \infty$. It is easily checked that

$$(1) \qquad (\alpha - \lambda)g^\alpha g^\lambda = \int_0^\infty (e^{-\lambda t} - e^{-\alpha t})p^t dt$$

and hence that

$$(2) \qquad g^\lambda\mu = g^\alpha\mu + (\alpha - \lambda)g^\alpha g^\lambda\mu.$$

Since $g^\lambda\mu$ is bounded on X, $(\alpha - \lambda)g^\alpha g^\lambda\mu$ is continuous on X and hence the restriction of $g^\alpha\mu$ to B is continuous on B by (2). Now $(\alpha - \lambda)g^\alpha g^\lambda \uparrow g^\lambda$ as $\alpha \uparrow \infty$ by (1), so $g^\alpha\mu \downarrow 0$ as $\alpha \uparrow \infty$ by (2). Consequently $g^\alpha\mu \downarrow 0$ uniformly on B as $\alpha \uparrow \infty$ by Dini's theorem. Thus $g^\alpha\mu \downarrow 0$ as $\alpha \uparrow \infty$ uniformly on X by the maximum principle for α-potentials. Hence by (2), $(\alpha - \lambda)g^{\alpha}g^{\lambda}\mu \to g^{\lambda}\mu$ uniformly on X as $\alpha \to \infty$ and therefore $g^{\lambda}\mu$ is continuous on X. It now follows easily from (3.6) that $k\mu$ is continuous on X as desired.

THEOREM 5.4. *Let* $\mu \in \mathcal{M}^{+}(B)$ *be such that* $k\mu < \infty$ *a.e.* (μ). *Then for every* $\varepsilon > 0$ *there is a compact subset* C *of* B *such that* $\mu(B \setminus C) \leq \varepsilon$ *and* $k\mu|_C$ *is continuous on* X.

Proof. Choose $\varepsilon > 0$. By Lusin's theorem there is a compact subset C of B such that $\mu(B \setminus C) \leq \varepsilon$ and the restriction of $k\mu$ to C is continuous on C. Now $k\mu = k\mu|_C + k\mu|_{B \setminus C}$ and $k\mu|_C$ and $k\mu|_{B \setminus C}$ are both lower semicontinuous, so the restriction of $k\mu|_C$ to C is continuous on C. By the continuity principle, $k\mu|_C$ is continuous on X.

THEOREM 5.5. *Let* B *be nonpolar. Then there is a nonzero measure* $\mu \in \mathcal{M}^{+}(B)$ *such that* $k\mu$ *is continuous on* X.

Proof. It can be assumed that B is compact. Now $\mu_B \in \mathcal{M}^{+}(B)$, $\mu_B(B) = 1$ and $k\mu_B \leq R(B) < \infty$ on X. By Theorem 5.4 there is a compact subset C of B such that $\mu_B(B \setminus C) \leq 1/2$ and $k\mu$ is continuous on X, where $\mu = \mu_B|_C$. Clearly μ satisfies the conclusion of the theorem.

The next result is called the *complete maximum principle*.

THEOREM 5.6. *Let* B *be nonpolar and let* $\nu \in \mathcal{M}^{+}(B^r)$. *Suppose that* $k\nu \leq k\mu + \alpha$ *q.e. on* B, *where* $\mu \in \mathcal{M}^{+}$, $\mu(X) \leq \nu(X)$

<u>and</u> $\alpha \in \mathbb{R}$. <u>Then</u> $k\nu \leq k\mu + \alpha$ <u>on</u> X.

<u>Proof</u>. Without loss of generality it can be assumed that B is relatively compact. Let $\{B_n\}$ be an increasing sequence of compact nonpolar subsets of B such that $P_{\cdot}(\tau_{B_n} \downarrow \tau_B) = 1$ on X. Then

$$(3) \qquad h_{B_n} k\nu \leq h_{B_n} k\mu + \alpha$$

on B_n^c and (3) holds q.e. on B_n^r. Thus (3) holds q.e. on X and hence a.e. on X. It now follows from Theorem 3.8 that

$$h_B k\nu \leq h_B k\mu + \alpha \quad \text{a.e. on } X.$$

Consequently by the fundamental identity

$$k\nu = h_B k\nu - \nu(X)W_B \leq h_B k\mu - \mu(X)W_B + \alpha \leq k\mu + \alpha$$

a.e. on X. Therefore $k\nu \leq k\mu + \alpha$ on X by Theorem 3.1.

The next result is called the <u>uniqueness principle</u>.

<u>THEOREM 5.7</u>. <u>Let</u> $\mu, \nu \in \mathcal{M}^+(B^r)$ <u>be such that</u> $\nu(X) = \mu(X)$ <u>and</u> $k\nu = k\mu + \alpha$ <u>q.e. on</u> B <u>for some</u> $\alpha \in \mathbb{R}$. <u>Then</u> $\nu = \mu$.

<u>Proof</u>. Without loss of generality it can be assumed that B is nonpolar. Then $k\nu = k\mu + \alpha$ on X by the complete maximum principle, so $\nu = \mu$ by Theorem 3.1.

The uniqueness principle leads to alternative formulations of the balayage and equilibrium problems.

<u>THEOREM 5.8</u>. <u>Let</u> B <u>be relatively compact and nonpolar and</u>

let $\mu \in \mathcal{M}^+$. Then μh_B is the unique measure $\nu \in \mathcal{M}^+(B^r)$ such that $\nu(X) = \mu(X)$ and $k\nu = k\mu + \alpha$ q.e. on B for some $\alpha \in \mathbb{R}$.

Proof. Since $\mu h_B \in \mathcal{M}^+(B^r)$, $\mu h_B(X) = \mu(X)$ and $k(\mu h_B) = k\mu + \alpha$ on B for some $\alpha \in \mathbb{R}$ by (3.37), the desired conclusion follows from the uniqueness principle.

THEOREM 5.9. Let B be relatively compact and nonpolar. Then μ_B is the unique probability measure $\mu \in \mathcal{M}^+(B^r)$ such that $k\mu$ is constant q.e. on B.

Proof. Since μ_B is a probability measure in $\mathcal{M}^+(B^r)$ and $k\mu_B$ is constant q.e. on B, the desired result follows from the uniqueness principle.

THEOREM 5.10. Let $\mu, \nu \in \mathcal{M}^+$ and let B be a relatively compact polar set such that $\mu(B) = 0$. Suppose there is an open set U containing $\overline{B}$ such that $k\nu \leq k\mu + \alpha$ on U for some $\alpha \in \mathbb{R}$. Then $\nu(B) = 0$.

Proof. By Theorem 4.2 there is a measure $\gamma \in \mathcal{M}^+(U)$ such that $k\gamma = \infty$ on B and $\int k\gamma d\mu < \infty$. Now

$$\int k\gamma d\nu = \int k\nu d\gamma \leq \int (k\mu + \alpha)d\gamma = \int k\gamma d\mu + \alpha\gamma(X) < \infty,$$

so $\nu(B) = 0$ as desired.

THEOREM 5.11. Suppose $\mu, \nu \in \mathcal{M}^+$, μ does not charge polar sets and $k\nu \leq k\mu + \alpha$ on X for some $\alpha \in \mathbb{R}$. Then ν does not charge polar sets.

Proof. Let A be polar. By Theorem 5.10, $\nu(B) = 0$ if B is a relatively compact subset of A, so $\nu(A) = 0$. Thus ν does not charge polar sets.

THEOREM 5.12. Let B be a compact nonpolar set, let $\mu \in \mathcal{M}^+$ be such that $\mu(B \setminus B^r) = 0$, and let $\nu \in \mathcal{M}^+(B)$ satisfy $\nu(X) = \mu(X)$. Then $\nu = \mu h_B$ if and only if there is an $\alpha \in \mathbb{R}$ such that $k\nu \leq k\mu + \alpha$ on X and $k\nu = k\mu + \alpha$ q.e. on B.

Proof. Suppose first that $\nu = \mu h_B$. Then by (3.39) and (3.37), $k\nu \leq k\mu + \alpha$ on X and $k\nu = k\mu + \alpha$ q.e. on B, where $\alpha = \int W_B d\mu$. Suppose conversely that $k\nu \leq k\mu + \alpha$ on X and $k\nu = k\mu + \alpha$ q.e. on B for some $\alpha \in \mathbb{R}$. Then $\nu(B \setminus B^r) = 0$ by Theorem 5.10, so $\nu \in \mathcal{M}^+(B^r)$. Consequently $\nu = \mu h_B$ by the uniqueness principle.

6. Energy

Let $\mu,\nu \in \mathcal{M}^+$ and $\lambda > 0$. It follows easily from (3.6) that $\int k\mu d\nu < \infty$ if and only if $\int g^\lambda \mu d\nu < \infty$. The *energy* $I(\mu)$ and *λ-energy* $I^\lambda(\mu)$ of μ are defined by $I(\mu) = \int k\mu d\mu$ and $I^\lambda(\mu) = \int g^\lambda \mu d\mu$. Thus μ has finite energy if and only if it has finite λ-energy. The equilibrium measure μ_B of a relatively compact nonpolar set B has energy $I(\mu_B) = R(B)$.

THEOREM 6.1. *Let $\mu \in \mathcal{M}^+$ have finite energy. Then μ does not charge polar sets. If $\nu \in \mathcal{M}^+$ also has finite energy, then $\int k\mu d\nu < \infty$.*

Proof. To see that μ does not charge polar sets, let B be a compact polar set. Set $B_1 = \{x \in B : k\mu(x) = \infty\}$ and $B_2 = \{x \in B : k\mu(x) < \infty\}$. Since μ has finite energy, $k\mu < \infty$ a.e. (μ) and hence $\mu(B_1) = 0$. Now $\mu|_{B_2} \in \mathcal{M}^+(B_2)$ and $k\mu|_{B_2} < \infty$ on B_2, so $\mu(B_2) = 0$ by Theorem 5.1 and hence $\mu(B) = 0$, which proves the desired result.

Suppose that ν also has finite energy. Now

$$g^\lambda(x,y) = \int_0^\infty e^{-\lambda t} p(t,x,y)dt = \int_0^\infty e^{-\lambda t}\left(\int p(t/2,x,z)p(t/2,z,y)dz\right)dt.$$

Thus by Schwarz's inequality and the symmetry of $p(t/2,x,y)$ in x and y

$$\begin{aligned}\int g^\lambda \mu d\nu &= \int_0^\infty e^{-\lambda t}\left(\int p^{t/2}\mu(z)p^{t/2}\nu(z)dz\right)dt \\ &\le \left[\int_0^\infty e^{-\lambda t}\left(\int (p^{t/2}\mu(z))^2 dz\right)dt\right]^{1/2}\left[\int_0^\infty e^{-\lambda t}\left(\int (p^{t/2}\nu(z))^2 dz\right)dt\right]^{1/2} \\ &= [I^\lambda(\mu)I^\lambda(\nu)]^{1/2} < \infty\end{aligned}$$

Consequently $\int k\mu d\nu < \infty$, which completes the proof of the theorem.

THEOREM 6.2. *A set B is nonpolar if and only if there is a nonzero measure $\mu \in \mathcal{M}^+(B)$ having finite energy.*

Proof. Suppose B is nonpolar and let A be a relatively compact nonpolar subset of B. Then μ_A is a nonzero measure in $\mathcal{M}^+(B)$ having finite energy. The converse result follows from Theorem 6.1.

It follows from the Theorem 6.1 that if $\mu \in \mathcal{M}^+(B)$ and $I(\mu) < \infty$, then $\mu \in \mathcal{M}^+(B^r)$.

THEOREM 6.3. *Let $\nu \in \mathcal{M}^+$ be a nonzero measure having finite energy. Suppose $k\nu \leq k\mu + \alpha$ a.e. (ν) where $\mu \in \mathcal{M}^+, \mu(X) \leq \nu(X)$ and $\alpha \in \mathbb{R}$. Then $k\nu \leq k\mu + \alpha$ on X.*

Proof. Set $B = \{x \in X : k\nu(x) \leq k\mu(x) + \alpha\}$. Then $\nu \in \mathcal{M}^+(B)$ and $k\nu \leq k\mu + \alpha$ on B. Since ν has finite energy, $\nu \in \mathcal{M}(B^r)$, so $k\nu \leq k\mu + \alpha$ on X by the complete maximum principle.

Let $\mathcal{E}$ denote the vector space of signed measures $\mu \in \mathcal{M}$ such that $|\mu|$ has finite energy. If $\mu \in \mathcal{E}$, then $k\mu = k\mu^+ - k\mu^-$ is well defined whenever $k|\mu| < \infty$ and hence is well defined q.e. on X. Set $\mathcal{E}(B) = \mathcal{M}(B) \cap \mathcal{E}$, $\mathcal{E}^+ = \mathcal{M}^+ \cap \mathcal{E}$ and $\mathcal{E}^+(B) = \mathcal{M}^+(B) \cap \mathcal{E}$.

Let C be a fixed compact nonpolar set and let $c > 0$. For $\mu,\nu \in \mathcal{E}(C)$,

$$(\mu,\nu) = \int k(\mu - \mu(X)\mu_C)d(\nu - \nu(X)\mu_C) + c\mu(X)\nu(X)$$
$$= \int k\mu d\nu - (R(C) - c)\mu(X)\nu(X)$$

is well defined and finite by Theorem 6.1. If either $\mu(X) = 0$ or $\nu(X) = 0$, then $(\mu,\nu) = \int k\mu d\nu$ independently of C and c. In particular if $\mu(X) = 0$, then $(\mu,\mu) = \int k\mu d\mu$. Note that (μ,ν) is symmetric in μ and ν and defines a bilinear form on $\mathcal{E}(C)$. It follows from the next result, called the <u>energy principle</u>, that (μ,ν) determines an inner product on $\mathcal{E}(C)$.

<u>THEOREM 6.4</u>. <u>Let</u> $\mu \in \mathcal{E}(C)$. <u>Then</u> $(\mu,\mu) \geq 0$ <u>with equality holding if and only if</u> $\mu = 0$.

<u>Proof</u>. Set $\nu = \mu(X)\mu_C$. It follows from the definition of k^λ in Section 3 that

$$\lim_{\lambda \to 0} \int k^\lambda \mu_1 d\nu_1 = \int k\mu_1 d\nu_1, \quad \mu_1, \nu_1 \in \mathcal{M}^+.$$

Consequently

$$\lim_{\lambda \to 0} \int k^\lambda(\mu - \nu)d(\mu - \nu) = \int k(\mu - \nu)d(\mu - \nu).$$

In other words

(1) $$(\mu,\mu) = \lim_{\lambda \to 0} \int k^\lambda(\mu - \nu)d(\mu - \nu) + c(\mu(X))^2.$$

Now $\nu(X) = \mu(X)$, so

(2) $$\int k^\lambda(\mu - \nu)d(\mu - \nu) = \int g^\lambda(\mu - \nu)d(\mu - \nu).$$

Observe that $|\mu|$ and $|\nu|$ have finite λ-energy. It follows as in the proof of Theorem 6.1 that

(3) $\int g^{\lambda}(\mu - \nu)d(\mu - \nu) = \int_0^{\infty} e^{-\lambda t}(\int (p^{t/2}(\mu - \nu)(z))^2 dz)dt,$

where $p^{t/2}(\mu - \nu) = p^{t/2}\mu - p^{t/2}\nu$. By (1)-(3)

(4) $(\mu,\mu) = \int_0^{\infty}(\int p^{t/2}(\mu - \nu)(z))^2 dz)dt + c(\mu(X))^2.$

Consequently $(\mu,\mu) \geq 0$. Suppose $(\mu,\mu) = 0$. By (4), $\mu(X) = 0$, so $\nu = 0$ and hence by another application of (4)

$$p^t\mu(z) = 0 \text{ a.e. on } (0,\infty) \times X.$$

Therefore for $t > 0$

$$\int_0^t p^s\mu ds = 0 \text{ a.e. on } X.$$

It now follows as in the proof of Theorem 3.1 that $\mu = 0$, which completes the proof of the theorem.

The norm $\|\mu\|$ for $\mu \in \mathcal{E}(C)$, the Schwarz and triangle inequalities and the notions of strong convergence, weak convergence, Cauchy sequence, weak Cauchy sequence, denseness, completeness and weak completeness are defined as usual in an inner product space. A weakly complete subset of $\mathcal{E}(C)$ is complete.

THEOREM 6.5. *The set of elements* $\nu \in \mathcal{E}(C)$ *such that* $k\nu$ *is continuous on* X *is dense in* $\mathcal{E}(C)$.

Proof. Choose $\mu \in \mathcal{E}^+(C)$ and $\varepsilon > 0$. Since $\iint |k(x,y)|\mu(dx)\mu(dy) < \infty$, there is a δ, $0 < \delta < \varepsilon$, such that if $A \subset X \times X$ and

(5) $$\int_A\int \mu(dx)\mu(dy) \leq \delta,$$

then

$$(6) \qquad \iint_A |k(x,y)|\mu(dx)\mu(dy) \leq \varepsilon.$$

By Theorem 5.4 there is a compact subset B of C such that $\mu(C \setminus B) \leq \delta^{1/2}$ and $k\nu$ is continuous on X, where $\nu = \mu|_B$. Also (5) and (6) hold with $A = (C \setminus B) \times (C \setminus B)$. Consequently

$$\|\mu - \nu\|^2 = \int_{B\setminus C} \int_{B\setminus C} k(x,y)\mu(dx)\mu(dy) - (R(C) - c)(\mu(B \setminus C))^2$$
$$\leq \varepsilon(1 + c + |R(C)|).$$

Thus the indicated set of elements is dense in $\mathcal{E}^+(C)$. It is therefore dense in $\mathcal{E}(C)$.

The next result implies in particular that <u>if</u> μ_n $n \geq 1$, <u>and</u> μ <u>are elements of</u> $\mathcal{E}^+(C)$ <u>and</u> μ_n <u>converges strongly to</u> μ, <u>then</u> μ_n <u>converges completely to</u> μ.

<u>THEOREM 6.6</u>. <u>Let</u> μ_n, $n \geq 1$, <u>be elements of</u> $\mathcal{E}^+(C)$ <u>such that</u> $\|\mu_n\|$ <u>is bounded in</u> n. <u>Then there is a strictly increasing sequence</u> $\{n_j\}$ <u>of positive integers and an element</u> $\mu \in \mathcal{E}^+(C)$ <u>such that</u> μ_{n_j} <u>converges completely to</u> μ. <u>Also</u> μ_n <u>converges weakly to</u> μ <u>if and only if</u> μ_n <u>converges completely to</u> μ.

<u>Proof</u>. Suppose first that μ_n converges completely to some finite measure μ. Then $\mu_n(C)$ is bounded in n. Now $\|\mu_n\|^2 = \int k\mu_n d\mu_n - (R(C) - c)(\mu_n(C))^2$ and hence $\int k\mu_n d\mu_n$ is bounded in n. By (3.6)

$$\int k\mu d\mu \leq \underline{\lim}_n \int k\mu_n d\mu_n < \infty$$

and hence $\mu \in \mathcal{E}^+(C)$. Choose $\nu \in \mathcal{E}(C)$ such that $k\nu$ is continuous on X. Then

$$(\mu_n,\nu) = \int k\nu d\mu_n - (R(C) - c)\mu_n(C)\nu(C)$$
$$\to \int k\nu d\mu - (R(C) - c)\mu(C)\nu(C) = (\mu,\nu).$$

Thus μ_n converges weakly to ν by Theorem 6.5.

To prove the first conclusion of the theorem it is enough to show that $\mu_n(C)$ is bounded in n. Assume the contrary. Then there exist probability measures ν_n, $n \geq 1$, and ν all concentrated on C such that $\nu_n \to \nu$ completely and $\|\nu_n\| \to 0$. By the first paragraph of this proof, $\nu \in \mathcal{E}^+(C)$ and ν_n converges weakly to ν. Consequently by Schwarz's inequality

$$\|\nu\|^2 = (\nu,\nu) = \lim_n (\nu_n,\nu) = 0$$

and hence $\nu = 0$ by the energy principle. But this contradicts the fact that $\nu(C) = 1$.

Suppose finally that μ_n converges weakly to $\mu \in \mathcal{E}^+(C)$. Now $\mu_n(C)$ is bounded in n as was just shown. Let $\{n_j\}$ be a strictly increasing sequence of positive integers such that μ_{n_j} converges completely to some element μ_0 of $\mathcal{E}^+(C)$. Then μ_{n_j} converges weakly to μ_0 and hence $\mu_0 = \mu$. Consequently μ_n converges completely to μ, which completes the proof of the theorem.

THEOREM 6.7. *The set* $\mathcal{E}^+(C)$ *is weakly complete.*

Proof. Let $\{\mu_n\}$ be a weak Cauchy sequence of elements of $\mathcal{E}^+(C)$. By definition, $\|\mu_n\|$ is bounded in n and $(\mu_n - \mu_m,\nu) \to 0$ as $m,n \to \infty$ for all $\nu \in \mathcal{E}$. By Theorem 6.6 (or the proof of that theorem) $\mu_n(C)$ is bounded in n. According to that theorem, to show that μ_n coverges weakly to a measure $\mu \in \mathcal{E}^+(C)$, it is enough to show that μ_n converges completely to μ. It suffices to show that if $\{m_j\}$ and $\{n_j\}$ are strictly increasing sequences of positive integers such that $\mu_{m_j} \to \gamma_1 \in \mathcal{E}^+(C)$ completely and $\mu_{n_j} \to \gamma_2 \in \mathcal{E}^+(C)$ completely, then $\gamma_1 = \gamma_2$. Choose $\nu \in \mathcal{E}(C)$. Then $(\mu_{m_j},\nu) \to (\gamma_1,\nu)$ and $(\mu_{n_j},\nu) \to (\gamma_2,\nu)$ by Theorem 6.6. Now $(\mu_{m_j} - \mu_{n_j},\nu) \to 0$, so $(\gamma_1,\nu) = (\gamma_2,\nu)$ for $\nu \in \mathcal{E}(C)$. Consequently $\|\gamma_2 - \gamma_1\|^2 = 0$ and hence $\gamma_1 = \gamma_2$ by the energy principle. This completes the proof of the theorem.

THEOREM 6.8. *Let B be relatively compact. Then B is polar if and only if there is a measure* $\mu \in \mathcal{E}^+$ *such that* $k\mu = \infty$ *on* B.

Proof. If there is a measure $\mu \in \mathcal{E}^+$ such that $k\mu = \infty$ on B, then B is polar by Theorem 4.3. Suppose conversely that B is polar. Without loss of generality it can be assumed that B is nonempty. Since $R(B) = \infty$ and B is capacitable by Theorem 3.10, there is a decreasing sequence $\{U_n\}$ of relatively compact nonempty open sets each containing B such that $R(U_n) \uparrow \infty$. It can be assumed (take a subsequence if necessary) that $R(U_n) > 0$ for each n and that $\Sigma_n R(U_n)^{-1/2} < \infty$. Now $\mu_{U_n}(X) = 1$ and

$$k\mu_{U_n} = R(U_n) - W_{U_n} = R(U_n) \text{ on } U_n.$$

Set $\mu_n = R(U_n)^{-1}\mu_{U_n}$. Then $k\mu_n = 1$ on B, $\mu_n(X) = R(U_n)^{-1}$ and $I(\mu_n) = R(U_n)^{-1}$. It can be assumed that $U_1 \subset C$. Then

$$\|\mu_n\|^2 = I(\mu_n) - (R(C) - c)(\mu_n(X))^2 = R(U_n)^{-1} - (R(C) - c)R(U_n)^{-2},$$

so $\Sigma_{n \geq 1} \|\mu_n\| < \infty$. Consequence $\{\Sigma_1^n \mu_m\}$ is a Cauchy sequence in n. This sequence clearly converges completely to $\mu = \Sigma_1^\infty \mu_n$, and $\mu \in \mathcal{E}^+$

by Theorem 6.6. It follows easily from (3.6) that $k\mu = \sum_n k\mu_n = \infty$ on B, which completes the proof of the theorem.

Let B be a nonpolar subset of C and let $\mu \in \mathcal{E}^+(C)$. By (3.39)

$$\int k(\mu h_B)d(\mu h_B) \leq \int (k\mu + \int W_B d\mu)d(\mu h_B) = \int k(\mu h_B)d\mu + \mu(X)\int W_B d\mu,$$

so by another application of (3.39)

$$\int k(\mu h_B)d(\mu h_B) \leq \int k\mu d\mu + 2\mu(X) \int W_B d\mu.$$

Thus $\mu h_B \in \mathcal{E}^+(C)$. Suppose now that $\mu \in \mathcal{E}(C)$. Set $\mu h_B = \int \mu(dy)h_B(y,\cdot) = \mu^+ h_B - \mu^- h_B$. Then $\mu h_B \in \mathcal{E}(C)$. It follows from (3.38) that

$$(\mu - \mu h_B, \mu h_B) = \int (g_B\mu - \int W_B d\mu)d(\mu h_B) = \int g_B(\mu h_B)d\mu - \mu(X) \int W_B d\mu;$$

now $g_B(\mu h_B) = 0$, so

(7) $$(\mu - \mu h_B, \mu h_B) = -\mu(X) \int W_B d\mu.$$

Similary it follows from (3.38) that

(8) $$(\mu - \mu h_B, \mu) = \int g_B\mu d\mu - \mu(X) \int W_B d\mu.$$

By (7) and (8)

(9) $$\|\mu - \mu h_B\|^2 = \int g_B \mu d\mu$$

and

(10) $$\|\mu h_B\|^2 = \|\mu\|^2 + 2\mu(X) \int W_B d\mu - \int g_B\mu d\mu \leq \|\mu\|^2 + 2\mu(X) \int W_B d\mu.$$

THEOREM 6.9. *Let $\mu \in \mathcal{E}$ and let B be relatively compact and nonpolar. Then*

$\inf[\|\mu h_A - \mu h_B\| : A$ *is a nonpolar compact subset of* $B] = 0$

and

$$\inf\left[\|\mu h_U - \mu h_B\| : U \text{ is a relatively compact open set containing } B\right] = 0$$

Proof. It can be assumed that $\overline{B} \subset \mathring{C}$ and (see the proof below) that $\mu \in \mathcal{E}^+(C)$. Let $\{B_n\}$ be an increasing sequence of nonpolar compact subsets of B such that $P.(\tau_{B_n} \downarrow \tau_B) = 1$ on X. By (9)

$$\|\mu h_B - \mu h_{B_n}\|^2 = \int g_{B_n}(\mu h_B) d(\mu h_B).$$

Now $q_{B_n}(t,x,y) \downarrow q_B(t,x,y)$ for $t > 0$ and $x,y \in X$, so by the dominated convergence theorem

$$\lim_n \int g_{B_n}(\mu h_B) d(\mu h_B) = \int g_B(\mu h_B) d(\mu h_B) = 0$$

and hence $\|\mu h_B - \mu h_{B_n}\|^2 \to 0$. This proves the first result.

Let $\{U_n\}$ be a decreasing sequence of open sets such that $B \subset U_n \subset C$ for each n and $P.(\tau_{U_n} \uparrow \tau_B) = 1$ a.e. (μ). Set $\mu_n = \mu h_{U_n}$. Then

$$\lim_n \int W_B d\mu_n = 0. \tag{11}$$

To verify (11) observe first that

$$\int W_B d\mu_n = \int W_B(z) \int \mu(dy) h_{U_n}(y,dz) = \int E.W_B(\xi(\tau_{U_n}))d\mu.$$

Now W_B is bounded on compacts and upper semicontinuous and $W_B = 0$ on B^r. Since $\xi(\tau_B) \in B^r$ a.s. $(P.)$

$$\lim_n E_{\cdot}W_B(\xi(\tau_{U_n})) = 0 \text{ a.e. } (\mu),$$

so (11) holds.

Choose $m > n$. Then $\mu_n h_{U_m} = \mu_m$. Since $W_{U_m} \leq W_B$ and $\mu_n(X) = \mu(X)$, it follows from (9) and the equality in (10) that

$$(12) \quad \|\mu_n - \mu_m\|^2 + \|\mu_m\|^2 \leq \|\mu_n\|^2 + 2\mu(X) \int W_B d\mu_n .$$

By (11) and (12),

$$(13) \qquad \lim_n \|\mu_n\|^2 \text{ exists and is finite.}$$

Equations (11)-(13) together imply that $\{\mu_n\}$ is a Cauchy sequence. Thus to prove that $\{\mu_n\}$ converges strongly to μh_B it suffices to prove that μ_n converges weakly to μh_B. By Theorem 6.6 it suffices to prove that μ_n converges completely to μh_B.

Let φ be a bounded continuous function on X. Then

$$\int \varphi d\mu_n = \int \varphi d\left(\int \mu(dy) h_{U_n}(y,\cdot)\right) = \int E_{\cdot}\varphi(\xi(\tau_{U_n}))d\mu$$

and similarly

$$\int \varphi d(\mu h_B) = \int E_{\cdot}(\varphi(\xi(\tau_B))d\mu,$$

so $\int \varphi d\mu_n \to \int \varphi d(\mu h_B)$. Therefore μ_n converges completely to μh_B, which yields the second result of the theorem.

A characterization of balayage in terms of energy will now be given.

THEOREM 6.10. *Let* B *be relatively compact and nonpolar, let* $\mu \in \mathcal{E}^+$ *and set* $\mu' = \mu h_B$. *Then*

$$\min[\|\nu - \mu\| : \nu \in \mathcal{E}^+(B^r) \text{ and } \nu(X) = \mu(X)] = \|\mu' - \mu\|$$

and the minimum occurs uniquely at $\nu = \mu'$. *Also*

$$\inf[\|\nu - \mu\| : \nu \in \mathcal{E}^+(B) \text{ and } \nu(X) = \mu(X)] = \|\mu' - \mu\|;$$

if $\nu_n \in \mathcal{E}^+(B)$, $\nu_n(X) = \mu(X)$ *and* $\|\nu_n - \mu\| \to \|\mu' - \mu\|$,

then ν_n *converges strongly to* μ'.

Proof. Now $\mu' \in \mathcal{E}^+(B^r)$, $\mu'(X) = \mu(X)$ and $k\mu' = k\mu + \int W_B du$ on B^r by (3.37); so $(\mu' - \mu, \nu - \mu') = 0$ for $\nu \in \mathcal{E}^+(B^r)$ with $\nu(X) = \mu(X)$. Consequently

(14) $\|\nu - \mu\|^2 = \|\nu - \mu'\|^2 + \|\mu' - \mu\|^2$ for $\nu \in \mathcal{E}^+(B^r)$ with $\nu(X) = \mu(X)$.

The first conclusion of the theorem now follows from the energy principle.

Observe that $\mathcal{E}^+(B) \subset \mathcal{E}^+(B^r)$. By Theorem 6.9 there is a sequence $\{B_n\}$ of nonpolar compact subsets of B such that $\nu_n = \mu h_{B_n}$ converges strongly to μ' and hence $\|\nu_n - \mu\| \to \|\mu' - \mu\|$. Now $\nu_n \in \mathcal{E}^+(B)$ and $\nu_n(X) = \mu(X)$, so the second result follows from (14).

A characterization of the equilibrium measure and Robin constant in terms of energy will now be given.

THEOREM 6.11. *Let* B *be relatively compact and nonpolar. Then*

$$\min[I(\nu) : \nu \in \mathcal{E}^+(B^r) \text{ and } \nu(X) = 1] = R(B)$$

and the minimum occurs uniquely at $\nu = \mu_B$. *Also*

$$\inf[I(\nu) : \nu \in \mathcal{E}^+(B) \text{ and } \nu(X) = 1] = R(B);$$

if $\nu_n \in \mathcal{E}^+(B)$, $\nu_n(X) = 1$ *and* $I(\nu_n) \to R(B)$, *then* ν_n *converges strongly to* μ_B.

Proof. This result can be obtained directly. Alternatively let μ be the equilibrium measure of a compact set C containing B. Then $\mu \in \mathcal{E}^+$ and $\mu h_B = \mu_B$. Choose $\nu \in \mathcal{E}^+(B^r)$ such that $\nu(X) = 1 = \mu(X)$. Then $\int k\nu d\mu = \int k\mu d\nu = R(C)$ and hence

$$\|\nu - \mu\|^2 = \int k(\nu - \mu)d(\nu - \mu) = \int (k\nu - R(C))d(\nu - \mu) = I(\nu) - R(C).$$

In particular

$$\|\mu_B - \mu\|^2 = I(\mu_B) - R(C) = R(B) - R(C).$$

The theorem now follows from Theorem 6.10.

THEOREM 6.12. *Let* B *be relatively compact and nonpolar. Then*

$$\min[\sup_x k\mu(x) : \mu \in \mathcal{M}^+(B^r) \text{ and } \mu(X) = 1] = R(B)$$

and the minimum occurs uniquely at $\mu = \mu_B$. *Also*

$$\inf[\sup_x k\mu(x) : \mu \in \mathcal{M}^+(B) \text{ and } \mu(X) = 1] = R(B);$$

if $\mu_n \in \mathcal{M}^+(B)$, $\mu_n(X) = 1$ *and* $\sup_x k\mu_n(x) \to R(B)$, *then* μ_n *converges strongly to* μ.

Proof. Now $\mu_B \in \mathcal{M}^+(B^r)$ and $\mu_B(X) = 1$. Since $W_B \geq 0$ on X and $W_B = 0$ on the nonempty set B^r, $\sup_x k\mu_B(x) = \sup_x(R(B) - W_B(x)) = R(B)$.

Let μ be a probability measure in $\mathcal{M}^+(B^r)$. Now $I(\mu) \leq \sup_x k\mu(x)$; so it follows from Theorem 6.11 that $\sup_x k\mu(x) \geq R(B)$, with equality holding only if $\mu = \mu_B$. This completes the proof of the first conclusion.

Since B is capacitable by Theorem 3.10, there is an increasing sequence $\{B_n\}$ of compact nonpolar subsets of B such that $R(B_n) \downarrow R(B)$. Now $\mu_{B_n} \in \mathcal{M}^+(B)$, $\mu_{B_n}(X) = 1$ and

$$\sup_x k\mu_{B_n}(x) = R(B_n) \downarrow R(B).$$

Let $\{\mu_n\}$ be a sequence of probability measures in $\mathcal{M}^+(B)$. Then $M_n = \sup_x k\mu_n(x) \geq I(\mu_n)$. Thus by Theorem 6.11, $\underline{\lim}_n M_n \geq R(B)$ and if $\lim_n M_n = R(B)$, then μ_n converges strongly to μ_B. Therefore the second conclusion of the theorem is valid.

The next result is a direct application of Theorem 6.12.

THEOREM 6.13. *Let B be a relatively compact set such that $0 < R(B) < \infty$. Then*

$$\max[\mu(X) : \mu \in \mathcal{M}^+(B^r) \text{ and } k\mu \leq 1] = R(B)^{-1}$$

and the maximum occurs uniquely at $\mu = \mu_B$. Also

$$\sup[\mu(X) : \mu \in \mathcal{M}^+(B) \text{ and } k\mu \leq 1] = R(B)^{-1};$$

if $\mu_n \in \mathcal{M}^+(B)$, $k\mu_n \leq 1$ and $\mu_n(X) \to R(B)^{-1}$, then μ_n converges strongly to $R(B)^{-1}\mu_B$.

The final result yields another characterization of the equilibrium measure.

THEOREM 6.14. *Let B be relatively compact and nonpolar. Then μ_B is the unique probability measure $\mu \in \mathcal{M}^+$ such that*

$k\mu = R(B)$ q.e. on B and $k\mu \leq R(B)$ on X.

Proof. Now μ_B satisfies the indicated properties and $\int k\mu_B d\mu_B = R(B)$. Suppose that μ also satisfies the indicated properties. Then $\int k\mu d\mu \leq R(B)$. By the complete maximum principle

$$R(B) - W_B = k\mu_B \leq k\mu \leq R(B) \quad \text{on } X.$$

In particular $k\mu = R(B)$ on B^r and hence $\int k\mu d\mu_B = R(B)$. Therefore

$$\|\mu - \mu_B\|^2 = \int k\mu d\mu - 2\int k\mu d\mu_B + \int k\mu_B d\mu_B \leq 0,$$

so $\mu = \mu_B$ by the energy principle.

BIBLIOGRAPHY

[1] R.M. BLUMENTHAL and R.K. GETOOR, Markov Processes and Potential Theory, Academic Press, New York, 1968.

[2] E. HEWITT and K.A. ROSS, Abstract Harmonic Analysis I, Springer-Verlag, Berlin, 1963.

[3] S.C. PORT and C.J. STONE, Potential theory of random walks on Abelian groups, Acta Math., 122 (1969), 19-114.

[4] S.C. PORT and C.J. STONE, Infinitely divisible processes and their potential theory, Ann. Inst. Fourier, 21(1971), (2) 157-275 and (4) 179-265.

[5] S.C. PORT and C.J. STONE, Brownian Motion and Classical Potential Theory, Academic Press, New York, 1978.

LOIS DE ZERO-UN ET LOIS SEMI-STABLES DANS UN GROUPE.

A. TORTRAT

1. Introduction

L'extension de la "loi de zéro ou un" connue dans un espace vectoriel pour les lois stables, à l'initiative de MM. Louie et Rajput, nous incite à étudier des lois μ semi-stables, dans un groupe, définies comme vérifiant pour un couple d'entiers n et K($n.\mu$ désigne l'image de μ pour $x \to x^n$, $\bar{\mu}$ celle pour $x \to x^{-1}$ et e est l'unité du groupe).

(1) $n.\mu = \mu^K \delta(b)$, n et K entiers ≥ 2, $\delta(b)$ loi de Dirac en b.

En ce qui concerne la loi de zéro-un :

(2) $G \in \mathcal{B}$, $\mu(Gx) = 0$ ou 1, G sous-groupe quelconque (normal dans le cas non abélien),

le remarquable travail [4] de A. Janssen nous permet par un résultat particulier (cf. le "lemme fondamental" ci-après) de tirer bien meilleur parti de nos résultats de [6] , [7] , lorsque μ appartient à un demi-groupe continu μ^t (que nous noterons aussi μ_t). Nous obtenons un résultat plus général que celui que donnent les méthodes propres de [4] ; mais ces dernières - celles de l'exemple 3 de [4] - peuvent fournir un résultat presque équivalent, et meilleur dans certains cas (cf. le théorème 3).

Auparavant nous allons réexprimer, sous une forme plus générale, les résultats en fait démontrés dans [6] et [7] . En ce § 1, X n'est pas nécessairement abélien.

Nous supposons que $(X, \mathcal{B})$ est un groupe mesurable, ou un groupe topologique muni de sa tribu borélienne (en ce cas μ est prise τ-régulière). G est un sous groupe normal et $\dot{X}$ désigne le groupe quotient X/G muni de la tribu induite par $\mathcal{B}$.

"ν divise μ au sens large signifie que $\mu=\rho\nu\rho'$ ", ce qui contient les cas $\mu=\rho\nu$ et $\mu=\rho\nu'$.

Lemme 1

Si μ ou $\overline{\mu}$ divise $n.\mu$ au sens large, et si μ ou $\overline{\mu}$ divise (au sens large) le cofacteur précédent, alors $\mu(Ga) > 0$ implique que :

Dans $\dot{X}$, μ se réduit à I atomes égaux à $1/I$, portés par $A = \{a_i\}$. Dans le cas vectoriel, même résultat avec l'homothétie $x \to cx$ au lieu de nx, si $c > 1$ et $cG \subset G$.

Lemme 2

i) Si $\mu\,\overline{\mu}$ et $\overline{\mu}\,\mu$ divisent $n.\mu$ au sens large, ou si $n.\mu = \mu^n$ et μ^{n+1} divise un $n'.\mu$ au sens large, on a, $\dot{G}$ étant un sous-groupe fini :

$$A\overline{A} = \overline{A}A=\dot{G},\ Aa^{-1}=a^{-1}\ A=\dot{G}\ (\text{tout } a \text{ de } A).$$

ii) Si μ^2 ou $\overline{\mu}^2$ divisent $n.\mu$ au sens large, on a $A=\dot{G}$ si $Ga=G$, ou, si μ est symétrique $A^2 = \dot{G} = Aa = aA$ (tout a de A).

iii) Si $n.\mu = \mu^K$, $n < K$, on $A^{K-n} = \dot{G}$.

Même extension qu'au lemme 1 au cas de l'homothétie $x \to cx$ dans X vectoriel.

Remarque 1

Dans [7], on prouve $A^{n-K} = \dot{G}$ si $n.\mu = \mu^K$ pour $K < n$, sous une condition supplémentaire donnée en iii) de l'énoncé qui suit, énoncé qui est une conséquence immédiate de ces résultats : il suffit d'assurer que $n'.\dot{G}$ a I éléments pour un n' multiple de I, ou d'un de ses facteurs premiers, dans le cas abélien, soit, dans le cas non abélien, que μ ou $\overline{\mu}$ divise $n'.\mu$ pour un n' multiple de kI avec $k = |K-n| \neq 0$, si $A \neq \dot{G}$, ou $k=1$ si $A=\dot{G}$. Dans le cas abélien on raisonne avec $\mu\,\overline{\mu}$, ce qui ramène au cas $A=\dot{G}$.

Théorème 1

(2) a lieu si μ ou $\overline{\mu}$ divisent au sens large des $n'.\mu$ pour des n' multiples de tout entier (de tout entier premier dans le cas abélien) et si

i) μ^2 ou $\mu\,\overline{\mu}$ ou $\overline{\mu}^2$ divise un $n.\mu$ dans le cas abélien,

ou

ii) - cas non abélien - l'une des conditions ci-après est vérifiée (a) ou b) ou c)) a). $\mu\bar{\mu}$ <u>et</u> $\bar{\mu}\mu$ devisent $n.\mu$ au sens large et $Gx=G$ si le support de μ dans $\dot{X}$ n'est pas symétrique,

b). μ^2 ou $\bar{\mu}^2$ divisent $n.\mu$ au sens large et $Gx=G$ si le support de μ dans $\dot{X}$ n'est pas symétrique,

c). $n.\mu = \mu^K$ $n < K$,

ou

iii) c). vaut pour un couple $n > K$ et $\mu^{K'}$ divise un $n'.\mu$ pour un $n-K < K' \leq \min(n,2n-2K)$. Pour $n=K$ il faut ajouter $Gx=G$ ou "le support de μ est symétrique", et supposer seulement que μ^{n+1} divise un $n'.\mu$ au sens large.

<u>2</u>. X demeure non nécessairement abélien.

Nous supposons maintenant que μ^t, $t \geq 0$, est un demi-groupe continu de lois de Radon.

Dans le cas de X vectoriel séparé par son dual, <u>il suffit</u> suivant le théorème de Siebert de supposer μ convexement de Radon.

<u>Lemme fondamental (A. Janssen)</u>

L'hypothèse $\mu^t \underset{t\to 0}{\to} \delta(e)$ interdit que, dans $\dot{X}$, les lois projection soient pour tout t équiréparties sur un même sous-groupe fini $\dot{G}$ non réduit à l'unité (: interdit $I > 1$). Il faut pour la preuve remplacer G par une $\cup(K_n)^n = G'$ avec des K_n compacts symétriques et $\mu(G-G') = 0$, ce qui est loisible si l'on sait que les μ_t restent équiréparties dans X/G'.

Cette preuve est alors celle, simplifiée, de la partie 1 de celle du théorème 6 de [4] . Dans ce cas elle s'étend sans changement à un groupe non abélien, et sans supposer les μ_t symétriques.

<u>Théorème 2</u>

i) On suppose que le demi-groupe continu μ^t de lois de Radon vérifie pour une infinité d'entiers k, dont k=1, la relation

(3) $$\mu_{2/k} \text{ divise } n.\mu_{1/k}, \quad n \text{ entier } \geq 2.$$

ii) Si X n'est pas abélien on suppose les μ^t symétriques, ou, sinon, que $\bar{\mu}_t\,\mu_t$ égale $\mu_t\,\bar{\mu}_t$ et est un demi-groupe symétrique (noté ν_t) vérifiant (3).

iii) Si X est abélien, on opère avec $\mu\,\overline{\mu} = \nu$. Si X est (abélien) localement compact, ou vectoriel séparé par son dual, on peut se borner à la seule relation (3), au sens de Parthasarathy $^{(+)}$, pour $k=1$ (les autres s'ensuivent). Dans ces mêmes conditions, il suffit que μ^r divise $n.\mu$ au sens de Parthasarathy, avec $r > 1$, (et $n=c > 1$ dans le cas vectoriel) car ce fait vaut alors aussi avec les n^ℓ et r^ℓ (ℓ entier, cf. § 4 dans le cas où r n'est pas entier, et la remarque 5).

Alors la loi de zéro-un, soit (2), vaut pour μ.

Preuve : Il résulte du lemme 2 que dans $\dot{X}$, chaque $\nu_{1/k}$ (raisonnons avec ν_t, si les μ_t ne sont pas symétriques) est équirépartie sur des sous-groupes finis $\dot{G}_{1/k}$. Il faut noter que si un $\mu(Gx)$ est > 0, il existe un $\mu_{1/k}(Gx') > 0$, pour tout k.

Puisque $\nu = (\nu_{1/k})^k$, on a $\dot{G}_t = \dot{G}_1$ pour tout $t=1/k$.

Alors, ν_t, comme facteur symétrisé de $\mu_{t/2}$, de ν, est tendu suivant les $2.K_{\varepsilon}$, si ν l'est suivant les compacts symétriques K_i contenus dans le groupe $\sum_1^I G\,a_i$ qui porte ν. C'est dire que dans $\dot{X}$, les ν_t sont portées par $\dot{G}=\dot{G}_1$, donc y sont toutes de Haar puisque égales à $\nu_{t-1/k}\ \nu_{1/k}(t > 1/k)$. Le lemme fondamental assure la conclusion.

Il est intéressant de noter d'autres idées de A. Janssen et d'améliorer son exemple 3, retrouvant exactement presque tous les résultats du théorème 2 et un autre plus fort, dans le cas d'un espace de Banach.

1e. Considérons dans X vectoriel, la relation, au sens de Parthasarathy

(4) $\qquad \mu^r$ divise $c.\mu$, c et $r > 1$.

Dans les conditions d'unicité de mesure de Lévy F, (4) équivaut à

(4') $\qquad c.F \geq rF$.

L'auteur suppose $cG=G$ pour conclure pour une loi stable $F(G^c) = 0$ ou ∞. Or il suffit de

---

(+) ainsi renforcé : le cofacteur est une loi (de Radon) qui se plonge elle aussi dans un demi-groupe continu.

(5) $$cG^c \supset G^c$$

pour obtenir

$$rF(G^c) \leq F(c^{-1}G^c) \leq F(G^c) \quad \text{donc} \quad F(G^c) = 0 \text{ ou } \infty.$$

Alors les théorèmes 9 et 10 de [4] donnent $\mu(G+x) = 0$ ou 1, sous la condition (5).

On note que pour $c=n$ la condition (5) équivalente à $nG \subset G$ est triviale ; sinon le résultat est plus faible que celui du théorème 2 en ce que (5) est nécessaire (mais c réel quelconque).

2e. Si maintenant X est un groupe abélien localement compact, dans les conditions du lemme 2, on obtient pareillement

$$\mu^r \text{ divise } n.\mu \implies (6)\ n.F \geq rF \implies F(G^c) = 0 \text{ ou } \infty.$$

Alors le corollaire 7 de [4] donne (1).

3e. Donnons d'abord deux exemples d'application du théorème 2.

Exemple 1 : Si dans X groupe topologique abélien, une infinité de $\mu^{1/k}$ sont gaussiennes au sens de Bernstein, on a $2.\mu = \mu^3\,\bar{\mu}$ (pour toutes ces $\mu^{1/k}$), (2) vaut si ces $\mu^{1/k}$ se plongent dans un demi-groupe continu de lois de Radon. Il en est de même dans le cas non abélien avec la relation $2.\mu_t = \mu_t^2\,\bar{\mu}_t\,\mu_t$ pour chaque t, par exemple si un vecteur aléatoire ζ_t de loi μ_t vérifie (pour chaque t et une réplique indépendante ζ_t' de ζ_t)

$$\zeta_t\,\zeta_t' \text{ est indépendant de } \zeta_t'^{-1}\,\zeta_t.$$

Exemple 2 : Dans X vectoriel soit μ convexement de Radon telle $\frac{1}{n}.\mu$ divise $\mu' = \mu^{1-\frac{1}{m}}$, ce qui équivaut à $\mu'^{m/(m-1)}$ divise $n.\mu'$. Alors (1) vaut pour tout G : $\mu_*(G+x) = 0$ ou 1. La conjecture de Kanter (cf. [6]) était que c rationnel et "$c.\mu$ divise μ" implique (1) pour tout module sur les rationnels.

Le Théorème I.3.2. de [6] donne ce résultat pour tout $0 < c < 1$, en ce sens que

$$G+x \in \mathcal{B}_\mu \implies \mu(G+x) = 0 \text{ ou } 1$$

vaut pour tout module G sur le corps engendré par c (dans X vectoriel mesurable, ou avec μ seulement τ-régulière dans le cas topologique) à condition que $c.\mu$ divise $\mu^{1-\frac{1}{m}}$. Ici nous obtenons mieux (G groupe quelconque) mais en supposant μ de Radon et indéfiniment divisible, et seulement pour $c=1/n$. Pour $c > o$ quelconque, le théorème 2 ne contient pas le résultat noté en, (4), (4') (5) qui fait intervenir pleinement [4]. Le théorème de Janssen donne en fait un résultat plus fort, correspondant à r=1 dans (4), lorsque X est un espace de Banach :

Théorème 3

(2) vaut pour tout G tel que $cG^c \subset G^c$ et toute loi μ inféfiniment divisible dans un espace de Banach séparable, telle que $c.\mu$ divise μ (pour un réel $c < 1$ et au sens de Parthasarathy). La restriction $r > 1$ de (4) (ici $r < 1$) disparaît et le résultat (2) vaut en particulier pour toute loi semi-stable (cf. [6]), avec la restriction $cG^c \subset G^c$.

Preuve : Posons $A_n = G^c \cap \{c^n < |x| \leq c^{n-1}\}$, on a

$$\frac{1}{c} A_{n+1} \supset A_n \quad \text{donc} \quad F(A_{n+1}) \geq c.F(A_{n+1}) = F(\frac{1}{c} A_{n+1}) \geq F(A_n),$$

et si $F(G^c) > 0$, un $F(A_n)$ est > 0, donc $F(G^c) = \sum_{-\infty}^{\infty} F(A_i) \geq \sum_{n}^{\infty} F(A_i) = \infty$.

Remarque 2 : La preuve s'étend au cas où (notations du § 4) μ_α divise μ et il existe des parties B_n disjointes telles que $\sum_{-\infty}^{\infty} B_n \supset G^c$, $\alpha^{-1} B_{n+1} \supset B_n$ et $\alpha^{-1} G^c \supset G^c$ (poser $A_n = G^c \cap B_n$). C'est le cas des lois V-décomposables dans un espace de Hilbert(cf. [4'], une seule relation (2.1) suffit).

3. Théorème 4

On suppose que le groupe topologique abélien (séparé) X n'a pas d'élément d'ordre n (autre que l'unité) et que la loi μ vérifiant (1) est de Radon (et propre : non de Dirac).

Alors

i) $\nu = \mu\tilde{\mu}$ s'intègre à un demi-groupe ν^r défini sur l'ensemble D des rationnels $r = k/_{K^I}$ à developpement de base K fini. Ces ν^r sont portées par un sous-groupe G vérifiant G=nG(1).

Ce demi-groupe est continu et se prolonge tel à R_+ si K est pair et ν apériodique (2).

ii) Si μ est apériodique et symétrique, <u>et</u> K pair, ν est apériodique.

iii) Si ν est apériodique on a $\hat{\nu}(y) > 0$, pour tout caractère (continu) y, et si X est localement compact, μ est portée par un translaté de C composante connexe de l'unité de X(3), et ν est indéfiniment divisible au sens faible (pour chaque k, il existe une loi λ_k avec $\nu = \lambda_k^k\,\delta(b_k)$).

iv) Si μ a un presque-caractère de loi propre, l'exposant p est ≤ 2.

v) Si X est un sous-groupe additif de R, (1) (avec $p \leq 2$) nécessite $\gamma^*X=1$, pour la loi normale réduite γ dans R. Alors $d\nu = f(.)d\nu$, sur X, avec f(.) définie et indéfiniment dérivable dans R. Sinon X est de mesure nulle pour γ, et (1) est impossible (nous donnons une famille de tels X ayant la puissance du continu).

---

(1) Dans ce cas abélien nous remplaçons la notation n.G par nG, tout en gardant éventuellement celle de x^n au lieu de nx.

(2) : sans facteur idempotant, ce qui équivaut à : ν a translatée de ν par a (à droite = ici à gauche) n'égale ν que trivialement, si a est l'unité, $\hat{\nu}$ désignera la transformée de Fourier de ν.

(3) Comme pour une loi gaussienne au sans de Parthasarathy, mais $2.\nu = \nu^4$ suffit, si X n'a pas d'élément d'ordre 2.

Remarque 3

Ainsi il sera naturel de chercher les lois $\mu \in (1)$, soit dans un groupe compact connexe, avec $p \leq 2$, si X est sans éléments d'ordre n (si $\mu' = \mu b$ est dans C, $n.\mu'' = \mu'^{K}\delta(ab^{n-k})$), soit dans un groupe compact possédant (tel Γ) des éléments d'ordre n (cf. [7]).

Preuve de i). Ses éléments sont dus à Byczkowski (pour une loi gaussienne, cf. [1] et la remarque finale de [6]).

On peut réduire X au groupe $\cup jK_i^!$ engendré par une $\lim\uparrow_j$ de compacts symétriques $K_i^!$ (avec $\mu K_i^! \uparrow 1$), soit $X = \lim\uparrow K_j$ avec $K_j = \bigcup_1 iK_j^!$.

Puisque tout fermé F égale $\cup F \cap K_i$, on a $nF \in \mathcal{B}$ donc $nB \in \mathcal{B}$ si $B \in \mathcal{B}$, car la classe des B ayant cette propriété est stable pour toutes les opérations d'une tribu (ceci parce que $\theta(x) = x^n$ étant injective, $B \cap B'$ vide $\Longrightarrow$ $nB \cap nB'$ vide).

(1) s'itère en

$$(1') \qquad n^\ell.\mu = \mu^{K^\ell}\delta(a_\ell),\ n^\ell. \doteq \nu^{K^\ell}.$$

Puisque $n^\ell.\mu$ est portée par $G_\ell = n^\ell X$, on a

$$\mu(G_\ell - b_\ell) = 1 \text{ pour un } b_\ell \text{ de } X \text{ au moins, donc } \nu G_\ell = 1.$$

Puisque la trace de $\mathcal{B}$ sur G_ℓ a pour image (ponctuelle), par $\theta^{-\ell}$, $\mathcal{B}$ elle-même, $\nu_\ell = \theta^{-\ell}\nu$ est une loi sur $\mathcal{B}$ qui vérifie $\theta^\ell(\nu_\ell^K) = (\theta^\ell \nu_\ell)^K = n^\ell.\nu$, c'est à dire $\theta^\ell\nu$. Ainsi on peut poser

$$\nu_\ell = \nu^{K^{-\ell}},\ \ell \text{ entier relatif},$$

on a cohérence entre les ν_ℓ ce qui définit ν^r comme demi-groupe sur $D = \{k/K^I\}$. La famille $\{\nu^r\}$ est tendue $(r \leq 1)$ si K est pair, car

$$\nu^{k/K^I} = \nu^{\frac{K}{2}\frac{2k}{K^{I+1}}},$$

et les $\nu^{2k/K^{I+1}}$ le sont comme symétrisées des facteurs $\nu^{k/K^{I+1}}$ de ν.

La continuité en 0, sur D, s'en suit si ν n'a pas de facteur idempotent, comme dans [1]. Classiquement (notons le, pour ne pas multiplier la dépendance de cette preuve) ν^t est univoquement défini sur R_+ par $\lim\limits_{r_i \uparrow t} \nu^{r_i}$: deux limites ν' et ν'' distinctes seraient reliées par

$\nu'_\rho = \nu'', \rho$ étant une limite de ν^r relatives à des exposants $r_{i_\beta} - r_{i_\alpha} \to 0$, et $\rho = \delta(0)$ par la continuité en 0.

Enfin si $t_\alpha \to t$, il suffit que ν^{t-t_α} ou $\nu^{t_\alpha - t}$ (suivant le cas) $\to \delta(0)$ pour que $\nu^{t_\alpha} \to \nu^t$. Or $\nu^r \xrightarrow[r \to 0]{} \delta(0)$ et $\{\nu^t, t \leq \eta\}$ est pour η assez petit, continu en même temps que $\{\nu^r, r \leq \eta\}$ dans un voisinage fermé de $\delta(0)$ donné arbitrairement petit.

Preuve de ii). Si $\mu = \bar{\mu}$ et K pair, on a $n.\mu = \nu^{K/2}\delta(a)$.

Or $\nu\delta(b) = \nu$ impliquerait $(n.\mu)\ \delta(b) = n.\mu$, donc $b \in nX$ (si non $nX+b$ serait disjoint de nX et ne pourrait porter μ).

On aurait (b/n existe et) $n.(\mu\delta(b/n)) = n.\mu$, donc $\mu = \mu\delta(b/n)$ vu $\mathcal{B} \xleftrightarrow{\theta} \mathcal{B}_{nX}$.

Preuve de iii). La preuve que $\hat{\nu}(y) = 0$ implique un facteur idempotent pour ν est celle du théorème 5.1.5 page 339 de [2], adaptée : si λ est une loi limite pour la famille $\{\nu_\ell \delta(b_\ell)\}$ (tendue pour des b_ℓ convenables), les λ^k divisent ν. Il existe donc une famille $\{\lambda^k \delta(a_k)\}$ tendue, c'est une "convolution dénombrable"

$\rho_1 \dots \rho_k$, avec $\rho_k = a_{k-1}^{-1}\ \lambda a_k$ ($\rho_1 = \lambda a_1$, notation simplifiée).

Suivant la proposition 5 de [5], les lois adhérentes σ à cette dernière famille sont translatées les unes des autres, et suivant la preuve du lemme 5.1.1. de [2], on a $\sigma = m\delta(a)$, m étant une loi de Haar. Pour toutes ces lois, telle σ, on aurait $\hat{\sigma}(y)=0$, donc m propre <u>et</u> divisant ν.
Que $\nu C=1$ si X est localement compact admet la même preuve que pour une loi gaussienne : Les y valant 1 sur C (sous-groupe ouvert et fermé de X) sont exactement les éléments compacts du dual X (: y^n est relativement compact).
Puisque (1) implique $\hat{\nu}(y^{n^\ell}) = (\hat{\nu}(y))^{K^\ell}$, ces nombres, ℓ variant, ont une borne inférieure > 0, donc $\hat{\nu}(y) = 1$ si $y=1$ sur C : $y=1$ p.s. Mais ces y prolongent à X les caractères de X/C, dont ils séparent les points (C étant fermé). Ainsi $C = \cap\{x : y(x) = 1\}$ (intersection prise sur ces y) est de ν mesure 1 avec chacun de ces fermés.

De plus ν est indéfiniment divisible au sens susdit suivant le théorème des lois accompagnantes (cf. 5.1.16 et 5.1.17 de [2]).

Preuve de iv). Nous appelons presque caractère $\mathcal{X}$ un homomorphisme de G (sous-groupe $\epsilon\ \nu G=1$) dans Γ. Sa loi ρ vérifie

(7) $$n.\rho = \rho^K.$$

Ou ρ est apériodique, elle est alors suivant iii) indéfiniment divisible dans Γ, on a donc $p \leq 2$ suivant [7].

Ou ρ admet la période minimale $e^{2\pi i/I}$ (: I maximal), alors la loi ρ' de moments

$$u_k(\rho') : u_{kI}(\rho)$$

est apériodique et vérifie (2), on retombe sur le cas précédent pour interdire $p > 2$.

Preuve de v). Elle s'étend au cas où ν symétrisée de μ vérifie ($c > 0$ quelconque)

(8) $$n.\nu=\nu^c, \text{ soit } (8')\ \ \phi(nt)=\phi^c(t) \text{ avec } \phi \geq 0 \text{ f.c. de la loi } \hat{\nu} \text{ induite par } \nu \text{ dans } \mathbb{R}.$$

(8') s'itère ici sans problème pour assurer

$$\phi(n^\ell t) = \phi^{c^\ell}(t).$$

Par changement d'échelle, on peut supposer $\phi \neq 1$ sur $]0,1]$, donc a fortiori $\sup_{\Delta_o} \phi = b < 1$, avec $\Delta_k =]n^{k-1},n^k]$.

Sur Δ_k on a

$$\phi(t) = \phi_{n^k.\nu}(\theta) = \phi^{c^k}(\theta), \text{ avec : } \theta = n^{-k}t \text{ décrit } \Delta_o.$$

Ainsi ϕ décroit exponentiellement, et ν a une densité indéfiniment dérivable, dans $\mathbb{R}$, donc une densité telle $p(x)$ par rapport à la loi normale γ.

Pour un groupe additif X, on a $\gamma^* X=0$ ou 1 (le corollaire du théorème 12 de [6] s'applique à ce cas de dimension 1, et une preuve directe de ce fait est facile et sans doute connue de longue date !). Le 1er cas est clairement exclu, et $\gamma^* X=1$ suffit à assurer $\int_X p(x) d\gamma = 1$: toute loi ν symétrique vérifiant (8) définit ν telle sur X.

Remarque 4

Voici une classe (peut être bien connue ?) de sous-groupes additifs X de R, de puissance du continu, et de mesure de Lebesgue nulle, qui ne peuvent donc (comme les X dénombrables) porter aucune loi semi-stable.

Soit

$$x = n + \sum_{1}^{\infty} \frac{\alpha_i}{3^i}, \quad \alpha_i = 0 \text{ ou } \pm 1$$

une représentation des réels, excluant celles qui après un rang fini ne comportent que des $\alpha_i=1$, ou que des $\alpha_i=-1$.

Posant $S_I(x) = \sum_{1}^{I} |\alpha_i|$, on voit assez aisément qu'on a

$$S_I(x+x') \leq S_I(x) + S_I(x') + 1.$$

Alors $G = \{x : S_I(x)/I \xrightarrow[I \to \infty]{} 0\}$ définit un groupe additif, dont la puissance est celle du continu (prendre $\alpha_{2i+1}=1$, $\alpha_{2i+1} = -1$, ou -1 et $+1$, alternativement), de mesure de Lebesgue nulle, puisque sur chaque intervalle $[n,n+1[$, $S_I(x)/I \to 2/3$ p.s. pour cette mesure. En prenant une base $2k+1$ au lieu de 3 õur le développement, on aurait le même résultat.

4. Dans le cas où μ vérifie une seule relation (1') l'itération n'est plus assurée. D'où la proposition beaucoup plus générale qui suit.

Proposition

Soit X un groupe abélien localement compact métrisable (la restriction métrisable est peut être inutile), μ une loi de radon plongeable dans un demi-groupe continu μ^t, et α un homorphisme continu de X dans X. μ_α^t et F_α désignent les images par α de μ^t et de F mesure de Lévy de μ (μ_α^t est un demi-groupe continu).

Alors F_α supposée restreinte à $X^*=X-e$ (e unité de X) égale la mesure de Lévy F' de μ_α^t. En particulier dire que "μ_α divise μ au sens de Parthasarathy" (cf. ci-dessus au théorème 2), équivaut à $F_\alpha \leq F$.

Remarque 5

Dans le cas d'un espace vectoriel localement convexe complet (avec μ de Radon indéfiniment divisible, la preuve est dans [6] (théorème II.2.4). Pour un espace de Banach elle suit aussi de ce que F est pleinement le générateur infinitésimal de μ^t : $\frac{1}{t}\mu^t \xrightarrow[t \to 0]{} F$, au sens de la convergence en loi des restrictions à $|x| > r$, $r > 0$ étant de continuité pour F (on peut se borner à $t=1/n$, cf. le corollaire 1.11 de [O]).

Dans le cas localement compact non abélien, la preuve immédiate est la suivante, si $\alpha^{-1}(K)$ est compact avec K (pour tout K ne contenant pas e). En effet, en ce cas, si f est continue à support dans K, $f\circ\alpha = f_\alpha$ l'est également (à support dans $\alpha^{-1}(K)$ ne contenant pas e), et la mesure de Lévy F' de μ_α^t est donnée par

$$F'f = \lim \frac{1}{t}(\mu_\alpha^t)\ f = \lim \frac{1}{t}\ \mu^t f_\alpha = Ff_\alpha = F_\alpha f.$$

Dans le cas général, si par exemple μ_α divise μ, nous ne savons pas si la relation $F_\alpha f \leq Ff$ pour les seules $f \geq 0$ telles que f et f_α soient à support compact, assure $F_\alpha \leq F$.

Preuve de la proposition : Il suffit de se borner au cas où μ est sans composante gaussienne.

a). $y_\alpha(.) = y\circ\alpha$ est un caractère (continu) si y en est un.

b). Soit $g(x,y)$ la fonction de Parthasarathy (cf. [2] p. 340) assurant la représentation

$$\mathrm{Log}\ \hat{\mu}(y) = \int [y(x)-1-i(x,y)]dF.$$

On prouve par l'absurde que

$$\sup_{x\in X}\ \sup_{V} g(x,y) \downarrow 0$$

suivant le filtre des voisinages V de l'unité ε du dual de X.

c). Soit $|g(x,y)| \leq 1$ pour tout $y \in V$, et $F(\alpha^{-1} U^c - K_\eta) \leq \eta$ pour un voisinage U de e dans X, et un compact K_η.

Pour $x'\times y \in U_x \times V_x$ convenable, on a

$$|g(x',y)-g(x,\varepsilon)| = |g(x',y)| < \eta,$$

donc pour un recouvrement fini de K_η par $\bigcup_1^I U_{x_i}$, avec $V_o = \cap V_{x_i}$,

on a

$$|g(x,y)| < \eta \quad \text{dans} \quad K_\eta \times V_o$$

d'où

$$\int_{\alpha^{-1} U^c} g(x,y)\,dM < 2\eta \quad \text{pour} \quad y_\alpha \in V \cap V_o,$$

soit pour tout $y \in$ un V' convenable. Ceci prouve la continuité, peut être connue, de ce terme.

Alors $\mathrm{Log}\{\hat{\mu}_\alpha(y) = \hat{\mu}(y_\alpha)$ s'écrit

$$\int [y_\alpha(x)-1-ig(x,y_\alpha)]\, dF,$$

et on a

$$(*) \int_{\alpha^{-1}U^c} [y_\alpha(x)-1-i\, g(x,y_\alpha)]dF = \int_{U^c} [y(x)-1]d\, F_\alpha - i\int_{\alpha^{-1}U^c} g(x,y_\alpha)dF.$$

Dans $(*)$ le 2ème terme à droite représente une translation, s'écrit $\mathrm{Log}\ a_U(y)$, prouvant que $\exp(F_{\alpha,U^c})$ est un facteur de μ_α (car F restreinte à U est une mesure de Lévy, dont l'image par α est le cofacteur de $\exp(F_{\alpha,U^c})\delta(a_U)$, donc que $F_{\alpha,U^c} \leq F'$ (mesure de Lévy de μ_α) et $F_\alpha \leq F'$.

d). Si pour un U on avait $\sigma = F'_{U^c} - F_{\alpha,U^c} > 0$, on aurait pour le filtre des $U' \subset U$,

$$\mu_\alpha = \lim\{\exp(F_{\alpha,U'^c})\delta(a_{U'})\}.\nu_{U'},$$

avec $\nu_{U'}$ divisible par $\exp \sigma$, et le premier terme à droite tendant vers μ_α, cela exigerait (cette convergence étant tendue, pour une suite $U'_n \downarrow e$) une limite non dégénérée pour $\nu_{U'}$, donc μ_α périodique, donc non plongeable.

Remarque 6

Dans le cas où on a $n.\mu = \mu^c\delta(a)$, on peut prouver que a se plonge dans un demi-groupe a^t et que, pour tout entier ℓ on a

$$n^\ell.\mu = \mu^{c^\ell}\delta(a_\ell), \text{ avec } a^t_{\ell+1} = a^{(c^\ell+c^{\ell-1}n+\ldots+n^\ell)t}.$$

Remarque 7. La preuve dite à la remarque 5, vaut dans le cas général si (9) :

$$(9) \qquad \frac{1}{t}\, \mu^t g \to Fg$$

vaut pour les fonctions $g=f\circ\alpha$, avec f à support compact dans $X-e$.

Il suffit donc que (9) vaille pour <u>toutes</u> les fonctions g continues bornées nulles dans un voisinage de l'unité. Suivant une lettre de A. JANSSEN, les travaux de HAZOD (cf. Lecture notes n° 594) peuvent fournir la réponse, au moins pour $\alpha=x^n$).

Notons d'autre part que le théorème 12 du texte de A. JANSSEN dans ces mêmes "Proceedings" est une forme très affaiblie du théorème 3 (de notre texte), déduit par nous des théorèmes 9 et 10 de [4] : il suffit suivant ce théorème 3, <u>de la seule hypothèse</u> que $\frac{1}{K}.\mu$ divise μ, pour <u>un seul</u> entier $K \geq 2$, pour que $\mu(G+x) = 0$ ou 1. En effet pour $c=\frac{1}{K}$, la condition $cG^c \subset G^c$ est ici triviale, et suivant la remarque 2, en prenant une seule semi-norme p et les $B_n = \{x : \frac{1}{K^{n+1}} < p(x) \leq \frac{1}{K^n}\}$, on obtient le résultat (p attachée à un voisinage convexe symétrique de 0).

Références

[0] DE ACOSTA, ARAUJO &GINE : On Poisson measures, Gaussian measures and the central limit theorem in Banach spaces. Advances in Probability, 4 (1978), pp. 1-68.

[1] T. BYCZKOWSKI : Zero-on laws for gaussian measures on metric abelian groups. Studia mathematica, tome 69 (1981), pp. 159-188.

[2] H. HEYER : Probability measures on locally compact groups. Springer, Ergebnisse der Math. n° 94 (1977).

[3] PARTHASARATHY K.R., RANGA RAO, VARADHAN S.R.S. : Probability distributions on locally compact abelian groups. Illinois J. of Math. 7 (1963) pp. 337-369.

[4] Arnold JANSSEN : Zero-un laws for infinitely divisible probability measures on groups, cette "Proceedings".

[4'] R. JAJTE : V-decomposable measures on Hilbert-spaces. Lecture notes n° 828, pp. 108-127.

[5] A. TORTRAT : Lois tendues et convolutions dénombrables dans un groupe topologique X. Annales Inst. Henri Poincaré 2 (1966), pp. 279-298.

[6] A. TORTRAT : Lois de zéro-un pour des probabilités semi-stables ou plus générales, dans un espace vectoriel ou un groupe (abélien ou non). Colloque de Saint-Flour 1980 (Publications du C.N.R.S.).

[7] A. TORTRAT : Lois stables dans un groupe. Annales Inst. Henri Poincaré XVII (1981) pp. 51-61.

A. TORTRAT
85, rue de Paris
92190 MEUDON

A. TORTRAT
UNIVERSITE PARIS VI
Laboratoire de Probabilités
Tour 56
4 Place Jussieu
75230 PARIS CEDEX 05

A LOCAL LIMIT THEOREM FOR RANDOM WALKS ON CERTAIN DISCRETE GROUPS

Wolfgang WOESS

1. *Introduction*

Let G be a discrete countable group with identity e and μ a probability measure on G, i.e. a function $\mu: G \to [0,1]$ with $\sum_{x\in G} \mu(x) = 1$.
If (Y_n) is a sequence of independent identically distributed random variables taking their values in G with common distribution μ, then the random variables $X_n = Y_1 \cdot Y_2 \cdots Y_n$, $n=1,2,\ldots$, form a Markov chain with state space G and transition probability $p_{x,y} = \mu(x^{-1}y)$ called (right) *random walk on* G *with law* μ and starting point e.

We are interested in the asymptotic behaviour of $\mathrm{Prob}[X_n = x] = \mu^{(n)}(x)$ for $x \in G$ as $n \to \infty$ ($\mu^{(n)}$ denotes the n-th convolution power of μ).

Definition : μ is said to satisfy a *local limit theorem* if there exists a sequence (a_n) of positive real numbers such that

$$\lim_{n\to\infty} a_n \mu^{(n)}(x) = \nu(x) \quad \forall\, x \in G ,$$

where ν is a measure on G (i.e. a real valued function on G) with $\nu(x) > 0 \ \ \forall\, x \in G$.

In particular, we want to know the sequence (a_n) explicitely.
In this paper a local limit theorem is calculated for certain probability measures on

$$G = \langle a , b \mid a^r = b^s = e \rangle ,$$

where r and s are positive integers ≥ 2 and not $r=s=2$.
For $r=2$ and $s=3$ this group has been attracting particular interest:
- as a group of rotations of the unit sphere in three-dimensional space to prove the "*Hausdorff* paradox", thus being the first group to be considered with respect to nonamenability,
- as the *modular group* $SL_2(\mathbb{Z})/\{\pm I\}$.

2. *Some basic definitions and facts*

The *support* of μ is the set $S_\mu = \{ x \in G \mid \mu(x) > 0 \}$.
μ is called *irreducible* if $\bigcup_{n\geq 1}^{\infty} S_\mu^n = G$.

(2.1) For irreducible μ the number $\limsup \mu^{(n)}(x)^{1/n} = \rho(\mu)$ is independent of $x \in G$ and $0 < \rho(\mu) \leq 1$. Furthermore, $\rho(\mu)$ is minimal among the numbers ρ admitting a positive measure ν on G such that $\mu*\nu(x) \leq \rho\cdot\nu(x) \quad \forall\, x\in G$. ([6],[11])

(2.2) If μ is irreducible and $\rho(\mu)=1$ then G is amenable.([2])

μ is called *aperiodic* , if S_μ is not contained in any proper coset of a normal subgroup of G . For irreducible μ this is equivalent to: $\mu^{(n)}(e)>0$ for almost all $n\in\mathbf{N}$.

(2.3) If μ is irreducible and aperiodic then $\lim \frac{\mu^{(n+1)}(x)}{\mu^{(n)}(x)} = \rho(\mu)$ $\forall\, x\in G$.([3])

(2.4) If μ is irreducible and $\lim \mu^{(n+1)}(x)/\mu^{(n)}(x) = \rho(\mu) \quad \forall\, x\in G$ and if S_μ is finite, then the sequence of measures $(\mu^{(n)}/\mu^{(n)}(e))$ is relatively compact in the space of measures on G equipped with the vague topology, and every accumulation point ν of the sequence satisfies :

$$\mu*\nu = \nu*\mu = \rho(\mu)\cdot\nu$$
$$\nu(e) = 1 \quad \text{and} \quad \nu(x) > 0 \quad \forall\, x\in G . \qquad \text{(compare [3],[8])}$$

If μ is irreducible, the number $r(\mu)=1/\rho(\mu)$ is the radius of convergence of the series $\Sigma_n\, \mu^{(n)}(x)z^n$ without depending on $x\in G$, and it is easy to verify that for $z=r(\mu)$ the series is either convergent or divergent for all $x\in G$ simultaneously.
In the second case μ is called $r(\mu)$-*recurrent* .

(2.5) If μ is irreducible and $r(\mu)$-recurrent then G is a recurrent group (and thus amenable) : There exists a 1-recurrent probability measure on G (which means that the random walk given by this probability is recurrent in the usual sense). ([6])

3. *Survey of results on nonamenable discrete groups*

(3.1) If $G = F_N$ is the *free group* with N generators ($N\geq 2$, finite) and if the random walk given by μ is *"isotropic"* , i.e. $\mu(x)$ depends just on the word-length of $x\in G$ then

$$\frac{n^{3/2}\mu^{(n)}}{\rho(\mu)^n} \quad \text{converges pointwise to a positive measure on}$$

G, provided μ is irreducible and aperiodic.
This was proved in [10] using methods of harmonic analysis. Isotropy is important for these methods.

(3.2) If $G = F_I$ is the *free group* with free generators $\{\, x_j \mid j\in I \,\}$ where $I = \{\, 1,\dots,N \,\}$, $N\geq 2$ or $I = \{\, 1,2,3,\dots \,\}$ and if the support of μ is $S_\mu = \{\, x_j, x_j^{-1} \mid j\in I \,\}$, then

$\frac{(2n)^{3/2}\mu^{(2n)}}{\rho(\mu)^{2n}}$ converges vaguely to a positive measure on the subgroup of all words of even length (the odd convolution powers of μ are 0 on this subgroup - μ has period 2). ([4] for $N=2$,[12] for $N>2$ and for the infinite case)

No isotropy is needed here. It should be remarked that for infinite index set I this gives a local limit theorem on a group which is not finitely generated. No isotropy can hold in this case. The nonexponential factor of the convolution powers is still $n^{-3/2}$ in the infinitely generated case.

(3.3) If $G = < a , b \mid a^r = b^s = e >$ is the group with two generators a , b satisfying the defining relations $a^r = b^s = e$, where r , s are integers ≥ 2 , and if the support of μ is the set $S_\mu = \{ a, a^2, \ldots, a^{r-1}, b, b^2, \ldots, b^{s-1} \}$ and if furthermore μ is uniformly distributed on $\{ a, \ldots, a^{r-1} \}$ and on $\{ b, \ldots, b^{s-1} \}$ (this is a kind of isotropy condition) then :

(a) G is amenable if and only if $r=s=2$, and in this case

$\lim_{n\to\infty} (2n)^{1/2}\mu^{(2n)}(x) = \sqrt{2/\pi}$ for each $x \in G$ in the subgroup of words of "even length".([5])

(b) If $r>2$ or $s>2$ then we can determine $\rho(\mu)$ and a positive measure ν_0 on G such that

$\lim_{n\to\infty} \frac{n^{3/2}\mu^{(n)}(x)}{\rho(\mu)^n} = \nu_0(x) \quad \forall\, x \in G$. ([5] for $r=s$, the general case is going to be proved in this paper)

(3.2) and (3.3) are proved by combinatorical methods using the results given above in section 2. .

The following sections are devoted to the proof of (3.3.b).-

A local limit theorem on $G = < a , b \mid a^r = b^s = e >$

Let G and μ as given above in (3.3.b), that is, $\mu = \mu_{r,s,\theta}$ for some θ $(0<\theta<1)$ is given by

$$\mu(a^i) = \frac{\theta}{r-1} \ , \ i=1,\ldots,r-1 \quad \text{and} \quad \mu(b^j) = \frac{1-\theta}{s-1} \ , \ j=1,\ldots,s-1 \ .$$

4. *Calculation of* $\rho(\mu)$ *and of ratio limits*

We are going to determine $\nu(x) = \lim_{n\to\infty} \mu^{(n)}(x)/\mu^{(n)}(e) \quad \forall\, x \in G$.-

Each $x \in G$ different from e has a unique reduced representation

(4.1) $x = x_1 \cdots x_m$ for some $m \in \mathbb{N}$ such that for $k \le m$

$x_k \in \{ a^i, b^j \mid i=1,\dots,r-1 \;,\; j=1,\dots,s-1 \}$ and for $k=1,\dots,m-1$

$x_k = a^i \;,\; i \in \{1,\dots,r-1\} \iff x_{k+1} = b^j \;,\; j \in \{1,\dots,s-1\}$

For each $x \in G$ different from e let

(4.2) $\overline{a}(x)$ be the number of powers of a and
$\overline{b}(x)$ the number of powers of b in the reduced representation (3.1) of x, furthermore let $\overline{a}(e) = \overline{b}(e) = 0$.

Then $\overline{a}(x)$ and $\overline{b}(x)$ are nonnegative integers satisfying $\overline{a}(x)-1 \le \overline{b}(x) \le \overline{a}(x)+1$, and for each pair (α,β) of nonnegative integers satisfying $\alpha-1 \le \beta \le \alpha+1$ there is at least one $x \in G$ such that $\overline{a}(x) = \alpha$ and $\overline{b}(x) = \beta$. For each of these "admitted" pairs (α,β) we define

(4.3) $K = K(\alpha,\beta) = \{ x \in G \mid \overline{a}(x)=\alpha \;,\; \overline{b}(x)=\beta \}$,
$K_a = K_a(\alpha,\beta) = \{ x \in K \mid x$ starts with a power of $a \}$,
$K_b = K_b(\alpha\ \beta) = \{ x \in K \mid x$ starts with a power of $b \}$.

Proposition :
(4.4) $\mu^{(n)}(x)$ is constant on each $K(\alpha,\beta)$: $\mu^{(n)}(x) = f_n(\overline{a}(x),\overline{b}(x))$.

Proof : By induction on n.-
(4.4) is true for $n=1$. Let us assume it is true for n. We use

$$\mu^{(n+1)}(x) = \frac{\theta}{r-1}\sum_{i=1}^{r-1} \mu^{(n)}(a^i x) + \frac{1-\theta}{s-1}\sum_{j=1}^{s-1} \mu^{(n)}(b^j x)$$

$$= \frac{\theta}{r-1}\sum_{i=1}^{r-1} \mu^{(n)}(x a^i) + \frac{1-\theta}{s-1}\sum_{j=1}^{s-1} \mu^{(n)}(x b^j) \quad .$$

Let $x \in K(\alpha,\beta)$. For $\alpha=\beta=0$, $x=e$, (4.4) is clear. Therefore, let $x \neq e$ and $K(\alpha,\beta) \neq \{e\}$.

Case 1 : $\alpha=\beta+1$, $\beta \ge 0$, $K=K_a$, $K_b=\emptyset$.- In the representation (4.1) of x we have $x_1=a^{i_0}$ with $i_0 \in \{1,\dots,r-1\}$.

$$\overline{a}(a^i x) = \begin{cases} \alpha \;, & i+i_0 \neq r \\ \alpha-1, & i+i_0 = r \end{cases} \;,\quad \overline{b}(a^i x) = \beta \;,\quad i=1,\dots,r-1 \text{ and}$$

$$\overline{a}(b^j x) = \alpha \;,\quad \overline{b}(b^j x) = \beta+1 \;,\quad j=1,\dots,s-1 \;.$$

$$\mu^{(n+1)}(x) = \frac{\theta}{r-1}((r-2)f_n(\alpha,\beta) + f_n(\alpha-1,\beta)) + (1-\theta)f_n(\alpha,\beta+1) = f_{n+1}(\alpha,\beta)$$

Case 2 : $\alpha+1=\beta$, $\alpha \ge 0$, $K=K_b$, $K_a=\emptyset$.- $x_1=b^{j_0}$ with $j_0 \in \{1,\dots,s-1\}$.

$$\overline{a}(a^i x) = \alpha+1 \;,\quad \overline{b}(a^i x) = \beta \;,\quad i=1,\dots,r-1 \text{ and}$$

$$\overline{a}(b^j x) = \alpha \;,\quad \overline{b}(b^j x) = \begin{cases} \beta \;, & j+j_0 \neq s \\ \beta-1, & j+j_0 = s \end{cases} \;,\quad j=1,\dots,s-1 \;.$$

$$\mu^{(n+1)}(x) = \theta f_n(\alpha+1,\beta) + \frac{1-\theta}{s-1}((s-2)f_n(\alpha,\beta) + f_n(\alpha,\beta-1)) = f_{n+1}(\alpha,\beta) \;.$$

Case 3 : $\alpha=\beta>0$. If x starts with a power of a then it ends with a power of b and conversely. By subdividing into the cases : $x_1=a^{i_0}$, $x_1=b^{j_0}$ and by using the right and left- sided convolution formulas given above for $\mu^{(n+1)}$ respectively, it is easy to see that *both* of the formulas from case 1 and case 2 hold.-

μ is irreducible and aperiodic ($\mu^{(n)}(e)>0 \quad \forall\ n\geq 2$), therefore by (2.1) and (2.3) we have

$$\lim_{n\to\infty} \frac{\mu^{(n+1)}(x)}{\mu^{(n)}(x)} = \rho(\mu) \quad \forall\ x\in G \quad .$$

As furthermore the support of μ is finite, by (2.4) the sequence $(\mu^{(n)}/\mu^{(n)}(e))$ is relatively compact in the vague topology. Any measure ν which is an accumulation point of this sequence satisfies the following relations for $\rho=\rho(\mu)$:

<u>(4.5)</u> $\mu*\nu = \nu*\mu = \rho\cdot\nu$, $\nu(e) = 1$ and - due to (4.4) - $\nu(x) = g(\overline{a}(x),\overline{b}(x))$ where g is a function defined on all "admitted" pairs of integers (α,β) taking positive real values.

Because of (2.1) $\rho(\mu)$ is just the smallest positive number ρ such that the equation (4.5) admits a *positive* solution. If for $\rho=\rho(\mu)$ (4.5) has a *unique* solution ν , this is the only accumulation point, i.e. $\lim\limits_{n\to\infty} \mu^{(n)}(x)/\mu^{(n)}(e) = \nu(x) \quad \forall\ x\in G$.

Therefore the next step will be to solve (4.5) for arbitrary parameter ρ , and by discussing the solutions we will then determine $\rho(\mu)$ and the ratio limit.-

(4.5) gives :

$$\rho\nu(x) = \frac{\theta}{r-1}\sum_{i=1}^{r-1} \nu(a^i x) + \frac{1-\theta}{s-1}\sum_{j=1}^{s-1} \nu(b^j x)$$

$$= \frac{\theta}{r-1}\sum_{i=1}^{r-1} \nu(xa^i) + \frac{1-\theta}{s-1}\sum_{j=1}^{s-1} \nu(xb^j) \quad .$$

$$\nu(e) = 1 \quad \text{and} \quad \nu(x) = g(\overline{a}(x),\overline{b}(x)) \quad .$$

Regarding the same cases as in the proof of (4.4) one obtains :

Case 1 : $\alpha=\beta+1$, $\beta\geq 0$

$$\rho g(\alpha,\beta) = \frac{\theta}{r-1}((r-2)g(\alpha,\beta) + g(\alpha-1,\beta)) + (1-\theta)g(\alpha,\beta+1)$$

Case 2 : $\alpha+1=\beta$, $\alpha\geq 0$

$$\rho g(\alpha,\beta) = \theta g(\alpha+1,\beta) + \frac{1-\theta}{s-1}((s-2)g(\alpha,\beta) + g(\alpha,\beta-1))$$

Case 3 : $\alpha=\beta\geq 1$

Both of the formulas (from case 1 and case 2) hold.

Now the equation (4.5) may be written in the following way :

$$\underline{(4.6)}\quad \begin{aligned}
&\text{a)}\quad g(0,0) = 1\\
&\text{b)}\quad \rho g(0,0) = \theta g(1,0) + (1-\theta)g(0,1)\\
&\text{c)}\quad \left(\rho-\frac{r-2}{r-1}\theta\right)g(n+1,n) = \frac{\theta}{r-1}g(n,n) + (1-\theta)g(n+1,n+1)\\
&\text{d)}\quad \left(\rho-\frac{s-2}{s-1}(1-\theta)\right)g(n,n+1) = \theta g(n+1,n+1) + \frac{1-\theta}{s-1}g(n,n)\\
&\text{e)}\quad \left(\rho-\frac{r-2}{r-1}\theta\right)g(n+1,n+1) = \frac{\theta}{r-1}g(n,n+1) + (1-\theta)g(n+1,n+2)\\
&\text{f)}\quad \left(\rho-\frac{s-2}{s-1}(1-\theta)\right)g(n+1,n+1) = \theta g(n+2,n+1) + \frac{1-\theta}{s-1}g(n+1,n) \qquad \forall\, n\in\mathbb{N}_0
\end{aligned}$$

This gives the following recursion :

$$\underline{(4.7)}\quad (r-1)(s-1)\theta(1-\theta)\cdot g(n+2,n+2) - A_\theta(\rho)\cdot g(n+1,n+1) + \theta(1-\theta)\cdot g(n,n) = 0$$

$$g(0,0) = 1\ ,$$

$$g(1,1) = \frac{\rho A_\theta(\rho) + \theta(1-\theta)\big((s-2)\theta + (r-2)(1-\theta)\big)}{\theta(1-\theta)\ \frac{d}{d\rho}A_\theta(\rho)}\ , \text{ where}$$

$$A_\theta(\rho) = (r-1)(s-1)\rho^2 - \big((s-1)(r-2)\theta + (r-1)(s-2)(1-\theta)\big)\rho -$$
$$- \big((s-1)\theta^2 + (r-1)(1-\theta)^2 - (r-2)(s-2)\theta(1-\theta)\big)\ .$$

Let $\sigma = \frac{1}{2}\left(\frac{r-2}{r-1}\theta + \frac{s-2}{s-1}(1-\theta)\right)$ denote the solution of $\frac{d}{d\rho}A_\theta(\rho) = 0$. For $\rho = \sigma$ we are not interested in $g(1,1)$, as it will turn out that for this parameter no *positive* solution of the recursion can be found. The characteristic polynomial of the recursion is

$$\underline{(4.8)}\quad P_\theta(z) = (r-1)(s-1)\theta(1-\theta)z^2 - A_\theta(\rho)z + \theta(1-\theta)\ .$$

$P_\theta(z)$ vanishes at

$$\underline{(4.9)}\quad z_{1,2} = \left(A_\theta(\rho) \pm \sqrt{D_\theta(\rho)}\right)/\big(2(r-1)(s-1)\theta(1-\theta)\big)\ , \text{ where}$$
$$D_\theta(\rho) = A_\theta(\rho)^2 - 4(r-1)(s-1)\theta^2(1-\theta)^2\ .$$

$D_\theta(\rho) = 0$ has the following four solutions :

$$\underline{(4.10)}\quad \begin{aligned}
\rho_1 &= \sigma + \tfrac{1}{2}\sqrt{\left(\tfrac{r-2}{r-1}\theta - \tfrac{s-2}{s-1}(1-\theta)\right)^2 + 4\left(\tfrac{\theta}{\sqrt{r-1}} + \tfrac{1-\theta}{\sqrt{s-1}}\right)^2}\\
\rho_2 &= \sigma + \tfrac{1}{2}\sqrt{\left(\tfrac{r-2}{r-1}\theta - \tfrac{s-2}{s-1}(1-\theta)\right)^2 + 4\left(\tfrac{\theta}{\sqrt{r-1}} - \tfrac{1-\theta}{\sqrt{s-1}}\right)^2}\\
\rho_3 &= \sigma - \tfrac{1}{2}\sqrt{\left(\tfrac{r-2}{r-1}\theta - \tfrac{s-2}{s-1}(1-\theta)\right)^2 + 4\left(\tfrac{\theta}{\sqrt{r-1}} - \tfrac{1-\theta}{\sqrt{s-1}}\right)^2}\\
\rho_4 &= \sigma - \tfrac{1}{2}\sqrt{\left(\tfrac{r-2}{r-1}\theta - \tfrac{s-2}{s-1}(1-\theta)\right)^2 + 4\left(\tfrac{\theta}{\sqrt{r-1}} + \tfrac{1-\theta}{\sqrt{s-1}}\right)^2}
\end{aligned}$$

We have $\rho_4 < \rho_3 \le \sigma \le \rho_2 < \rho_1$. ρ_1 and ρ_4 are the solutions of $A_\theta(\rho) = 2\sqrt{(r-1)(s-1)}\theta(1-\theta)$, ρ_2 and ρ_3 are the solutions of $A_\theta(\rho) = -2\sqrt{(r-1)(s-1)}\theta(1-\theta)$.

Proposition : $\rho(\mu)$ cannot be less than ρ .

Proof : We are looking for solutions of the recursion such that

$g(n,n) > 0 \ \forall\, n \in \mathbf{N}$, furthermore $\rho(\mu)$ has to be strictly positive.

a) If $\rho \le \rho_4$ and $\rho > 0$ then $A_\theta(\rho) > 0$ and $\frac{d}{d\rho}A_\theta(\rho) < 0$, thus by (4.7) $g(1,1) < 0$.

b) For $\rho_4 < \rho < \rho_3$ and for $\rho_2 < \rho < \rho_1$ we have $D_\theta(\rho) < 0$, the solutions (4.9) of $P_\theta(z) = 0$ have nonvanishing imaginary parts and are conjugates, $z_1 = \overline{z_2}$. The solution of the recursion (4.7) is : $g(n,n) = C_1 z_1^n + C_2 z_2^n$. C_1 and C_2 are uniquely determined by $C_1 + C_2 = g(0,0) = 1$ *and* $C_1 z_1 + C_2 \overline{z_1} = g(1,1)$ which is a real number. Therefore, $C_2 = \overline{C_1}$ and $g(n,n) = 2\cdot \mathrm{Re}(C_1 z_1^n)$ (real part). If $C_1 = r_1 exp(i\phi_1)$, $0 < \phi_1 < 2\pi$, $r_1 > 0$ and $z_1 = r\cdot exp(i\phi)$, $0 < \phi < 2\pi$, $\phi \neq \pi$, $r > 0$, then $C_1 z_1^n = r_1 r\cdot exp(i(\phi_1 + n\phi))$ and it is easy to see that there is an n such that $\phi_1 + n\phi$ modulo 2π lies in the open interval $(\frac{\pi}{2}, \frac{3\pi}{2})$ which means that $\mathrm{Re}(C_1 z_1^n) < 0$, $g(n,n) < 0$.

c) $\rho_3 < \rho < \rho_2$, $\rho > 0$: The solutions of $P_\theta(z) = 0$ satisfy $z_2 < z_1 < 0$, as $A_\theta(\rho) < 0$ and $0 < D_\theta(\rho) < A_\theta(\rho)^2$.
If we suppose that $g(n,n) > 0 \ \forall\, n \in \mathbf{N}$, then C_1 is real and $g(n,n) = C_1 z_1^n + (1-C_1) z_2^n$.
For $n=1$ we obtain $C_1 z_1 + (1-C_1) z_2 > 0$, therefore $C_1 > 1$ because of $z_2/z_1 > 1$. By asumption we have $C_1 z_1^{2n} + (1-C_1) z_2^{2n} > 0$. This implies $0 < (C_1 - 1)/C_1 < (z_1/z_2)^{2n} \to 0$ as $n \to \infty$, contradiction.

d) For $\rho = \rho_2$, resp. $\rho = \rho_3$, $P_\theta(z) = 0$ has the (double) solution $z_1 = -1/\sqrt{(r-1)(s-1)}$, $g(n,n) = (1 + C_2 n) z_1^n$, where C_2 is real, and we see that there exist negative $g(n,n)$. -

So now we know that $\rho(\mu) \ge \rho_1$.
For $\rho = \rho_1$ the recursion (4.7) and thus also the convolution equation given in (4.5) have a unique positive solution :
$P_\theta(z) = 0$ has the double solution $z_1 = 1/\sqrt{(r-1)(s-1)}$,

(4.11) $g(n,n) = (C_1 + C_2 n)/\sqrt{(r-1)(s-1)}^{\,n}$, where (by (4.7), $n=0$ and 1)

$$C_1 = 1 \quad \text{and} \quad C_2 = (2(\rho_1 - \sigma))^{-1}\cdot\left(\frac{r-2}{\sqrt{r-1}} + \frac{s-2}{\sqrt{s-1}}\right)\cdot\left(\frac{\theta}{\sqrt{r-1}} + \frac{1-\theta}{\sqrt{s-1}}\right)$$

As $g(n,n) > 0$ and as $\rho_1 > \frac{r-2}{r-1}\theta$ and $\rho_1 > \frac{s-2}{s-1}(1-\theta)$, by (4.6) c) and d) $g(n+1,n)$ and $g(n,n+1)$ are also positive and uniquely determined. Therefore ρ_1 is the smallest positive parameter admitting a positive solution of (4.5), and $\rho(\mu) = \rho_1$.
(2.4) and the remark after (4.5) imply

(4.12) $$\lim_{n\to\infty} \frac{\mu^{(n)}(x)}{\mu^{(n)}(e)} = \frac{1 + C_2 \overline{a}(x)}{\sqrt{(r-1)(s-1)}^{\,\overline{a}(x)}} \quad \text{for each } x \in G \text{ with } \overline{a}(x) = \overline{b}(x)$$

(C_2 as given in (4.11)).

For $\bar{a}(x) = \bar{b}(x) \pm 1$ the ratio limit is obtained by use of the formulas (4.6) c) and d) with $\rho = \rho_1$.

5. *Asymptotic behaviour of* $\mu^{(n)}(e)$ *as* $n \to \infty$

The asymptotic behaviour of the return probabilities $\mu^{(n)}(e)$ is determined by use of the *method of Darboux* ([1],[4],[9]). -

For this purpose the random walk is reduced to a Markov chain on the integers :

By (4.4) the convolution powers of μ are constant on each of the sets $K_a(\alpha,\beta)$ and $K_b(\alpha,\beta)$. As $K_a(\alpha,\alpha+1) = K_b(\beta+1,\beta) = \emptyset$, we can enumerate the nonvoid sets K_a and K_b and $K(0,0) = \{e\}$ as follows :

$$K(0,0) \leftrightarrow 0 \quad , \quad K_a(\alpha,\beta) \leftrightarrow \alpha+\beta \quad , \quad K_b(\alpha,\beta) \leftrightarrow -(\alpha+\beta) \quad ,$$

and to each integer corresponds exactly one nonvoid class.

For $x \in G$ and a nonvoid class K_l (l integer) the one-step transition probability of the *right* random walk from x to K_l , $p_{x,K_l} = \mu(x^{-1}K_l)$ depends only on the class of x . We obtain a Markov chain on the nonvoid classes (i.e. the integers) with transition probability $p_{K_l,K_m} = \mu(x^{-1}K_m)$ where $x \in K_l$ is arbitrary. This Markov chain is illustrated by the following graph :

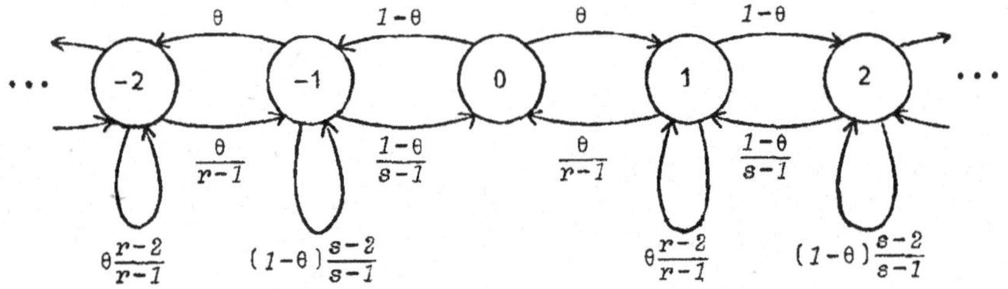

Fig. 1

Now $\mu^{(n)}(e)$ is the probability of returning to 0 at the n-th step after having started in 0 . Let a_n be the probability of returning to 0 for the first time at the n-th step after the first step from 0 to 1 and let b_n be the probability of returning to 0 for the first time at the n-th step after first step from 0 to -1 $(a_0=b_0=0)$.

We now define

(5.1) $$E = E(z) = \sum_{n=0}^{\infty} \mu^{(n)}(e) z^n ,$$

$$A = A(z) = \sum_{n=0}^{\infty} a_n z^n \quad , \quad B = B(z) = \sum_{n=0}^{\infty} b_n z^n \quad (z \text{ complex}).$$

From the formula $\mu^{(n)}(e) = \sum_{k=0}^{n} (a_k+b_k)\mu^{(n-k)}(e)$ we obtain
$E = 1/(1 - (A+B))$.
Equations for A and B are found by analysing the graph of the Markov chain given in fig.1 (using similar formulas as above, see [7]):

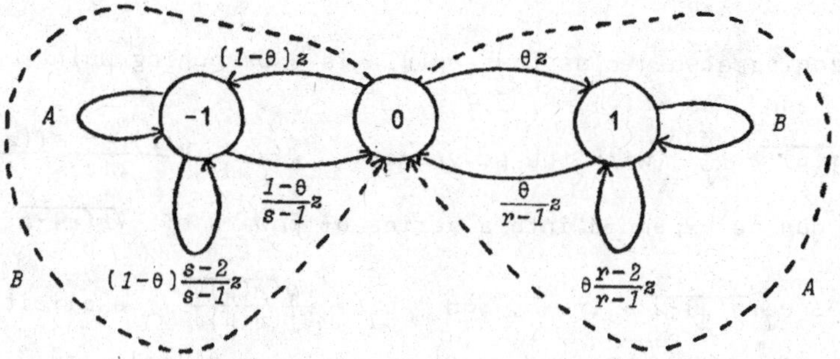

Fig. 2

Let $t = t(z) = \frac{1}{r-1}\theta^2 z^2$, $u = u(z) = \frac{1}{s-1}(1-\theta)^2 z^2$,

$v = v(z) = 1 - \theta\frac{r-2}{r-1}z$, $w = w(z) = 1 - (1-\theta)\frac{s-2}{s-1}$.

Then fig.2 gives $A = t/(v - B)$, $B = u/(w - A)$.

After some calculation the following equation for E is found :

(5.2) $f(z)E^2 + g(z)E + h(z) = 0$, where

$$f = -vw(v+w-1) + (v-w)(t-u) + (v+w)(wt+vu) - (t-u)^2 ,$$

$$g = vw(v+w-2) - (v-w)(t-u) \quad \text{and} \quad h = vw .$$

The branch of the complex function E given by (5.2) in which we are interested is the power series defined in (5.1). As we know from §4 , this series has the radius of convergence around 0 $r(\mu) = 1/\rho_1$, where $\rho_1 = \rho(\mu)$ is given by (4.10). ρ_1 satisfies the equation $D_\theta(\rho) = 0$ (see (4.9)). In terms of $r(\mu)$:

$r(\mu)$ satisfies the equation $\hat{D}(z) = 0$, where $\hat{D} = (vw - (t+u))^2 - 4tu$.

Furthermore, by a well known theorem of *Pringsheim* , $r(\mu)$ is a singularity of E and no pole, as G is not amenable and therefore μ cannot be $r(\mu)$-recurrent (by (2.5)), i.e. $\sum_{n=0}^{\infty} r(\mu)^n \mu^{(n)}(e) < \infty$.

Proposition :

(5.3) $r(\mu)$ is the only singularity of E on the circle of convergence.

Proof : Singularities of E are solutions of $f(z) = 0$ or of $g(z)^2 - 4f(z)h(z) = 0$. We find : $g^2 - 4fh = (v+w)^2\hat{D}$. As $r(\mu)$ is a simple zero of $\hat{D}$ (compare (4.10)) and no zero of $(v+w)^2$, $r(\mu)$ is a simple solution of $g(z)^2 - 4f(z)h(z) = 0$. Calculation also shows that $r(\mu)$ is not a solution of $f(z) = 0$.

The zeroes of $g^2 - 4fh$ different from $r(\mu)$ have all an absolute value different from $r(\mu)$ (compare (4.10)). If we suppose that there exists a singularity z_0 of E such that $|z_0| = r(\mu)$ and $z_0 \neq r(\mu)$ then $f(z_0) = 0$ and $g(z_0) \neq 0$, z_0 is a *pole* of E contradicting

$|\sum_{n=0}^{\infty} z_0^n \mu^{(n)}(e)| \leq \sum_{n=0}^{\infty} r(\mu)^n \nu^{(n)}(e) < \infty$. -

The above reasoning also yields that $r(\mu)$ is a branching point of order 1 of E and , as $E(0) = 1$,

$$E = -\frac{g}{2f} - \sqrt{r(\mu)-z}\,\frac{\sqrt{Q}}{2f} \quad \text{with the polynomial} \quad Q(z) = \frac{g(z)^2 - 4f(z)h(z)}{r(\mu) - z} .$$

Therefore E can be expanded into a series of powers of $\sqrt{r(\mu)-z}$ near $r(\mu)$:

(5.4) $E = c_0 - c_1\sqrt{r(\mu)-z} + \ldots$, and $c_1 = \frac{\sqrt{Q(r(\mu))}}{2f(r(\mu))}$, explicitely

$$c_1 = \rho(\mu)^2 \cdot \left(\frac{r-2}{\sqrt{r-1}} + \frac{s-2}{\sqrt{s-1}}\right)^{-2} \cdot (2(\rho(\mu)-\sigma))^{3/2} \cdot \left(\frac{\theta(1-\theta)}{\sqrt{(r-1)(s-1)}}\right)^{-3/2}$$

after some calculation. ($\rho(\mu) = \rho_1$ as given in (4.10), σ as defined just after (4.7)).

Proceeding exactly as in [5], the *method of Darboux* (see also [1],[4], [9],[12]) yields

$$\mu^{(n)}(e) = \frac{c_1\sqrt{r(\mu)}}{2\sqrt{\pi}} \cdot r(\mu)^{-n} n^{-3/2} + O(r(\mu)^{-n} n^{-2}) \text{ and thus (as } \frac{1}{r(\mu)} = \rho(\mu))$$

(5.5) $$\lim_{n\to\infty} \mu^{(n)}(e)/(\rho(\mu)^n n^{-3/2}) = c_1/2\sqrt{\pi\rho(\mu)} .$$

6. Conclusion

Combining the results of §4 and §5 , (4.12) and (5.5), we obtain :

$$\lim_{n\to\infty} \frac{\mu^{(n)}(x)}{\rho(\mu)^n n^{-3/2}} = \frac{c_1}{2\sqrt{\pi\rho(\mu)}} \cdot \frac{1 + C_2\bar{a}(x)}{\sqrt{(r-1)(s-1)}^{\bar{a}(x)}} \quad (= \nu_0(x))$$

for each $x \in G$ with $\bar{a}(x) = \bar{b}(x)$. c_1 is given in (5.4), C_2 is given in (4.11), $\bar{a}(x)$ and $\bar{b}(x)$ are defined in (4.2).

For $\bar{a}(x) = \bar{b}(x) \pm 1$ the limit formulas are a bit more complicated and may be obtained by using (4.6) c) and d). -

References

[1] BENDER,E.A.: Asymptotic methods in enumeration. Siam Review 16, 485-515 (1974).

[2] BERG,CH. and J.P.R.CHRISTENSEN : On the relation between amenability of locally compact groups and the norms of convolution operators. Math. Ann.208, 149-153 (1974).

[3] GERL,P.: Wahrscheinlichkeitsmaße auf diskreten Gruppen. Archiv Math. 31, 611-619 (1978).

[4] GERL,P.: Eine asymptotische Auswertung von Faltungspotenzen. Sitzungsber.Öst.Akad.Wiss.,Math.-Naturw.Klasse,Abt.II, 186, 385-396 (1978).

[5] GERL,P.: A local limit theorem on some groups. Arbeitsber.Math.Inst. Univ.Salzburg 1/1980, 1-10 (preprint,1980).

[6] GUIVARC'H,Y.: Sur la loi des grands nombres et le rayon spectral d' une marche aléatoire sur un groupe de Lie. Astérisque 74, 47-98 (1980).

[7] HOWARD,R.A.: Dynamic Probabilistic Systems.Vol.I. New York-London-Sidney-Toronto : John Wiley & Sons. 1971.

[8] LE PAGE,E.: Théorèmes quotients pour les marches aléatoires. C.R. Acad.Sc.Paris 279 A, 69-72 (1974).

[9] PLOTKIN,J.M. and J.ROSENTHAL : Some asymptotic methods in combinatorics. J.Austral.Math.Soc. Ser.A 28, 452-460 (1979).

[10]SAWYER,S.: Isotropic random walks in a tree. Z.Wahrscheinlichkeitsth.verw.Gebiete 42, 279-292 (1978).

[11]VERE-JONES,D.: Geometric ergodicity in denumerable Marcov chains. Quarterly J.Math.13, 7-28 (1962).

[12]WOESS,W.: Puissances de convolution sur les groupes libres ayant un nombre quelconque de générateurs. To appear,Astérisque (preprint in Arbeitsber.Math.Inst.Univ.Salzburg 1-2/1981, 13-22).

Wolfgang WOESS
Institut für Mathematik der
Universität Salzburg
Petersbrunnstraße 19
A-5020 Salzburg
Austria

Vol. 787: Potential Theory, Copenhagen 1979. Proceedings, 1979. Edited by C. Berg, G. Forst and B. Fuglede. VIII, 319 pages. 1980.

Vol. 788: Topology Symposium, Siegen 1979. Proceedings, 1979. Edited by U. Koschorke and W. D. Neumann. VIII, 495 pages. 1980.

Vol. 789: J. E. Humphreys, Arithmetic Groups. VII, 158 pages. 1980.

Vol. 790: W. Dicks, Groups, Trees and Projective Modules. IX, 127 pages. 1980.

Vol. 791: K. W. Bauer and S. Ruscheweyh, Differential Operators for Partial Differential Equations and Function Theoretic Applications. V, 258 pages. 1980.

Vol. 792: Geometry and Differential Geometry. Proceedings, 1979. Edited by R. Artzy and I. Vaisman. VI, 443 pages. 1980.

Vol. 793: J. Renault, A Groupoid Approach to C*-Algebras. III, 160 pages. 1980.

Vol. 794: Measure Theory, Oberwolfach 1979. Proceedings 1979. Edited by D. Kölzow. XV, 573 pages. 1980.

Vol. 795: Séminaire d'Algèbre Paul Dubreil et Marie-Paule Malliavin. Proceedings 1979. Edited by M. P. Malliavin. V, 433 pages. 1980.

Vol. 796: C. Constantinescu, Duality in Measure Theory. IV, 197 pages. 1980.

Vol. 797: S. Mäki, The Determination of Units in Real Cyclic Sextic Fields. III, 198 pages. 1980.

Vol. 798: Analytic Functions, Kozubnik 1979. Proceedings. Edited by J. Ławrynowicz. X, 476 pages. 1980.

Vol. 799: Functional Differential Equations and Bifurcation. Proceedings 1979. Edited by A. F. Izé. XXII, 409 pages. 1980.

Vol. 800: M.-F. Vignéras, Arithmétique des Algèbres de Quaternions. VII, 169 pages. 1980.

Vol. 801: K. Floret, Weakly Compact Sets. VII, 123 pages. 1980.

Vol. 802: J. Bair, R. Fourneau, Etude Géometrique des Espaces Vectoriels II. VII, 283 pages. 1980.

Vol. 803: F.-Y. Maeda, Dirichlet Integrals on Harmonic Spaces. X, 180 pages. 1980.

Vol. 804: M. Matsuda, First Order Algebraic Differential Equations. VII, 111 pages. 1980.

Vol. 805: O. Kowalski, Generalized Symmetric Spaces. XII, 187 pages. 1980.

Vol. 806: Burnside Groups. Proceedings, 1977. Edited by J. L. Mennicke. V, 274 pages. 1980.

Vol. 807: Fonctions de Plusieurs Variables Complexes IV. Proceedings, 1979. Edited by F. Norguet. IX, 198 pages. 1980.

Vol. 808: G. Maury et J. Raynaud, Ordres Maximaux au Sens de K. Asano. VIII, 192 pages. 1980.

Vol. 809: I. Gumowski and Ch. Mira, Recurences and Discrete Dynamic Systems. VI, 272 pages. 1980.

Vol. 810: Geometrical Approaches to Differential Equations. Proceedings 1979. Edited by R. Martini. VII, 339 pages. 1980.

Vol. 811: D. Normann, Recursion on the Countable Functionals. VIII, 191 pages. 1980.

Vol. 812: Y. Namikawa, Toroidal Compactification of Siegel Spaces. VIII, 162 pages. 1980.

Vol. 813: A. Campillo, Algebroid Curves in Positive Characteristic. V, 168 pages. 1980.

Vol. 814: Séminaire de Théorie du Potentiel, Paris, No. 5. Proceedings. Edited by F. Hirsch et G. Mokobodzki. IV, 239 pages. 1980.

Vol. 815: P. J. Slodowy, Simple Singularities and Simple Algebraic Groups. XI, 175 pages. 1980.

Vol. 816: L. Stoica, Local Operators and Markov Processes. VIII, 104 pages. 1980.

Vol. 817: L. Gerritzen, M. van der Put, Schottky Groups and Mumford Curves. VIII, 317 pages. 1980.

Vol. 818: S. Montgomery, Fixed Rings of Finite Automorphism Groups of Associative Rings. VII, 126 pages. 1980.

Vol. 819: Global Theory of Dynamical Systems. Proceedings, 1979. Edited by Z. Nitecki and C. Robinson. IX, 499 pages. 1980.

Vol. 820: W. Abikoff, The Real Analytic Theory of Teichmüller Space. VII, 144 pages. 1980.

Vol. 821: Statistique non Paramétrique Asymptotique. Proceedings, 1979. Edited by J.-P. Raoult. VII, 175 pages. 1980.

Vol. 822: Séminaire Pierre Lelong–Henri Skoda, (Analyse) Années 1978/79. Proceedings. Edited by P. Lelong et H. Skoda. VIII, 356 pages, 1980.

Vol. 823: J. Král, Integral Operators in Potential Theory. III, 171 pages. 1980.

Vol. 824: D. Frank Hsu, Cyclic Neofields and Combinatorial Designs. VI, 230 pages. 1980.

Vol. 825: Ring Theory, Antwerp 1980. Proceedings. Edited by F. van Oystaeyen. VII, 209 pages. 1980.

Vol. 826: Ph. G. Ciarlet et P. Rabier, Les Equations de von Kármán. VI, 181 pages. 1980.

Vol. 827: Ordinary and Partial Differential Equations. Proceedings, 1978. Edited by W. N. Everitt. XVI, 271 pages. 1980.

Vol. 828: Probability Theory on Vector Spaces II. Proceedings, 1979. Edited by A. Weron. XIII, 324 pages. 1980.

Vol. 829: Combinatorial Mathematics VII. Proceedings, 1979. Edited by R. W. Robinson et al.. X, 256 pages. 1980.

Vol. 830: J. A. Green, Polynomial Representations of GL_n. VI, 118 pages. 1980.

Vol. 831: Representation Theory I. Proceedings, 1979. Edited by V. Dlab and P. Gabriel. XIV, 373 pages. 1980.

Vol. 832: Representation Theory II. Proceedings, 1979. Edited by V. Dlab and P. Gabriel. XIV, 673 pages. 1980.

Vol. 833: Th. Jeulin, Semi-Martingales et Grossissement d'une Filtration. IX, 142 Seiten. 1980.

Vol. 834: Model Theory of Algebra and Arithmetic. Proceedings 1979. Edited by L. Pacholski, J. Wierzejewski, and A. J. Wilkie. VI, 410 pages. 1980.

Vol. 835: H. Zieschang, E. Vogt and H.-D. Coldewey, Surfaces and Planar Discontinuous Groups. X, 334 pages. 1980.

Vol. 836: Differential Geometrical Methods in Mathematical Physics. Proceedings, 1979. Edited by P. L. García, A. Pérez-Rendón, and J. M. Souriau. XII, 538 pages. 1980.

Vol. 837: J. Meixner, F. W. Schäfke and G. Wolf, Mathieu Functions and Spheroidal Functions and their Mathematical Foundations. Further Studies. VII, 126 pages. 1980.

Vol. 838: Global Differential Geometry and Global Analysis. Proceedings 1979. Edited by D. Ferus et al. XI, 299 pages. 1981.

Vol. 839: Cabal Seminar 77 – 79. Proceedings. Edited by A. S. Kechris, D. A. Martin and Y. N. Moschovakis. V, 274 pages. 1981.

Vol. 840: D. Henry, Geometric Theory of Semilinear Parabolic Equations. IV, 348 pages. 1981.

Vol. 841: A. Haraux, Nonlinear Evolution Equations- Global Behaviour of Solutions. XII, 313 pages. 1981.

Vol. 842: Séminaire Bourbaki vol. 1979/80. Exposés 543–56[illegible] IV, 317 pages. 1981.

Vol. 843: Functional Analysis, Holomorphy, and Approximation Theory. Proceedings. Edited by S. Machado. VI, 636 pages. 198[illegible]

Vol. 844: Groupe de Brauer. Proceedings. Edited by M. Kervaire and M. Ojanguren. VII, 274 pages. 1981.

Vol. 845: A. Tannenbaum, Invariance and System Theory: Algebraic and Geometric Aspects. X, 161 pages. 1981.

Vol. 846: Ordinary and Partial Differential Equations, Proceedings. Edited by W. N. Everitt and B. D. Sleeman. XIV, 384 pages. 1981.

Vol. 847: U. Koschorke, Vector Fields and Other Vector Bundle Morphisms – A Singularity Approach. IV, 304 pages. 1981.

Vol. 848: Algebra, Carbondale 1980. Proceedings. Ed. by R. K. Amayo. VI, 298 pages. 1981.

Vol. 849: P. Major, Multiple Wiener-Itô Integrals. VII, 127 pages. 1981.

Vol. 850: Séminaire de Probabilités XV. 1979/80. Avec table générale des exposés de 1966/67 à 1978/79. Edited by J. Azéma and M. Yor. IV, 704 pages. 1981.

Vol. 851: Stochastic Integrals. Proceedings, 1980. Edited by D. Williams. IX, 540 pages. 1981.

Vol. 852: L. Schwartz, Geometry and Probability in Banach Spaces. X, 101 pages. 1981.

Vol. 853: N. Boboc, G. Bucur, A. Cornea, Order and Convexity in Potential Theory: H-Cones. IV, 286 pages. 1981.

Vol. 854: Algebraic K-Theory. Evanston 1980. Proceedings. Edited by E. M. Friedlander and M. R. Stein. V, 517 pages. 1981.

Vol. 855: Semigroups. Proceedings 1978. Edited by H. Jürgensen, M. Petrich and H. J. Weinert. V, 221 pages. 1981.

Vol. 856: R. Lascar, Propagation des Singularités des Solutions d'Equations Pseudo-Différentielles à Caractéristiques de Multiplicités Variables. VIII, 237 pages. 1981.

Vol. 857: M. Miyanishi. Non-complete Algebraic Surfaces. XVIII, 244 pages. 1981.

Vol. 858: E. A. Coddington, H. S. V. de Snoo: Regular Boundary Value Problems Associated with Pairs of Ordinary Differential Expressions. V, 225 pages. 1981.

Vol. 859: Logic Year 1979–80. Proceedings. Edited by M. Lerman, J. Schmerl and R. Soare. VIII, 326 pages. 1981.

Vol. 860: Probability in Banach Spaces III. Proceedings, 1980. Edited by A. Beck. VI, 329 pages. 1981.

Vol. 861: Analytical Methods in Probability Theory. Proceedings 1980. Edited by D. Dugué, E. Lukacs, V. K. Rohatgi. X, 183 pages. 1981.

Vol. 862: Algebraic Geometry. Proceedings 1980. Edited by A. Libgober and P. Wagreich. V, 281 pages. 1981.

Vol. 863: Processus Aléatoires à Deux Indices. Proceedings, 1980. Edited by H. Korezlioglu, G. Mazziotto and J. Szpirglas. V, 274 pages. 1981.

Vol. 864: Complex Analysis and Spectral Theory. Proceedings, 1979/80. Edited by V. P. Havin and N. K. Nikol'skii. VI, 480 pages. 1981.

Vol. 865: R. W. Bruggeman, Fourier Coefficients of Automorphic Forms. III, 201 pages. 1981.

Vol. 866: J.-M. Bismut, Mécanique Aléatoire. XVI, 563 pages. 1981.

Vol. 867: Séminaire d'Algèbre Paul Dubreil et Marie-Paule Malliavin. Proceedings, 1980. Edited by M.-P. Malliavin. V, 476 pages. 1981.

Vol. 868: Surfaces Algébriques. Proceedings 1976–78. Edited by J. Giraud, L. Illusie et M. Raynaud. V, 314 pages. 1981.

Vol. 869: A. V. Zelevinsky, Representations of Finite Classical Groups. IV, 184 pages. 1981.

Vol. 870: Shape Theory and Geometric Topology. Proceedings, 1981. Edited by S. Mardešić and J. Segal. V, 265 pages. 1981.

Vol. 871: Continuous Lattices. Proceedings, 1979. Edited by B. Banaschewski and R.-E. Hoffmann. X, 413 pages. 1981.

Vol. 872: Set Theory and Model Theory. Proceedings, 1979. Edited by R. B. Jensen and A. Prestel. V, 174 pages. 1981.

Vol. 873: Constructive Mathematics, Proceedings, 1980. Edited by F. Richman. VII, 347 pages. 1981.

Vol. 874: Abelian Group Theory. Proceedings, 1981. Edited by R. Göbel and E. Walker. XXI, 447 pages. 1981.

Vol. 875: H. Zieschang, Finite Groups of Mapping Classes of Surfaces. VIII, 340 pages. 1981.

Vol. 876: J. P. Bickel, N. El Karoui and M. Yor. Ecole d'Eté de Probabilités de Saint-Flour IX – 1979. Edited by P. L. Hennequin. XI, 280 pages. 1981.

Vol. 877: J. Erven, B.-J. Falkowski, Low Order Cohomology and Applications. VI, 126 pages. 1981.

Vol. 878: Numerical Solution of Nonlinear Equations. Proceedings, 1980. Edited by E. L. Allgower, K. Glashoff, and H.-O. Peitgen. XIV, 440 pages. 1981.

Vol. 879: V. V. Sazonov, Normal Approximation – Some Recent Advances. VII, 105 pages. 1981.

Vol. 880: Non Commutative Harmonic Analysis and Lie Groups. Proceedings, 1980. Edited by J. Carmona and M. Vergne. IV, 553 pages. 1981.

Vol. 881: R. Lutz, M. Goze, Nonstandard Analysis. XIV, 261 pages. 1981.

Vol. 882: Integral Representations and Applications. Proceedings, 1980. Edited by K. Roggenkamp. XII, 479 pages. 1981.

Vol. 883: Cylindric Set Algebras. By L. Henkin, J. D. Monk, A. Tarski, H. Andréka, and I. Németi. VII, 323 pages. 1981.

Vol. 884: Combinatorial Mathematics VIII. Proceedings, 1980. Edited by K. L. McAvaney. XIII, 359 pages. 1981.

Vol. 885: Combinatorics and Graph Theory. Edited by S. B. Rao. Proceedings, 1980. VII, 500 pages. 1981.

Vol. 886: Fixed Point Theory. Proceedings, 1980. Edited by E. Fadell and G. Fournier. XII, 511 pages. 1981.

Vol. 887: F. van Oystaeyen, A. Verschoren, Non-commutative Algebraic Geometry, VI, 404 pages. 1981.

Vol. 888: Padé Approximation and its Applications. Proceedings, 1980. Edited by M. G. de Bruin and H. van Rossum. VI, 383 pages. 1981.

Vol. 889: J. Bourgain, New Classes of $\mathcal{L}^p$-Spaces. V, 143 pages. 1981.

Vol. 890: Model Theory and Arithmetic. Proceedings, 1979/80. Edited by C. Berline, K. McAloon, and J.-P. Ressayre. VI, 306 pages. 1981.

Vol. 891: Logic Symposia, Hakone, 1979, 1980. Proceedings, 1979, 1980. Edited by G. H. Müller, G. Takeuti, and T. Tugué. XI, 394 pages. 1981.

Vol. 892: H. Cajar, Billingsley Dimension in Probability Spaces. III, 106 pages. 1981.

Vol. 893: Geometries and Groups. Proceedings. Edited by M. Aigner and D. Jungnickel. X, 250 pages. 1981.

Vol. 894: Geometry Symposium. Utrecht 1980, Proceedings. Edited by E. Looijenga, D. Siersma, and F. Takens. V, 153 pages. 1981.

Vol. 895: J.A. Hillman, Alexander Ideals of Links. V, 178 pages. 1981.

Vol. 896: B. Angéniol, Familles de Cycles Algébriques – Schéma de Chow. VI, 140 pages. 1981.

Vol. 897: W. Buchholz, S. Feferman, W. Pohlers, W. Sieg, Iterated Inductive Definitions and Subsystems of Analysis: Recent Proof-Theoretical Studies. V, 383 pages. 1981.

Vol. 898: Dynamical Systems and Turbulence, Warwick, 1980. Proceedings. Edited by D. Rand and L.-S. Young. VI, 390 pages. 1981.

Vol. 899: Analytic Number Theory. Proceedings, 1980. Edited by M.I. Knopp. X, 478 pages. 1981.